BIOLOGICAL COMPLEXITY
AND THE DYNAMICS OF LIFE PROCESSES

New Comprehensive Biochemistry

Volume 34

General Editor

G. BERNARDI
Paris

ELSEVIER
Amsterdam · Lausanne · New York · Oxford · Shannon · Singapore · Tokyo

Biological Complexity and the Dynamics of Life Processes

Jacques Ricard

Institut Jacques Monod
CNRS, Universités Paris 6 et Paris 7
2 Place Jussieu
75251 Paris Cedex 05, France

1999
ELSEVIER
Amsterdam · Lausanne · New York · Oxford · Shannon · Singapore · Tokyo

Sara Burgerhartstraat 25
P.O. Box 211, 1000 AE Amsterdam, The Netherlands

First edition 1999

Library of Congress Cataloging in Publication Data
A catalog record from the Library of Congress has been applied for.

ISBN: 0 444 50081 2
ISBN: 0 444 80303 3 (series)

♾ The paper used in this publication meets the requirements of ANSI/NISO Z39.48-1992 (Permanence of Paper)

Printed in The Netherlands

Preface

Classical molecular biology attempts at understanding the logic of biological events through a study of the structure and function of certain biological macromolecules. In essence, molecular biology is reductionist since its aim is to reduce a macroscopic system (a biological property) to the structure and properties of microscopic elements of the system (nucleic acids and proteins). This approach to biological phenomena has been extremely successful.

Most cell biologists working with complex eukaryotic cells, however, are using molecular biology as a tool to unravel some of the intricacies of living processes, but they do not necessarily believe that the reductionist approach relies upon firm epistemological grounds. On the contrary, they are aware that the knowledge of the structure and function of individual macromolecules is necessary but not sufficient to understand the internal logic of the living world, and that the supramolecular organization of the eukaryotic cell plays an essential role in the expression of biological functions. This means that many biological functions are emergent with respect to the individual properties of macromolecules involved in the expression of these functions, and a major problem of present day biology is to understand in physical terms the mechanisms of this emergence.

We feel that the best illustration of this idea comes, perhaps, from the chemiosmotic theory, which offers a physical explanation of the energy storage under the form of adenosine triphosphate (ATP) in mitochondria and chloroplasts. For decades, biochemists have been looking for a molecule that could have been responsible for the phosphorylation of adenosine diphosphate (ADP) into ATP. This search was in vain for this molecule did not exist, but an enzyme was discovered that catalysed the reverse process, namely the hydrolysis of ATP into ADP. Mitchell was the first to realize that ATP synthesis in mitochondria could be explained in terms of nonequilibrium thermodynamics if it was assumed that the scalar process of ATP synthesis was coupled to a vectorial event, namely proton transfer across the inner mitochondrial membrane. Then the same enzyme that catalysed ATP hydrolysis *in vitro* could catalyse its synthesis if anchored *in vivo* in the mitochondrial membrane, and if protons are transferred across this membrane. The predictions of this physical theory have been tested experimentally with success. The enzyme that allows ATP synthesis *in vivo* has thus been called ATP synthase and has recently been shown to be a motor protein. It is thus clear that the individual properties of this isolated enzyme are not sufficient to explain ATP synthesis. This process is the result of the action of a system (the ATP synthase–membrane system), and not of an isolated molecule.

We are convinced that there is much to be done in order to understand the physical laws that govern complex biological systems but we feel that this interdisciplinary approach of biological problems will be rewarding. The contribution of physics to biology is often considered to be exclusively exerted through advanced technologies such as X-ray diffraction, electron microscopy, nuclear magnetic resonance, ... In the present case this contribution is more conceptual, more precisely it is the introduction in biology of physical concepts, and not of physical techniques. The consequence of this approach is the use of a mathematical tool that becomes an essential part of the reasoning.

The title of this book puts the emphasis on two fundamental aspects of biological systems, namely their complexity and the fact that they are in a dynamic state. A great deal of interest has been devoted to a science of complexity, more precisely to a science that aims at understanding how complex systems have properties that are emergent with respect to those of their elements. The settlement in Santa Fè (New Mexico, USA) of an Institute devoted to these studies may be considered a testimony of this interest. There has been a dispute about the definition of complexity and the existence of general laws that would govern complex systems independently of their nature, but we shall not get involved in this dispute. We shall solely be concerned, in this book, with the complexity of the living cell and with the analysis of how this complexity may offer a physical approach to important biological problems. Moreover, as the living cell is a dynamic system, its study, in this perspective, has to be effected through nonequilibrium thermodynamics and kinetics.

The aim of this book is not to present the latest experimental data, but to discuss experimental results, whether recent or not, in an integrated dynamic physical perspective. This perspective, at the border between cell biology, physics and physical chemistry, might well be considered unexpected and disturbing, but we are confident, however, this is an important new field of research. This book is thus partly theoretical and partly experimental. It describes why the living cell may be considered a complex system; how enzyme reactions may be viewed as elementary dynamic life processes; how coupling between scalar and (or) vectorial dynamic processes may act as signaling devices; how metabolism is controlled; how cell compartmentalization may explain energy storage and active transport; how information and small molecules are transferred within multienzyme complexes; how complexity of the cell envelope may modulate catalytic activity of cell wall bound enzymes; how free energy stored in the cell may generate motility; how complexity of cell organelles can generate temporal organization of metabolic cycles, such as oscillations and chaos; how diffusion of morphogens in the young embryo can induce spatio-temporal organization of biochemical processes and emergence of patterns; and, last but not least, how it is possible to conceive the evolution of complexity of living systems. All these topics are traditionally viewed as independent. Considered in a mechanistic pespective, however, they appear as different aspects of the same fundamental problem, namely how genetic information and biological complexity take part in the emergence of complex functions that stretch far beyond the individual properties of biological macromolecules.

We do hope this book will be of interest to physicists and physical chemists interested in biological complexity, and to biologists interested in the physical interpretation of dynamic biological processes.

We are extremely grateful to our former collaborators and colleagues who have contributed to various results presented in this book. We have a special debt to Dick D'Ari who has read and corrected the manuscript and who has spent hours discussing with me the topics of the book. Brigitte Meunier has been extremely helpful on many occasions. Last but not least, my wife, Käty, has kept us going during the exciting but difficult task of preparing the manuscript of this book.

Contents

Other volumes in the series

Volume 18a. *Hormones and Their Actions, Part 1* (1988)
B.A. Cooke, R.J.B. King and H.J. van der Molen (Eds.)

Volume 18b. *Hormones and Their Actions, Part 2 – Specific Action of Protein Hormones* (1988)
B.A. Cooke, R.J.B. King and H.J. van der Molen (Eds.)

Volume 19. *Biosynthesis of Tetrapyrroles* (1991)
P.M. Jordan (Ed.)

Volume 20. *Biochemistry of Lipids, Lipoproteins and Membranes* (1991)
D.E. Vance and J. Vance (Eds.) – Please see Vol. 31 – revised edition

Volume 21. *Molecular Aspects of Transport Proteins* (1992)
J.J. de Pont (Ed.)

Volume 22. *Membrane Biogenesis and Protein Targeting* (1992)
W. Neupert and R. Lill (Eds.)

Volume 23. *Molecular Mechanisms in Bioenergetics* (1992)
L. Ernster (Ed.)

Volume 24. *Neurotransmitter Receptors* (1993)
F. Hucho (Ed.)

Volume 25. *Protein Lipid Interactions* (1993)
A. Watts (Ed.)

Volume 26. *The Biochemistry of Archaea* (1993)
M. Kates, D. Kushner and A. Matheson (Eds.)

Volume 27. *Bacterial Cell Wall* (1994)
J. Ghuysen and R. Hakenbeck (Eds.)

Volume 28. *Free Radical Damage and its Control* (1994)
C. Rive-Evans and R.H. Burdon (Eds.)

Volume 29a. *Glycoproteins* (1995)
J. Montreuil, J.F.G. Vliegenthart and H. Schachter (Eds.)

Volume 29b. *Glycoproteins II* (1997)
J. Montreuil, J.F.G. Vliegenthart and H. Schachter (Eds.)

Volume 30. *Glycoproteins and Disease* (1996)
J. Montreuil, J.F.G. Vliegenthart and H. Schachter (Eds.)

Volume 31. *Biochemistry of Lipids, Lipoproteins and Membranes* (1996)
D.E. Vance and J. Vance (Eds.)

Volume 32. *Computational Methods in Molecular Biology* (1998)
S.L. Salzberg, D.B. Searls and S. Kasif (Eds.)

Volume 33. *Biochemistry and Molecular Biology of Plant Hormones* (1999)
P.J.J. Hooykaas, M.A. Hall and K.R. Libbenga (Eds.)

CHAPTER 1

Complexity and the structure of the living cell

For a biologist, there is little doubt that the living cell has a complex structure and that this structure is at the origin of the complex functions of the cell. Although everyone understands the meaning of the word complexity, as used in everyday life, a precise definition of the corresponding concept is difficult to offer. One may even wonder whether a widely acceptable definition of this concept will ever be proposed. Nevertheless, biological systems are unquestionably complex, much more complex than man-made systems. Indeed the brain of higher vertebrates, including man, is considered by many to be the most complex object on earth. It is therefore an important matter to know how the complexity of the interactions that exist between the elements of a living system contributes to the expression of the function of this system. In other words, the question at stake, is to know what is the respective importance of the "parts and the whole" [1] in the global behaviour of the system. The aim of this chapter is thus twofold: first, to state clearly what we intend by complexity; and second, to present a brief overview of the cell structure which may allow one to appreciate the extent of its own complexity.

1.1. What do we mean by complexity?

There has been, in recent years, a great interest in the notion of complexity and its developments. Many scientists, coming from different fields [2–7], have offered tentative definitions of complexity and have attempted to discover the general laws that govern complex systems. Thus, fluid turbulence, neural and metabolic networks, languages, animal population dynamics, to cite but a few, are complex systems and should, in that unified perspective, follow common laws ... In these chapters, we will not develop a new general theory of complexity, but rather stick to definite biological problems. Although the attempts at defining complexity in a rational and synthetic manner, as well as the efforts to construct a unified theory of complexity are extremely stimulating, there is no general agreement as to the possibility of developing such a general theory [8]. What we are going to do here is to present some features of complex systems and see how living cells display precisely these features.

- A complex system, like any system, is made up of a number of elements in interaction. These interactions may be physical, but this is not indispensable. They may also express the existence of information transfer between the elements of the system.
- The system displays a certain degree of order. It is neither strictly ordered, as atoms in a crystal, nor fully random, as molecules in a gas. It displays a fuzzy structural and functional organization.
- The system must display nonlinear effects and often exhibits feedback loops. Nonlinearity can generate thresholds, that is, small causes can have large effects.

- A complex system is thermodynamically open and operates away from equilibrium. It is thus in interaction with the external world and requires matter and energy to maintain its organization.
- A complex system is in a dynamic state and has a history. This means that the present behaviour of the system is, in part, determined by its past behaviour.
- Most of interactions between elements of the system take place over a rather short range. This means that most of these interactions are local and that each element of the system does not "know" what is happening to the system as a whole. This is an important condition, indeed, for if an element had a "knowledge" of the behaviour of the system, the complexity would be present in that element and not in the system.

These features, which have been in part expressed by Cilliers [7], do not always allow a clear-cut distinction between systems that are complex and the others that are not. Nevertheless an important feature of complex systems is that their properties are emergent. This means that these properties cannot be predicted from the individual study of the elements of the system. The dynamic properties of the system should also require precise knowledge of the interactions that exist between these elements. Physical chemistry should be thus able to predict and explain the nature and extent of these emergent properties.

1.2. The living cell

The aim of this section is to present an overview of the structural organization of the living cell in relation to the concept of complexity. This section can be left skipped by biologists and biochemists but may be of interest to physicists and physical chemists.

1.2.1. The bacterial cell

The typical, time-honoured example of bacterial cell is *Escherichia coli*. These cells are rod-shaped (Fig. 1.1). Depending on the external conditions, their size is between two and four micrometers in length. The cell is surrounded by several envelopes. Most of the cell compartment is occupied by the cytoplasm which contains, among many molecular components, a double-stranded deoxyribonucleic acid molecule, or DNA, which is the bacterial chromosome [9,10].

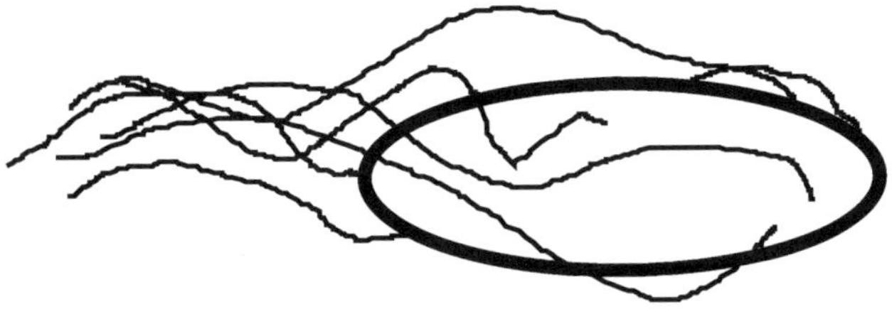

Fig. 1.1. Schematic representation of *Escherichia coli* cell.

1.2.1.1. The cell envelopes

The cell envelopes are composed of several layers. The outer membrane is a lipid bilayer associated with polysaccharides that constitute most of the outer region of the cell. This liposaccharidic membrane is interrupted in many different places by proteins referred to as porins that form holes in the membrane. Through these holes, nutrients may enter the cell. Immediately inside the outer membrane is the periplasm containing a layer of peptidoglycan which constitutes the wall and gives the cell its proper shape. The outer membrane is anchored to the peptydoglycan cell wall by lipoproteins. Inside the peptidoglycan cell wall is the inner region of the periplasmic space. Some protein molecules are located in this periplasm. These proteins include enzymes involved in the degradation of nutrients so to allow their transport to the cytoplasm. Other proteins sense the concentration of nutrients such as aminoacids and sugars. In its internal region, the periplasmic space is bound by the cytoplasmic membrane. This membrane is a lipid bilayer studded with proteins that span the membrane. Most of these proteins are carriers. Some of them carry ions to the cytoplasm, others carry ions to the periplasm. Yet others allow the exchange of different ions between the cytoplasm and the periplasm, or take part in the transport of neutral substances to the cytoplasm. A chain of electron transfer processes exists in the cytoplasmic membrane, thus leading to the oxidation of different substances. Part of the free energy released is converted into an electrochemical proton gradient. This gradient, which is a consequence of energy dissipation, results in proton extrusion to the periplasm and part of the corresponding energy is used to convert adenosine bisphosphate (ADP) into adenosine triphosphate (ATP). Thus part of the energy released by the electron transfer process is stored as ATP molecules. This ATP is used as a fuel that allows synthetic reactions to take place in the cell. It allows also the motion of this cell in a given milieu. In fact, about ten flagella are present at the surface of the bacterium. Each rises from a complex supramolecular structure that spans the cell envelopes and plays the part of a motor. This motor turns the flagellum in either direction, clockwise or counterclockwise. This allows the bacterium to swim [9–13].

1.2.1.2. The cytoplasm

Most of the cell volume is occupied by the cytoplasm [9]. About 70% of the cytoplasmic volume is occupied by a solution containing many different solutes. The remaining 30% of the volume is occupied by proteins, ribosomes, transfer ribonucleic acids (tRNAs), etc.

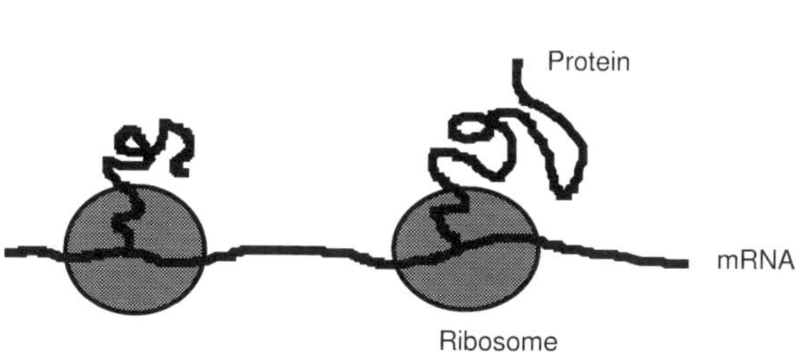

Fig. 1.2. Schematic picture of protein synthesis on a ribosome.

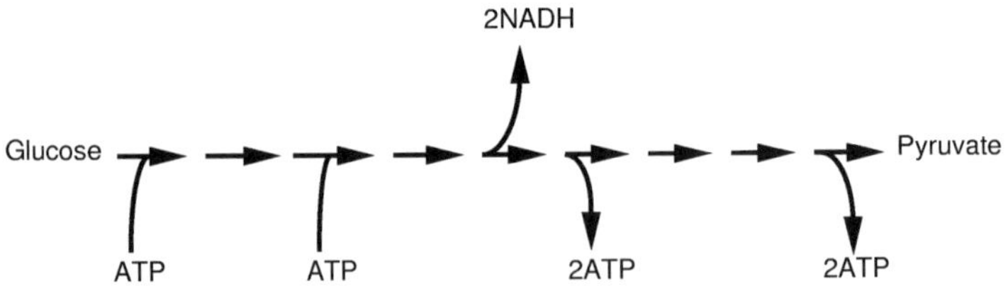

Fig. 1.3. Schematic representation of glycolysis.

Two major processes take place in the cytoplasm: the synthesis of new molecules and energy production through complex metabolic networks. Protein synthesis takes place on ribosomes where messenger RNAs (mRNAs), which are gene transcripts, are translated into proteins. Several ribosomes, each synthesizing a polypeptide chain, can be connected by a twisted mRNA molecule (Fig. 1.2). The aminoacids are transported to the ribosomes by tRNAs and are polymerized on the ribosomes. The association of tRNAs and aminoacids is specific. Thus, as proteins are made up of twenty different aminoacids, there exists twenty different tRNAs. The whole process of gene translation is carried in several steps each requiring a specific enzyme molecule.

The main metabolic networks involved in energy production are glycolysis and the citric acid cycle (Fig. 1.3). Glycolysis is a sequence of nine enzyme reactions that convert glucose to pyruvate. Starting with six carbon molecules, the process ends up with three-carbon molecules. Part of the available energy is stored as two ATP and two reduced nicotinamide adenine dinucleotide (NADH) molecules [13]. The available experimental data strongly suggests that the enzyme molecules involved in the overall process, and specific for each reaction step, are physically distinct entities. Completion of the whole glycolytic process implies that each reaction intermediate, which is both the substrate and the product of an enzyme reaction, has to be desorbed from the active site of an enzyme and to collide with the active site of the enzyme that comes next in the reaction sequence. This represents, indeed, a physical limitation to the efficiency of the overall reaction flow.

Under aerobic conditions, glycolysis is followed by the citric acid cycle. Pyruvate (3 C) is converted into acetyl coenzyme A (2 C) which can be coupled with oxaloacetate (4 C) and therefore enters the cycle as citrate (6 C). After eight enzymatic steps, oxaloacetate is regenerated. Two molecules of carbon dioxide are formed per turn of the Krebs cycle (Fig. 1.4). More importantly, three molecules of reduced nicotinamide adenine dinucleotide (NADH) and one molecule of reduced flavine adenine dinucleotide ($FADH_2$) are formed per turn. Similarly, two molecules of NADH are also generated during glycolysis and one additional NADH molecule is formed during the conversion of pyruvate to acetyl coenzyme A. These NADH molecules are reoxidized through an electron transfer chain which takes place, as already outlined, in the cytoplasmic membrane. As previously mentioned, the energy released by the oxidation process is converted into ATP. Thus, the free energy originating from glucose degradation may be converted into a form of chemical energy that can easily be used by the cell to perform its own syntheses [13–16].

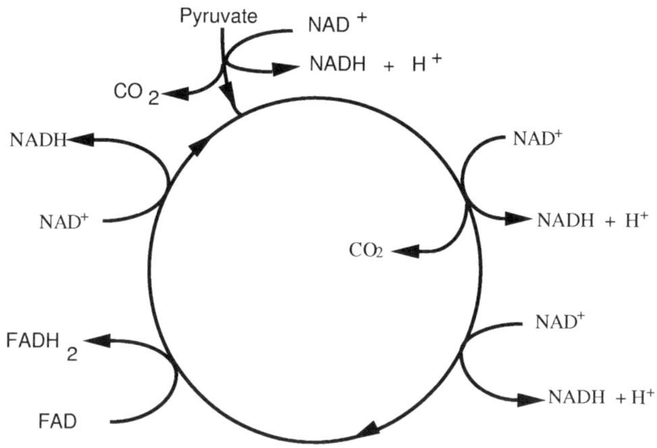

Fig. 1.4. Schematic representation of the Krebs cycle.

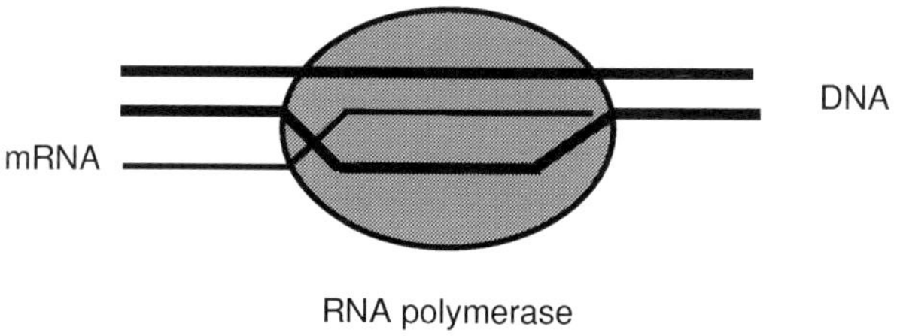

Fig. 1.5. Schematic representation of DNA transcription.

1.2.1.3. The nuclear region

A region of the cytoplasm is occupied by a circular DNA molecule which is attached to the cell envelope by two specific regions of its molecule. All but a few proteins of the bacterium have their informational sequence stored in this circular chromosome [9]. The circular DNA is twisted and folded around protein cores. This nuclear region of the cytoplasm is relatively free of proteins. As mentioned above, regions of this DNA are transcribed as messenger RNA molecules. For a given gene, this transcription process involves one strand only of the DNA molecule (Fig. 1.5).

The circular chromosome is replicated during the cell cycle. The replication process starts and ends at the two points of attachment of the DNA to the cell envelope. The whole process of cell division takes less than half-an-hour to proceed. In addition to the chromosome, bacteria contain circular plasmids that code for several proteins each. Plasmids often confer resistance to antibiotics. These plasmids can be transferred from cell to cell. As the nuclear region is not physically separated from the cytoplasm, *Escherichia coli* is referred to as a prokaryote. All bacteria are prokaryotes.

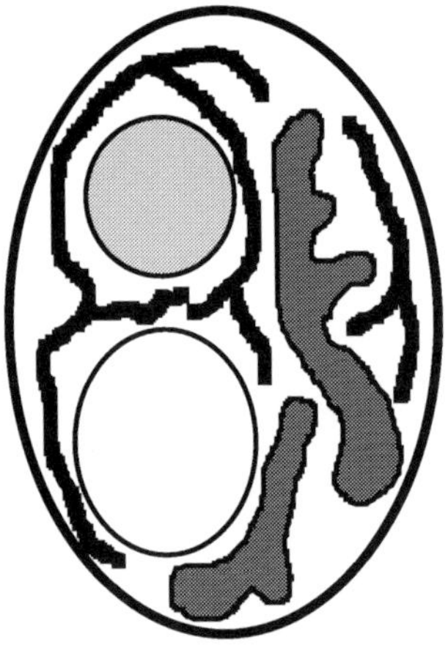

Fig. 1.6. Schematic representation of a yeast cell. In the cell are represented: the nucleus, a vacuole, the mitochondria and the endoplasmic reticulum.

1.2.2. The eukaryotic cell

Contrary to prokaryotes, more evolved organisms, either unicellular or multicellular, have their genetic material physically separated from the rest of the cell. They are called eukaryotes. For about two billion years, prokaryotes were the only inhabitants of our earth and, after this long period, some of these primitive cells evolved and generate eukaryotic cells, and this event was certainly a major step in the history of living organisms. A brief overview of the main features of eukaryotic cells is presented below, taking as a first example baker's yeast, *Saccharomyces cerevisiae*, and as a second example sycamore, *Acer pseudoplatanus*.

1.2.2.1. Baker's yeast cell

Baker's yeast is one of the simplest eukaryotes [9]. It is a unicellular organism but, unlike bacteria, it is highly compartmentalized. Bounded by their membrane, baker's yeast cells have, in their cytoplasm, an internal cytoskeleton which maintains the cell shape. The nucleus contains the chromosomes and is surrounded by a double membrane. Mitochondria are the organelles that generate energy. The Golgi apparatus is a complex of flattened sacs involved in sorting and packaging macromolecules. The endoplasmic reticulum, which is a system of flattened sheets and tubes, is involved in the transport of lipids and membrane proteins. A yeast cell is schematized in Fig. 1.6.

1.2.2.1.1. The cytoplasm. Yeast cytoplasm is crisscrossed by three types of filaments that play the part of a cytoskeleton: actin filaments, intermediate filaments, and microtubules. Actin is the commonest structural element. It constitutes a network of filaments that run through the cell. Proteins such as filamin and fimbrin crosslink actin filaments, forming an elastic mesh of fibers. Intermediate filaments are larger than actin filaments and usually perpendicular to them. They are very strong and give shape to the cell. Microtubules are the largest of these structural elements. Their networks are used by motor proteins, such

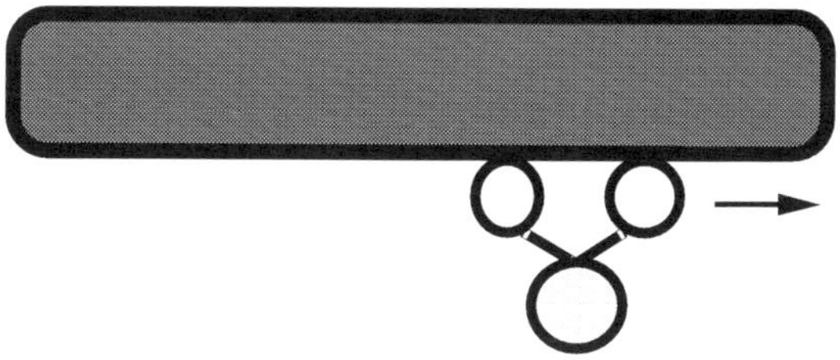

Fig. 1.7. A motor protein moves along a microtubule.

as kinesin and dynein, that move along these microtubules thanks to the energy originating from ATP hydrolysis [9,13] (Fig. 1.7). They may haul cell material, such as vesicles, along the microtubules. Both actin filaments and microtubules are unstable structures of the cell that require hydrolysis of adenosine triphosphate (for actin filaments) or of guanosine triphosphate (for microtubules) in order to maintain, or extend, their network. These networks may thus be considered as dissipative structures [6].

All the chemical reactions occurring in the cytoplasm take place in between the network of filaments and microtubules. The gene translation machinery is present in the cytoplasm. It is nearly identical to that occurring in bacteria, namely ribosomes, tRNAs, mRNAs and the required enzymes. The mechanism of translation is also similar to that taking place in bacteria. The main difference is that mRNAs are synthesized in the nucleus and, after some transformations, transported to the cytoplasm. Moreover the ribosomes are larger than those of bacteria, but their function is the same.

The cytoplasm is crowded with proteins and most of these are associated as supramolecular edifices. This is a major difference with respect to bacterial cells. If the cytoplasm of bacteria has often been referred to as an "enzyme bag", this is certainly not the case for the cytoplasm of eukaryotic cells which has a fuzzy organisation. Enzymes present in the cytoplasm may occur as multienzyme complexes, or as supramolecular associations with the cytoskeleton, or with the membranes. As we shall see later, one may expect these enzymes, present in a fuzzy organized state, to act in a manner that could be different from that occurring in bacteria, where most of the enzymes are soluble and act as physically distinct entities. The reactions of glycolysis take place in the cytoplasm of yeast cells, whereas the Krebs cycle and the electron transfer processes associated with the reoxidation of NADH take place in mitochondria. Many other chemical reactions also take place in the cytoplasm. These reactions do not fit the specific tasks of the organelles. They have been named "the general housekeeping reactions of the cell" [9].

1.2.2.1.2. The mitochondria. Mitochondria are the organelles that store, in a form available to the cell, part of the free energy released from metabolic processes, and in particular from the Krebs cycle and the associated electron transfer chains. Mitochondria are about the size of bacteria. They are surrounded by a double membrane (Fig. 1.8). The outer membrane is traversed by protein molecules similar to bacterial porins. The intermembrane space is filled with proteins similar to periplasmic proteins of bacteria. The inner membrane is folded and form cristae (Fig. 1.8). It is studded with supramolecular

Fig. 1.8. Schematic representation of a mitochondrion. The internal membrane displays cristae.

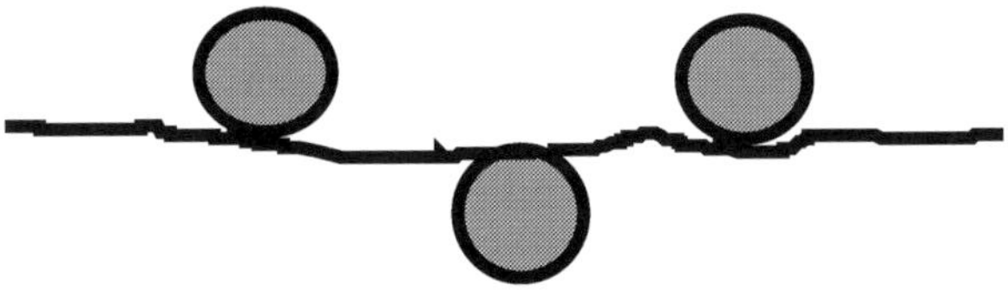

Fig. 1.9. A sequence of nucleosomes.

complexes embedded in the membrane and involved in electron transfer and energy conversion processes [13]. The internal region of the mitochondrion is called the matrix. It is packed with proteins and, surprisingly, contains a circular DNA molecule, ribosomes and tRNAs. These ribosomes are of the bacterial type, that is, smaller than regular cytoplasmic ribosomes. The presence, in mitochondria, of a machinery that allows the storage and translation of information, and which is clearly of the bacterial type, has suggested that present-day eukaryotic cell is the result of symbiosis. A bacterium has perhaps entered another cell as a parasite some 1.5 billion years ago, and the two partners took advantage of this situation. The enzymes of the citric acid cycle are located in the mitochondrial matrix [14–16]. The available experimental evidence suggests that these enzymes occur as multimolecular complexes. As a matter of fact, protein concentration in mitochondria is such that these proteins must be in physical contact.

1.2.2.1.3. The nucleus. Most of the information of the cell is stored in DNA molecules arranged in chromosomes and located in the nucleus [9,13]. The organelle is surrounded by a double membrane continuous with the endoplasmic reticulum. The nuclear membrane is traversed by pores. On the internal face of the membrane, protein filaments give shape to the nucleus. The DNA is stored in several chromosomes. Each chromosome is a DNA molecule wrapped around histone proteins, thus constituting a sequence of nucleosomes (Fig. 1.9). Nucleosomes protect DNA and help compact the molecule in a small volume.

Transcription takes place in the nucleus. Most of the potential information contained in eukaryotic DNA is never expressed. Moreover protein coding in eukaryotic DNA is not a continuous process, as it is the case in bacteria. Some regions of the DNA, called exons, are expressed and are separated by others, called introns, which are not. Most of the cell mRNAs originate from a maturation of heterogeneous nuclear RNAs (hnRNAs). hnRNAs are synthesized in the nucleus. They undergo some transformations (capping and polyadenylation) and then a splicing process edits out the introns. Editing is effected thanks to large protein complexes called spliceosomes. Most of the RNA which is synthesized is thus degraded inside the nucleus. The resulting mRNAs associated to proteins are transported to the nuclear membrane. The protein-mRNA complex then unfolds and the mRNA is expelled through a pore to the cytoplasm [17–22].

1.2.2.1.4. Transport of proteins within the cell. Since the eukaryotic cell contains different compartments and most proteins are synthesized in the cytoplasm, there must exist sorting and transporting devices that allow the proper final location of proteins. It is striking that, although mitochondria possess their specific machinery allowing storage and expression of information, most mitochondrial proteins are in fact synthesized in the cytoplasm and transported, across the mitochondrial membrane, in the matrix. This is effected by specific proteins, called chaperones, that bind to the polypeptide chain to be transported, unfolds this chain, which may then pass across the membrane. Other chaperones inside the matrix may help the polypeptide chain to regain its initial conformation [23–35].

Most of the protein transport within the cell, however, is effected by the endoplasmic reticulum and the Golgi apparatus. The endoplasmic reticulum is a network of tubes and sheets enclosed by a single membrane. Multiprotein complexes embedded in the membrane can bind ribosomes that are precisely in the process of synthesizing a polypeptide chain [36–42]. The polypeptide chain can then enter the endoplasmic reticulum, before its correct folding is completed and fold afterwards. The endoplasmic reticulum then expels vesicles containing proteins to be transported. These vesicles travel to the Golgi apparatus. This apparatus is a set of membrane-bound sacs that may be roughly divided into three regions: the *cis*-Golgi network, the Golgi stack and the *trans*-Golgi network. The vesicles, originating in the endoplasmic reticulum and carrying proteins, coalesce with the Golgi. These proteins are transported within the network and differentially tagged depending on their destination [43–48]. Then they leave the Golgi network via vesicles that carry them to their final location.

1.2.2.2. The sycamore cell

Clumps of plant cells, for instance sycamore cells (*Acer pseudoplatanus*), can be cultured *in vitro* under sterile conditions. When transferred to fresh medium, the cell first divides, then elongates. Plant cells therefore display different aspects depending on whether they are studied at the beginning or end of their growth period. There are two main differences between a young plant cell and an animal cell: the existence of a cell wall, which plays the part of an external skeleton for the plant cell; the presence, in the cytoplasm of the plant cell, of a new type of organelle, the chloroplast. As the cell extends, one may observe an additional difference, namely the presence of vacuoles in the plant cell that tend to coalesce and occupy most of the cell compartment (Fig. 1.10).

1.2.2.2.1. The cell wall. The wall of sycamore cells cultured *in vitro* is called the primary cell wall. It is made up of cellulose microfibrils interconnected by xyloglucan molecules. Xyloglucans are associated with cellulose microfibrils through hydrogen bonds. Moreover the primary cell wall contains acidic compounds, called pectins. Some of the carboxylate groups of pectins are methylated. Thus, the cell wall behaves like an insoluble polyanion. Moreover, structural proteins and enzymes are also present in the wall. Cell wall enzymes play at least three different roles. Trans glucanases are involved in local wall loosening, and peroxidases in wall stiffening. As we shall see later, both are involved in the control of the growth process. Others, for instance phosphatases, are involved in the hydrolysis of organic compounds that have to be split before they can enter the cell. Last but not least, pectin methyl esterases help control the local pH. It is important to stress that the cell wall

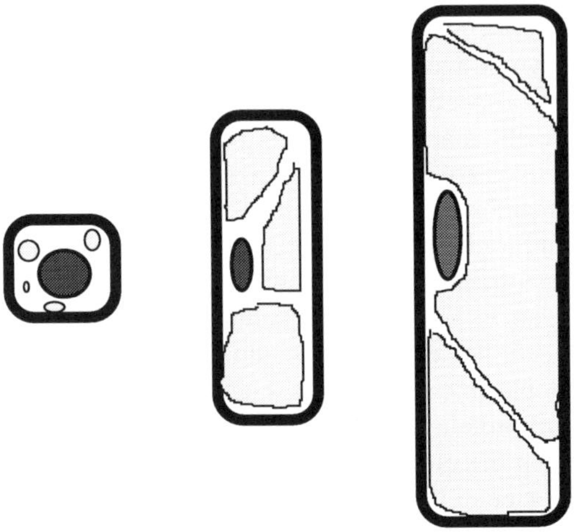

Fig. 1.10. A young growing plant cell. As the cell grows the vacuoles tend to fuse and occupy most of the internal volume. The nucleus then occupies a lateral position in the cell.

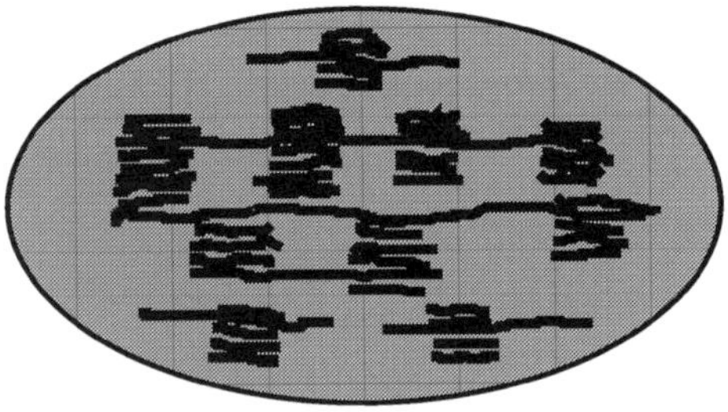

Fig. 1.11. Schematic representation of a chloroplast. The thylakoids are represented in black.

is not an inert envelope of the plant cell, it can grow and respond to external or internal signals. Plant cell wall extension, which is, to a significant extent, unidirectional, requires the participation of trans glucanases that break $\beta(1 \rightarrow 4)$ bonds of xyloglucans. Under the influence of turgor pressure exerted by the vacuoles, the cell wall extends and new $\beta(1 \rightarrow 4)$ bonds of xyloglucans are formed after the cell has extended. New glucidic material may be transported to the wall within Golgi vesicles and become incorporated in the cell envelope. These polysaccharides are methylated, and therefore neutral [49–56]. Plant cell wall enzymes are extremely sensitive to the local pH. Pectin methyl esterases may demethylate methylated pectins, thereby releasing negatively charged compounds that, depending on the pH, tend to attract protons and cations. The local pH is thus under the control of pectin methyl esterases. Within plant tissues, the cell wall may become extremely rigid owing to the presence of different polysaccharide compounds, such as lignin.

1.2.2.2.2. The chloroplasts. Chloroplasts are organelles that resemble mitochondria. Like mitochondria, they are surrounded by a double membrane with a small intermembrane space. But, unlike mitochondria, the inner membrane does not display cristae (Fig. 1.11). Inside the inner membrane is a region called the stroma, which is somewhat similar to the matrix of mitochondria. In the central region of the chloroplast is a system of flattened disclike sacs referred to as thylakoids (Fig. 1.11).

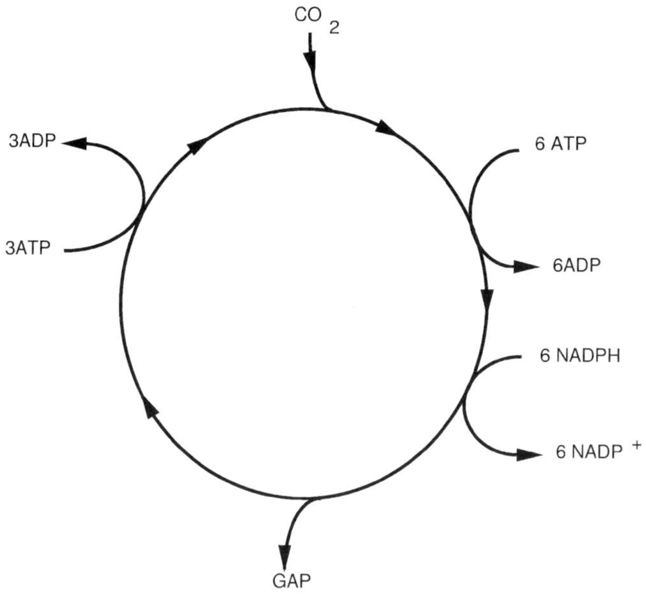

Fig. 1.12. Schematic representation of the Benson–Calvin cycle. See text.

Electrons taken from a water molecule thanks to the light excitation of two different chlorophyll pigments, are transferred to a series of transporters and ultimately to nicotinamide adenine dinucleotide phosphate (NADP), which becomes reduced (NADPH). Part of the energy is converted, during the electron transfer process, to ATP. Thus ATP represents a form of storage of light energy. All these processes, which are the counterpart of those already alluded to for mitochondria, take place in the thylakoids. Reduced NADP (NADPH) and ATP are then used in a metabolic process, the so-called Benson–Calvin cycle, which takes place in the chloroplast stroma. This cycle allows the fixation and reduction of carbon dioxide. It starts with the fixation of three CO_2 molecules on three molecules of ribulose 1, 5-bisphosphate (5 C) to give six molecules of 3-phosphoglycerate (3 C). The cycle comprises six steps (Fig. 1.12). Two of these require the participation of ATP and one that of NADPH. Exactly as for mitochondria, the chloroplast stroma is crammed with proteins [57–60]. Several of these proteins have been shown to exist as multienzyme complexes [61,62]. Moreover, circular DNA, bacterial-type ribosomes and tRNAs are present in the stroma. Again, this suggests that present day plant cells are the result of symbiosis between two prokaryotic cells.

1.3. The living cell is a complex system

It has now become easy to decide whether living cells meet the requirements of a complex system, as defined in section 1.1.

- A living cell can be viewed as a system that encompasses many elements (the biomolecules, the biochemical reactions, etc.) in a complex network.
- A living cell displays a general pattern of organization. This pattern, however, is not strictly defined but displays fuzz, in terms of both structure and functional organization.
- One may expect biochemical networks to be nonlinear. This nonlinearity may originate from the complexity of some biochemical reactions, or from the coupling between vectorial (diffusional) and scalar (chemical) processes, or from the electrostatic interactions between the cell milieu (the cell wall for instance) and some charged reaction intermediates.
- A living cell is an open system, in the thermodynamic sense of the word. It dissipates matter and energy in order to maintain its structural and functional organization.
- Since scalar processes (biochemical reactions) and vectorial processes (diffusion of molecules and ions) rely upon different laws, their coupling may be expected to generate nonlinearity and multistability of the coupled system. Multistability, therefore, means that the system will react differently depending on whether the present intensity of a signal is reached through an increase or a decrease of intensity. The system is thus endowed with a history.
- Any biomolecule, in a network within a cell, receives, or gives, inputs from, or to, a limited number of other biomolecules. This means that this biomolecule cannot "know" what is happening over the entire system. The biomolecule can only establish local structural and (or) functional contacts with its neighbours. One may therefore expect the properties of the system as a whole to be emergent relative to the properties of the biomolecule.

Although some of the above propositions may not appear obvious at first sight, we are confident, however, that further reading of the following chapters will make them obvious to any reader, and will prompt him to conclude that a living cell is indeed a complex system. The complexity of the eukaryotic cell, however, is by far larger than that of a prokaryotic organism. There are two reasons for this difference of complexity: compartmentalization of the eukaryotic cell and its molecular crowding. Compartmentalization generalizes the concept of vectorial-scalar coupling which is often associated with nonlinearity of the response. Because of molecular crowding and the existence of multienzyme complexes, different enzymes may functionally interact, and these interactions considerably increase the complexity of the system. The forthcoming chapters will offer illustrations of these views.

References

[1] Heisenberg, W. (1972) La Partie et le Tout. Le monde de la Physique Atomique. Albin Michel, Paris.

[2] Gell-Mann, M. (1995) Le Quark et le Jaguar. Voyage au Coeur du Simple et du Complexe. Albin Michel, Paris.

[3] Kauffman, S.A. (1993) The Origins of Order. Self-organization and Selection in Evolution. Oxford University Press, Oxford, New York.

[4] Kauffman, S.A. (1995) At Home in the Universe. The Search for Laws of Complexity. Viking Press, London.

[5] Nicolis, G. and Prigogine, I. (1989) Exploring Complexity. Freeman, New York.

[6] Prigogine, I. and Stengers, I. (1979) La Nouvelle Alliance. Gallimard, Paris.

[7] Cilliers, P. (1998) Complexity and Postmodernism. Understanding Complex Systems. Routledge, London and New York.
[8] Horgan, J. (1995) From complexity to perplexity. Scientific American, June, 104–109.
[9] Goodsell, D.S. (1993) The Machinery of Life. Springer-Verlag, New York.
[10] Stryer, L. (1981) Biochemistry. Second edition. Freeman, San Francisco.
[11] Block, S.M. (1997) Real engines of creation. Nature 386, 217–219.
[12] Adler, J. (1979) The Sensing of chemicals by bacteria. Scientific American, April, 40–47.
[13] Alberts, B., Bray, D., Lewis, J., Raff, M., Roberts, K. and Watson, J. (1994) The Molecular Biology of the Cell. Third edition. Garland Publishing, New York and London.
[14] Kornberg, H.L. (1987) Tricarboxylic acid cycle. Bioessays 7, 236–238.
[15] Krebs, H.A. (1970) The History of the tricarboxylic acid cycle. Perspect. Biol. Med. 14, 154–170.
[16] Krebs, H.A. and Martin, A. (1981) Reminiscences and Reflections. Oxford University Press, Oxford.
[17] Kornberg, R.D. and Klug, A. (1981) The Nucleosome. Scientific American, February, 52–64.
[18] McGhee, J.D. and Felsenfeld, G. (1980) Nucleosome structure. Annu. Rev. Biochem. 49, 1115–1156.
[19] Richmond, T.J., Finch, J.T., Rushton, B., Rhodes, D. and Klug, A. (1984) Structure of the nucleosome core particle at 7 Å resolution. Nature 311, 532–537.
[20] Hansen, J.C. and Ausio, J. (1992) Chromatin dynamics and the modulation of genetic activity. Trends Biochem. Sci. 17, 187–191.
[21] Pederson, D.S., Thoma, F. and Simpson, R. (1986) Core particle fiber, and transcriptionally active chromatin structure. Annu. Rev. Cell Biol. 2, 117–147.
[22] Swedlow, J.R., Agard, D.A. and Sedat, J.W. (1993) Chromosome structure inside the nucleus. Curr. Opin. Cell Biol. 5, 412–416.
[23] Pfanner, N., Rassow, J. and Wienhues, U. (1990) Contact sites between inner and outer membranes: structure and role in protein translocation into the mitochondria. Biochim. Biophys. Acta 1018, 239–242.
[24] Pon, L., Moll, T., Vestveber, D., Marshallay, B. and Shatz, G. (1989) Protein inport into mitochondria: ATP-dependent protein translocation activity in a submitochondrial fraction enriched in membrane contact sites and specific proteins. J. Cell Biol. 109, 2306–2316.
[25] Schleyer, M. and Neupert, W. (1985) Transport of proteins into mitochondria. Translocational intermediates spanning contact sites between outer and inner membranes. Cell 43, 339–350.
[26] Deshaies, R.J., Koch, B.D., Werner-Washburne, M., Craig, E.A. and Schekman, R.A. (1988) A subfamily of stress proteins facilitates translocation of secretory and mitochondrial precursor polypeptides. Nature 332, 800–805.
[27] Eilers, M. and Schatz, G. (1988) Protein unfolding and the energetics of protein translocation across biological membranes. Cell 52, 481–483.
[28] Wienhues, U., Becker, K. and Schleyer, M. (1991) Protein folding causes an arrest of preprotein translocation into mitochondria *in vivo*. J. Cell Biol. 115, 1601–1609.
[29] Hendrick, J.P. and Hartl, F.U. (1993) Molecular chaperone functions of heat-shock proteins. Annu. Rev. Biochem. 62, 349–384.
[30] Kelley, W.L. and Georgopoulos, C. (1992) Chaperones and protein folding. Curr. Opin. Cell Biol. 4, 984–991.
[31] Koll, H., Guiard, B. and Rossow, J. (1992) Antifolding activity of hsp 60 couples protein inport into the mitochondrial matrix with export to the intramembrane space. Cell 68, 1163–1175.
[32] Scherer, P.E., Krieg, U.C., Hwang, S.T., Westweber, D. and Schatz, G. (1990) A precursor protein partly translocated into yeast mitochondria is bound to a 70 kD mitochondrial stress protein. EMBO J. 9, 4315–4322.
[33] Glick, B.S., Beasley, E.M. and Schatz, G. (1992) Protein sorting in mitochondria. Trends Biochem. Sci. 17, 453–459.
[34] Hartl, F.U., Ostermann, J., Guiard, B. and Neupert, W. (1987) Successive translocation into and out of the mitochondrial matrix: targeting of proteins to the intermembrane space by a bipartite signal peptide. Cell 51, 1027–1037.
[35] Blobel, G. and Doberstein, B. (1975) Transfer of proteins across membranes. J. Cell Biol. 67, 835–851.
[36] Kaiser, C.A., Preuss, D., Grisafi, P. and Botstein, D. (1987) Many random sequences functionally replace the secretion signal sequence of yeast invertase. Science 235, 312–317.

[37] Simon, K., Perara, E. and Lingappa, V. (1987) Translocation of globin fusion proteins across the endoplasmic reticulum membrane in *Xenopus laevis* oocytes. J. Cell Biol. 104, 1165–1172.

[38] von Heijne, G. (1985) Signal sequences: the limit of variation. J. Mol. Biol. 184, 99–105.

[39] Gilmore, R. (1991) The protein translocation apparatus of the rough endoplasmic reticulum, its associated proteins and the mechanism of translocation. Curr. Opin. Cell Biol. 3, 580–584.

[40] Meyer, D.I., Krause, E. and Doberstein, B. (1982) Secretory protein translocation across membranes. The role of the docking protein. Nature 297, 647–650.

[41] Simon, S. (1993) Translocation of proteins across the endoplasmic reticulum. Curr. Opin. Cell Biol. 5, 581–588.

[42] Walter, P. and Lingappa, V. (1986) Mechanisms of protein translocation across the endoplasmic reticulum membrane. Annu. Rev. Cell Biol. 2, 499–516.

[43] Rambourg, A. and Clermont, Y. (1990) Three-dimensional electron microscopy: structure of the Golgi apparatus. Eur. J. Cell Biol. 51, 189–200.

[44] Hauri, H.P. and Schweizer, A. (1992) The endoplasmic reticulum-Golgi intermediate compartment. Curr. Opin Cell Biol. 4, 600–608.

[45] Lippincott-Schwartz, J. (1993) Bidirectional membrane traffic between the endoplasmic reticulum and Golgi apparatus. Trends Cell Biol. 3, 81–88.

[46] Pelham, H.R. (1991) Recycling of proteins between the endoplasmic reticulum and Golgi complex. Curr. Opin. Cell Biol. 3, 585–591.

[47] Balch, W.E., Dupuy, W.G., Braell, W.A. and Rothman, J.E. (1984) Reconstitution of the transport of proteins between successive compartments of the Golgi measured by the coupled incorporation of N-acetylglucosamine. Cell 39, 405–416.

[48] Kornfeld, R. and Kornfeld, S. (1985) Assembly of asparagine-linked oligosaccharides. Annu. Rev. Biochem. 54, 631–664.

[49] Fry, S.C. (1993) Loosening the ties. Current Biol. 3, 355–357.

[50] Fry, S.C., Smith, R.C., Renwick, K.F., Martin, D.J., Hodge, S.K. and Matthews, K.J. (1992) Xyloglucan endotransglycosylase, a new wall loosening activity from plants. Biochem. J. 282, 821–828.

[51] Smith, R.C. and Fry, S.C. (1991) Endotransglycosylation of xyloglucans in plant cell suspension cultures. Biochem. J. 279, 529–535.

[52] McDougall, G. and Fry, S.C. (1980) Xyloglucan oligosaccharides promote growth and activate cellulase: evidence for a role of cellulase in cell suspensions. Plant Physiol. 93, 1042–1048.

[53] McQueen-Mason, S. and Cossgrove, D.J. (1994) Disruption of hydrogen bonding between plant cell wall polymers by proteins that induce wall extension. Proc. Natl. Acad. Sci. USA 91, 6574–6578.

[54] Nishiani, K. and Tominaga, R. (1992) Endo-xyloglucanase transferase, a novel class of glycosyl transferase that catalyzes transfer of a segment of xyloglucan molecule to another xyloglucan molecule. J. Biol. Chem. 267, 21058–21064.

[55] Ricard, J. (1987) Enzyme regulation. In: P.K. Stumpf and E.E. Conn (Eds.), The Biochemistry of Plants. Academic Press, New York, Vol. 11, pp. 69–105.

[56] Ricard, J. and Noat, G. (1988) Electrostatic effects and the dynamics of multienzyme reactions at the surface of plant cells. In: P.B. Chock, C.Y. Huang, C. Tsou and J.K. Wang (Eds.), Enzyme Dynamics and Regulation. Springer-Verlag, New York, pp. 235–246.

[57] Cramer, W.A., Widger, W.R., Herrmann, R.G. and Trebst, A. (1985) Topography and function of the thylakoid membrane proteins. Trends Biochem. Sci. 10, 125–129.

[58] Bogorad, L. (1981) Chloroplasts. J. Cell Biol. 91, 256s–270s.

[59] Clayton, R.K. (1980) Photosynthesis. Physical Mechanisms and Chemical Patterns. Cambridge University Press, Cambridge.

[60] Haliwell, B. (1981) Chloroplast Metabolism. The Structure and Functions of Chloroplasts in Green Leaf Cells. Clarendon Press, Oxford.

[61] Gontéro, B., Cardenas, M.L. and Ricard, J. (1988) A functional five-enzyme complex of chloroplasts involved in the Calvin cycle. Eur. J. Biochem. 173, 437–443.

[62] Avilan, L., Gontéro, B., Lebreton, S. and Ricard, J. (1997) Memory and imprinting effects in multienzyme complexes. I. Isolation, dissociation and reassociation of a phosphoribulokinase-glyceraldehyde-3-phosphate dehydrogenase complex from *Chlamydomonas reinhardtii* chloroplasts. Eur. J. Biochem. 246, 78–84.

CHAPTER 2

Elementary life processes viewed as dynamic physicochemical events

Elementary biological events can be understood in terms of molecules, of chemical reactions and of the interconnections that exist between them. The rationale of these biological events should thus be looked for in terms of the laws of physics and chemistry. The chemical reactions, however, take place in the living cell which, from the chemist's point of view, is far from a simple standard medium. As already discussed in the previous chapter, it is rather a fuzzily organized system whose functional and structural complexity can generate emergent properties. However, isolated enzymes in free solution can already display some unexpected emergent properties. The aim of this chapter is precisely to review these properties.

2.1. *General phenomenological description of dynamic processes*

Dynamic processes, which take place in a test tube and in a living cell, may be scalar or vectorial. Enzyme catalysed chemical reactions are typical scalar processes, whereas the transport of molecules and ions is a vectorial process. Even though the vectorial process of diffusion can often be neglected in a dilute stirred solution where an enzyme catalysed reaction is taking place, because it is much faster than any step of the chemical reaction, these vectorial processes still exist and, as will be seen later, they may become rate-limiting in the living cell. It is thus mandatory to develop a formal quantitative language that can be applied to both chemical reactions and transport processes. This language is the language of nonequilibrium thermodynamics [1–7].

Let us consider a process

$$\cdots + \nu_i X_i + \cdots \rightarrow \cdots + \nu_j X_j + \cdots \tag{2.1}$$

whether scalar or vectorial, where ν_i and ν_j are the stoichiometric coefficients and X_i and X_j are different or identical (but localized in different regions of space) molecules. If n_i and n_j are the corresponding mole numbers, the advancement of the reaction may be defined as

$$\mathrm{d}\xi = -\frac{\mathrm{d}n_i}{\nu_i} = \frac{\mathrm{d}n_j}{\nu_j}. \tag{2.2}$$

If the reaction occurs at constant volume, thermodynamics allows one to write

$$\mathrm{d}S = -\frac{1}{T}\,\mathrm{d}G = -\frac{1}{T}\left(\sum_{i=1}^{k}\mu_i\,\mathrm{d}n_i + \sum_{j=k+1}^{n}\mu_j\,\mathrm{d}n_j\right), \tag{2.3}$$

where S, G, T and μ are the entropy, the Gibbs free energy, the absolute temperature and the chemical potentials, respectively. Substituting for $\mathrm{d}n_i$ and $\mathrm{d}n_j$ their expression derived from (2.2) leads to

$$\mathrm{d}S = -\frac{1}{T}\mathrm{d}G = \frac{1}{T}\left(\sum_{i=1}^{k} \nu_i \mu_i - \sum_{j=k+1}^{n} \nu_j \mu_j\right)\mathrm{d}\xi. \tag{2.4}$$

The affinity, A, of this process is defined as

$$A = \sum_i \nu_i \mu_i - \sum_j \nu_j \mu_j = -\left(\frac{\partial G}{\partial \xi}\right). \tag{2.5}$$

The rate of entropy production is then

$$\frac{\mathrm{d}S}{\mathrm{d}t} = \frac{1}{T} A \frac{\mathrm{d}\xi}{\mathrm{d}t}. \tag{2.6}$$

The process reaches equilibrium when $\mathrm{d}S/\mathrm{d}t = 0$ and therefore when $A = 0$. The affinity may thus be viewed as the force that pulls the physicochemical process towards equilibrium. The rate of this process per unit volume, v, may be defined as

$$v = J_{\mathrm{v}} = \frac{1}{V}\frac{\mathrm{d}\xi}{\mathrm{d}t} = -\frac{1}{V}\frac{1}{\nu_i}\frac{\mathrm{d}n_i}{\mathrm{d}t} = \frac{1}{V}\frac{1}{\nu_j}\frac{\mathrm{d}n_j}{\mathrm{d}t} \tag{2.7}$$

and is called the flow, J_{v}, per unit volume. This rate (or this flow) is expressed in $\mathrm{M\,s^{-1}}$. The rate in the forward direction, v_+, may be written as

$$v_+ = k \prod_{i=1}^{n} [X_i]^{\alpha_i}, \tag{2.8}$$

where α_i are numbers, in general different from the stoichiometric coefficients, which express the order of the dynamic process relative to the various reagents, and k is the corresponding rate constant. The overall order in the forward direction is

$$\omega = \sum_{i=1}^{n} \alpha_i. \tag{2.9}$$

If the rate is expressed in $\mathrm{M\,s^{-1}}$, the corresponding rate constant will be expressed in $\mathrm{M^{1-\omega}\,s^{-1}}$.

Equation (2.6) is a relationship between entropy production, the flow and the force that generates this flow. It is thus of interest to derive a more direct relationship between flows and forces. Moreover, the present formalism should apply to both scalar and vectorial processes. Therefore, this relationship will be derived below for a simple chemical reaction and for a diffusion process.

Let there be the simple chemical process

$$\mathrm{A} + \mathrm{B} \underset{k'}{\overset{k}{\longleftrightarrow}} \mathrm{C} \tag{2.10}$$

The flow per unit volume, or the net reaction rate, may be expressed as

$$v = J_{\mathrm{v}} = \frac{1}{V}\frac{\mathrm{d}\xi}{\mathrm{d}t} = k[\mathrm{A}][\mathrm{B}] - k'[\mathrm{C}]. \tag{2.11}$$

Let

$$Q = \frac{[\mathrm{C}]}{[\mathrm{A}][\mathrm{B}]} \quad \text{and} \quad K = \frac{k}{k'} = \frac{[\bar{\mathrm{C}}]}{[\bar{\mathrm{A}}][\bar{\mathrm{B}}]}, \tag{2.12}$$

where $[\bar{\mathrm{A}}]$, $[\bar{\mathrm{B}}]$, $[\bar{\mathrm{C}}]$ are equilibrium concentrations and K is the equilibrium constant, whereas Q is a concentration ratio which varies as the reaction proceeds towards equilibrium. Equation (2.11) can be rewritten as

$$v = J_{\mathrm{v}} = \frac{1}{V}\frac{\mathrm{d}\xi}{\mathrm{d}t} = k[\mathrm{A}][\mathrm{B}]\left(1 - \frac{Q}{K}\right). \tag{2.13}$$

The chemical potential of a reagent having a concentration c is

$$\mu = \mu^0 + RT \ln c, \tag{2.14}$$

where μ^0 is the standard chemical potential. Therefore, the affinity of the chemical reaction may be expressed as

$$A = \mu_{\mathrm{A}} + \mu_{\mathrm{B}} - \mu_{\mathrm{C}} = \mu_{\mathrm{A}}^0 + \mu_{\mathrm{B}}^0 - \mu_{\mathrm{C}}^0 - RT \ln Q. \tag{2.15}$$

Moreover, the standard free energy difference, ΔG^0, is defined as

$$\Delta G^0 = \mu_{\mathrm{C}}^0 - \left(\mu_{\mathrm{A}}^0 + \mu_{\mathrm{B}}^0\right) = -RT \ln K. \tag{2.16}$$

Therefore, the expression for the affinity is now

$$A = -RT \ln \frac{Q}{K}, \tag{2.17}$$

and the net reaction rate, or the flow per unit volume, assumes the form

$$v = J_{\mathrm{v}} = k[\mathrm{A}][\mathrm{B}]\left\{1 - \exp(-A/RT)\right\}. \tag{2.18}$$

This equation shows that there is an exponential relationship between the flow and the force that generates the flow. Moreover, close to equilibrium, one has approximately

$$\mathrm{e}^{-A/RT} \approx 1 - \frac{A}{RT}, \tag{2.19}$$

and eq. (2.18) becomes

$$J_v = \frac{k[\mathrm{A}][\mathrm{B}]}{RT}A. \tag{2.20}$$

The flow-force relationship is thus linear close to thermodynamic equilibrium.

This conclusion can be extended to diffusion processes. Fick's first law predicts that the diffusion flow of molecules that pass across a surface should be proportional to the concentration gradient. One has thus

$$J = \frac{\partial n}{\partial t} = -DA_\mathrm{d}\frac{\partial c}{\partial x}, \tag{2.21}$$

where D is the transport coefficient (in $\mathrm{cm^2\,s^{-1}}$), A_d is the area (in $\mathrm{cm^2}$) through which the diffusion occurs, c the concentration (in M) and x the distance (in cm). Similarly Fick's second law describes the diffusion of molecules through a volume element and assumes the form

$$\frac{\partial c}{\partial t} = D\frac{\partial^2 c}{\partial x^2}. \tag{2.22}$$

If there is a steady state within this volume element, then

$$\frac{\partial^2 c}{\partial x^2} = 0, \tag{2.23}$$

and, therefore, the concentration gradient must be linear. If c_o and c_i are the respective concentrations in two different regions of space ($c_\mathrm{o} > c_\mathrm{i}$) separated by the distance l, the concentration gradient of eq. (2.21) assumes the form

$$\frac{\partial c}{\partial x} = -\frac{c_\mathrm{o} - c_\mathrm{i}}{l}, \tag{2.24}$$

therefore, under steady state conditions, the diffusion flow can be expressed as

$$J = \frac{DA_\mathrm{d}}{l}(c_\mathrm{o} - c_\mathrm{i}). \tag{2.25}$$

DA_d/l is expressed in $\mathrm{cm^3\,s^{-1}}$, it has the dimension of a volume (D_v) times a first order rate constant (h_d). Therefore

$$\frac{DA_\mathrm{d}}{l} = D_\mathrm{v}h_\mathrm{d}, \tag{2.26}$$

and the steady state diffusion flow per unit volume is then

$$J_\mathrm{v} = \frac{J}{D_\mathrm{v}} = h_\mathrm{d}(c_\mathrm{o} - c_\mathrm{i}). \tag{2.27}$$

This equation is written in the "quasi-chemical" formalism since J_v is a reaction rate, h_d is a first order rate constant and c_o and c_i are concentrations. We shall see in the forthcoming chapter the importance of this formalism.

This equation, however, does not make obvious a simple relationship between flow and force (or affinity). Let us assume, for instance, that eq. (2.27) describes the diffusion of molecules through a porous membrane. c_o is then the concentration of a substance on the cis side and c_i is the concentraton of the same substance on the trans side of this membrane. Expression (2.27) can be rewritten as

$$J_v = h_d \left\{ \exp\left(\frac{\mu_o - \mu^0}{RT} \right) - \exp\left(\frac{\mu_i - \mu^0}{RT} \right) \right\}, \tag{2.28}$$

where μ_o and μ_i are the chemical potentials on the cis and on the trans sides and μ^0 is still the standard potential. The affinity that drives the flow of molecules from the cis to the trans side of the membrane is thus

$$A = \mu_o - \mu_i, \tag{2.29}$$

and eq. (2.28) should be rewritten to express the variation of the flow, J_v, as a function of this affinity. This can be done by algebraic manipulation of eq. (2.28). One has

$$J_v = h_d \exp\left(-\frac{\mu^0}{RT} \right) \left\{ \exp\left(\frac{\mu_o}{RT} \right) - \exp\left(\frac{\mu_i}{RT} \right) \right\}. \tag{2.30}$$

Setting

$$\langle \mu \rangle = \frac{\mu_o + \mu_i}{2} \tag{2.31}$$

one finds

$$J_v = h_d \exp\left(-\frac{\mu^0}{RT} \right) \exp\left(\frac{\langle \mu \rangle}{RT} \right) \left\{ \exp\left(\frac{A}{2RT} \right) - \exp\left(-\frac{A}{2RT} \right) \right\}, \tag{2.32}$$

which is a typical flow-force relationship. Close to thermodynamic equilibrium

$$\exp\left(\frac{A}{2RT} \right) \approx 1 + \frac{A}{2RT}, \tag{2.33a}$$

$$\exp\left(-\frac{A}{2RT} \right) \approx 1 - \frac{A}{2RT}, \tag{2.33b}$$

and eq. (2.32) reduces to

$$J_v = h_d \exp\left(\frac{\langle \mu \rangle - \mu^0}{RT} \right) \frac{A}{RT}, \tag{2.34}$$

which is a linear flow-force relationship. One is thus led to conclude that the same type of relationship applies whether the dynamic process is scalar or vectorial.

This thermodynamic formulation may be presented in an even more general way. Let us consider a flow J_i generated by different forces, $X_1, \ldots, X_j, \ldots$. Let

$$J_i = f(X_1, \ldots, X_j, \ldots). \tag{2.35}$$

Close to equilibrium this function can be expanded in MacLaurin series giving

$$J_i = \overline{J}_i + \sum_{j=1}^{n} \frac{\partial J_i}{\partial X_j} X_j + \cdots, \tag{2.36}$$

where $\overline{J}_i$ is the equilibrium flow which is indeed nil. Thus, neglecting the nonlinear terms

$$J_i = \sum_{j=1}^{n} L_{ij} X_j. \tag{2.37}$$

The L_{ij} are coupling coefficients, defined as

$$L_{ij} = \frac{\partial J_i}{\partial X_j}. \tag{2.38}$$

Close to equilibrium, a flow is thus a linear combination of forces. This conclusion can be extended to a set of flows through the matrix equation

$$\begin{bmatrix} J_1 \\ \vdots \\ J_n \end{bmatrix} = \begin{bmatrix} L_{11} & \cdots & L_{1n} \\ \vdots & \ddots & \vdots \\ L_{n1} & \cdots & L_{nn} \end{bmatrix} \begin{bmatrix} X_1 \\ \vdots \\ X_n \end{bmatrix}. \tag{2.39}$$

Moreover, Onsager's reciprocity relationships state that

$$L_{ij} = L_{ji} \quad (\forall i, j = 1, \ldots, n). \tag{2.40}$$

If the matrix relation (2.39) is expressed as

$$\mathbf{X} = \mathbf{L}^{-1}\mathbf{J}, \tag{2.41}$$

where $\mathbf{X}$ and $\mathbf{J}$ are the column vectors of forces and flows and $\mathbf{L}^{-1}$ is the inverse matrix of coupling coefficients, it displays an obvious similarity to a generalized Ohm's law. Hence there is a parallel between linear nonequilibrium thermodynamics and the study of electrical circuits. This parallel represents the very basis of so-called network thermodynamics [4,7].

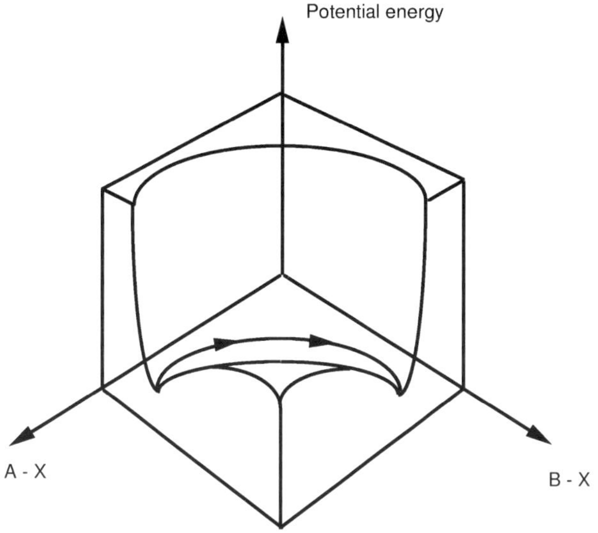

Fig. 2.1. Potential energy surface of a chemical reaction. A–X and B–X represent the distance that separates X from A and B, respectively. The reaction follows the "valley" on the surface (arrows).

2.2. *Enzyme reactions under simple standard conditions*

Usually, enzyme reactions are studied in stirred liquid media, under highly dilute conditions. These conditions are considered "standard" for most enzyme processes. The aim of this section is to review briefly how catalysed chemical reactions occur under these "standard" conditions.

2.2.1. Simple transition state theory and enzyme reactions

It was realized as early as 1889 by Arrhenius that one or several energy barrier(s) should exist between the initial and final stages of a chemical reaction, whether catalysed or not. Let us consider the simple uncatalysed chemical process

$$\mathrm{A} + \mathrm{B{-}X} \rightleftharpoons \mathrm{A{-}X} + \mathrm{B}. \tag{2.42}$$

If one plots the corresponding potential energy E as a function of B–X and A–X distances, one obtains a potential energy surface such as the one shown in Fig. 2.1. The reaction will follow a path associated with the lowest permissible potential energy, that is the reaction will follow the bottom of a valley on the potential energy surface. The corresponding pass will be associated with the transition state, $T_{\neq}$, of the reaction (Fig. 2.1). Thus the chemical reaction will have to overcome, along its reaction coordinate, one or several, energy barrier(s) associated with this (or these) transition state(s).

One of the effets of a catalyst, an enzyme for instance, on the reaction is to increase the number of steps involved in this process and to decrease the height of the energy barriers.

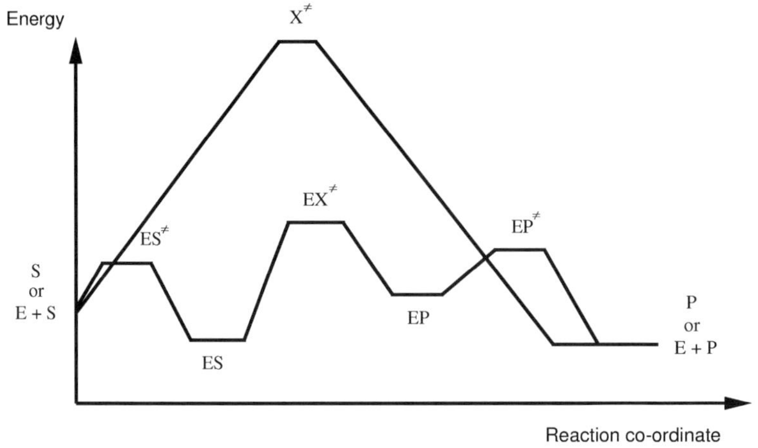

Fig. 2.2. Uncatalysed and enzyme-catalysed chemical reaction. The upper curve represents the energy profile of the uncatalysed reaction. The chemical process has to overcome the energy barrier associated with the transition state $X^{\neq}$. The lower curve represents the energy profile of the enzyme-catalysed reaction. The chemical process has to overcome three barriers associated with three transition states.

In the case of the transfer reaction (2.42), the corresponding enzyme-catalysed process could be

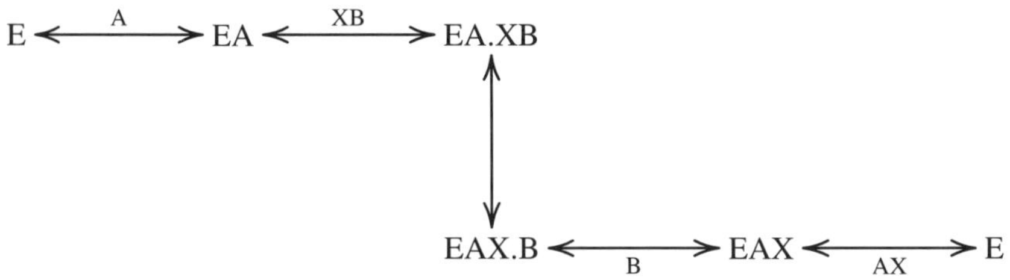

$$E \xleftrightarrow{A} EA \xleftrightarrow{XB} EA.XB \updownarrow EAX.B \xleftrightarrow{B} EAX \xleftrightarrow{AX} E \tag{2.43}$$

E is the enzyme. By convention the substrates, A and BX, are positioned above the arrows and the products AX and B below. The reaction is assumed to follow a compulsory order for substrate binding and product release. The catalytic step refers to the transfer of X from BX to A. For a two-substrate, two-product enzyme-catalysed reaction there may exist many *a priori* different possible reaction mechanisms. The aim of enzyme kinetics is precisely to screen amongst these *a priori* possible processes. Figure 2.2 shows possible energy profiles for the same uncatalysed and enzyme catalysed reaction.

There are different theories that aim at explaining the mechanism of a chemical reaction [8]. Probably one of the simplest and most convenient of these theories, as applied to enzyme reactions, is the so-called transition state theory [9,10] which is based on simple considerations of statistical mechanics and, more precisely, on the concept of partition function. A molecule may be distributed over different energy states, ε_n. A molecular partition function, f', precisely describes how the molecules are distributed over the allowed energy states, namely

$$f' = \sum_{n=0}^{k} \exp(-\varepsilon_n / k_B T), \tag{2.44}$$

where k_B is the Boltzmann constant and T is the absolute temperature. If reaction (2.42) is at equilibrium, its transition state must also be at equilibrium with the reagents (or substrates) A and B–X. The corresponding equilibrium constant, $K_{\neq}$, then takes the form

$$K_{\neq} = \frac{[T_{\neq}]}{[A][BX]}, \tag{2.45}$$

where [A], [BX] and $[T_{\neq}]$ are the concentration of the reagents and of the transition state. If ξ is the advancement of the reaction in the forward direction, one has

$$d\xi = -dn_A = -dn_{BX} = dn_{AX} = dn_B, \tag{2.46}$$

where n_A, n_{BX}, n_{AX} and n_B are the number of moles of substrates and products. The reaction rate in the same direction, v, is then defined by the expression

$$v = \frac{1}{V}\frac{d\xi}{dt} = k[A][BX], \tag{2.47}$$

where V is still the reaction volume and k is the rate constant in the forward direction.

The usual thermodynamic functions may be expressed in terms of partition functions [3] and indeed the same applies to equilibrium constants. Thus

$$K_{\neq} = \frac{f'_{\neq}}{f'_A f'_{BX}}, \tag{2.48}$$

where $f'_{\neq}$ is the molecular partition function of the transition state, f'_A and f'_{BX} are the molecular partition functions of the substrates. In these molecular partition functions, f', the energies are expressed from the zero level. As will be seen later, there is an advantage in expressing these energies from the lowest permissible level, ε_0, which is, in general, different from zero. Then one has

$$\sum_{n=0}^{k} \exp(-\varepsilon_n / k_B T) = \sum_{n=0}^{k} \exp\{-(\varepsilon_n - \varepsilon_0)/k_B T\} \exp(-\varepsilon_0 / k_B T), \tag{2.49}$$

or

$$f' = f\, e^{-\varepsilon_0 / k_B T}, \tag{2.50}$$

where f is now this new partition function of which the energies are expressed from that of the lowest level, ε_0. The previous equilibrium constant, $K_{\neq}$, can then be rewritten as

$$K_{\neq} = \frac{f_{\neq}}{f_A f_{BX}} \exp\{-(\varepsilon_{0\neq} - \varepsilon_{0A} - \varepsilon_{0BX})/k_B T\}, \tag{2.51}$$

where $\varepsilon_{0\neq}$, ε_{0A} and ε_{0BX} are the lowest levels in the transition state and in the substrates A and BX. Equation (2.51) above gives physical grounds to the empirical concept of energy of activation [11]. From this equation, the energy of activation, E_0, is defined as

$$E_0 = N(\varepsilon_{0\neq} - \varepsilon_{0A} - \varepsilon_{0BX}), \tag{2.52}$$

where N is the Avogadro number. Thus eq. (2.51) assumes the form

$$K_{\neq} = \frac{f_{\neq}}{f_A f_{BX}} e^{-E_0/RT}, \tag{2.53}$$

where R is the gas constant ($R = Nk_B$).

Within the framework of this theory, the transition state can be viewed as an unstable molecule a vibration of which becomes very loose and is converted into a translation, thus leading to the formation of the products. The partition function of this transition state, $f_{\neq}$, may thus be expressed as a product of two partition functions f_ν and $f^{\neq}$. The partition function f_ν refers to the set of atoms involved in the conversion process of the vibration into the translation. The partition function $f^{\neq}$ refers to the rest of the molecule. Thus one has

$$f_{\neq} = f_\nu f^{\neq}. \tag{2.54}$$

In the transition state theory, the partition function f_ν is identical to the partition function of the harmonic oscillator [3]

$$f_\nu = \sum_{n=0}^{\infty} \exp\{-(n + 1/2)h\nu/k_B T\}, \tag{2.55}$$

where n is now the quantum number of the vibration, h is the Planck's constant and ν is the frequency of the vibration. When the vibration is converted into a translation, $\nu \to 0$ and

$$f_\nu \approx \frac{k_B T(1 - h\nu/2k_B T)}{h\nu} \approx \frac{k_B T}{h\nu}. \tag{2.56}$$

Inserting this expression into eq. (2.54) yields

$$f_{\neq} = \frac{k_B T}{h\nu} f^{\neq}. \tag{2.57}$$

Similarly, eq. (2.53) becomes equivalent to

$$K_{\neq} = \frac{k_B T}{h\nu} \frac{f^{\neq}}{f_A f_{BX}} e^{-E_0/RT}. \tag{2.58}$$

Setting

$$K^{\neq} = \frac{f^{\neq}}{f_A f_{BX}} e^{-E_0/RT}, \tag{2.59}$$

eq. (2.17) becomes

$$K_{\neq} = \frac{k_B T}{h\nu} K^{\neq}. \tag{2.60}$$

An important idea of the transition state theory is to consider that the reaction rate depends on both the concentration of the transition state, $[T_{\neq}]$, and the frequency, ν, of the vibration which is being converted into a translation. The basic postulate of this theory is precisely to assume that the forward rate is equal to the product of these two variables. Thus

$$v = \frac{1}{T} \frac{d\xi}{dt} = \nu[T_{\neq}] = \nu K_{\neq}[A][BX]. \tag{2.61}$$

Comparing this relationship to eq. (2.47) leads to

$$k = \nu K_{\neq}, \tag{2.62}$$

and substituting the expression of $K_{\neq}$, derived from eq. (2.19), into expression (2.62) leads to

$$k = \frac{k_B T}{h} K^{\neq}, \tag{2.63}$$

which is the fundamental equation of the transition state theory. As $K^{\neq}$ is an equilibrium constant, one may define, from this equilibrium constant, a free energy of activation, $\Delta G^{\neq}$, as

$$K^{\neq} = e^{-\Delta G^{\neq}/RT}, \tag{2.64}$$

and, from this free energy of activation, an enthalpy, $\Delta H^{\neq}$, and an entropy, $\Delta S^{\neq}$, of activation can be defined as

$$\Delta G^{\neq} = \Delta H^{\neq} - T\Delta S^{\neq}. \tag{2.65}$$

Thus, through the transition state theory, a rate constant, which is considered as an operational quantity in chemical kinetics, gains physical significance. From eqs. (2.63)–(2.65) one can write

$$k = \frac{k_B T}{h} e^{-\Delta G^{\neq}/RT} \tag{2.66}$$

and

$$k = \frac{k_B T}{h} e^{\Delta S^{\neq}/R} e^{-\Delta H^{\neq}/RT}. \tag{2.67}$$

Transition state theory gives also physical support to the empirical Arrhenius equation

$$k = A_a e^{-E_0/RT}, \tag{2.68}$$

where A_a is a constant called the frequency factor and E_0 is still the activation energy. Combining eqs. (2.59) and (2.63) yields

$$k = \frac{k_B T}{h} \frac{f^{\neq}}{f_A f_{BX}} e^{-E_0/RT}, \tag{2.69}$$

which shows that the empirical frequency factor of the Arrhenius equation can be expressed in terms of partition functions

$$A_a = \frac{k_B T}{h} \frac{f^{\neq}}{f_A f_{BX}}. \tag{2.70}$$

It is now important to realize what the transition state theory has offered to the understanding of enzyme action. If we consider the simple chemical process

$$S \longleftrightarrow P \tag{2.71}$$

and if we now assume the process to be catalysed by an enzyme, the simplest of these enzyme processes should be made up of, at least, three elementary steps: a substrate fixation–desorption step, a catalytic step and a product desorption–fixation step, namely

$$E \overset{S}{\longleftrightarrow} ES \longleftrightarrow EP \underset{P}{\longleftrightarrow} E \tag{2.72}$$

Indeed, thermodynamics imposes that the overall equilibrium constant of processes (2.71) and (2.72) be the same. If it is possible to measure each of the rate constants, it then becomes possible, through eq. (2.66), to express the free energy of activation associated with each of these steps, and one can thus obtain the free energy profile of the enzyme reaction.

A possible energy profile of this reaction is shown in Fig. 2.2 and implies that a free energy barrier has to be overcome for each of the reaction steps. This type of energy profile raises a question that has long been a matter of debate. If, as was formely believed, the enzyme is viewed as a rigid molecule whose active site is complementary to the substrate, one may perfectly understand that the substrate is bound to the enzyme, but it becomes impossible to explain, in the framework of this lock and key hypothesis, how the high energy barrier associated with catalysis can be overcome. Moreover, as the enzyme reaction is more or less reversible, both S and P may be considered as the "substrate" of the reaction, depending on the direction chosen for following the dynamics of the chemical process. It is obvious that the enzyme cannot be complementary to both S and P. In order to solve this difficulty and to explain the origin of the energy required to overcome the energy barrier associated with catalysis, Pauling [12,13], as early as 1946, suggested that the active site of the enzyme be complementary, at least in an approximate manner, to the transition state, $T^{\neq}$, of the reaction.

2.2.2. *"Complementarity" between the active site of the enzyme and the transition state*

Over the last few decades, numerous results have been obtained that lead to the conclusion that Pauling's predictions are indeed valid. This evidence comes from simple thermodynamic considerations, from the use of transition state analogues as enzyme inhibitors, from structural studies on enzymes and from the discovery of catalytic antibodies ("abzymes").

2.2.2.1. *Enzyme-transition state "complementarity" derived from thermodynamic considerations*

Let us consider the simple uncatalysed and enzyme-catalysed reactions (2.71) and (2.72). If one aims at comparing the forward rate constant of the uncatalysed reaction with the catalytic constant of the corresponding enzyme-catalysed process, one may derive an ideal "thermodynamic box" that precisely allows this comparison. This "thermodynamic box" is shown in Fig. 2.3.

From this scheme, the first principle of thermodynamics requires that

$$\Delta G^{\neq}_{\mathrm{ne}} - \Delta G^{\neq}_{\mathrm{e}} = \Delta G_{\mathrm{S}} - \Delta G_{\mathrm{T}}. \tag{2.73}$$

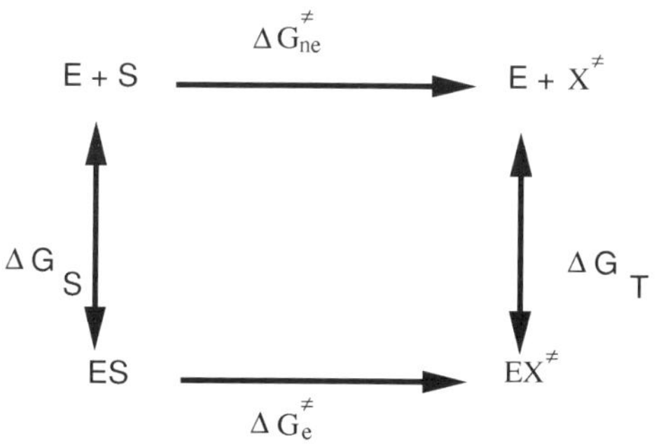

Fig. 2.3. Thermodynamic box that associates binding free energies of substrate and transition state to the enzyme.

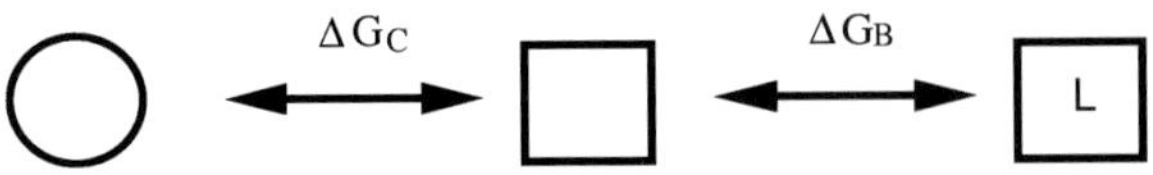

Fig. 2.4. The two ideal energy steps of an induced-fit process. The first step is associated with a conformation change of the protein, the second with the binding of the ligand L.

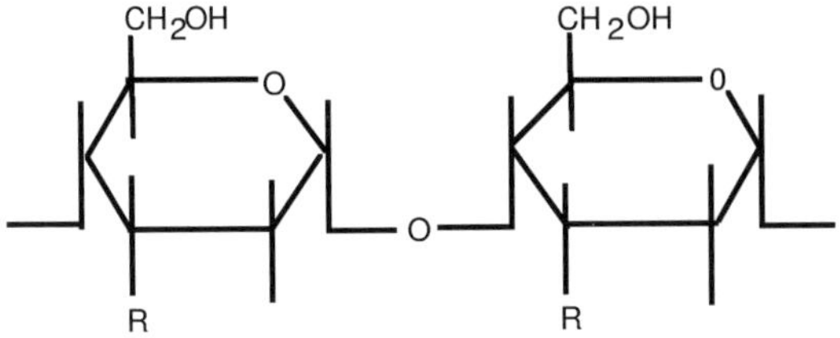

Fig. 2.5. A polymer of N-acetylglucosamine.

In this expression, $\Delta G^{\neq}_{\mathrm{ne}}$ and $\Delta G^{\neq}_{\mathrm{e}}$ represent the non-enzymatic and the enzymatic free energies of activation of the reaction. ΔG_{S} and ΔG_{T} are the binding energies of the substrate and of the transition state $\mathrm{X}^{\neq}$ to the enzyme. $\Delta G^{\neq}_{\mathrm{ne}}$ and $\Delta G^{\neq}_{\mathrm{e}}$ are of necessity positive whereas ΔG_{S} and ΔG_{T} may be positive or negative. If the protein behaves as a real catalyst, the free energy barrier of the uncatalyzed process, $\Delta G^{\neq}_{\mathrm{ne}}$, must be much higher than that of the corresponding catalyzed step, $\Delta G^{\neq}_{\mathrm{e}}$. Thus,

$$\Delta G^{\neq}_{\mathrm{ne}} > \Delta G^{\neq}_{\mathrm{e}}, \tag{2.74}$$

which, in turn, implies from eq. (2.73) that

$$\Delta G_{\mathrm{S}} > \Delta G_{\mathrm{T}}. \tag{2.75}$$

The larger the ΔG_{S} or ΔG_{T} value, the weaker is the binding of the substrate or transition state to the enzyme. Thus, as expected by Pauling, thermodynamics imposes that the transition state be bound much more tightly to the enzyme than the substrate itself [12,14].

One must stress that the tight binding of a ligand to an enzyme implies some sort of "complementarity" between the ligand and the active site of the enzyme. The concept of induced fit, developed by Koshland and associates [15–17], provides the rationale for this relationship between tight binding and "complementarity". Let us assume that a ligand requires a conformation change of the enzyme in order to be bound at the active site (Fig. 2.4).

The overall free energy change, ΔG_{L}, associated with the binding process may be split into two contributions. The first pertains to a conformation change that will give the enzyme the proper three-dimensional structure for "correct" ligand binding, i.e. a structure "complementary" to that of the ligand. In Fig. 2.4 above, this process is pictured by the conversion of a circle into a square. As the free enzyme occurs in its "circle" and not in its "square" state, one is led to the conclusion that the free energy of this ideal conformational transition, ΔG_{C}, can only be positive. The second ideal process corresponds to the binding of the ligand to the "complementary", distorted, active site. Its free energy is ΔG_{B} and may be positive or negative. It is the binding itself which induces the conformation change. The overall free energy ΔG_{L} that can be measured experimentally is thus

$$\Delta G_{\mathrm{L}} = \Delta G_{\mathrm{C}} + \Delta G_{\mathrm{B}}. \tag{2.76}$$

It is obvious that if the free enzyme is already "complementary" to the ligand, the binding process can take place without any conformation change of the enzyme. The energy contribution, ΔG_{C}, will then be equal to zero and the overall free energy, ΔG_{L}, will be smaller than one would have expected if an induced fit were taking place. The binding of the ligand, which can be measured experimentally, will then be tighter. There is thus a relationship between tight binding and enzyme-ligand "complementarity". In a way, one can conclude that thermodynamics imposes that the free enzyme be more "complementary" to the transition state than to the substrate itself.

It is now possible to understand the origin of the free energy required to overcome the energy barrier associated with catalysis. The central idea of the induced fit theory is that the enzyme is in a "strained" state when bound to the substrate. This "strained", or "activated", state has been termed "entatic" state by Williams [18]. It spontaneously tends to relapse to the conformation that has a lower energy level, and this is achieved at the top of the barrier, when the substrate has been converted into the transition state and when the enzyme adopts the conformation it had in the free, unliganded, state. As expected by Pauling [12], enzyme-transition state "complementarity" is one of the driving forces of catalysis [19,20]. As a transition state is some sort of unstable molecule, it is indeed impossible to test experimentally the tight affinity it must have for the enzyme. This difficulty can be circumvented, however, by using transition state analogues, i.e. non-reactive stable molecules that possess some features of real transition states, and by measuring their binding to the enzyme. An example of this approach will be described in the next section.

The induced movements of the enzyme may be small or large. In the case of kinases, and in particular of phosphoglycerate kinase, there is direct experimental evidence that these movements are of large amplitude. This enzyme is made up of two halves, each having a subsite that accomodates a substrate molecule. The binding of the two substrates on the enzyme induces a hinge motion of the domains that become into contact [21]. Since these domains are made of α-helical strands, they move relatively easily relative to one another.

Thus, according to the induced fit theory, an enzyme may undergo a conformation change corresponding to a "strain", when it binds a substrate. But one may also perfectly well assume that the enzyme behaves as a relatively rigid entity and that it is the substrate which undergoes this strain. The above reasoning about "complementarity" between the enzyme and the transition state still holds. There is little doubt that, for some enzymes, the strain is exerted on the substrate. This is precisely the case for lysozyme [22–24].

2.2.2.2. *Enzyme-transition state "complementarity" derived from structural considerations and from the use of transition state analogues*

The archetype of structural studies aimed at discovering some "complementarity" between the active site of an enzyme and the transition state of the corresponding reaction were done

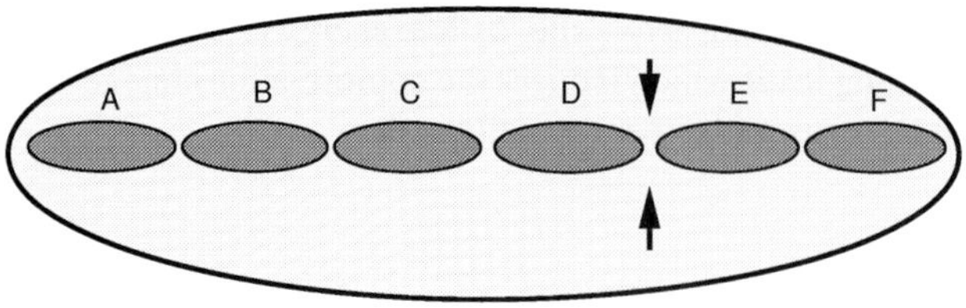

Fig. 2.6. Schematic picture of lysozyme. The picture represents the six subsites. Hydrolysis of the polymer is effected between subsites D and E (arrows).

on lysozyme [22,23]. The enzyme catalyses the hydrolysis of $\beta(1 \rightarrow 4)$ bonds of various polysaccharides of bacterial cell wall and, in particular, polymers of N-acetylglucosamine (Fig. 2.5).

The three-dimensional structure of the enzyme has been solved by crystallography [22,23]. Roughly speaking, the enzyme has an ellipsoidal shape and its active site is comprised of six subsites A, B, C, D, E, F (Fig. 2.6). Each subsite can accomodate a N-acetylglucosamine unit. Model building, however, has shown that whereas subsites A, B, C, E, F can accomodate sugar units, it is impossible to fit N-acetylglucosamine into subsite D, unless the pyranose ring is distorted.

This observation indeed suggests that the substrate is strained at subsite D. Moreover, direct measurements of the binding free energies of sugar residues to subsites A, B, C, D provide the following values [25]:

subsite A: -7 kJ/mole,

subsite B: -15 kJ/mole,

subsite C: -23 kJ/mole,

subsite D: $+12$ kJ/mole.

Measurements of the relative rate of hydrolysis of polymers of N-acetylglucosamine varies from 1 for the trimer to 4000 for the pentamer and 30 000 for the hexamer. A sharp increase of the rate takes place between the degrees of polymerization 4 and 5. This result strongly suggests that hydroysis takes place between the subsites D and E (Fig. 2.6). Crystallographic studies show that, in this region of the protein molecule, aspartic 52 and glutamic 35 are sitting on the two sides of the enzyme's cleft. These two aminoacid residues have markedly different environments. Whereas aspartic 52 is in a polar environment, glutamic 35 is in a nonpolar region of the enzyme molecule. This means that, under optimum pH conditions for lysozyme activity (about pH 5), aspartic 52 must be ionized and glutamic 35 un-ionized. In O^{18} water, the hydrolysis of a polymer of N-acetylglucosamine leads to smaller molecules of which some bear the C_1–O^{18} bond. Taken together, these results suggest that the hydroysis of the $\beta(1 \rightarrow 4)$ bond occurs according to the following mechanism [26]. The –COOH group of glutamic 35 donates a proton to the glycosidic oxygen atom and this leads to the cleavage of the C_1–O bond. The breakage of this bond generates a carbonium ion on the C_1 carbon of the glycosidic ring. This carbonium ion reacts with

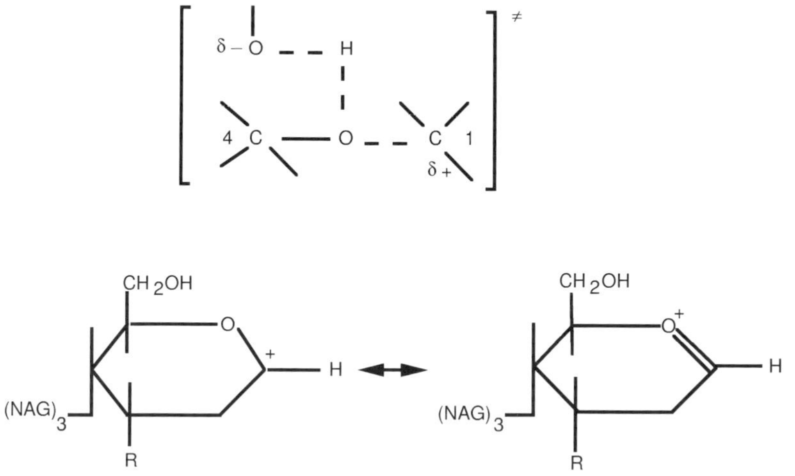

Fig. 2.7. Possible structure for the transition state of the lysozyme reaction. The lower part of the figure represents the resonance between the carbonium and oxonium states.

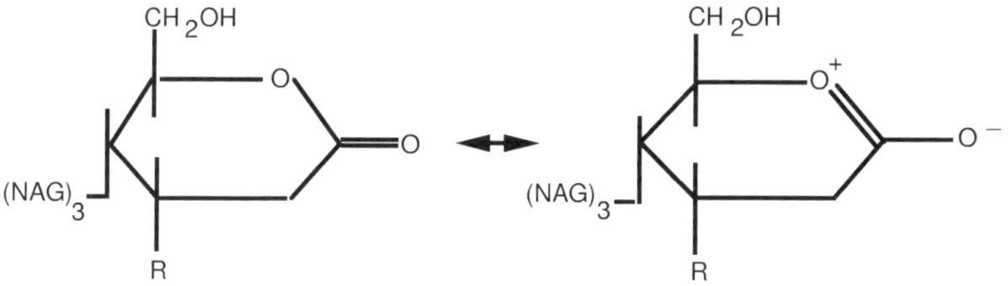

Fig. 2.8. A transition state analogue for the lysozyme reaction. The analogue displays the resonance between carbonium and oxonium states.

OH^- from the solvent and glutamic 35 becomes protonated. The basic features of this reaction mechanism are thus an acid catalysis exerted by glutamic 35 sitting 3 Å away from the glycosidic oxygen atom and the stabilization of the carbonium ion by the aspartic 52 located 3 Å from the ring D [27–30]. The likely transition state of the process is shown in Fig. 2.7. Moreover this transition state must be in resonance between the carbonium and the oxonium states (Fig. 2.7).

As mentioned previously, rings A, B, C fit more or less in the corresponding subsites whereas ring D does not. In order to fit nicely, this ring must be strained so as to have the half-chair conformation [26]. There is little doubt that this configuration is precisely that of the transition state. This may be confirmed by the result that the lactone analogue of tetra-N-acetylglucosamine carries the oxonium ion exactly as the transition state of the reaction does (Fig. 2.8).

In fact, this lactone is a stable analogue of the transition state and has an apparent affinity for lysozyme which is about 3600 times that of the corresponding tetra-N-acetylglucosamine. This situation is not specific for the lysozyme reaction; it has also been found for many other enzyme catalysed reactions. Transition state analogs have been synthesized for glucosidase, proline racemase, cytidine deaminase, etc., and behave as strong competitive inhibitors of the corresponding reactions [31–34]. The rationale for their high affinity is their "complementarity" to the active sites of the corresponding enzymes.

2.2.2.3. Enzyme-transition state "complementarity" derived from the study of catalytic antibodies

Probably the most direct and elegant evidence that an enzyme has to be "complementary" to the transition state of the reaction it catalyses comes from the study of catalytic antibodies, or abzymes [35,36]. If the idea that the active site of an enzyme has to be "complementary" to the transition state were correct, one should expect that, by using as hapten a synthetic substance that mimics a transition state structure, the immune response should provide specific antibodies with the presumed binding and catalytic activities of enzymes. This is precisely what has been observed experimentally and these antibodies have been called catalytic antibodies, or abzymes.

Of particular interest in this respect is the study of the hydrolysis of carboxylic esters by these catalytic antibodies. A schematic picture of the corresponding transition state is shown in Fig. 2.9. Phosphonate esters have been used as transition state analogs of this reaction. Monoclonal antibodies have been raised against protein conjugates of these phosphonate esters (Fig. 2.10). Depending on the substrate used, the antibodies obtained may take part in a stoichiometric, or in a catalytic reaction. In Fig. 2.10 are shown a stoichiometric and a catalytic reaction. In a stoichiometric reaction the substrate tends to inactivate the monoclonal antibody for, after breakage of the substrate molecule, part of it remains bound to the active site of the protein. The free antibody is therefore not released at the end of the chemical reaction and so it does not behave as a catalyst (Fig. 2.10). Alternatively, a catalytic substrate does not inactivate the antibody which is released in a free state at the end of the reaction (Fig. 2.10).

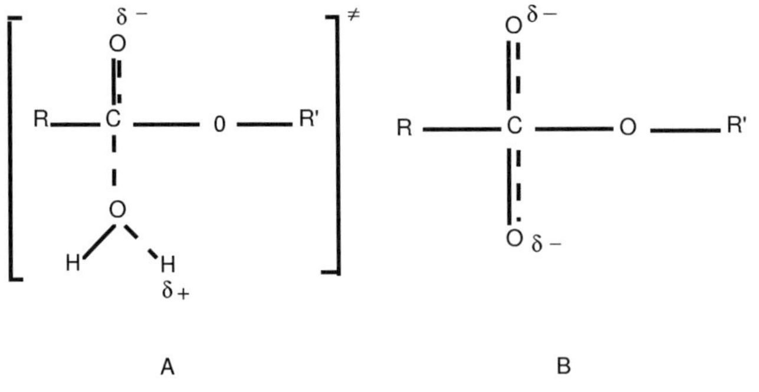

Fig. 2.9. Transition state for carboxylic ester hydrolysis. A – Transition state for the hydrolysis of carboxylic esters. B – Transition state analogue.

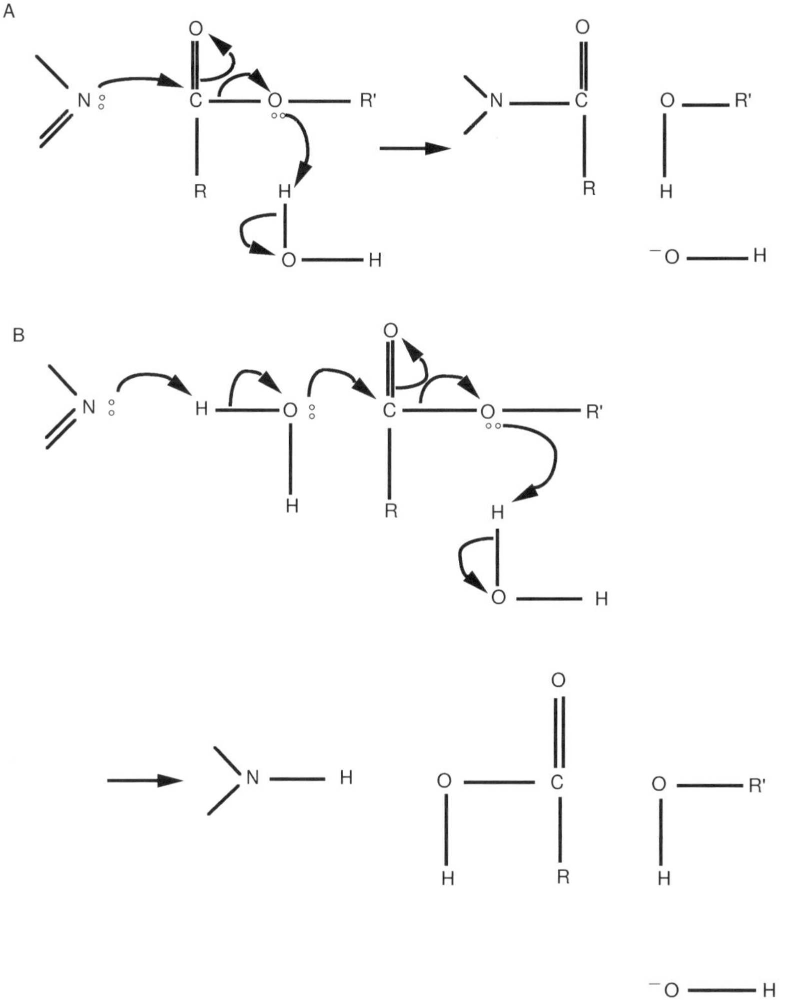

Fig. 2.10. Hydrolysis of carboxylic esters by "abzymes". A – Hydrolytic process involving a "stoichiometric" antibody. B – Hydrolytic process involving a "catalytic" antibody.

The study of catalytic antibodies is now expanding at a very rapid pace. But its main interest for our purpose is to show that the catalytic activity of a protein originates from its "complementarity" to the transition state of the reaction.

2.2.3. The time-course of an enzyme reaction

Let us consider, as an example, the following chemical process

$$A \longleftrightarrow P + Q. \tag{2.77}$$

If this reaction is catalysed by an enzyme and if it is studied during its early phase, its simplest kinetic scheme may be written as

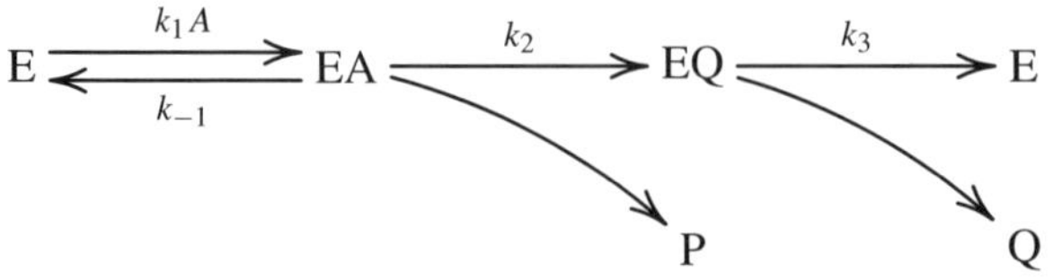

(2.78)

The steps of product release are considered irreversible because the product concentrations are small. Moreover, the substrate concentration, $[A] = a$, is very large relative to the total enzyme concentration. It will thus be considered as approximately constant during the early stages of the reaction. Then the differential equations associated with system (2.78) become linear and can be solved analytically. If e_0 and e are the total and the free enzyme concentrations, if m and n are the concentrations of the two enzyme intermediates, one must have the conservation equation

$$e_0 = e + m + n, \tag{2.79}$$

and the system of linear differential equations assume the form

$$\begin{aligned} \frac{dm}{dt} &= k_1 a e_0 - (k_1 a + k_{-1} + k_2) m - k_1 a n, \\ \frac{dn}{dt} &= k_2 m - k_3 n. \end{aligned} \tag{2.80}$$

This system is readily solved using, for instance, the Laplace transform (see Appendix). The system of differential equations can then be transformed into a system of linear algebraic equations

$$\begin{bmatrix} s(s + k_1 a + k_{-1} + k_2) & s k_1 a \\ -k_2 & s + k_3 \end{bmatrix} \begin{bmatrix} L\{m\} \\ L\{n\} \end{bmatrix} = \begin{bmatrix} k_1 a e_0 \\ 0 \end{bmatrix}, \tag{2.81}$$

where s is a complex variable (see Appendix), $L\{m\}$ and $L\{n\}$ are the Laplace transforms of m and n. Solving this system for $L\{m\}$ and $L\{n\}$ and rearranging yields

$$\begin{aligned} L\{m\} &= \frac{k_1 k_3 a e_0}{\lambda_1 \lambda_2} \frac{1}{s} - \frac{(k_3 - \lambda_1) k_1 a e_0}{\lambda_1 (\lambda_2 - \lambda_1)} \frac{1}{s + \lambda_1} - \frac{(k_3 - \lambda_2) k_1 a e_0}{\lambda_2 (\lambda_1 - \lambda_2)} \frac{1}{s + \lambda_2}, \\ L\{n\} &= \frac{k_1 k_2 a e_0}{\lambda_1 \lambda_2} \frac{1}{s} - \frac{k_1 k_2 a e_0}{\lambda_1 (\lambda_2 - \lambda_1)} \frac{1}{s + \lambda_1} - \frac{k_1 k_2 a e_0}{\lambda_2 (\lambda_1 - \lambda_2)} \frac{1}{s + \lambda_2}, \end{aligned} \tag{2.82}$$

and these expressions allow one to determine the variation of m and n as a function of time

$$m = \frac{k_1 k_3 a e_0}{\lambda_1 \lambda_2} + \frac{(k_3 - \lambda_1) k_1 a e_0}{\lambda_1 (\lambda_1 - \lambda_2)} e^{-\lambda_1 t} + \frac{(k_3 - \lambda_2) k_1 a e_0}{\lambda_2 (\lambda_2 - \lambda_1)} e^{-\lambda_2 t},$$

$$n = \frac{k_1k_2ae_0}{\lambda_1\lambda_2} + \frac{k_1k_2ae_0}{\lambda_1(\lambda_1-\lambda_2)}\,\mathrm{e}^{-\lambda_1 t} + \frac{k_1k_2ae_0}{\lambda_2(\lambda_2-\lambda_1)}\,\mathrm{e}^{-\lambda_2 t}. \tag{2.83}$$

In these expressions λ_1 and λ_2 represent the Encke's roots of the determinant of the square matrix of system (2.81). The rate of product appearance $\mathrm{d}p/\mathrm{d}t$ and $\mathrm{d}q/\mathrm{d}t$ are

$$\frac{\mathrm{d}p}{\mathrm{d}t} = k_2m, \qquad \frac{\mathrm{d}q}{\mathrm{d}t} = k_3n, \tag{2.84}$$

and from eq. (2.82) it is obvious that, at the beginning of the reaction, these rates are different, but, as time increases, the exponential terms vanish and the two rates of product appearance become identical

$$\frac{\mathrm{d}p}{\mathrm{d}t} = \frac{\mathrm{d}q}{\mathrm{d}t} = \frac{k_1k_2k_3}{\lambda_1\lambda_2}ae_0. \tag{2.85}$$

The concentations of the enzyme intermediates, m and n, are now time-independent. The reaction has reached an initial steady state. This expression can be given a more conventional form. The determinant Δ of the square matrix of system (2.81) is

$$\Delta = s\left\{s^2 + (k_1a + k_{-1} + k_2 + k_3)s + k_1(k_2+k_3)a + k_3(k_{-1}+k_2)\right\} \tag{2.86}$$

or

$$\Delta = s(s+\lambda_1)(s+\lambda_2), \tag{2.87}$$

and the product of the two roots λ_1 and λ_2 is thus

$$\lambda_1\lambda_2 = k_1(k_2+k_3)a + k_3(k_{-1}+k_2), \tag{2.88}$$

and the steady state rate takes the more conventional form

$$\frac{\mathrm{d}p}{\mathrm{d}t} = \frac{\mathrm{d}q}{\mathrm{d}t} = \frac{k_1k_2k_3ae_0}{k_1(k_2+k_3)a + k_3(k_{-1}+k_2)}. \tag{2.89}$$

The progress curves of the two products can be derived from the eq. (2.83) which can be written in condensed form as

$$\begin{aligned} m &= \psi_0 + \psi_1\,\mathrm{e}^{-\lambda_1 t} + \psi_2\,\mathrm{e}^{-\lambda_2 t}, \\ n &= \psi_0' + \psi_1'\,\mathrm{e}^{-\lambda_1 t} + \psi_2'\,\mathrm{e}^{-\lambda_2 t}, \end{aligned} \tag{2.90}$$

where the ψ and ψ' are groupings of rate constants equivalent to relaxation amplitudes. Integration of these expressions in the interval $[0, t]$ yields

$$\begin{aligned} p &= k_2\psi_0 t - k_2\frac{\psi_1}{\lambda_1}\left(\mathrm{e}^{-\lambda_1 t} - 1\right) - k_2\frac{\psi_2}{\lambda_2}\left(\mathrm{e}^{-\lambda_2 t - 1}\right), \\ q &= k_3\psi_0' t - k_3\frac{\psi_1'}{\lambda_1}\left(\mathrm{e}^{-\lambda_1 t} - 1\right) - k_3\frac{\psi_2'}{\lambda_2}\left(\mathrm{e}^{-\lambda_2 t - 1}\right). \end{aligned} \tag{2.91}$$

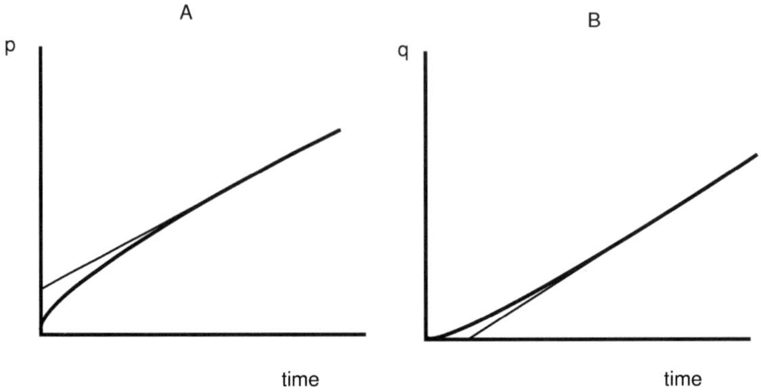

Fig. 2.11. Burst phase and lag phase of an enzyme reaction. A – Burst phase. B – Lag phase.

These equations allow one to determine the behaviour of the system during the pre-steady state phase. In order to do so, one can first derive the equations of the progress curve during the initial steady state phase. These equations are

$$\begin{aligned} p &= k_2\psi_0 t + \frac{k_2\psi_1}{\lambda_1} + \frac{k_2\psi_2}{\lambda_2}, \\ q &= k_3\psi_0' t + \frac{k_3\psi_1'}{\lambda_1} + \frac{k_3\psi_2'}{\lambda_2}. \end{aligned} \tag{2.92}$$

If these straight lines intersect the ordinate axis for negative values of p and q, the reaction displays a lag. Alternatively, if the intersect has a positive value, the reaction has a burst (Fig. 2.11). It may be shown (see Appendix) that the progress curve of product Q can only display a lag phase whereas the progress curve of product P can have either a lag, or a burst phase.

A burst phase will be obtained if (see Appendix)

$$\lambda_1\lambda_2 > k_3(\lambda_1 + \lambda_2), \tag{2.93}$$

and a lag will be obtained if this inequality is reversed. This mathematical reasoning is rigorous for the early stages of the reaction, when the concentrations of the products are small enough to allow one to neglect the reverse process of product binding to the enzyme. When this is not the case, the differential equations become nonlinear and cannot be solved analytically. When the products accumulate in the medium, the reaction rate falls off and becomes equal to zero when the equilibrium between substrate and products is reached. Thus, when the reaction reaches a plateau, this does not usually mean that all the substrate has been converted into the products, but rather that an equilibrium between substrate and products is reached. A typical progress curve of an enzyme-catalysed reaction is shown in Fig. 2.12.

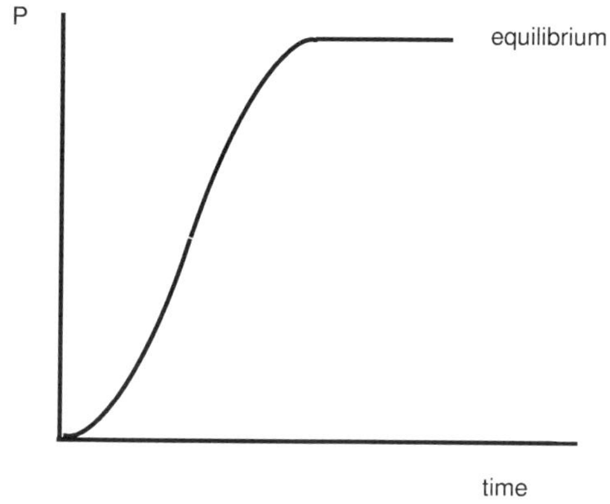

Fig. 2.12. The time course of an enzyme reaction. The curve shows the succession of: a pre-steady state phase, a steady state, the approach to equilibrium, and the equilibrium.

2.2.4. Simple enzymes that catalyse simple reactions

Some enzymes are relatively simple. They are made up of one polypeptide chain and bear only one active site. Their steady state kinetic properties may also be very simple. Let us consider a one-substrate, one-product enzyme catalyzed reaction. The simplest relevant kinetic model may be represented in cyclic form as

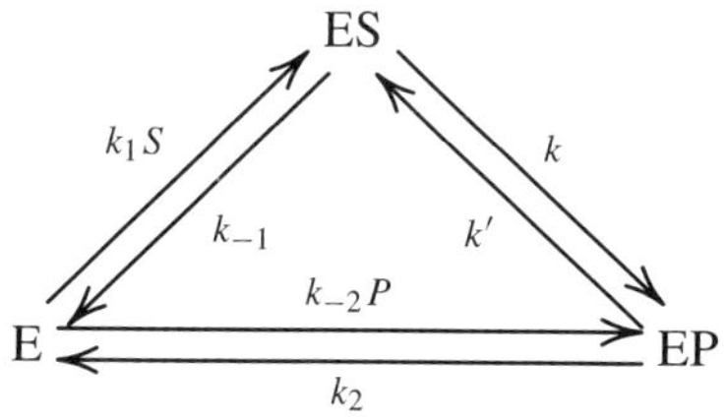

Under steady state conditions, one has the set of algebraic equations

$$-\big(k_1[\mathrm{S}]+k_{-2}[\mathrm{P}]\big)\frac{[\mathrm{E}]}{[\mathrm{E}]_0}+k_{-1}\frac{[\mathrm{ES}]}{[\mathrm{E}]_0}+k_2\frac{[\mathrm{EP}]}{[\mathrm{E}]_0}=0,$$

$$k_1[\mathrm{S}]\frac{[\mathrm{E}]}{[\mathrm{E}]_0}-(k_{-1}+k)\frac{[\mathrm{ES}]}{[\mathrm{E}]_0}+k'\frac{[\mathrm{EP}]}{[\mathrm{E}]_0}=0,$$

$$k_{-2}[\mathrm{P}]\frac{[\mathrm{E}]}{[\mathrm{E}]_0}+k\frac{[\mathrm{ES}]}{[\mathrm{E}]_0}-(k_2+k')\frac{[\mathrm{EP}]}{[\mathrm{E}]_0}=0, \tag{2.94}$$

and a conservation equation

$$\frac{[\mathrm{E}]}{[\mathrm{E}]_0}+\frac{[\mathrm{ES}]}{[\mathrm{E}]_0}+\frac{[\mathrm{EP}]}{[\mathrm{E}]_0}=1, \tag{2.95}$$

where $[\mathrm{E}]_0$ is the total enzyme concentration. If any of the steady state equations (2.94) is replaced by the conservation equation (2.95) and if the resulting system of simultaneous equations is solved for $[E]/[E]_0$, $[ES]/[E]_0$ and $[EP]/[E]_0$. The steady state concentrations of enzyme intermediates can be obtained. For example,

$$\frac{[ES]}{[E]_0} = \frac{\begin{vmatrix} 1 & 1 & 1 \\ k_1[S] & 0 & k' \\ k_{-2}[P] & 0 & -(k_2+k') \end{vmatrix}}{\begin{vmatrix} 1 & 1 & 1 \\ k_1[S] & -(k_{-1}+k) & k' \\ k_{-2}[P] & k & -(k_2+k') \end{vmatrix}}. \tag{2.96}$$

The net reaction rate (the difference between the reaction rates in the forward and reverse directions) is then

$$\frac{v}{[E]_0} = k_2\frac{[EP]}{[E]_0} - k_{-1}\frac{[ES]}{[E]_0}. \tag{2.97}$$

Several short-hand methods [37–40] have been proposed that allow the automatic derivation of steady state rate equations and therefore avoid computation of determinants. Probably the most classical is that of King and Altman [37]. This method, like the others, can be formulated as simple rules. In the case of a one-substrate, one-product reaction, these rules are the following:

- the numerator of the rate equation is made up of two cyclical patterns of three arrows that describe the forward and the reverse enzyme processes;
- the first cyclical pattern is positive, the second negative;
- the denominator of the rate equation collects all the possible non-cyclical patterns of two arrows;
- patterns made up of two arrows originating from the same point are not allowed;
- all the patterns of the denominator are positive. The patterns of the numerator and of the denominator are shown in Fig. 2.13.

The net reaction rate that may be derived is

$$v = \frac{\{k_1kk_2[S] - k_{-1}k'k_{-2}[P]\}[E]_0}{kk_2 + k_{-1}k_2 + k_{-1}k' + (k+k'+k_2)k_1[S] + (k+k'+k_{-1})k_{-2}[P]}, \tag{2.98}$$

which can be rewritten as

$$v = \frac{V\overline{K}[S] - V'\overline{K'}[P]}{1 + \overline{K}[S] + \overline{K'}[P]} \tag{2.99}$$

with

$$\overline{K} = \frac{k_1(k+k'+k_2)}{kk_2 + k_{-1}k_2 + k_{-1}k'}, \qquad V = \frac{kk_2}{k+k'+k_2}[E]_0,$$

$$\overline{K'} = \frac{k_{-2}(k+k'+k_{-1})}{kk_2 + k_{-1}k' + k_{-1}k_2}, \qquad V' = \frac{k_{-1}k'}{k+k'+k_{-1}}[E]_0. \tag{2.100}$$

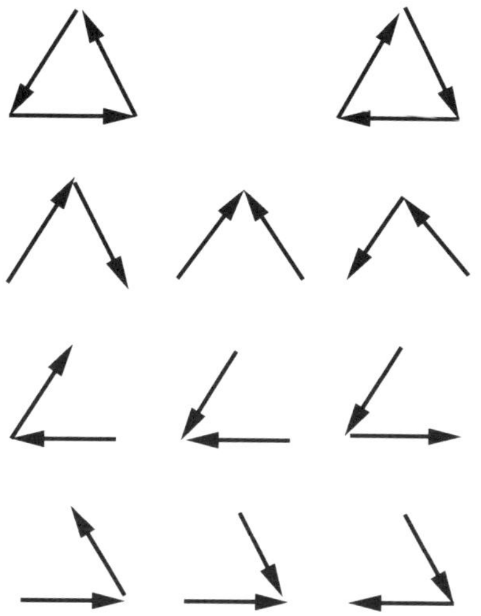

Fig. 2.13. The cyclical and non-cyclical patterns of a one-substrate, one-product enzyme reaction.

V and V' are the maximum reaction rates in the forward and in the reverse directions, respectively. Similarly, $\overline{K}$ and $\overline{K'}$ are the apparent affinity constants of the enzyme for the substrate and the product. When the equilibrium between the substrate and the product is reached, the net rate is nil and therefore

$$V\overline{K}[\overline{\mathrm{S}}] = V'\overline{K'}[\overline{\mathrm{P}}], \tag{2.101}$$

where $[\overline{\mathrm{S}}]$ and $[\overline{\mathrm{P}}]$ are equilibrium concentrations. From this expression, one can derive the so-called Haldane relationship

$$\frac{[\overline{\mathrm{P}}]}{[\overline{\mathrm{S}}]} = K = \frac{V\overline{K}}{V'\overline{K'}}, \tag{2.102}$$

which associates kinetic parameters in either direction and the thermodynamic equilibrium constant, K.

If the steady state is measured under initial conditions, the corresponding equation reduces to

$$v = \frac{V\overline{K}[\mathrm{S}]}{1 + \overline{K}[\mathrm{S}]}, \tag{2.103}$$

and its graphical representation, in v *versus* [S] space, is a rectangular hyperbola. This hyperbola may be transformed into a straight line (see, for instance, refs. [41] and [42]). The most classical of these transformations is no doubt the Lineweaver–Burk transformation, in which $1/v$ is plotted against $1/[\mathrm{S}]$. If the net reaction rate is plotted against the affinity

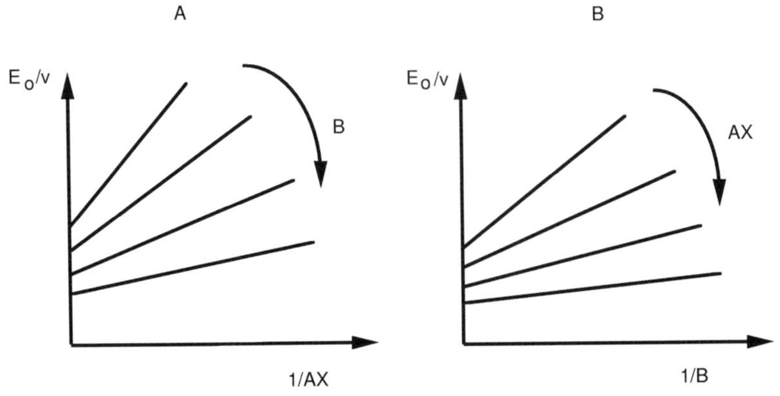

Fig. 2.14. The reciprocal plots for a sequential ordered mechanism. The plots pertain to eq. (2.106).

of the reaction, the curve thus obtained is sigmoidal and will therefore look linear about the inflection point [43]. Thus there is an apparent linear flow-force relationship even if the reaction is far away from thermodynamic equilibrium.

Most enzyme reactions involve the participation of several substrates and several products. For instance, a group transfer reaction

$$AX + B \longleftrightarrow A + BX \tag{2.104}$$

may take place according to different sequences of chemical events. Thus the three different sequences below are different but lead to the same overall result

$$1 \quad E \xleftrightarrow{AX} EAX \xleftrightarrow{B} EAX.B \longleftrightarrow EA.XB \xleftrightarrow[XB]{} EA \xleftrightarrow[A]{} E \tag{2.105}$$

$$2 \quad E \xleftrightarrow{B} EB \xleftrightarrow{AX} EB.XA \longleftrightarrow EBX.A \xleftrightarrow[A]{} EBX \xleftrightarrow[BX]{} E$$

$$3 \quad E \xleftrightarrow{AX} EAX \xleftrightarrow[A]{} EX \xleftrightarrow{B} EXB \xleftrightarrow[BX]{} E$$

The question that is now raised is to know whether it is possible to screen amongst these different reaction mechanisms in order to know which of them is compatible with the rate data. If the steady state rate is measured under initial conditions and in the absence of any product, the reciprocal rate equations are

$$\frac{[E]_0}{v} = \phi_0 + \frac{\phi_1}{[AX]} + \frac{\phi_2}{[B]} + \frac{\phi_{12}}{[AX][B]} \tag{2.106}$$

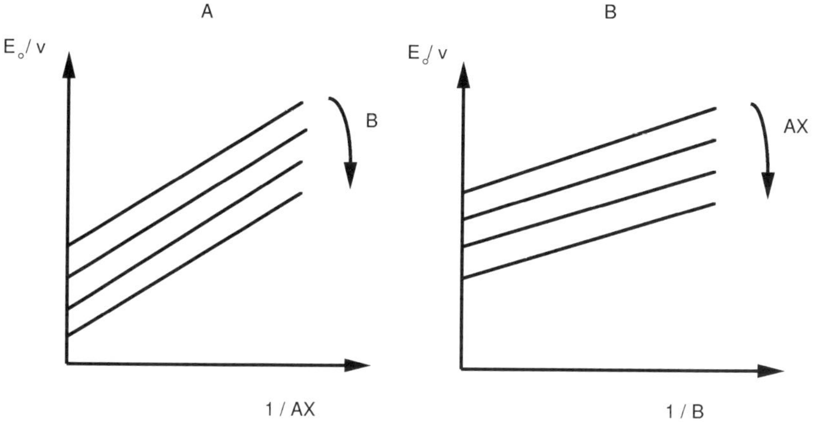

Fig. 2.15. The reciprocal plots for a substitution mechanism. The plots pertain to eq. (2.107).

for models 1 and 2 and

$$\frac{[\mathrm{E}]_0}{v} = \phi_0 + \frac{\phi_1}{[\mathrm{AX}]} + \frac{\phi_2}{[\mathrm{B}]} \tag{2.107}$$

for model 3. In these equations, the ϕs are groupings of rate constants [44]. On the basis of this analysis alone, models 1 and 2 are thus indistinguishable, but model 3 can be distinguished from the two others. Equation (2.106) predicts that the reciprocal steady state rate varies linearly with the reciprocal of either substrate concentration and that both the slope and the intercept depend on the concentration of the non-varied substrate (Fig. 2.14). In the case of eq. (2.107), the plots are parallel; only the intercept depends on the non-varied substrate (Fig. 2.15).

To distinguish between models 1 and 2, the variation of the reaction rate has to be studied as a function of the substrate concentrations, but in the presence of one of the products. For instance, in the presence of product A, the rate equation (2.106) becomes

$$\frac{[\mathrm{E}]_0}{v} = \phi_0 + \frac{\phi_1}{[\mathrm{AX}]} + \frac{\phi_2}{[\mathrm{B}]} + \frac{\phi_{12}}{[\mathrm{AX}][\mathrm{B}]} + \frac{\phi_1^{\mathrm{I}}[\mathrm{A}]}{[\mathrm{AX}]} + \frac{\phi_{12}^{\mathrm{I}}[\mathrm{A}]}{[\mathrm{AX}][\mathrm{B}]} \tag{2.108}$$

for model 1 and

$$\frac{[\mathrm{E}]_0}{v} = \phi_0 + \frac{\phi_1}{[\mathrm{AX}]} + \frac{\phi_2}{[\mathrm{B}]} + \frac{\phi_{12}}{[\mathrm{AX}][\mathrm{B}]} + \phi_0^{\mathrm{I}}[\mathrm{A}] + \frac{\phi_2^{\mathrm{I}}[\mathrm{A}]}{[\mathrm{B}]} + \frac{\phi_{12}^{\mathrm{I}}[\mathrm{A}]}{[\mathrm{AX}][\mathrm{B}]} \tag{2.109}$$

for model 2. These two equations can now be distinguished on the basis of experimental results. The product A is competitive relative to AX (i.e. A will influence the slope of the plots but not their intercept) for eq. (2.108) and non-competitive relative to the same substrate for model 2 (A will affect both the slope and the intercept of the plots obtained *versus* $1/[\mathrm{AX}]$). This kind of behaviour is depicted in Fig. 2.16.

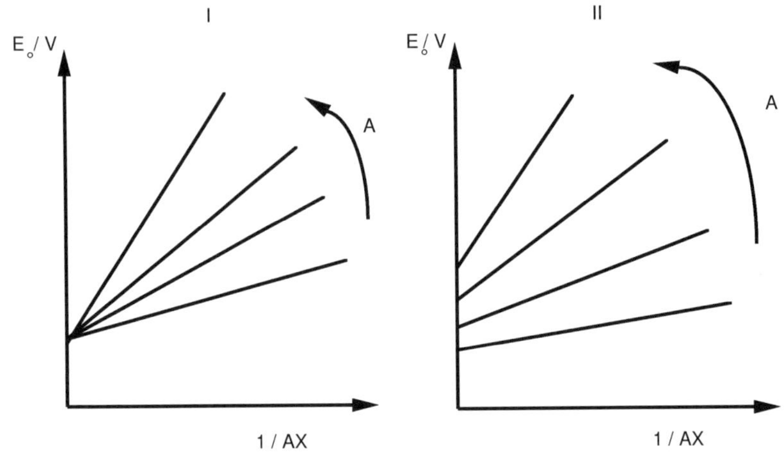

Fig. 2.16. Different types of inhibition exerted by a product on an enzyme reaction. I – The product behaves as a competitive inhibitor. II – The product behaves as a non competitive inhibitor. The type of inhibition allows to distinguish amongst different sequences of substrate binding and product release.

The situation will be exactly symmetrical when studying the effect of product BX on the reciprocal rate. In the case of model 3, the presence of either product will affect not only the intercept of the plots, but also the slope. The plots will then fail to remain parallel. The aim of steady state kinetics is thus to find an experimental strategy that allows the screening of *a priori* possible reaction schemes. The reader is referred to the various books of steady state kinetics that have been published over the past decades [41,42,45–47].

2.2.5. Simple enzymes that catalyse complex reactions

2.2.5.1. Kinetic co-operativity

Over the last decades, it has become more and more evident that some simple enzymes, bearing only one active site for their substrate(s), may catalyse very complex kinetic processes. This kinetic complexity appears to be the consequence of "slow" conformation changes of the enzyme during the reaction. As already discussed in the previous section, the usual response of an enzymatic process in steady state to a change of substrate concentration is hyperbolic. The rectangular hyperbola expresses the progressive saturation of the active site as the substrate concentration is increased. This so-called Michaelis–Menten behaviour is observed for a large number of enzymes. Other enzymes display significant departures from hyperbolic behaviour. They are called co-operative. There exists two extreme cases of a co-operative response. They are called positive and negative co-operativity. These two types of response generate deviations from linearity of their reciprocal plots. Positive co-operativity generates an upward, and negative co-operativity a downward curvature of the reciprocal plots.

In classical biochemistry, positive co-operativity is viewed as a positive interaction between identical active sites. The occupancy of a first site by the substrate facilitates the

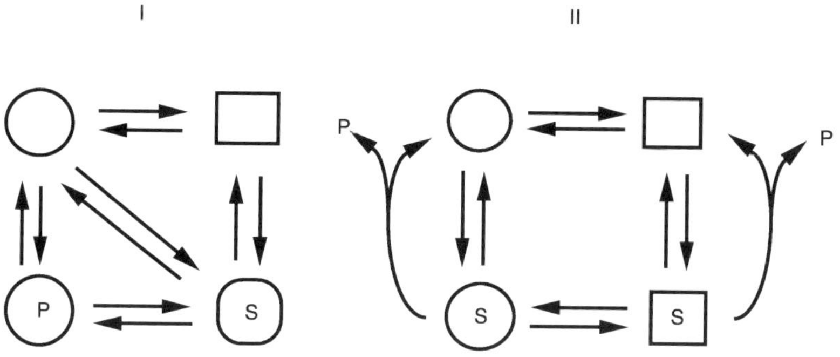

Fig. 2.17. Enzyme memory and enzyme hysteresis. I – Enzyme memory for a one substrate–one product enzyme. II – Enzyme hysteresis.

binding of another substrate molecule on a second active site and so on. This situation often leads to a sigmoidal curve, when the steady state rate is plotted against the substrate concentration. One may speculate that there exists a functional advantage for this type of behaviour, for the enzyme may display an all-or-none type of response to small changes of substrate concentration. Alternatively, the occupancy of a first active site by the substrate may lead to a decrease of affinity of a second site for the same substrate and so on. The co-operativity is then negative. The steady state rate increases sharply at low substrate concentration and does not vary significantly afterwards. Positively and negatively co-operative enzymes are thus made up of several identical polypeptide chains in interaction. These co-operative effects, whether positive or negative, imply that information is transferred from site to site within the polymeric enzyme, and therefore that the protein changes its conformation when the substrate concentration is varied [15,48]. It has long been implicitly considered that these conformation changes are fast events. Although this is certainly the case for many ligand-induced conformation changes, there are a number of enzymes, called hysteretic [49–53], that display conformation changes whose time-scale is similar to, or even slower than the other reaction steps. These "slow" conformation changes may be due to different types of molecular events: the rotation, or the sliding, of α-helices; the cis-trans isomerisation about a proline-imide bond; the segmental movement of a polypeptide chain; the hinge bending of two protein domains [51].

2.2.5.2. *Enzyme hysteresis and enzyme memory*

If an enzyme bears one active site only but displays these "slow" conformational changes, it may potentially exhibit co-operativity. This co-operativity, however, is only kinetic and cannot occur under thermodynamic equilibrium conditions. It originates from the occurrence of different catalytic paths having different relative importance as the substrate concentration is varied. Two models of this kinetic co-operativity have been particularly studied. The first one postulates that the free enzyme exists in two different conformations, each able to take part, at different rates, in two catalytic pathways (Fig. 2.17).

The second is simpler and postulates that, upon desorption of the last product, the enzyme retains, or "recalls", for a while the conformation stabilised by the last product of the

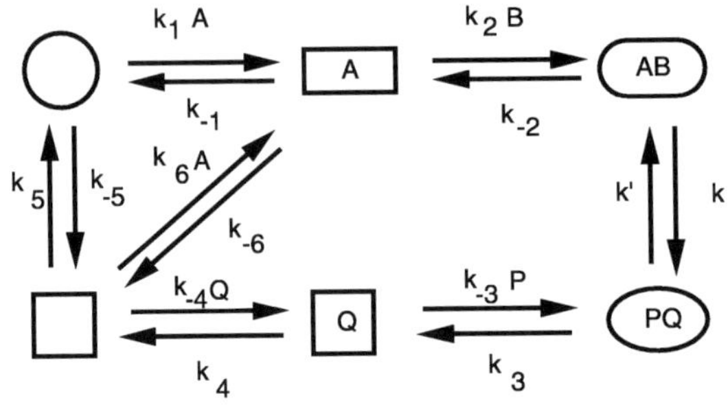

Fig. 2.18. Enzyme memory for a two substrate–two product enzyme. See text.

reaction sequence before relapsing to its initial conformation. For this reason this model is called "mnemonical" [52–56]. The two models are in fact compatible and, for both of them, kinetic co-operativity requires the isomerisation step of the free enzyme not to be much faster than the other reaction steps. In either of these models, co-operativity does not appear to be the consequence of information transfer between identical catalytic sites, but rather of co-operation between different conformations of the same enzyme molecule.

The two models cannot be experimentally distinguished for one-substrate, one-product enzymes, but this screening becomes possible for multisubstrate, multiproduct enzymes. In particular, the predictions of the mnemonical model are very straightforward and can be tested experimentally. These predictions are the following: the binding of the substrates should be ordered; kinetic co-operativity should be observed for the first substrate, but not for the second; the last product of the reaction sequence should enhance the kinetic co-operativity if this co-operativity is negative, but should reverse the co-operativity if it is positive. A mnemonical model for a two-substrate, two-product enzyme is presented in Fig. 2.18. In the case of hysteretic model II, there is no obvious reason to believe that all these effects will be encountered. In particular, it would be extremely difficult, without having recourse to *ad hoc* hypotheses, to explain that kinetic co-operativity may occur for the first, but not for the second substrate. Since the mnemonical model is simpler, most of the following discussion will be centered on it.

2.2.5.3. *Quantitative evaluation of the kinetic co-operativity*

In the absence of the products in the reaction medium, the reciprocal rate equation can be cast into either of the two following forms

$$\frac{[\mathrm{E}]_0}{v} = \frac{(\alpha_1 + \alpha_2[\mathrm{B}])\frac{1}{[\mathrm{A}]^2} + (\beta_1 + \beta_2[\mathrm{B}])\frac{1}{[\mathrm{A}]} + (\gamma_1 + \gamma_2[\mathrm{B}])}{\delta[\mathrm{B}]\frac{1}{[\mathrm{A}]} + \varepsilon[\mathrm{B}]}, \tag{2.110a}$$

$$\frac{[\mathrm{E}]_0}{v} = \frac{\alpha_2 + \beta_2[\mathrm{A}] + \gamma_2[\mathrm{A}]^2}{\delta[\mathrm{A}] + \varepsilon[\mathrm{A}]^2} + \frac{\alpha_1 + \beta_1[\mathrm{A}] + \gamma_1[\mathrm{A}]^2}{\delta[\mathrm{A}] + \varepsilon[\mathrm{A}]^2}\frac{1}{[\mathrm{B}]}, \tag{2.110b}$$

where α, β, ... are groupings of rate constants [53]. Under these two forms, it is obvious that the reciprocal plots should be nonlinear relative to $1/[\mathrm{A}]$ and linear with respect to $1/[\mathrm{B}]$. The simplest way to evaluate the extent of this co-operativity is probably to derive the expression of the second derivative of $[\mathrm{E}]_0/v$ with respect to $1/[\mathrm{A}]$ as $1/[\mathrm{A}] \to 0$. This is

$$\lim \frac{\partial^2([\mathrm{E}]_0/v)}{\partial(1/[\mathrm{A}])^2} = \Gamma = \frac{2k_5}{k_1 k_6^2}(k_6 - k_1). \tag{2.111}$$

This equation remains valid whatever the degree of complexty of the mnemonical model. That is, this expression would be unchanged for a one-substrate, one-product enzyme, or for a two-substrate, two-product enzyme, etc., provided that the concentration of the products is negligible in the reaction medium. The sign of the kinetic co-operativity depends solely on the respective values of k_6 and k_1. If $k_6 > k_1$, $\Gamma > 0$ and the kinetic co-operativity is positive. If $k_1 > k_6$, $\Gamma < 0$, the kinetic co-operativity is negative. The enzyme will display no co-operativity if $\Gamma = 0$, that is if $k_1 = k_6$. The Γ parameter has a dimension, and is expressed in $\mathrm{s\,M}^2$. If the last product, Q, is present in the reaction medium, the expression of Γ assumes the form

$$\Gamma = \frac{2k_5}{k_1 k_6^2}\{(k_6 - k_1) - k_1 K_4[\mathrm{Q}]\}, \tag{2.112}$$

where K_4 is the affinity constant of the last product Q for the enzyme. Again this equation is valid for any type of mnemonical model. From expression (2.112) it becomes obvious that, as the concentration of Q is increased, the value of Γ will become more and more negative (if Γ is already negative in the absence of product Q), or its sign will be reversed from positive to negative (if Γ is positive in the absence of product Q). This reversal of co-operativity by the last product of the reaction sequence is illustrated in Fig. 2.19. Moreover, the Γ value is independent of the concentration of the other substrate, B, or of the first product, P.

Although simple, the Γ parameter is not ideal, however, for it has a dimension and its value depends on the value of the reaction velocity. To avoid this potential difficulty, one can use the so-called extreme Hill coefficient, that is the maximum (for positive co-operativity), or the minimum (for negative co-operativity) of the function

$$h = \frac{\partial\{\log v/(V - v)\}}{\partial \log[\mathrm{A}]}, \tag{2.113}$$

where V is the maximum reaction rate. In the case of a one-substrate enzyme bearing several active sites displaying positive co-operativity, the maximum value of this function will be obtained for the substrate concentration, [A], that gives half-saturation of the enzyme by the substrate. In the case of a two-substrate mnemonical enzyme the situation is more complex and the extreme (maxium or minimum) co-operativity will be obtained for critical values of the concentration of A given by [57]

$$[\mathrm{A}]^2_{\mathrm{ext}} = \frac{\alpha_1 + \alpha_2[\mathrm{B}]}{\alpha_1 + (\alpha_2 - \varepsilon\Gamma/2)[\mathrm{B}]}\frac{\delta^2}{\varepsilon^2}, \tag{2.114}$$

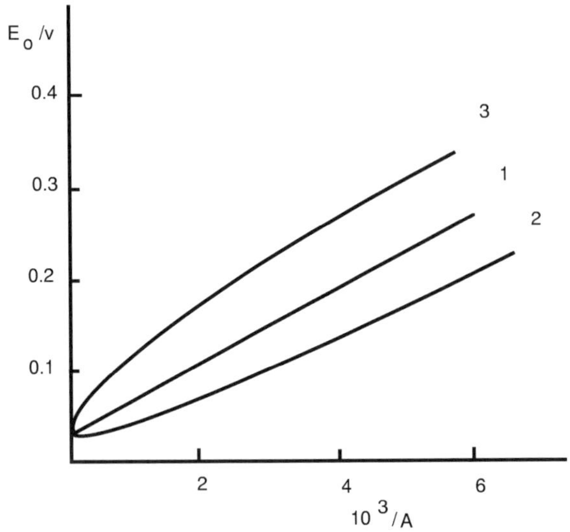

Fig. 2.19. Change of kinetic co-operativity exerted by the last product of the reaction sequence. In the absence of the last product (Q) the co-operativity is nil (curve 1) or positive (curve 2). This co-operativity becomes negative in the presence of Q (curve 3). Adapted from ref. [52].

which depends on the concentration of B. In this expression α, δ and ε are groupings of rate constants that have already appeared in eqs. (2.110a) and (2.110b). The extreme (maximum and minimum) value of the function h, $\tilde{h}_{ext}$, will be [57]

$$\tilde{h}_{ext} = \frac{2}{1 + \left\{ \frac{\alpha_1 + (\alpha_2 - \varepsilon\Gamma/2)[B]}{\alpha_1 + \alpha_2[B]} \right\}^{1/2}}, \tag{2.115}$$

and also depends upon the concentration of B. This extreme Hill coefficient is a dimensionless number equal to one when there is no co-operativity. If the co-operativity is positive, $1 < h < 2$, and if it is negative, $0 < h < 1$. As the substrate concentration, [B], is increased, positive and negative co-operativity, as revealed by the extreme Hill coefficient, are enhanced (Fig. 2.20).

The same applies to the extreme concentration of A, i.e. the concentration of A pertaining to the extreme Hill coefficient. When either of the products P or Q is present in the medium, the expression of the extreme Hill coefficient is more complex and assumes either of the forms [57]

$$\tilde{h}_{ext} = \frac{2}{1 + \left\{ \frac{\alpha_1 + (\alpha_2 - \varepsilon\Gamma/2)[B] + \alpha_3[P]}{\alpha_1 + \alpha_2[B] + \alpha_3[P]} \right\}^{1/2}}, \tag{2.116}$$

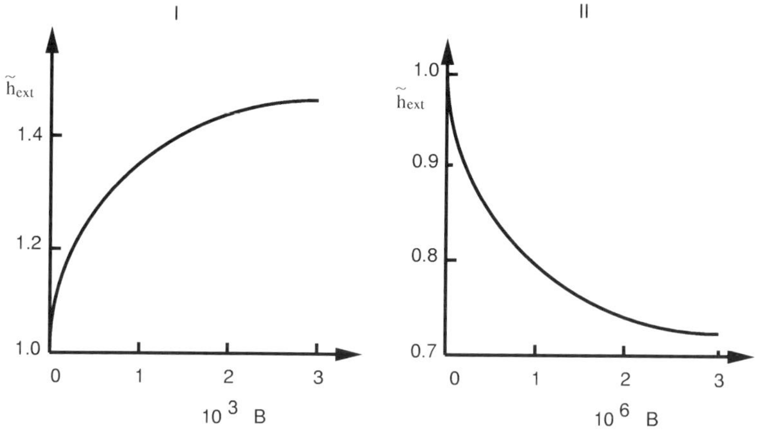

Fig. 2.20. Enhancement of co-operativity by the second substrate of a mnemonical enzyme. I – Enhancement of positive co-operativity. II – Enhancement of negative co-operativity. Adapted from ref. [57].

or [57]

$$\tilde{h}_{ext} = \frac{2}{1 + \left\{ \dfrac{\alpha_1 + (\alpha_2 - \varepsilon\Gamma/2)[B] + (\alpha_4 + \alpha_5[B])[Q]}{\alpha_1 + \alpha_2[B] + (\alpha_4 + \alpha_5[B])[Q]} \right\}^{1/2}}. \tag{2.117}$$

The corresponding critical concentrations of A pertaining to the extreme Hill coefficients are derived from

$$[A]^2_{ext} = \frac{\alpha_1 + \alpha_2[B] + \alpha_3[P]}{\alpha_1 + (\alpha_2 - \varepsilon\Gamma/2)[B] + \alpha_3[P]} \frac{\delta^2}{\varepsilon^2} \tag{2.118}$$

for eq. (2.116), and

$$[A]^2_{ext} = \frac{\alpha_1 + \alpha_2[B] + (\alpha_4 + \alpha_5[B])[Q]}{\alpha_1 + (\alpha_2 - \varepsilon\Gamma/2)[B] + (\alpha_4 + \alpha_5[B])[Q]} \frac{\delta^2}{\varepsilon^2} \tag{2.119}$$

for eq. (2.117). Equation (2.116) shows that the product P brings about a decrease of the extreme Hill coefficient, if the co-operativity is positive in the absence of this product, and an increase of this coefficient if the co-operativity is negative under the same conditions. In fact, increasing the concentration of P tends to suppress the co-operativity (Fig. 2.21). The effet of Q, through its effects on the extreme Hill coefficient, tends to enhance negative co-operativity, or to reverse positive into negative co-operativity (Fig. 2.21).

2.2.5.4. Wheat germ hexokinase as a mnemonical enzyme

Several enzymes display a behaviour that has been considered hysteretic or mnemonic. This is the case for instance for wheat germ hexokinase LI [52–54,56], rat liver glucoki-

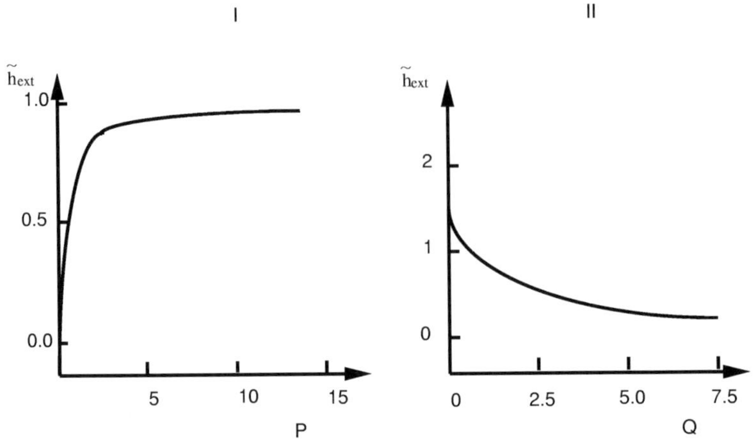

Fig. 2.21. Effect of the two products on the kinetic co-operativity of a mnemonical enzyme. I – The first product tends to suppress a negative co-operativity. II – The second product tends to reverse positive into negative co-operativity. Adapted from ref. [57].

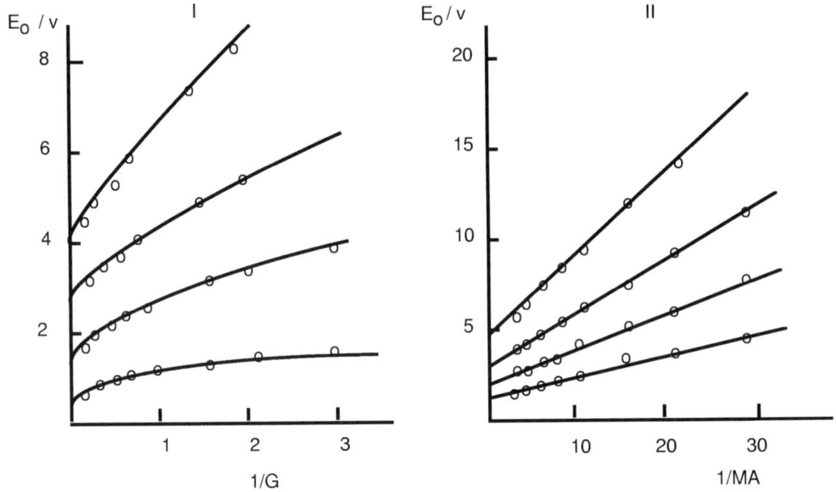

Fig. 2.22. Kinetic co-operativity of wheat germ hexokinase LI. I – The enzyme displays negative co-operativity with respect to the first substrate, glucose. II – The enzyme displays no co-operativity with respect to the second substrate MgATP. Adapted from ref. [53].

nase [58], octopine dehydrogenase [59], fumarase [60], plant cell wall glucosidase [61] and chloroplast fructose bisphosphatase [62,63]. Of particular interest is the case of wheat germ hexokinase LI for all the theoretical predictions above have been found experimentally to occur with this enzyme. This hexokinase is a monomeric protein of 50 000 molecular mass. It binds its substrates (glucose and MgATP) and releases its products (glucose-6-phosphate

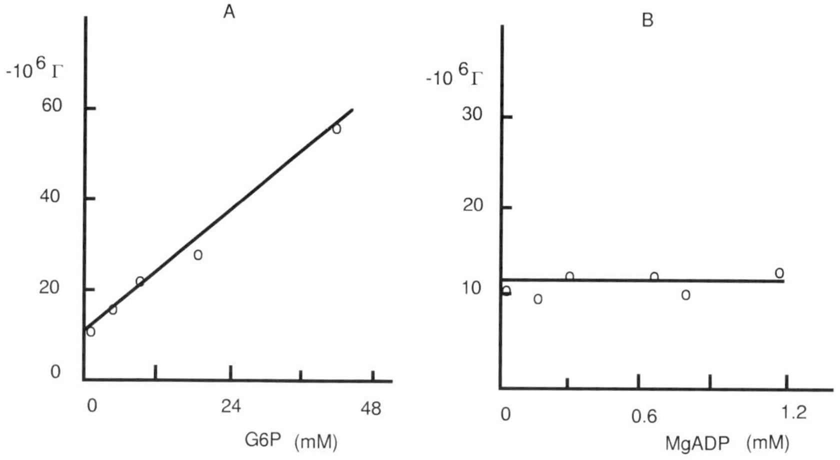

Fig. 2.23. Effect of products of wheat germ hexokinase LI reaction on the Γ coefficient. A – Glucose-6-phosphate (the last product of the reaction sequence) induces a decrease of the Γ coefficient. B – MgADP (the last product of the reaction sequence) has no effect on the Γ coefficient. Adapted from ref. [53].

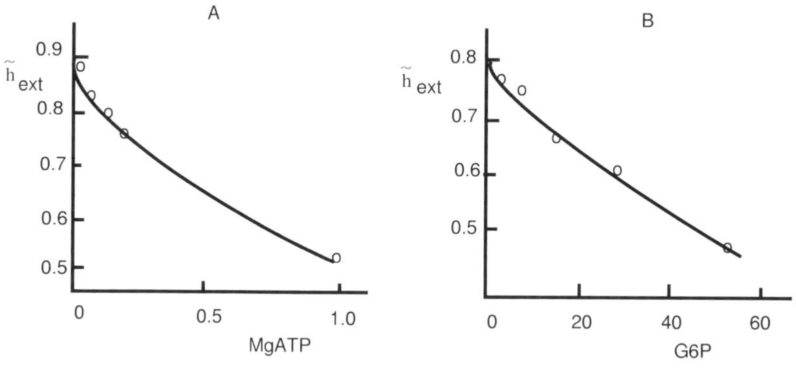

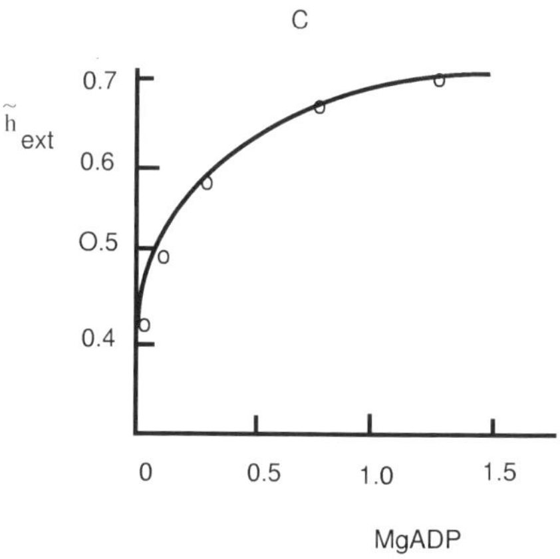

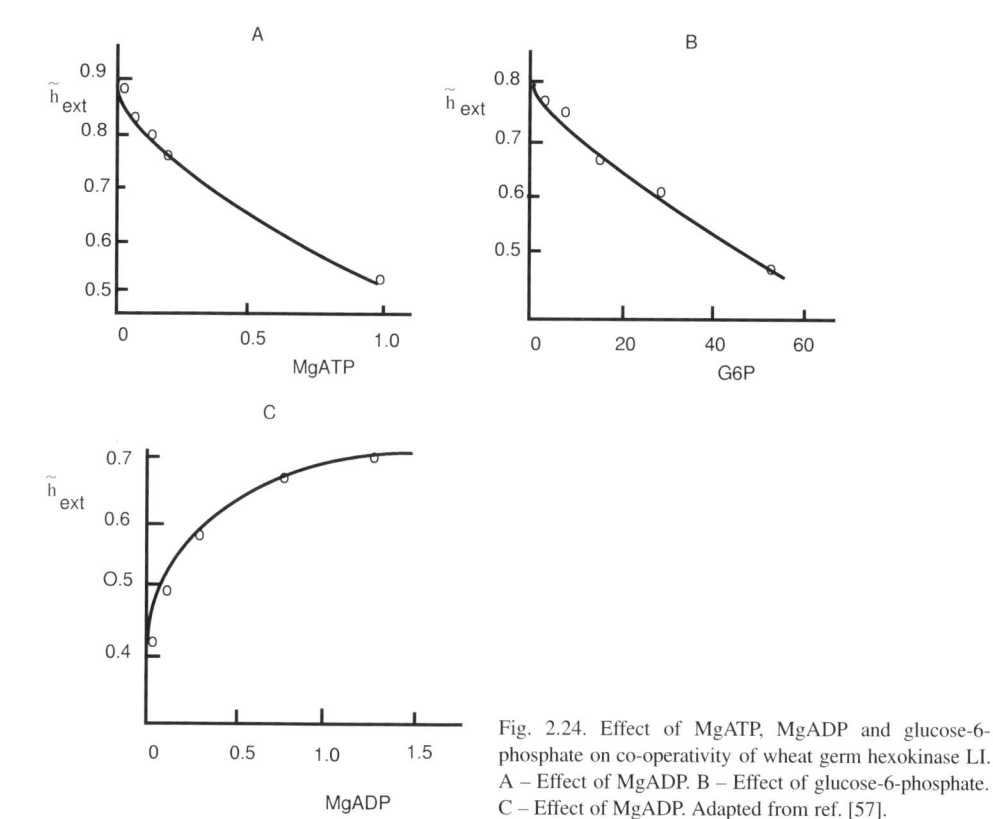

Fig. 2.24. Effect of MgATP, MgADP and glucose-6-phosphate on co-operativity of wheat germ hexokinase LI. A – Effect of MgADP. B – Effect of glucose-6-phosphate. C – Effect of MgADP. Adapted from ref. [57].

and MgADP) in a compulsory, or preferred, order pathway, glucose being bound first and glucose-6-phosphate released last. The enzyme has only one binding site for glucose. However, under steady state conditions, the enzyme displays negative co-operativity relative to glucose and no co-operativity with respect to ATP (Fig. 2.22). The Γ parameter is independent of the concentrations of MgATP and MgADP, but becomes more and more negative as the concentration of glucose-6-phosphate is increased (Fig. 2.23). The extreme Hill coefficient, however, decreases as the MgATP concentration is increased and the converse is observed for MgADP (Fig. 2.24). Moreover, the effect of glucose-6-phosphate is to generate a marked decrease of this Hill coefficient (Fig. 2.24).

2.2.5.5. Sensing chemical signals by hysteretic and mnemonic enzymes

An interesting property of hysteretic and mnemonic enzymes, which has for long escaped the attention of investigators, is the possible existence, in addition to the regular steady state, of a meta-steady state which is sensitive to the history of the system. This rather new concept of meta-steady state will be defined later and is an interesting feature of the enzyme reaction, for it gives such an enzyme the possibility to behave as a biosensor.

The integrated equation that describes the time-dependence of the concentration of an enzyme–substrate complex (see section 2.3) is, whatever the complexity of the reaction mechanism,

$$\frac{[\mathrm{ES}(t)]}{[\mathrm{E}]_0} = \frac{[\mathrm{ES}]_\sigma}{[\mathrm{E}]_0} + \sum_{i=1}^{n} \phi_i \mathrm{e}^{-\lambda_i t}, \tag{2.120}$$

where $[\mathrm{E}]_0$ is the total enzyme concentration, $[\mathrm{ES}]_\sigma$ the steady state concentration of the enzyme–substrate complex, ϕ_i and λ_i groupings of rate constants and ligand concentrations equivalent to relaxation amplitudes and time constants. As outlined previously, this type of equation is valid only if there is a large excess of substrate over the enzyme and if the initial conditions, referred to above, apply. Integration of eq. (2.120) leads to

$$\frac{[\mathrm{P}]}{[\mathrm{E}]_0} = \sum_{i=1}^{n} k\frac{\phi_i}{\lambda_i} + k\frac{[\mathrm{ES}]_\sigma}{[\mathrm{E}]_0}t - \sum_{i=1}^{n} k\frac{\phi_i}{\lambda_i}\mathrm{e}^{-\lambda_i t}, \tag{2.121}$$

which describes the progress curve, under initial conditions of the reaction. k is the catalytic rate constant. Setting

$$v_\sigma = k[\mathrm{ES}]_\sigma, \qquad \alpha = \sum_{i=1}^{n} k\frac{\phi_i}{\lambda_i}, \qquad \psi_i = -k\frac{\phi_i}{\lambda_i}, \tag{2.122}$$

eq. (2.121) takes the form

$$\frac{[\mathrm{P}]}{[\mathrm{E}]_0} = \alpha + \frac{v_\sigma}{[\mathrm{E}]_0}t + \sum_{i=1}^{n} \psi_i \mathrm{e}^{-\lambda_i t}. \tag{2.123}$$

It is usually considered, in classical steady state kinetics, that, when the progress curve becomes approximately linear, a steady state is reached and therefore all exponential terms of expression (2.123) have vanished. If this is not so and if the nth transient is very slow, then

$$e^{-\lambda_n t} \approx 1 - \lambda_n t, \tag{2.124}$$

and the equation of the progress curve now takes the form

$$\frac{[\mathrm{P}]}{[\mathrm{E}]_0} = \alpha + \psi_n + \left(\frac{v_\sigma}{[\mathrm{E}]_0} - \lambda_n \psi_n\right) t + \sum_{i=1}^{n-1} \psi_i e^{-\lambda_i t}, \tag{2.125}$$

so the steady state rate which is measured is only apparent and is called meta-steady state rate, v_σ^*. Its expression is

$$v_\sigma^* = v_\sigma + \lambda_n \psi_n [\mathrm{E}]_0. \tag{2.126}$$

For hysteretic and mnemonic models, as we shall see later, the expression of ψ_n, which depends on a ligand concentration, will adopt different values, for the same ligand concentration, depending on whether this concentration has been reached by an increase or a decrease from a previous concentration. Thus, for certain kinetic schemes, the value of ψ_n will depend upon the history of the system. This should give the meta-steady state some unique properties.

Probably the simplest kinetic scheme that can display the existence of a meta-steady state is the mnemonical model governing the conversion of the substrate S into two products P and Q. Simulation of this model shows that a meta-steady state indeed exists for a while before the real steady state is reached. The meta-steady state concentration of ES may be larger or smaller than the corresponding real steady state concentration. One can show after some algebra that the meta-steady state rate for this model assumes the form

$$\begin{aligned} \frac{v_\sigma^*}{k[\mathrm{E}]_0} &= \left\{ k_1[\mathrm{S}](1 + K_4[\mathrm{Q}]) \frac{[\mathrm{E}_1(0)]}{[\mathrm{E}]_0} + k_2[\mathrm{S}] \frac{[\mathrm{E}_2(0)] + [\mathrm{E}_2\mathrm{Q}(0)]}{[\mathrm{E}]_0} \right\} \\ &\quad \times \left\{ k + k_{-1} + k_{-2} + k_{-3} + k_3 + (k_1 + k_2)[\mathrm{S}] \right. \\ &\qquad \left. + (k + k_{-1} + k_{-2} + k_{-3}) K_4[\mathrm{Q}] + k_1 K_4 [\mathrm{S}][\mathrm{Q}] \right\}^{-1}. \end{aligned} \tag{2.127}$$

In this equation K_4 is the affinity constant of product Q for the square conformation of the enzyme, $[\mathrm{E}_1(0)]$, $[\mathrm{E}_2(0)]$ and $[\mathrm{E}_2\mathrm{Q}(0)]$ the initial concentrations (at the start of the reaction) of the corresponding enzyme forms. These concentrations are indeed not independent since

$$\frac{[\mathrm{E}_1(0)]}{[\mathrm{E}]_0} + \frac{[\mathrm{E}_2(0)]}{[\mathrm{E}]_0} + \frac{[\mathrm{E}_2\mathrm{Q}(0)]}{[\mathrm{E}]_0} = 1. \tag{2.128}$$

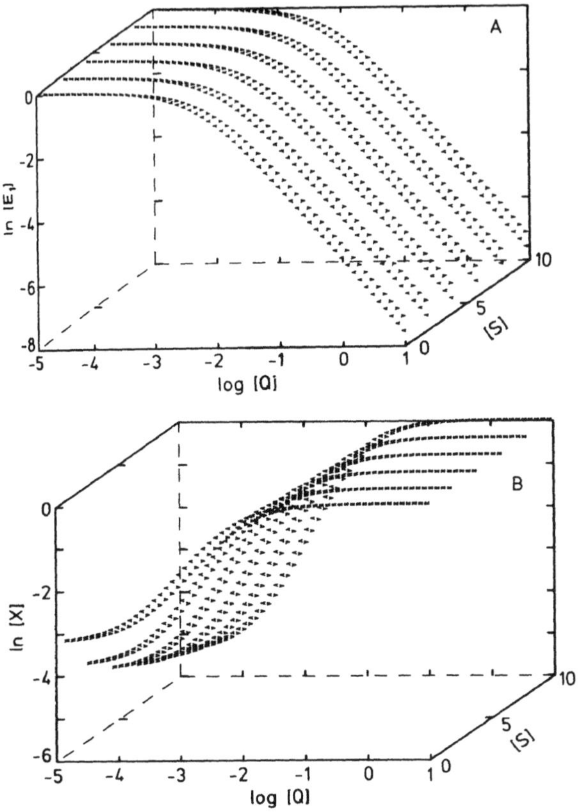

Fig. 2.25. Hysteresis of the two conformations of a mnemonical enzyme under meta steady state conditions. A – Hysteresis of the "free" conformation. B – Hysteresis of the conformation stabilized by the last product. From ref. [64].

These initial enzyme concentrations depend on the initial concentrations of Q. They are also different depending on whether the present concentration of Q is reached after an increase or a decrease from the previous concentration of this ligand (Fig. 2.25).

The meta-steady state rate equation thus displays two unique features: the product Q can act both as an activator and as an inhibitor of the meta-steady state rate; the meta-steady state rate can have, for the same values of [S] and [Q], two different values depending on whether the present concentration of Q is reached by an increase or a decrease from a previous concentration. The first property allows to distinguish the meta-steady state rate of a one-sited monomeric mnemonic enzyme from the corresponding steady state for, in the latter case, a product can only be inhibitory. The second property above gives a hysteretic, or mnemonic, enzyme the property to behave as a biosensor, i.e. to sense not only the concentration of a ligand, but also whether this concentration is reached by an increase or a decrease from the former concentration. The meta-steady state rate is thus sensitive to the history of the system.

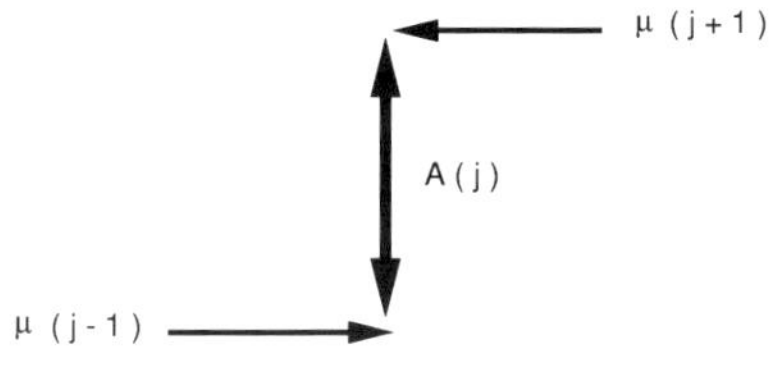

Fig. 2.26. The sensing force. See text.

It is important to understand the logic, at the molecular level, of the difference of the meta-steady state rate, depending on whether the final concentration of the product Q has been reached by an increase or a decrease from a former concentration. Let us assume, for instance, that the concentration of Q is decreased by a value $-\Delta Q$ (that is from a value Q_{j+1} down to a value Q_j) or, alternatively, that the concentration of Q is increased by a value $+\Delta Q$ (that is from a value Q_{j-1} up to the value Q_j). If the desorption and the binding of a ligand Q, from and to the enzyme, are nearly instantaneous relative to the corresponding conformation changes, one may expect the concentration of the enzyme states, E_1 for instance, to remain unchanged immediately after dilution of Q, or after an increase of its concentration. This means, for instance, that, for a concentration $[Q_j]$, the concentration of E_1 will be the concentration this enzyme form had at stage $j-1$, or will have at stage $j+1$, depending on whether the concentration of Q has been increased or decreased (Fig. 2.25). If $E_1(j-1)$ and $E_1(j+1)$ are these concentrations and $\mu_1(j-1)$ and $\mu_1(j+1)$ the corresponding chemical potentials one can define a force

$$A_1(j) = \mu_1(j-1) - \mu_1(j+1), \tag{2.129}$$

which allows the "perception" of a concentration change. This force may be called a sensing force. In the present case it is defined as a positive number (Figs. 2.25 and 2.26).

In the simple mnemonical model, the square conformation of the enzyme exists in a free and and in a complexed state (E_2Q). As is it assumed that there is a fast equilibrium between these two enzyme forms, let us call $[X_2]$ the sum of the concentrations of these two forms

$$[X_2] = [E_2] + [E_2Q]. \tag{2.130}$$

X_2 is thus the square conformation, whether free or bound to Q. The reasoning that has been applied to enzyme form E_1 can also be applied to X_2. One can thus define another sensing force

$$A_2(j) = \mu_2(j-1) - \mu_2(j+1), \tag{2.131}$$

which, in the present case has a negative value (Fig. 2.25).

As the concentrations of the various enzyme forms, for fixed substrate (S) and product (Q) concentrations, may be different depending on whether the actual concentration of Q is reached by an increase or a decrease from a former value, one may expect the meta-steady state rate to be different under these two types of experimental conditions. One can thus define a difference, Δv_j, between the two meta-steady state rates v_{j-1} (if Q_j has

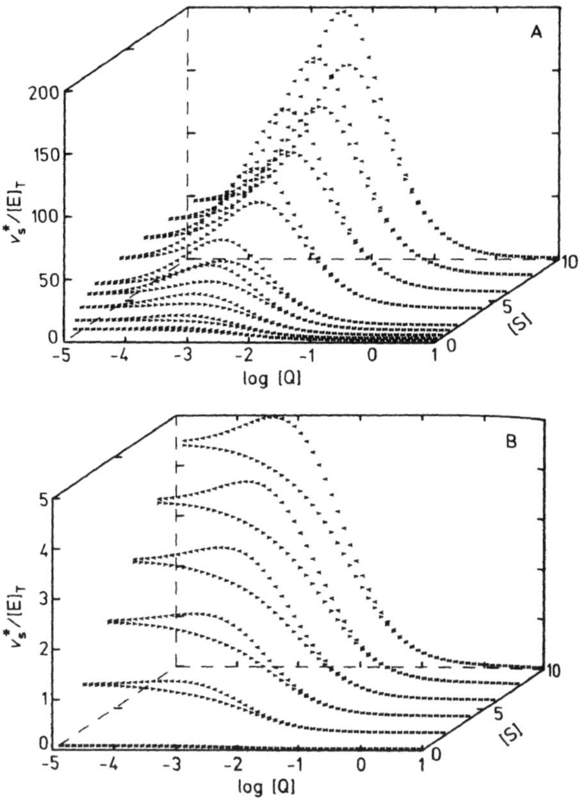

Fig. 2.27. Hysteresis of the meta steady state rate of a mnemonical enzyme as a function of the last product concentration. A and B – Activation by the last product and hysteresis under different conditions. From ref. [64].

been reached by an increase of concentration) and v_{j+1} (if Q_j is obtained by a decrease of concentration). The difference (see Fig. 2.27)

$$\Delta v_j = v_{j-1} - v_{j+1} \tag{2.132}$$

may be positive or negative. It represents how the meta-steady state reacts to the *direction* of a concentration change. Δv_j may thus be termed directional sensitivity of the meta-steady state rate.

There must exist a relationship between this sensitivity and the sensing forces. This relationship can be derived from eq. (2.127). As the denominator of this equation contains terms that are independent of the *direction* of a concentration change the expression of Δv_j is easily derived. One finds

$$\Delta v_j = L_1\{E_1(j-1) - E_1(j+1)\} + L_2\{X_2(j-1) - X_2(j+1)\}, \tag{2.133}$$

where L_1 and L_2 are collections of rate constants, substrate and product concentrations, which are not given here. In order to derive a relationship between Δv_j, $A_1(j)$ and $A_2(j)$ one can follow a calculation procedure identical to that used to derive eq. (2.32). One has thus

$$\begin{aligned} E_1(j-1) - E_1(j+1) &= \exp\left(-\frac{\mu_1^0}{RT}\right)\exp\left(\frac{\langle\mu_1\rangle}{RT}\right) \\ &\quad\times\left\{\exp\left[\frac{A_1(j)}{2RT}\right] - \exp\left[-\frac{A_1(j)}{2RT}\right]\right\}, \\ X_2(j-1) - X_2(j+1) &= \exp\left(-\frac{\mu_2^0}{RT}\right)\exp\left(\frac{\langle\mu_2\rangle}{RT}\right) \\ &\quad\times\left\{\exp\left[\frac{A_2(j)}{2RT}\right] - \exp\left[-\frac{A_2(j)}{2RT}\right]\right\}. \end{aligned} \tag{2.134}$$

Computer simulations show that the sensing forces, $A_1(j)$ and $A_2(j)$, cannot be very large [64,65]. Thus eq. (2.134) approximate to

$$\begin{aligned} E_1(j-1) - E_1(j+1) &= \exp\left(-\frac{\mu_1^0}{RT}\right)\exp\left(\frac{\langle\mu_1\rangle}{RT}\right)\frac{A_1(j)}{RT}, \\ X_2(j-1) - X_2(j+1) &= \exp\left(-\frac{\mu_2^0}{RT}\right)\exp\left(\frac{\langle\mu_2\rangle}{RT}\right)\frac{A_2(j)}{RT}. \end{aligned} \tag{2.135}$$

In expressions (2.134) and (2.135)

$$\begin{aligned} \langle\mu_1\rangle &= \frac{\mu_1(j-1) + \mu_1(j+1)}{2}, \\ \langle\mu_2\rangle &= \frac{\mu_2(j-1) + \mu_2(j+1)}{2}. \end{aligned} \tag{2.136}$$

Thus the directional sensitivity of the meta-steady state rate can be expressed by a linear combination of the sensing forces

$$\Delta v_j = L_1^* \frac{A_1(j)}{RT} + L_2^* \frac{A_2(j)}{RT}, \tag{2.137}$$

where

$$\begin{aligned} L_1^* &= L_1 \exp\left(-\frac{\mu_1^0}{RT}\right)\exp\left(\frac{\langle\mu_1\rangle}{RT}\right), \\ L_2^* &= L_2 \exp\left(-\frac{\mu_2^0}{RT}\right)\exp\left(\frac{\langle\mu_2\rangle}{RT}\right), \end{aligned} \tag{2.138}$$

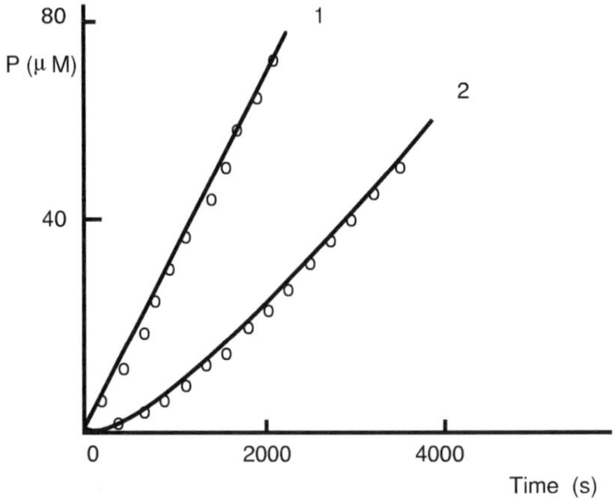

Fig. 2.28. Memory effects of fructose bisphosphatase. Curve 1: The enzyme is incubated in the presence of fructose 2,6-bisphosphate, this ligand is then chased away and the reaction rate is measured in the presence of the substrates of the enzyme. Curve 2: The reaction rate is measured under the same conditions as before but the enzyme has not been incubated with fructose 2,6-bisphosphate. See text. From ref. [63].

These theoretical considerations are not without experimental background. Biological macromolecules, such as nucleic acids and proteins, can occur under metastable states giving rise to hysteresis. Thus, nucleic acids may display metastable secondary structure upon their titration and therefore can exhibit hysteresis effects [66,67]. Of particular interest, in relation to the theoretical considerations above, is the kinetic behaviour of choroplast fructose 1,6-bisphosphatase. This enzyme is a tetramer made up of apparently identical subunits. It plays a key role in the regulation of the Benson–Calvin cycle, which results in the fixation and the reduction of carbon dioxide in plants and in other photosynthetic organisms. In its oxidized state, at pH 7.5, it is nearly totally devoid of activity. It can however regain full activity if it is incubated with low concentrations of a substrate analogue, fructose 2,6-bisphosphate.

If the inactive oxidized enzyme is incubated with a reaction medium containing the substrate (fructose 1,6-bisphosphate) and the analogue (fructose 2,6-bisphosphate) the progress curve of the reaction displays a slow lag (Fig. 2.28). In another experiment, the enzyme is incubated with fructose 2,6-bisphosphate and magnesium, then these ligands are removed by dilution and the enzyme is introduced into the same reaction mixture as in the previous experiment. Under these conditions, the progress curve displays no lag (Fig. 2.28). The enzyme retains the conformation stabilized by fructose 2,6-bisphosphate and this conformation is the active one. Fructose 2,6-bisphosphate thus exerts an imprinting effect on the enzyme which is associated with the existence of a meta-steady state.

Since the enzyme can retain the conformation stabilized by fructose 2,6-bisphosphate, the meta-steady state can display a hysteresis loop. This can be experimentally demonstrated by incubating fructose 2,6-bisphosphate and magnesium with fructose bisphosphatase. These ligands are then removed and the reaction rate (the meta-steady state rate) is measured at a given concentration of fructose 1,6-bisphosphate, fructose 2,6-bisphosphate

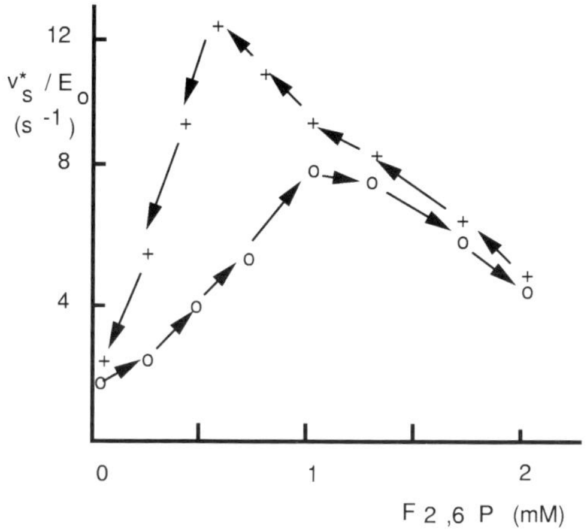

Fig. 2.29. Hysteresis of meta steady state rate of fructose bisphosphatase as a function of fructose 2,6-bisphosphate. The reaction rate depends on the history of the system and is different depending on the concentration of fructose 2,6-bisphosphate is increased or decreased. Adapted from ref. [63].

and magnesium. In general, the velocity will differ depending on whether the concentration of fructose 2,6-bisphosphate that has been removed, is higher or lower than the concentration of this ligand in the assay mixture (Fig. 2.29). The meta-steady state rate of fructose bisphosphatase thus depends on the direction of a concentration change (increase or decrease) of fructose 2,6-bisphosphate. These theoretical and experimental results thus clearly prove that, owing to "slow" conformation changes, simple enzymes can display, even in dilute solutions, an extremely complex behaviour. Of course the "slow" conformation changes that generate this complex kinetics are not restricted to simple enzymes, and occur with complex enzymes as well. But the aim of this section was to show that even simple enzymes, owing to "slow" conformation changes and imprinting effects, may possibily display, very subtle kinetic and dynamic properties.

2.3. *Does the complexity of the living cell affect the dynamics of enzyme-catalysed reactions?*

The general physical principles that govern the dynamics of enzyme reactions in stirred free solutions must no doubt be followed in all cicumstances. However, within the living cell, molecular crowding and fuzzy supramolecular organization may add other physical principles that must also be followed. One may thus wonder whether the dynamics of biological events, viewed at the chemical level, is similar to their chemical counterparts taking place in free stirred solutions. Answering this question is precisely the aim of this book. Without anticipating the content of future chapters, simple intuitive reasoning, nevertheless, leads us to the conclusion that the answer to this question can only be negative. Indeed, supramolecular complexity of the living cell must, of necessity, play a major part in the dynamics of elementary life processes.

The first reason that contributes to explain the difference of behaviour of an enzyme *in vitro* and *in vivo* is the viscosity and the compartmentalization of the cell milieu, which tends to decrease the rate of diffusion of molecules. When an enzyme reaction takes place in a stirred free solution, the diffusion of the substrate(s) to the active site is a fast process relative to any step of the catalysed reaction. Thus, under these conditions, the rate of free product appearance is the rate of the enzymatic process itself. However, within a living cell, in a viscous and compartmentalized medium, the rate of substrate diffusion and transport is considerably decreased, in such a way that the diffusion process may take place at about the same rate, or even at a lower rate, than that of the enzymatic process itself. Thus, under these conditions, the rate of product appearance does not necessarily match the rate of the enzyme process alone, but rather the rate of a coupled system involving slow diffusion and the enzyme reaction. Since the dynamics of diffusion is quite different from that of an enzyme reaction, one may expect the dynamics of the overall process to be different from that of either component of the coupled system.

Another reason that may contribute to the difference of behaviour of enzyme reactions *in vitro* and *in vivo* stems from the fact that *in vivo* an enzyme reaction is inserted into an enzyme network that constitutes a metabolic pathway. It is a time-honored idea, since the early 1960s, to consider that an enzyme network must be controlled by an enzyme located at the beginning of the network and retro-inhibited, or activated, by one of the last products of the enzyme sequence. This holds true indeed for a number of cases, but represents a reductionist view of metabolism, which is considered as a sort of molecular autocracy [68] in which a whole network would be under the sole control of a single enzyme, or of a small number of enzymes. One may wonder why many enzymes, or perhaps even all the enzymes of a network, would not take a part in the regulation of that nework. If this were so, one would replace the idea of molecular autocracy by that of molecular democracy [68]. Owing to the functional constraints that are exerted on an enzymatic process by the other chemical reactions of the same network, one may expect the steady state and the dynamic behaviour of this enzymatic process to be altered as a consequence of its insertion in a metabolic sequence. Moreover, an isolated enzyme in free solution may display one stable steady state only, and one may indeed wonder whether the same is true for enzymes cycles and networks, or whether there exists multiple steady states, or unstable steady states. If unstable steady states were occurring this would perhaps represent a molecular basis for biological clocks and excitability phenomena [69].

Many essential biological reactions, such as the synthesis of adenosine triphosphate (ATP), are thermodynamically disfavoured, if considered in isolation, and take place at the interface of different cell compartments. One may thus wonder whether this scalar chemical reaction is not coupled with the vectorial transfer of molecules, or ions, from one cell compartment to another. If this type of process were really taking place, one could wonder about the thermodynamics of coupling between these two types of processes.

Many enzymes are anchored in membranes or embedded in cell walls. Both biomembranes and cell walls are insoluble polyelectrolytes. Thus, the local pH within the cell walls, for instance, may be quite different from that prevailing outside this organelle. One may thus raise the question of the pH-dependence of these enzymes and how this pH-dependence is modified by their association with a biological polyelectrolyte. Moreover, the substrates of many membrane-bound or cell wall-bound enzymes are ions. One must therefore expect an electrostatic attraction or repulsion of the substrate to be exerted by the fixed charges of the polyelectrolyte. One may then wonder how this electrostatic attraction or repulsion alters the intrinsic kinetic properties of membrane-bound or cell wall-bound enzymes. One may also expect these effects to be modulated by the local ionic strength.

In eukaryotic cells most, perhaps all enzymes, are associated with other proteins, nucleic acids, membranes or cell walls. It appears highly probable that these associations represent a functional advantage for the eukaryotic cell, and one may again raise the question of the nature of this functional advantage. It has often been argued that, with large eukaryotic cells, the dilution of the reaction intermediates of a metabolic pathway and the random collision of a substrate with the competent enzyme would have resulted in poor efficiency of an enzyme network [70]. It is thus conceivable that the association of different enzymes, which catalyse consecutive reactions of the same pathway, allows the channelling of the reaction intermediates from active site to active site without a significant dilution in the cell milieu [70]. One may also wonder whether the physical association of an enzyme with other proteins, or membranes, does not change the intrinsic properties of this enzyme.

All these effects do exist in the eukaryotic cell and the aim of the next chapters of this book will be precisely to understand, on a physical basis, how they take part in important biological processes.

Appendix

Laplace transforms

The Laplace transform of a function $f(t)$, $L\{f(t)\}$, is defined as [71]

$$L\{f(t)\} = \int_0^\infty f(t)\,\mathrm{e}^{-st}\,\mathrm{d}t, \tag{A2.1}$$

where s is a complex variable. This function has several important properties, namely

$$\begin{aligned}
&L\left\{\frac{\partial}{\partial t} f(t)\right\} = sL\{f(t)\} + f(0),\\
&L\{af(t)\} = aL\{f(t)\},\\
&L\{a\} = \frac{a}{s},\\
&L\{\mathrm{e}^{-\lambda t}\} = \frac{1}{s+\lambda},\\
&L\{a_1 f_1(t) + a_2 f_2(t) + \cdots +\} = a_1 L\{f_1(t)\} + a_2 L\{f_2(t)\} + \cdots,
\end{aligned} \tag{A2.2}$$

where a and λ are constants. The properties (A2.2) allow one to convert differential equations (2.80) of the main text into the algebraic equations (2.81). The inverse conversion of the Laplace transforms into the corresponding functions, i.e. the conversion of equations (2.82) of the main text into eq. (2.83), requires the use of inverse Laplace transforms, namely

$$L^{-1}\left\{\frac{a}{s}\right\} = a, \qquad L^{-1}\left\{\frac{1}{s+\lambda}\right\} = \mathrm{e}^{-\lambda t}. \tag{A2.3}$$

From eq. (2.92) of the main text, one can show, after some algebra, that

$$k_2\left(\frac{\psi_1}{\lambda_1} + \frac{\psi_2}{\lambda_2}\right) = k_1 k_2 e_0 a \frac{\lambda_1\lambda_2 - k_3(\lambda_1 + \lambda_2)}{\lambda_1^2\lambda_2^2} \tag{A2.4}$$

and

$$k_3\left(\frac{\psi_1'}{\lambda_1} + \frac{\psi_2'}{\lambda_2}\right) = -\frac{k_1 k_2 k_3 e_0 a}{\lambda_1^2\lambda_2^2}. \tag{A2.5}$$

Expression (A2.4) may be positive or negative depending on the respective values of $\lambda_1\lambda_2$ and $k_3(\lambda_1 + \lambda_2)$ whereas the expression (A2.5) is of necessity negative. This means that if p is plotted as a function of time, one will observe a burst phase if

$$\lambda_1\lambda_2 > k_3(\lambda_1 + \lambda_2), \tag{A2.6}$$

and a lag phase if this inequality is reversed. But if q is plotted as a function of time one should obtain a lag phase only.

References

[1] Katchalsky, A. and Curran, P.F. (1975) Nonequilibrium Thermodynamics in Biophysics. Harvard University Press, Cambridge and London.
[2] Nicolis, G. and Prigogine, I. (1977) Self-Organization in Nonequilibrium Systems. From Dissipative Structures to Order through Fluctuations. John Wiley and Sons, New York.
[3] Castellan, G.W. (1977) Physical Chemistry. Addison-Wesley, Amsterdam and New York.
[4] Schnakenberg, J. (1981) Thermodynamic Network Analysis of Biological Systems. Springer-Verlag, Berlin, Heidelberg, New York.
[5] Haken, H. (1978) Synergetics. Springer-Verlag, Berlin, Heidelberg, New York.
[6] Prigogine, I. (1967) Introduction to Thermodynamics of Irreversible Processes. Wiley Interscience, New York.
[7] Peacocke, A.R. (1983) The Physical Chemistry of Biological Organization. Clarendon Press, Oxford.
[8] Laidler, K.J. (1969) Theories of chemical reaction rates. McGraw-Hill, New York.
[9] Glasstone, S., Laidler, K.J. and Eyring, H. (1941) The Theory of Rate Processes. McGraw-Hill, New York.
[10] Kraut, J. (1988) How do enzymes work? Science 242, 553–539.
[11] Laidler, K.J. (1965) Chemical Kinetics. McGraw-Hill, New York.
[12] Pauling, L. (1948) Nature of forces between large molecules of biological interest. Nature 161, 707–709.
[13] Pauling, L. (1946) Molecular architecture and biological reactions. Chem. Eng. News 24, 1375–1377.
[14] Kurz, J.L. (1963) Transition state characterization of catalyzed reactions. J. Amer. Chem. Soc. 85, 987–991.
[15] Koshland, D.E., Nemethy, G. and Filmer, D. (1966) Comparison of experimental binding data and theoretical models in proteins containing subunits. Biochemistry 5, 365–385.

[16] Koshland, D.E. (1970) The molecular basis of enzyme regulation. In: P.D. Boyer (Ed.), The Enzymes, 3rd edition. Academic Press, New York, Vol. 1, pp. 341–396.
[17] Koshland, D.E. and Neet, K.E. (1968) The catalytic and regulatory properties of enzymes. Annu. Rev. Biochem. 37, 359–410.
[18] Williams, R.J.P. (1972) The entatic state. Cold Spring Harbor Symp. Quant. Biol. 36, 53–62.
[19] Jencks, W.P. (1975) Binding energy, specificity and enzymic catalysis: the Circe effect. Adv. Enzymol. 43, 219–410.
[20] Jencks, W.P. (1969) Catalysis in Chemistry and Enzymology. McGraw-Hill, New York.
[21] Williams, R.J.P. (1993) Are enzymes mechanical devices? Trends Biochem. Sci. 18, 115–117.
[22] Phillips, D.C. (1967) The hen egg-white lysozyme molecule. Proc. Natl. Acad. Sci. USA 57, 484–495.
[23] Ford, L.O., Johnson, L.N., Mackin, P.A., Phillips, D.C. and Tian, R. (1974) Crystal structure of a lysozyme-tetrasaccharide lactone complex. J. Mol. Biol. 88, 349–371.
[24] Fersht, A. (1977) Enzyme Structure and Mechanism. Freeman, Reading and San Francisco.
[25] Chipman, D.M., Grisario, V. and Sharon, N. (1967) The binding of oligosaccharides containing N-acetylglucosamine and N-acetylmuramic acid to lysozyme. J. Biol. Chem. 242, 4388–4394.
[26] Stryer, L. (1991) Biochemistry. Freeman, San Francisco.
[27] Phillips, D.C. (1966) The three-dimensional structure of an enzyme molecule. Scientific American, May, 78–90.
[28] Imoto, T., Johnson, L.N., North, A.C.T., Phillips, D.C. and Rupley, J.A. (1972) Vertebrate lysozymes. In: P.D. Boyer (Ed.), The Enzymes. Academic Press, New York, Vol. 7, pp. 666–868.
[29] Dahlquist, F.W., Rand-Meir, T. and Raftery, M.A. (1968) Demonstration of carbonium ion intermediate during lysozyme catalysis. Proc. Natl. Acad. Sci. USA 61, 1194–1198.
[30] Rupley, J.A. (1967) The binding and cleavage by lysozyme of N-acetylglucosamine oligosaccharides. Proc. Roy. Soc. (B) 167, 416–428.
[31] Wolfenden, R. (1969) Transition state analogues for enzyme catalysis. Nature 223, 704–705.
[32] Wolfenden, R. (1972) Analog approaches to the structure of the transition state in enzyme reactions. Accounts Chem. Res. 5, 10–18.
[33] Wolfenden, R. (1976) Transition state analog inhibitors and enzyme catalysis. Annu. Rev. Biophys. Bioeng. 5, 271–306.
[34] Lienhard, G.E. (1979) Enzymatic catalysis and transition state theory. Transition state analogs show that catalysis is due to tighter binding of transition states than of substrates. Science 180, 149–154.
[35] Tramontano, A., Janda, K.D. and Lerner, R.A. (1986) Catalytic antibodies. Science 234, 1566–1570.
[36] Lerner, R.A. and Tramontano, A. (1987) Antibodies as enzymes. Trends Biochem. Sci. 143, 427–430.
[37] King, E.L. and Altman, C. (1956) A schematic method of deriving the rate laws for enzyme-catalyzed reactions. J. Phys. Chem. 60, 1375–1378.
[38] Laidler, K.J. (1968) The Chemical Kinetics of Enzyme Action. Clarendon Press, Oxford.
[39] Volkenstein, M.V. and Goldstein, B.N. (1966) A new method for solving the problems of the stationary kinetics of enzymological reactions. Biochim. Biophys. Acta 115, 471–477.
[40] Cha, S. (1968) A simple method for derivation of rate equations for enzyme-catalyzed reactions under rapid equilibrium assumption or combined assumptions of equilibrium and steady state. J. Biol. Chem. 243, 820–825.
[41] Wong, J.T. (1975) Kinetics of Enzyme Mechanisms. Academic Press, London.
[42] Cornish-Bowden, A. (1995) Fundamentals of Enzyme Kinetics. Portland Press, London.
[43] Westerhoff, H. and Van Dam, K. (1987) Thermodynamics and Control of Biological Free Energy Transduction. Elsevier, Amsterdam.
[44] Dalziel, K. (1957) Initial steady-state velocities in the evaluation of enzyme–coenzyme–substrate reactions mechanisms. Acta Chem. Scand. 114, 1706–1723.
[45] Ricard, J. (1973) Cinétique et Mécanismes d'Action des Enzymes. Doin, Paris.
[46] Fromm, H.J. (1975) Initial Rate Enzyme Kinetics. Springer-Verlag, Berlin, Heidelberg, New York.
[47] Keleti, T. (1986) Basic Enzyme Kinetics. Akademiai Kiado, Budapest.
[48] Monod, J., Wyman, J. and Changeux, J.P. (1965) On the nature of allosteric transitions: a plausible model. J. Mol. Biol. 12, 88–118.
[49] Frieden, C. (1970) Kinetic aspects of regulation of metabolic processes. The hysteretic enzyme concept. J. Biol. Chem. 245, 5788–5799.

[50] Ainslie, G.R., Shill, J.P. and Neet, K.E. (1972) Transients and cooperativity. A slow transition model for relating transients and cooperative kinetics of enzymes. J. Biol. Chem. 247, 7088–7096.
[51] Neet, K.E. and Ainslie, G.R. (1980) Hysteretic enzymes. In: D.L. Purich (Ed.), Methods Enzymol., Vol. 64, Part B, pp. 192–226.
[52] Ricard, J., Meunier, J.C. and Buc, J. (1974) Regulatory behavior of monomeric enzymes. I. The mnemonical enzyme concept. Eur. J. Biochem. 49, 195–208.
[53] Meunier, J.C., Buc, J. and Ricard, J. (1974) Regulatory behavior of monomeric enzymes. II. A wheat germ hexokinase as a mnemonical enzyme. Eur. J. Biochem. 49, 209–223.
[54] Ricard, J., Buc, J. and Meunier, J.C. (1977) Enzyme memory. I. A transient kinetic study of wheat germ hexokinase L I. Eur. J. Biochem. 80, 581–592.
[55] Buc, J., Ricard, J. and Meunier, J.C. (1977) Enzyme memory. II. Kinetics and thermodynamics of the slow conformation changes of wheat germ hexokinase L I. Eur. J. Biochem. 80, 593–601.
[56] Meunier, J.C., Buc, J. and Ricard, J. (1979) Enzyme memory. Effect of glucose-6-phosphate and temperature on the molecular transition of wheat germ hexokinase L I. Eur. J. Biochem. 97, 573–583.
[57] Ricard, J. and Noat, G. (1985) Kinetic co-operativity of monomeric mnemonical enzymes. The significance of the kinetic Hill coefficient. Eur. J. Biochem. 152, 557–564.
[58] Storer, A.C. and Cornish-Bowden, A. (1977) Kinetic evidence for a "mnemonical" mechanism for rat liver glucokinase. Biochem. J. 165, 61–69.
[59] Monneuse-Doublet, M.O., Olomucki, A. and Buc, J. (1978) Investigations on the kinetic mechanism of octopine dehydrogenase. A regulatory behavior. Eur. J. Biochem. 84, 441–448.
[60] Rose, I.A., Warms, J.V.B. and Yuan, R.G. (1993) Role of conformational change in the fumarase reaction cycle. Biochemistry 32, 8504–8511.
[61] Cheron, G., Noat, G. and Ricard, J. (1990) Hysteresis of plant cell wall glucosidase. Biochem. J. 269, 389–392.
[62] Soulié, J.M., Rivière, M. and Ricard, J. (1988) Enzymes as biosensors. II. Hysteretic response of chloroplastic fructose 1,6-bisphosphatase to fructose 2,6-bisphosphate. Eur. J. Biochem. 176, 111–117.
[63] Soulié, J.M., Rivière, M., Baldet, P. and Ricard, J. (1991) Kinetics of the conformational transition of the spinach chloroplast fructose 1,6-bisphosphatase. Eur. J. Biochem. 195, 671–678.
[64] Ricard, J. and Buc, J. (1988) Enzymes as biosensors. I. Enzyme memory and sensing chemical signals. Eur. J. Biochem. 176, 103–109.
[65] Ricard, J., Buc, J., Kellershohn, N. and Soulié, J.M. (1990) Sensing chemicals signals by enzymes. In: A. Cornish-Bowden and M.L. Cardenas (Eds.), Control of Metabolic Processes, pp. 291–296.
[66] Revzin, A., Neumann, E. and Katchalsky, A. (1973) Metastable secondary structures in ribosomal RNA. Molecular hysteresis in the acid-base titration of *Escherichia coli* ribosomal RNA. J. Mol. Biol. 79, 95–114.
[67] Schneider, F.W. (1976) Stability of steady states in nucleic acid poly(A-T) synthesis. Biopolymers 15, 1–14.
[68] Kacser, H. and Burns, J.A. (1972) The control of flux. Symp. Soc. Exp. Biol. 27, 65–104.
[69] Goldbeter, A. (1996) Biochemical Oscillations and Cellular Rhythms. The Molecular Bases of Periodic and Chaotic Behaviour. Cambridge University Press, Cambridge.
[70] Srere, P. (1994) Complexities of metabolic regulation. Trends Biochem. Sci. 19, 519–520.
[71] Kraut, E.A. (1967) Fundamentals of Mathematical Physics. McGraw-Hill, New York.

CHAPTER 3

Coupling between chemical and (or) vectorial processes as a basis for signal perception and transduction

Prokaryotic as well as eukaryotic cells perceive signals from the external world and react to them accordingly. Thus, cells belonging to the same multicellular organism receive signals originating from different cells and, in turn, emit signals that are perceived by other cells. Perception and transduction of signals is indeed compulsory if the cells are to behave in a coordinated manner, as multicellular organisms do. The term 'signaling' has been coined to designate the biochemical events involved in the processes of emission and transduction of signals. In a multicellular organism, perturbation of cell signaling may abolish the coordination between the cells, and may lead to cancer and ultimately to death. Cell signaling no doubt represents an important chapter in any modern textbook on cell biology. The present chapter is not devoted to the study of cell signaling *per se*, but is rather centred on the role played by the coupling between vectorial and chemical processes, or between chemical reactions, in the physical chemistry of cell signaling. As it is obvious to any biologist and biochemist that enzyme reactions occurring in the living cell, whether prokaryotic or eukaryotic, are not independent but coupled, this raises the question of whether novel properties emerge from this coupling.

We shall thus discuss in this chapter how the simplest type of coupling, namely the coupling between diffusion and an enzyme reaction, can generate unexpected effects. We shall also discuss how these emergent properties can be used as sensing devices. Since an essential feature of any sensing device is its sensitivity to a signal, we shall discuss this matter in some detail with antagonistic reactions. Last but not least, we shall consider the application of these general ideas to an important biochemical system, namely, bacterial chemotaxis.

3.1. Coupling between reagent diffusion and bound enzyme reaction rate as an elementary sensing device

Viscosity within a living cell may be rather high, such that the diffusion of small molecules within the cell is significantly decreased. The decline of the diffusion rate is called diffusional resistance. In the case of the substrate of an enzymatic process, the diffusion rate of this reagent may become of about the same magnitude, or even smaller, than the corresponding enzyme reaction rate. Diffusion and the enzyme reaction are then coupled. From this coupling, novel properties may emerge that are neither those of pure diffusion nor those of a pure enzyme reaction [1–11].

3.1.1. The basic equation of coupling

As already outlined in the previous chapter, the very existence of this coupling process relies on the possibility of representing a diffusion process with the so-called quasi-chemical

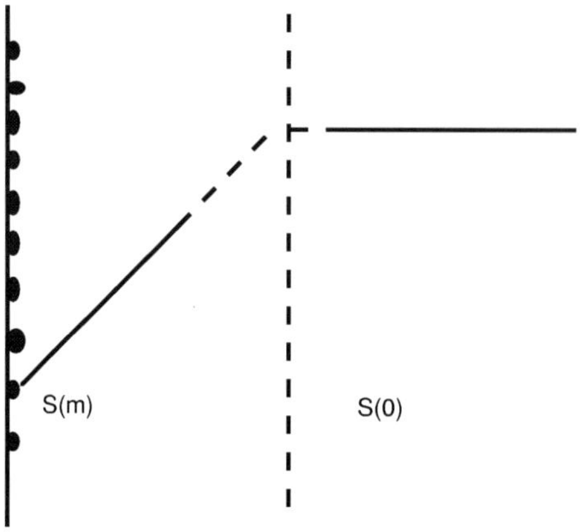

Fig. 3.1. A linear gradient in the vicinity of a surface. Enzyme molecules are bound to a surface. As there is a steady state between diffusion and enzyme reaction, a linear gradient is established in the vicinity of the surface.

formalism. If one considers, for instance, enzyme molecules that are bound to an impermeable surface, in the course of the enzyme reaction, a gradient of substrate and product appears in the vicinity of this surface (Fig. 3.1).

If the overall system is in steady state, the diffusion flow of substrate is

$$J_v = h_d\{S(0) - S(m)\}. \tag{3.1}$$

$S(0)$ is the substrate concentration "far" from the surface, in a region of space that can be considered as a reservoir and as the "origin" of the diffusion process. $S(m)$ is the substrate concentration in the immediate vicinity of the enzyme molecules, at a distance m from the origin. h_d is the substrate diffusion constant. If the enzyme follows Michaelis–Menten kinetics, the corresponding steady state rate assumes the form

$$v_e = \frac{VS(m)/K}{1 + S(m)/K}, \tag{3.2}$$

where V and K are the V_m and the K_m of the reaction, as already seen in the previous chapter. Overall steady state conditions demand that

$$h_d\{S(0) - S(m)\} = \frac{VS(m)/K}{1 + S(m)/K}. \tag{3.3}$$

As already outlined in chapter 2, these overall steady state conditions imply that the gradient of substrate be linear in the vicinity of the surface. Equation (3.3) represents the coupling equation of the system. There is an obvious advantage in expressing this equation in dimensionless form. Setting

$$S_\sigma = \frac{S(m)}{K}, \qquad s_0 = \frac{S(0)}{K}, \qquad h_d^* = \frac{h_d K}{V}, \tag{3.4}$$

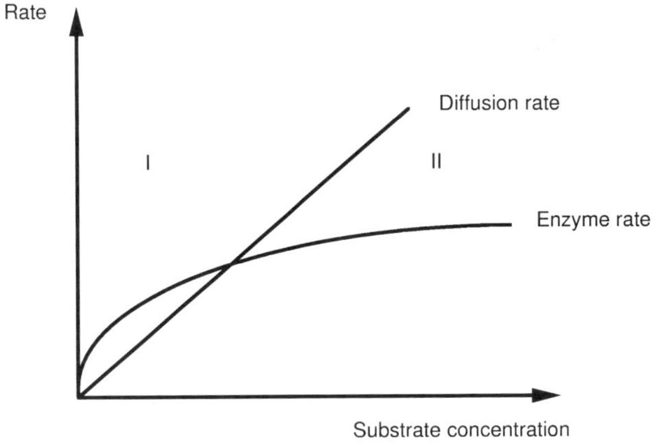

Fig. 3.2. A difference of behaviour between diffusion and enzyme reaction rate. Below the point of intersection of the two curves, the overall process is controlled by diffusion. Above this intersection, it is under the control of enzyme reaction.

eq. (3.3) becomes

$$h_{\mathrm{d}}^{*}(s_0 - s_\sigma) - \frac{s_\sigma}{1+s_\sigma} = 0. \tag{3.5}$$

This equation has two obvious advantages over eq. (3.3). The number of parameters has been decreased from three to one, and the new variables s_σ and s_0 are dimensionless. The dimensionless coupling equation may be reexpressed as

$$h_{\mathrm{d}}^{*} s_\sigma^2 + \left\{h_{\mathrm{d}}^{*}(1-s_0)+1\right\} s_\sigma - h_{\mathrm{d}}^{*} s_0 = 0, \tag{3.6}$$

and its positive root is

$$s_\sigma = \frac{1}{2h_{\mathrm{d}}^{*}}\left\{\sqrt{\left\{1+h_{\mathrm{d}}^{*}(1-s_0)\right\}^2 + 4h_{\mathrm{d}}^{*2} s_0} - \left\{h_{\mathrm{d}}^{*}(1-s_0)+1\right\}\right\}. \tag{3.7}$$

This equation expresses how the local dimensionless substrate concentration, s_σ, in the vicinity of the enzyme molecules, varies with respect to the bulk dimensionless substrate concentration, s_0. Although s_σ is the dimensionless substrate concentration "seen" by the enzyme molecules, s_0 is the only variable that can be determined directly by experiment. Moreover, eq. (3.7) shows that if, as assumed so far, the enzyme reaction follows Michaelis–Menten kinetics with respect to s_σ, which is not directly accessible to experiment, it cannot follow the same type of kinetics relative to s_0. As a matter of fact, the kinetics with respect to this substrate is apparently positively co-operative. Plotting $(1+s_\sigma)/s_\sigma$ as a function of $1/s_0$ yields plots that are concave upwards. This can be intuitively understood by simple inspection of Fig. 3.2. Plotting the bound enzyme reaction rate as a function of $S(0)$, it can be seen that, for low substrate concentrations, the rate is very similar to the rate of diffusion process but, as the bulk concentration increases, it becomes more and more similar to the rate process catalyzed by the free enzyme (Fig. 3.2). This means that, at low substrate concentrations, diffusion is the limiting process whereas at high substrate concentrations, it is the enzyme reaction that has become rate limiting.

3.1.2. Hysteresis loops and sensing chemical signals

Far more interesting for its biological implication is the situation where the enzyme reaction rate has at least one term that is nonlinear in substrate concentration. The simplest process that generates this type of equation is that of substrate inhibition. Inhibition by excess substrate can be generated in different ways: the simplest is the incorrect non-productive binding of two substrate molecules at the active site of the enzyme. The corresponding reaction rate is then

$$v_e = \frac{VS(m)/K}{1 + S(m)/K + K_2 S^2(m)/K}, \tag{3.8}$$

where K_2 is now the affinity constant of the second substrate molecule for the binary enzyme–substrate complex. If, as previously, the overall system is in steady state,

$$h_d\{S(0) - S(m)\} - \frac{VS(m)/K}{1 + S(m)/K + K_2 S^2(m)/K} = 0. \tag{3.9}$$

Defining the dimensionless variables and parameters as in eq. (3.4) and setting $\lambda = K_2 K$, the dimensionless coupling equation takes the form

$$h_d^*(s_0 - s_\sigma) - \frac{s_\sigma}{1 + s_\sigma + \lambda s_\sigma^2} = 0, \tag{3.10}$$

which can be rearranged to

$$h_d^* \lambda s_\sigma^3 - h_d^*(\lambda s_0 - 1)s_\sigma^2 + \{h_d^*(1 - s_0) + 1\}s_\sigma - h_d^* s_0 = 0. \tag{3.11}$$

This third degree coupling equation may have three real positive roots (Descartes rule of signs). This means that when plotting s_σ as a function of s_0 the resulting curve may have the S-shape shown in Fig. 3.3.

This is interesting, for the coupled system can work as a sensor. Its properties can be briefly described as follows. The coupled system, unlike isolated enzyme reactions, can display multiple steady states. Two of them (states 1 and 3) are stable whereas the last one (state 2) is unstable. Thus, two states can be detected experimentally. In other words, for a given bulk concentration, s_0, such that $s_0 \in [s_0^A s_0^B]$, the local concentration, s_σ, in the vicinity of the bound enzyme molecules will be different depending on whether the bulk concentration, s_0, was reached after an increase or a decrease of a former bulk concentration. This implies that the system is able to "recall" the values of the concentrations that were anterior to that occurring at time t. In other words the system displays short-term memory

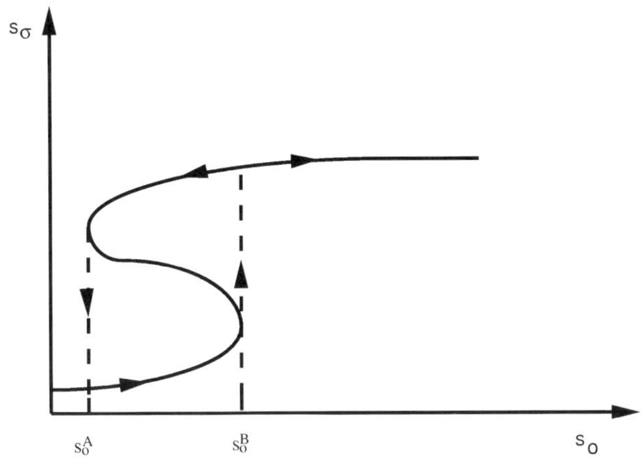

Fig. 3.3. Hysteresis of substrate concentration generated by the coupling between diffusion and enzyme reaction. See text.

and hysteresis effects. It can sense, not only the intensity of a signal, but also whether this intensity is increasing or decreasing, exactly as our eye coupled to our brain, for instance, is able to estimate the order of magnitude of a light intensity and is also able to know whether this intensity is increasing or decreasing. Moreover, these effects are solely due to a nonlinear term in the enzyme rate equation and to the fact that the coupled system occurs under nonequilibrium conditions. Last but not least, the system displays the ability of being extremely sensitive to slight changes of bulk substrate concentrations. For instance, in the vicinity of s_0^B, a slight increase of the bulk concentration will result in a dramatic increase of the local concentration, s_σ, whereas in the vicinity of s_0^A, a slight decrease of the bulk concentration will generate a large decrease of the local concentration, s_σ (Fig. 3.3).

3.1.3. Control of the substrate gradient

The systemic properties which we have considered so far require the existence of a substrate gradient in the vicinity of the surface. As this gradient is generated in part by the enzyme activity, one may expect that ligands that affect the reaction rate should also affect the steepness of the gradient. This can easily be demonstrated with an enzyme that follows Michaelis–Menten kinetics and is competitively inhibited by a ligand I [1–3]. The relevant rate equation is then

$$v_e = \frac{V S(m)/K_s}{1 + S(m)/K_s + I_0/K_i}. \tag{3.12}$$

As the inhibitor does not undergo any chemical transformation its concentration does not vary in space and is equal to I_0. Its dissociation constant from the enzyme is K_i. Setting

$$s_0 = \frac{S(0)}{K_s}, \qquad s_\sigma = \frac{S(m)}{K_s}, \qquad i_0 = \frac{I_0}{K_i}, \tag{3.13}$$

the dimensionless coupling equation assumes the form

$$h_{\rm d}^*(s_0 - s_\sigma) - \frac{s_\sigma}{1 + s_\sigma + i_0} = 0, \tag{3.14}$$

which can be rewritten as

$$h_{\rm d}^* s_\sigma^2 + \{h_{\rm d}^*(1 + i_0) - h_{\rm d}^* s_0 + 1\} s_\sigma - h_{\rm d}^* s_0 (1 + i_0) = 0. \tag{3.15}$$

The positive root of this equation is

$$s_\sigma = \frac{1}{2h_{\rm d}^*}\left\{\sqrt{\{h_{\rm d}^*(1 + i_0) - h_{\rm d}^* s_0 + 1\}^2 + 4h_{\rm d}^{*2} s_0 (1 + i_0)} - \{h_{\rm d}^*(1 + i_0) - h_{\rm d}^* s_0 + 1\}\right\}. \tag{3.16}$$

As i_0 increases,

$$h_{\rm d}^*(1 + i_0) - h_{\rm d}^* s_0 \gg 1. \tag{3.17}$$

Moreover, one has

$$\{h_{\rm d}^*(1 + i_0) - h_{\rm d}^* s_0\}^2 + 4h_{\rm d}^2 s_0 (1 + i_0) = \{h_{\rm d}^*(1 + i_0) + h_{\rm d}^* s_0\}^2. \tag{3.18}$$

Therefore, eq. (3.16) reduces to $s_\sigma = s_0$. Increasing the inhibitor concentration, tends to suppress the gradient of substrate and all the effects considered above vanish.

3.2. *Sensitivity amplification for coupled biochemical systems*

Many signaling mechanisms share common features. These common features are depicted in Fig. 3.4. A protein, P, may be converted into a modified form, P*, thanks to a converter

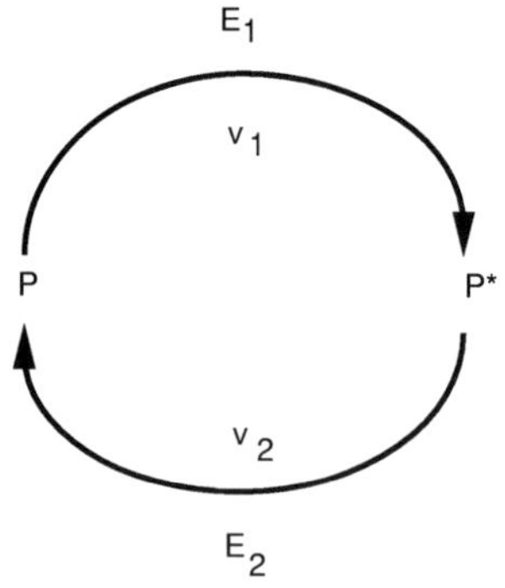

Fig. 3.4. A monocyclic cascade. Two enzymes, E_1 and E_2, are involved in phosphorylation–dephosphorylation, or methylation–demethylation, of a protein P. The corresponding rates are v_1 and v_2, respectively.

enzyme E_1. This modified protein P* may, in turn, be converted back to P thanks to another converter enzyme, E_2. The conversion P $\rightarrow$ P* may represent a phosphorylation or a methylation process, for example. Then, the reverse transition P* $\rightarrow$ P will be a dephosphorylation or a demethylation. The coupled process of Fig. 3.4 is defined as a monocyclic cascade. Moreover, as will be seen later, several cascades may be coupled. Their properties have been extensively studied [12–24].

3.2.1. Zero-order ultrasensitivity of a monocyclic cascade

What is important in the properties of a cascade is its sensitivity to small changes of the concentrations of either the converter enzymes or the effectors of these enzymes. We will study the first question first. The conservation equation of protein P is

$$P_{\mathrm{T}} = P + P^*, \tag{3.19}$$

where P_{T} is the total protein concentration. If, for simplicity, we assume that the converter enzymes follow Michaelis–Menten kinetics and that their concentration is small relative to that of the protein, we have

$$\frac{v_1}{P_{\mathrm{T}}} = \frac{V_1[1-(P^*/P_{\mathrm{T}})]}{(K_1/P_{\mathrm{T}}) + [1-(P^*/P_{\mathrm{T}})]} \tag{3.20}$$

for the first reaction and

$$\frac{v_2}{P_{\mathrm{T}}} = \frac{V_2(P^*/P_{\mathrm{T}})}{(K_2/P_{\mathrm{T}}) + (P^*/P_{\mathrm{T}})} \tag{3.21}$$

for the second. In these equations v_1 and v_2 are the steady state rates, V_1 and V_2 the corresponding maximum rates and K_1 and K_2 the Michaelis constants. Setting

$$\widetilde{K}_1 = \frac{K_1}{P_{\mathrm{T}}}, \qquad \widetilde{K}_2 = \frac{K_2}{P_{\mathrm{T}}}, \qquad \widetilde{P}^* = \frac{P^*}{P_{\mathrm{T}}}, \qquad \alpha = \frac{V_1}{V_2}, \tag{3.22}$$

and assuming that the whole system is in steady state, the relevant coupling equation assumes the form

$$\frac{V_1(1-\widetilde{P}^*)}{\widetilde{K}_1 + (1-\widetilde{P}^*)} = \frac{V_2\widetilde{P}^*}{\widetilde{K}_2 + \widetilde{P}^*}, \tag{3.23}$$

which can be rearranged to

$$(\alpha-1)\widetilde{P}^{*2} - \left[(\alpha-1) - \widetilde{K}_2\left(\alpha + \frac{\widetilde{K}_1}{\widetilde{K}_2}\right)\right]\widetilde{P}^* - \alpha\widetilde{K}_2 = 0. \tag{3.24}$$

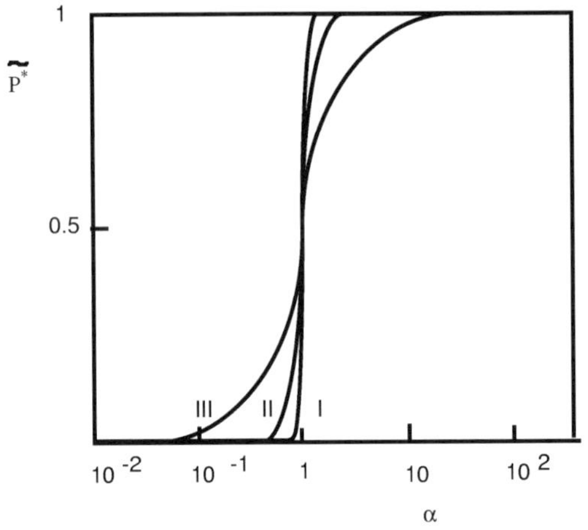

Fig. 3.5. Zero-order ultrasensitivity. The steepness of the curve decreases from I to III as the values of $\widetilde{K}_{1'}$ and $\widetilde{K}_{2'}$ decrease. Adapted from ref. [22].

The positive root of this second-degree equation is then

$$\widetilde{P}^* = \frac{1}{2(\alpha-1)}\left\{(\alpha-1) - \widetilde{K}_2\left(\alpha + \frac{\widetilde{K}_1}{\widetilde{K}_2}\right) + \sqrt{\left[(\alpha-1) - \widetilde{K}_2\left(\alpha + \frac{\widetilde{K}_1}{\widetilde{K}_2}\right)\right]^2 + 4(\alpha-1)\alpha\widetilde{K}_2}\right\}. \tag{3.25}$$

The variation of $\widetilde{P}^*$ can be studied as a function of α. The variation of α can be obtained by changing the respective concentrations of the two converter enzymes E_1 and E_2. Three examples of the variation of $\widetilde{P}^*$ as a function of α are shown in Fig. 3.5. As expected, $\widetilde{P}^*$ increases and reaches unity as α is increased but the steepness of the response depends markedly on the values of $\widetilde{K}_1$ and $\widetilde{K}_2$. Of particular interest is the response of the all-or-none type. This is obtained for very low values of $\widetilde{K}_1$ and $\widetilde{K}_2$. Since these two constants are normalized relative to the total protein concentration, low values of $\widetilde{K}_1$ and $\widetilde{K}_2$ mean that the overall system is located in a zero-order region. The steep reponse of the system has been referred to as zero-order ultrasensitivity by Goldbeter and Koshland [20,21].

3.2.2. Response of the system to changes in effector concentration

The two converter enzymes may be sensitive to various effectors. The effect exerted by such effectors on cascade systems has been extensively studied by Chock and Stadtman [12–19] and Goldbeter and Koshland [20–23]. A situation illustrating the possible effect of an effector L is shown in Fig 3.6.

There are many different variants of this scheme, but the assumption which is made here is that the same ligand can activate the inactive enzyme E'_1 into E_1 and, conversely, can de-activate the active enzyme E_2 into E'_2. This model, which is derived from real experimental situations, is interesting for it allows one to understand how a cascade system can display

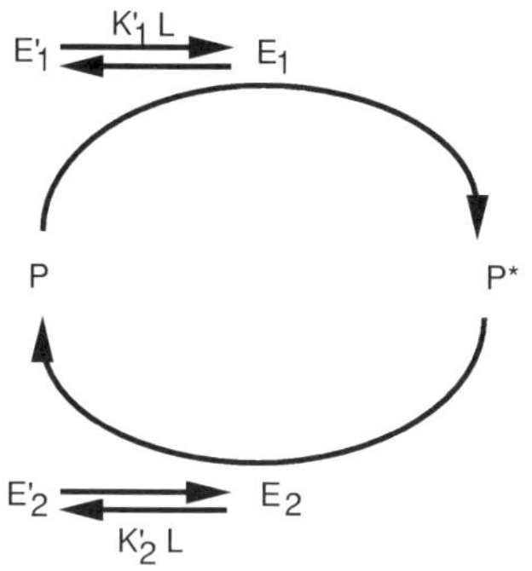

Fig. 3.6. Effect of a ligand L on the two converter enzymes of a monocyclic cascade. The ligand is assumed to activate enzyme E_1 and inhibit enzyme E_2.

an amplification of its response to slight changes of a signal. In this chapter we shall follow a method which is different from the strategies of Chock and Stadtman [12–19] and Goldbeter and Koshland [20–23]. The amplification factor, A_L, of the response to a change in effector concentration is defined as (see next chapter)

$$A_L = \left(\frac{\partial P^*}{\partial L}\right)\frac{L}{P^*} = \frac{\partial \ln P^*}{\partial \ln L}. \tag{3.26}$$

As we have already seen, $\widetilde{P}^*$ is a function of α (eq. (3.25)) and α must be a function of L, namely,

$$\widetilde{P}^* = f(\alpha), \qquad \alpha = g(L). \tag{3.27}$$

Thus the usual rules of differentiation of a function of function allow one to reexpress A_L as a product of two factors, namely,

$$A_L = \left(\frac{\partial \widetilde{P}^*}{\partial \alpha}\right)\frac{\alpha}{\widetilde{P}^*}\left(\frac{\partial \alpha}{\partial L}\right)\frac{L}{\alpha}. \tag{3.28}$$

The first factor, $(\partial \widetilde{P}^*/\partial \alpha)(\alpha/\widetilde{P}^*)$, measures the steepness of the curves of Fig. 3.5 and may indeed be very large for zero-order ultrasensitivity. The second factor, $(\partial \alpha/\partial L)/(L/\alpha)$, will enhance this amplification of the response if it assumes values larger than unity. It is therefore of interest to derive the expression of this factor. According to the model system of Fig. 3.6 the fractions, f_1 and f_2, of the converter enzymes that are active, and therefore in states E_1 and E_2, are equal to

$$f_1 = \frac{K_1' L}{1 + K_1' L}, \qquad f_2 = \frac{1}{1 + K_2' L}, \tag{3.29}$$

where K_1' and K_2' are the relevant equilibrium constants. The two maximum rates, V_1 and V_2, are expressed as

$$V_1 = k_1 E_1 = \frac{k_1 E_{1T} K_1' L}{1 + K_1' L}, \qquad V_2 = k_2 E_2 = \frac{k_2 E_{2T}}{1 + K_2' L}, \tag{3.30}$$

where k_1 and k_2 are the catalytic constants of the two antagonistic processes, E_{1T} and E_{2T} the total (active plus inactive) concentrations of the converter enzymes. Therefore, the expression of α is

$$\alpha = \frac{k_1 E_{1T}}{k_2 E_{2T}} \frac{K_1' L + K_1' K_2' L^2}{1 + K_1' L}, \tag{3.31}$$

and the expression of its derivative has the form

$$\frac{\partial \alpha}{\partial L} = \frac{k_1 E_{1T}}{k_2 E_{2T}} \frac{K_1'(1 + 2K_2' L + K_1' K_2' L^2)}{(1 + K_1' L)^2}. \tag{3.32}$$

Therefore, one finds

$$\left(\frac{\partial \alpha}{\partial L}\right)\frac{L}{\alpha} = \frac{1 + 2K_2' L + K_1' K_2' L^2}{1 + (K_1' + K_2')L + K_1' K_2' L^2}. \tag{3.33}$$

This expression will be larger than unity if $K_2' > K_1'$. Therefore, in the case considered, the effector may enhance the amplification of the response. The coupled effect exerted by the ultrasensitivity and the effector can be enormous, many orders of magnitude greater than any co-operative effect observed with an isolated macromolecule.

3.2.3. *Propagation of amplification in multicyclic cascades*

As we shall see in the next section, the internal logic of several biological problems relies on a sequence of cascades. Such a sequence is referred to as a polycyclic cascade. An example of a bicyclic cascade is shown in Fig. 3.7. In this scheme, it is assumed that protein P* serves as a converter enzyme for the reaction that converts protein Q into protein Q*.

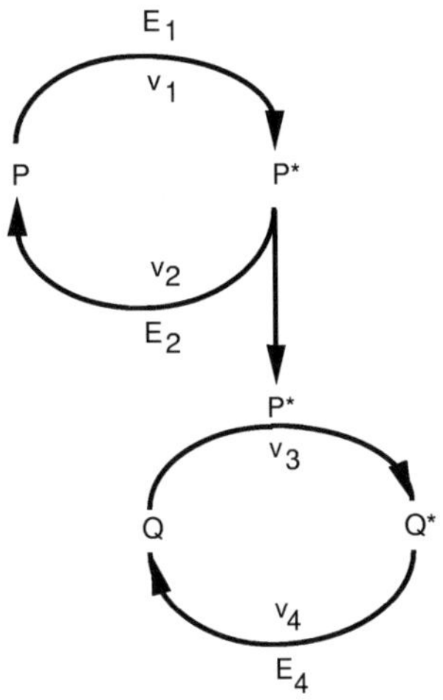

Fig. 3.7. A bicyclic cascade. The modified protein P* plays the part of a converter enzyme.

It is obvious that this process can be extended to a larger number of elementary cascades but the logic remains the same. A question which arises is to know whether a perturbation of the first elementary cascade, brought about by a change of a variable in this cascade, is transmitted to the other cascades, and whether the amplification of the response can be enhanced through the propagation process. More specifically, in the case of the bicyclic cascade of Fig. 3.7, an enhancement of the response will occur during propagation if

$$\left(\frac{\partial \widetilde{Q}^*}{\partial \alpha}\right)\frac{\alpha}{\widetilde{Q}^*} > \left(\frac{\partial \widetilde{P}^*}{\partial \alpha}\right)\frac{\alpha}{\widetilde{P}^*}. \tag{3.34}$$

The maximum reaction rates involved in the first cascade are V_1 and V_2 and those involved in the second, V_3 and V_4. One can thus define the variables α and α' of the two cascades as

$$\alpha = \frac{V_1}{V_2} = \frac{k_1 E_1}{k_2 E_2}, \qquad \alpha' = \frac{V_3}{V_4} = \frac{k_3 P^*}{k_4 E_4}. \tag{3.35}$$

Moreover, $\widetilde{P}^*$ is indeed a function of α and $\widetilde{Q}^*$ is a function of α', namely,

$$\widetilde{P}^* = f(\alpha), \qquad \widetilde{Q}^* = g(\alpha'), \tag{3.36}$$

as previously, these functions can be derived by solving the quadratic coupling equations

$$(\alpha - 1)\widetilde{P}^{*2} - \left[(\alpha - 1) - \widetilde{K}_2\left(\alpha - \frac{\widetilde{K}_1}{\widetilde{K}_2}\right)\right]\widetilde{P}^* - \alpha\widetilde{K}_2 = 0 \tag{3.37}$$

and

$$(\alpha' - 1)\widetilde{Q}^{*2} - \left[(\alpha' - 1) - \widetilde{K}_3\left(\alpha' - \frac{\widetilde{K}_3}{\widetilde{K}_4}\right)\right]\widetilde{Q}^* - \alpha'\widetilde{K}_4 = 0. \tag{3.38}$$

The constants $\widetilde{K}_3$ and $\widetilde{K}_4$ are defined as previously for $\widetilde{K}_1$ and $\widetilde{K}_2$, namely,

$$\widetilde{K}_3 = \frac{K_3}{Q_{\mathrm{T}}}, \qquad \widetilde{K}_4 = \frac{K_4}{Q_{\mathrm{T}}}, \tag{3.39}$$

where K_2 and K_3 are the Michaelis constants of the third and the fourth enzyme reactions, and Q_{T} is the total concentration of Q. Moreover, α' is a function of $\widetilde{P}^*$

$$\alpha' = h(\widetilde{P}^*). \tag{3.40}$$

The rule of differentiation of functions of functions now leads to

$$\left(\frac{\partial \widetilde{Q}^*}{\partial \alpha}\right)\frac{\alpha}{\widetilde{Q}^*} = \left(\frac{\partial \widetilde{Q}^*}{\partial \alpha'}\right)\frac{\alpha'}{\widetilde{Q}^*}\left(\frac{\partial \alpha'}{\partial \widetilde{P}^*}\right)\frac{\widetilde{P}^*}{\alpha'}\left(\frac{\partial \widetilde{P}^*}{\partial \alpha}\right)\frac{\alpha}{\widetilde{P}^*}. \tag{3.41}$$

This expression shows that changing the enzyme ratio of the first cascade, E_1/E_2, affects the behaviour of the second one. There is thus a popagation of the effect, generated by a perturbation of converter enzyme concentrations, within the bicyclic cascade, and this holds true for a more complex polycyclic cascade. Given such a polycyclic cascade, its response to a perturbation of the concentration of the converter enzymes of the first cycle can be viewed as the product of the responses of each elementary cycle to a perturbation of its own converter enzymes, connected by a number of converter elements. In the case of a bicyclic cascade there is one converter element, $(\partial\alpha'/\partial\widetilde{P}^*)(\widetilde{P}^*/\alpha')$, that couples the two elementary cycles. Comparison of expressions (3.26) and (3.41) shows that an enhancement of the response through a coupling of the elementary cascades is expected to occur if

$$\left(\frac{\partial\widetilde{Q}^*}{\partial\alpha'}\right)\frac{\alpha'}{\widetilde{Q}^*}\left(\frac{\partial\alpha'}{\partial\widetilde{P}^*}\right)\frac{\widetilde{P}^*}{\alpha'} > 1. \tag{3.42}$$

Moreover, the expression of the converter element may be derived. As $\alpha' = k_3P^*/(k_4E_4)$, one has

$$\frac{\partial\alpha'}{\partial\widetilde{P}^*} = P_\mathrm{T}\frac{\partial\alpha'}{\partial P^*} = \frac{k_3P_\mathrm{T}}{k_4E_4} \tag{3.43}$$

and

$$\left(\frac{\partial\alpha'}{\partial\widetilde{P}^*}\right)\frac{\widetilde{P}^*}{\alpha'} = \frac{k_3P_\mathrm{T}}{k_4E_4}\frac{P^*}{P_\mathrm{T}}\frac{1}{\alpha'} = \frac{k_3P^*}{k_4E_4}\frac{1}{\alpha'} = 1. \tag{3.44}$$

Expression (3.42) thus reduces to

$$\left(\frac{\partial\widetilde{Q}^*}{\partial\alpha'}\right)\frac{\alpha'}{\widetilde{Q}^*} > 1, \tag{3.45}$$

and eq. (3.41) reduces to

$$\left(\frac{\partial\widetilde{Q}^*}{\partial\alpha}\right)\frac{\alpha}{\widetilde{Q}^*} = \left(\frac{\partial\widetilde{Q}^*}{\partial\alpha'}\right)\frac{\alpha'}{\widetilde{Q}^*}\left(\frac{\partial\widetilde{P}^*}{\partial\alpha}\right)\frac{\alpha}{\widetilde{P}^*}. \tag{3.46}$$

Thus, the response of the overall bicyclic cascade to a perturbation of the converter enzymes of the first cycle is the product of the responses of the individual cycles to a perturbation of their own converter enzymes. The converter element, which couples the two cycles, is equal to unity. This important conclusion can be extended to a polycyclic cascade displaying a larger number of coupled elementary cycles.

3.2.4. Response of a polycyclic cascade to an effector

We are now in a position to study how an effector L, which affects the activity of the converter enzymes of the first cycle of a polycyclic cascade, may modulate the behaviour

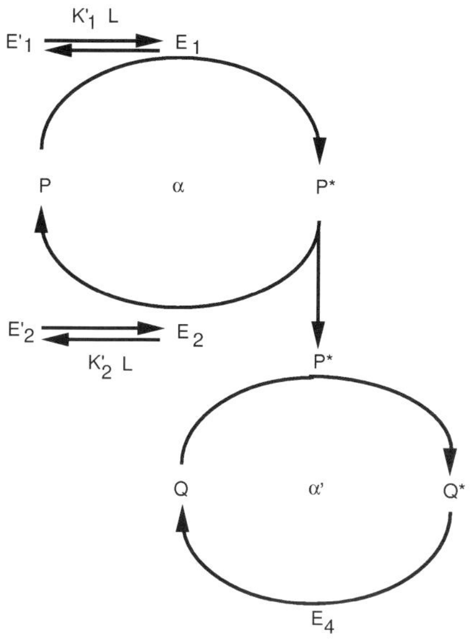

Fig. 3.8. Effect of a ligand L on two converter enzymes of a bicyclic cascade. As previously, the ligand L is assumed to activate enzyme E_1 and to inhibit enzyme E_2.

of the whole system. The model that is considered below is shown in Fig. 3.8. It represents an extension to the bicyclic cascade of the ideas already developed in section 3.2.2.

In this model, it is assumed, as previously, that an inactive converter enzyme E_1' is activated to a new form, E_1, and that the active enzyme form E_2 is converted into an inactive one E_2'. These two conversion processes require the binding of a ligand L to the two enzyme molecules. The response (and its potential amplification) of the polycyclic system to a perturbation of the ligand concentration is, as previously

$$A_{\mathrm{L}} = \left(\frac{\partial \widetilde{Q}^*}{\partial L}\right)\frac{L}{\widetilde{Q}^*}. \tag{3.47}$$

In the system of Fig. 3.8, $\widetilde{P}^*$ is a function of α which is itself a function of L, and $\widetilde{Q}^*$ is a function of α' which is a function of $\widetilde{P}^*$. Thus

$$\begin{aligned} \widetilde{P}^* &= f(\alpha), \qquad \alpha = g(L), \\ \widetilde{Q}^* &= h(\alpha'), \qquad \alpha' = i(\widetilde{P}^*), \end{aligned} \tag{3.48}$$

and the classical rules of differentiation of functions allow one to reexpress eq. (3.47) as

$$A_{\mathrm{L}} = \left(\frac{\partial \widetilde{Q}^*}{\partial \alpha'}\right)\frac{\alpha'}{\widetilde{Q}^*}\left(\frac{\partial \alpha'}{\partial \widetilde{P}^*}\right)\frac{\widetilde{P}^*}{\alpha'}\left(\frac{\partial \widetilde{P}^*}{\partial \alpha}\right)\frac{\alpha}{\widetilde{P}^*}\left(\frac{\partial \alpha}{\partial L}\right)\frac{L}{\alpha}. \tag{3.49}$$

As previously, one must have

$$\left(\frac{\partial \alpha'}{\partial \widetilde{P}^*}\right)\frac{\widetilde{P}^*}{\alpha'} = 1, \tag{3.50}$$

and the effector will generate an amplification of the response with respect to eq. (3.41) if

$$\left(\frac{\partial \alpha}{\partial L}\right)\frac{L}{\alpha} > 1. \tag{3.51}$$

The expression of $(\partial\alpha/\partial L)/(L/\alpha)$ has already been derived (eq. (3.33)) and can reach values larger than one if $K_2' > K_1'$. Therefore, the influence of an effector of the first cycle may propagate to the last one and may, for instance, enhance the overall response of the polycyclic cascade.

3.3. Bacterial chemotaxis as an example of cell signaling

Many different cell signaling phenomena bring into play cascade reactions and probably also multistability phenomena. Bacterial chemotaxis will be taken as an example of these types of events. This study will be limited to the most important biological results and to their likely explanation. Bacterial chemotaxis has been studied with *Escherichia coli* and *Salmonella enterica* [25–28]. Chemotaxis is the response of bacteria and other organisms to the introduction of certain chemicals in their environment. If a solution of a sugar or an aminoacid is introduced with a micropipette into a medium containing either of these bacteria, a concentration gradient builds up in the medium. The bacteria swim against the concentration gradient, towards a region of space that is rich in this chemical, which then plays the part of an attractant. Other substances, such as phenols, behave as repellents. In this case the bacteria swim down the concentration gradient towards a region that does not contain much phenol.

Bacterial swimming is brought about by their flagella. A bacterium such as *E. coli* bears six to ten flagella located on the cell surface. Each flagellum consists of a helical tube made up of a protein, referred to as flagellin, and is attached, through a short hook, to a complex molecular structure that looks like a protein disk embedded in the bacterial membrane. This complex structure allows osmomechanical coupling between the energy stored in the electrochemical gradient, which exists between the inside and the outside of the cell, and the movements of the flagellum (see chapter 5). In fact, the disk is part of a motor that converts the energy of the proton gradient into mechanical energy. A rotation of the flagellum of 360° is paid for at the cost of the diffusion of about a thousand protons. The angular velocity of the flagellum depends on the electrochemical potential difference of the protons. The maximum angular velocity that can be reached is about sixty revolutions per second. The flagella may rotate clockwise or counterclockwise. If the flagella rotate counterclockwise, they appear collected as a single bundle which plays the part of a propeller, and the bacterium swims in a definite, fixed, direction. Alternatively, if the flagella rotate clockwise, they do not bundle together and the cell tumbles chaotically without moving in a definite direction. In the absence of an environmental stimulus, the direction of rotation of flagella reverses every moment. This gives the bacterium a specific type of swimming that alternates smooth displacement in a definite direction and random changes of direction caused by tumbling. In the presence of an attractant, the frequency of tumbling decreases dramatically and the baceria are moving against the concentration gradient. Qualitatively,

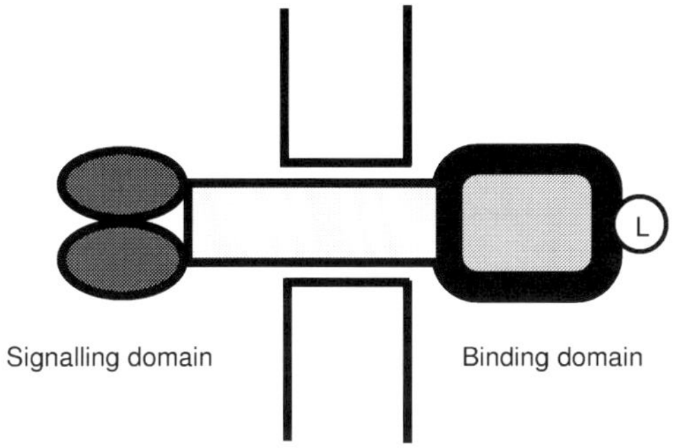

Fig. 3.9. Schematic picture of a receptor. See text.

the same result is observed with a repellent, but now the bacteria follow the concentration gradient and swim towards the region of space that are less concentrated.

As we shall see later, the presence of attractants and repellents can be detected by bacteria, for these substances bind to transmembrane receptors. An important question is to know how a bacterium detects a *concentration change* and swims against, or in the direction of, a concentration gradient. A bacterium is unable to detect, at a given time, a concentration difference over the distance that separates the receptors located in different regions of the cell surface. The bacterial cell is so small that it would be too minute a distance to sense a difference of concentration. But the bacteria can detect changes in time, not in space, of concentrations when they swim in a definite direction. Changes over time can be produced in the laboratory by increasing the concentration of attractant. Then, the frequency of tumbling decreases for a while, thus resulting in a smooth displacement of bacteria. But, after a short time period, the tumbling frequency returns to normal. The bacteria have become adapted to a higher concentration of the attractant.

All these experimental observations mean that the transmembrane receptors are able to sense not only the local concentration of an attractant or a repellent at time t, but also whether this concentration is increasing or decreasing, as the bacteria swim in a definite direction. In other words, the bacterial cell surface has a short-term memory of what is occurring in the external medium. This is precisely the situation that has been described theoretically in the first section of the present chapter. Although this matter has not been studied in great details so far, it is the present author's opinion that the sensivity of the cell surface to *changes of concentration* can be understood only through a coupling between dffusion of a ligand (attractant or repellent) and a series of chemical reactions that are initiated by the binding of this ligand to the transmembrane receptor. Qualitatively speaking, this is exactly the situation that was considered in the simple model of section 3.1, above.

It is therefore of interest to comment briefly on these receptors and on the complex reaction they are initiating [29–33]. There are four types of receptors for attractants in *E. coli* cells. Two of them bind aminoacids, one binds sugars and the last binds dipeptides. They are all transmembrane proteins. The ligands, attractants or repellents, should thus cross the external membrane and diffuse through the periplasmic space to the competent receptor. In the case of aminoacids, it is the free molecule that diffuses to the receptor. In the case of sugars and dipeptides, these ligands should first bind to a periplasmic protein and the resulting binary complex then binds to the receptor. As already mentioned, the

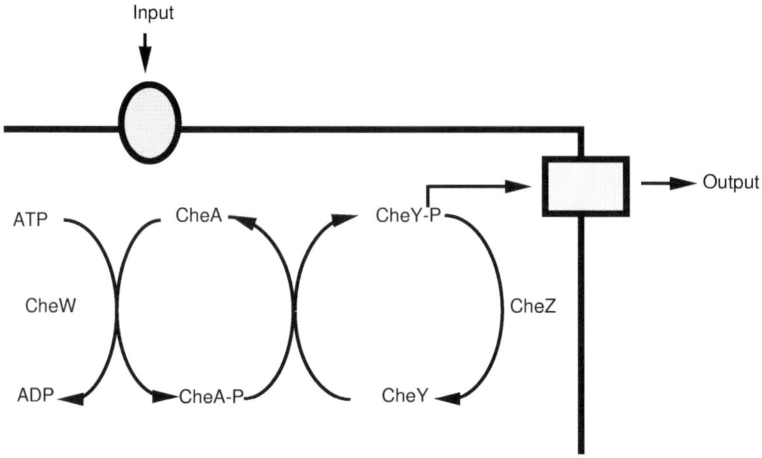

Fig. 3.10. Schematic picture of the cascade involved in bacterial chemotaxis. See text.

receptors are transmembrane proteins made up of one or two polypeptide chains (Fig. 3.9). The region of the polypeptide chain located in the periplasmic space is responsible for ligand binding. Both the N- and the C-terminus are located in the cytosol. In this region of the cell, the polypeptide chain is folded in two domains: a domain containing aminoacid residues that can be methylated and a signaling domain. When an attractant, or a repellent, is bound to the receptor domain located in the periplasmic space, a conformation change takes place and the signaling domain adopts a conformation that allows the interaction of this domain with proteins involved in the information transfer to the motor.

There are four cytoplasmic proteins that couple the receptors with the flagellar motor. They are referred to as CheA, CheW, CheY and CheZ. CheA is a histidine protein kinase that can bind to the signaling domain of the activated receptor and to CheW as well. It is then phosphorylated by ATP on a histidine residue (Fig. 3.10). The phosphorylated form of CheA, CheA-P, transfers its phosphate to another cytoplasmic protein, CheY. The phosphorylated form of CheY, CheY-P, binds to the flagellar motor. In the absence of CheY-P, the motor and the flagellum rotate counterclockwise, thus allowing the bacterium to swim smoothly in a definite direction. Once CheY-P is bound to the motor, the rotation becomes clockwise, which, in turn, induces tumbling. Another protein, CheZ, activates the dephosphorylation of CheY-P and antagonizes the effect of CheY-P on the flagellar motor (Fig. 3.10).

The binding of a repellent to a receptor increases the latter's activity, which leads in turn to an increased phosphorylation rate of both CheA and CheY. CheY-P is then synthesized in larger amounts and binds to the motor, inducing a clockwise rotation of the flagella and tumbling of the bacterium. An attractant behaves the opposite way. The receptor is in a less activated state. CheA and CheY are phosphorylated at a lower rate and the flagella continue to rotate counterclockwise, thus allowing the bacterium to swim smoothly. CheY thus plays the part of an on/off switch. Depending on whether it is phosphorylated or not, it induces a clockwise or counterclockwise rotation of the motor. Other proteins in different organisms, in animal cells for instance, also play the role of a switch, exactly like CheY. If the reaction scheme of Fig. 3.10 is compared to those of the previous sections, it is clear that it can be considered as a bicyclic cascade.

We have already outlined that bacterial chemotaxis displays adaptive responses. Thus, when an attractant, for instance, is introduced into the medium containing bacteria, tumbling is at first abolished, then after a while returns to its initial frequency. This return to the normal frequency of tumbling is called the adaptive response of bacteria to the attractant. It is associated with the methylation of the receptor. In fact, binding of an attractant to a receptor has two implications: deactivation of the receptor already mentioned and methylation of this receptor. The first event causes a decline of the phosphorylation rate of CheY, thus leading to a decrease in tumbling frequency. The specific methylation of the receptor catalysed by a methyltransferase restores the initial activity of the receptor with bound attractant. When the attractant is removed, the receptor is demethylated by a methylesterase. It is very likely that other regulation processes, yet to be discovered, play a role in the molecular control of bacterial chemotaxis.

3.4. General features of a signaling process

When considered at the macroscopic level, signaling processes appear extremely diverse. Neural activity, vision, bacterial chemotaxis, etc., may all be considered basically different biological phenomena. Nevertheless they display, at the molecular level, common features that will briefly be considered below.

The biological system, whatever its nature, receives a signal, often chemical, that interacts with a receptor. This signal is processed and converted into an output. In other words, the receptor transfers to the inside of the system a piece of information that comes from outside of the system. In most cases, information transfer through a receptor means a conformation change of the receptor on binding a chemical signal. This implies that the term "information" should be given a much broader meaning than that of nucleotide, or aminoacid, sequence which is encountered in classical molecular biology. This matter will be discussed in some details in a forthcoming chapter.

The output, or response, of the system is, in most cases, amplified. This means that slight changes of the signal result in dramatic changes of the response. The response is thus co-operative, or nonlinear, relative to the signal.

Coupled antagonostic chemical reactions, which constitute a cascade, represent a common basic feature of most, and perhaps all, signaling devices. These antagonistic reactions, for instance phosphorylation/dephosphorylation of a protein or binding/release of GTP to/from an enzyme, represent a molecular device that allows transduction of a signal. From a thermodynamic viewpoint the whole system must be open and requires expenditure of energy (hydrolysis of ATP or GTP, for instance).

The system must be able to sense not only the intensity of a signal but also whether this intensity is reached via an increase or a decrease from a previous intensity. This requires that the system possesses short-term memory of what has been happening in the external medium. From a physical viewpoint, this strongly suggests that the system displays several stable steady states. The system can then discriminate between these steady states on the basis of the direction of change of the signal. Coupling between chemical reactions, or between diffusion and enzyme reactions, can generate multistability. A number of experimental biochemical systems have been shown to display this multistable behaviour [7,34].

A constant and perhaps more general property, which may be encountered in coupled systems, is the emergence of novel properties from reaction coupling. Thus, the steep cooperativity of a cascade is a property of the system of coupled reactions, rather than the property of any individual reaction. Similarly, multistability and the ability to behave as a real biosensor does not exist for individual enzyme reactions and requires coupling of chemical reactions, or of vectorial and scalar processes. Individual enzyme reactions, if they display meta-steady states, behave as primitive biosensors, but this requires stringent conditions of "slow" conformation changes that do not appear to be encountered often in nature.

References

[1] Engasser, J.M. and Horvath, C. (1974a) Inhibition of bound enzymes. I. Antienergistic interaction of chemical and diffusional inhibition. Biochemistry 13, 3849–3854.

[2] Engasser, J.M. and Horvath, C. (1974b) Inhibition of bound enzymes. II. Characterization of product inhibition and accumulation. Biochemistry 13, 3849–3854.

[3] Engasser, J.M. and Horvath, C. (1974c) Inhibition of bound enzymes. III. Diffusion enhanced regulatory effect with substrate inhibition. Biochemistry 13, 3855–3859.

[4] Engasser, J.M. and Horvath, C. (1976) Diffusion and kinetics with immobilized enzymes. In: L.B. Wingard and E. Katchalski-Katzir (Eds.), Applied Biochemistry and Bioengeering, Vol. 1. Academic Press, New York, pp. 127–220.

[5] Hervagault, J.F. and Thomas, D. (1985) Theoretical and experimental studies on the behavior of immobilized multienzyme systems. In: G.R. Welch (Ed.), Organized Multienzyme Systems: Catalytic Properties, Academic Press, New York, pp. 381–418.

[6] Hervagault, J.F., Breton, J., Kernevez, J.P., Rajani, J. and Thomas, D. (1984) Photobiochemical memory. In: J. Ricard and A. Cornish-Bowden (Eds.), Dynamics of Biochemical Systems. Plenum Press, New York, pp. 157–169.

[7] Thomas, D., Barbotin, J.N., Hervagault, J.F. and Romette, J.L. (1977) Experimental evidence for a kinetic and electrochemical memory in enzyme membranes. Proc. Natl. Acad. Sci. USA 74, 5314–5317.

[8] Ricard, J. and Noat, G. (1984a) Enzyme reactions at the surface of living cells. I. Electric repulsion of charged ligands and recognition of signals from the external milieu. J. Theor. Biol. 109, 555–569.

[9] Ricard, J. and Noat, G. (1984b) Enzyme reactions at the surface of living cells. II. Destabilization in the membrane and conduction of signals. J. Theor. Biol. 109, 571–580.

[10] Ricard, J. (1989) Modulation of enzyme catalysis in organized biological systems. A physico-chemical approach. Catalysis Today 5, 275–384.

[11] Ricard, J., Mulliert, G., Kellershohn, N. and Giudici-Orticoni, M.T. (1994) Dynamics of enzyme reactions and metabolic networks in living cells. A physico-chemical approach. Progress Mol. Subcel. Biol. 13, 1–80.

[12] Chock, P.B., Rhee, S.G. and Stadtman, E.R. (1980a) Covalently interconvertible enzyme cascade systems. In: D.L. Purich (Ed.), Methods of Enzymology, Vol. 64, Part B. Academic Press, New York, pp. 297–325.

[13] Chock, P.B., Rhee, S.G. and Stadtman, E.R. (1980b) Inerconvertible enzyme cascades in cellular regulation. Annu. Rev. Biochem. 49, 813–843.

[14] Chock, P.B. and Stadtman, E.R. (1977) Superiority of interconvertible enzyme cascades in metabolic regulation: analysis of multicyclic systems. Proc. Natl. Acad. Sci. USA 74, 2766–2770.

[15] Rhee, S.G., Park, R., Chock, P.B. and Stadtman, E.R. (1978) Allosteric regulation of monocyclic interconvertible enzyme cascade systems: use of *E. coli* glutamine synthetase as an experimental model. Proc. Natl. Acad. Sci. USA 75, 3138–3142.

[16] Mura, U., Chock, P.B. and Stadtman, E.R. (1981) Allosteric regulation of the state of adenylation of glutamine synthetase in permeabilized cell preparations of *Escherichia coli*. J. Biol. Chem. 256, 13022–13029.
[17] Rhee, S.G., Chock, P.B. and Stadtman, E.R. (1989) Regulation of *Escherichia coli* glutamine synthetase. Adv. Enzymol. 62, 37–92.
[18] Stadtman, E.R. and Chock, P.B. (1977) Superiority of interconvertible enzyme cascades in metabolic regulation: analysis of monocyclic cascades. Proc. Acad. Sci. USA 74, 2761–2766.
[19] Chock, P.B., Rhee, S.G. and Stadtman, E.R. (1990) Metabolic control by the cyclic cascade mechanism: a study of *E. coli* glutamine synthetase. In: A. Cornish-Bowden and M.L. Cardenas (Eds.), Control of Metabolic Processes. Plenum Press, New York.
[20] Goldbeter, A. and Koshland, D.E. (1981) An amplified sensitivity arising from covalent modification in biological systems. Proc. Natl. Acad. Sci. USA 78, 6840–6844.
[21] Goldbeter, A. and Koshland, D.E. (1984) Ultrasensitivity in biochemical systems controlled by covalent modification. J. Biol. Chem. 259, 14441–14447.
[22] Goldbeter, A. and Koshland, D.E. (1982) Sensitivity amplification in biochemical systems. Quarterly Rev. Biophys. 15, 555–591.
[23] Goldbeter, A. and Koshland, D.E. (1990) Zero-order ultrasensitivity in interconvertible enzyme systems. In: A. Cornish-Bowden and M.L. Cardenas (Eds.), Control of Metabolic Processes. Plenum Press, New York.
[24] Cardenas, M.L. and Cornish-Bowden, A. (1990) Properties needed for the enzymes of an interconvertible cascade to generate a highly sensitive response. In: A. Cornish-Bowden and M.L. Cardenas (Eds.), Control of Metabolic Processes. Plenum Press, New York.
[25] Adler (1976) The sensing of chemicals by bacteria. Sci. Amer., April, 40–47.
[26] Berg, H. (1975) How bacteria swim. Sci. Amer., February, 36–44.
[27] Koshland, D.E. (1981) Biochemistry of sensing and adaptation in a simple bacterial system. Annu. Rev. Biochem. 50, 765–782.
[28] Alberts, B., Bray, D., Lewis, J., Raff, M., Roberts, K. and Watson, J. (1994) Molecular Biology of the Cell. Garland, New York and London.
[29] Bourret, R.B., Borkovitch, K.A. and Simon, M.I. (1991) Signal transduction pathways involving protein phosphorylation in prokaryotes. Annu. Rev. Biochem. 60, 401–441.
[30] Hazelbauer, G.L. (1992) Bacterial chemoreceptors. Curr. Opin. Struct. Biol. 2, 505–510.
[31] Parkinson, J.S. (1993) Signal transduction schemes of bacteria. Cell 73, 857–871.
[32] Stock, J.B., Lukat, G.S. and Stock, A.M. (1991) Bacterial chemotaxis and the molecular logic of intracellular signal transduction networks. Annu. Rev. Biophys. Biophys. Chem. 20, 109–136.
[33] Stoddard, B.L., Bui, J.D. and Koshland, D.E. (1992) Structure and dynamics of transmembrane signaling by the *Escherichia coli* transmembrane receptor. Biochemistry 31, 11978–11983.
[34] Naparstek, A., Romette, J.L., Kernevez, J.P. and Thomas, D. (1974) Memory in enzyme membranes. Nature 249, 490–491.

CHAPTER 4

Control of metabolic networks under steady state conditions

Enzyme reactions which take place in the living cell are usually included in a metabolic sequence. In order to understand the behaviour of the whole system, it is certainly not sufficient to have a complete knowledge of the structure and dynamics of the individual enzymes that take part in this metabolic network. What matters most is to understand the kinetics and the dynamics of the metabolic process considered as a whole.

This matter has long been disregarded and the control of metabolism has been studied in a reductionist manner. It was in fact considered that metabolic pathways were each controlled by a small number of enzymes, or even by a single enzyme, located at the beginning of the reaction sequence and retro-inhibited by one of the last products of the sequence. This was considered an example of a "molecular autocracy" in which the behaviour of a system could be reduced to that of one enzyme only. There are, however, numerous examples which show that one should not stick too closely to these reductionist views. By altering in succession, through site-directed mutagenesis for instance, the individual properties of the enzymes that partake in a reaction flux, the flux may well not be altered. This clearly shows that, in this case, no enzyme can specifically be considered as "regulatory", but rather that this function is shared by different enzymes of the pathway. The understanding, in quantitative terms, of the conrol of a metabolic network is achieved thanks to two formal models called the Metabolic Control Theory (MCT) and the Biochemical System Theory (BST). Both will be considered now.

4.1. Metabolic control theory

The bases of the Metabolic control theory were formulated independently by Kacser and Burns [1–4] and Heinrich and Rapoport [5,6]. The theory has been developed thanks to several contributions [7–19]. In spite of its name it is not really a theory, but rather a formal model that allows the expression of parameters that describe, in quantitative terms, the global behaviour of an enzyme system.

4.1.1. The parameters of Metabolic control theory

Let us consider a linear sequence of enzyme reactions

$$\mathrm{X_0} \xrightarrow[v_1]{\mathrm{E_1}} \mathrm{S_1} \xrightarrow[v_2]{\mathrm{E_2}} \mathrm{S_2} \cdots \mathrm{S}_{n-1} \xrightarrow[v_n]{\mathrm{E}_n} \mathrm{X_f} \tag{4.1}$$

X_0 and X_f are a source and a sink. Both are infinitely large and are thus considered reservoirs. $S_1, S_2, \ldots$ are the intermediates of the metabolic network, E_1, $E_2, \ldots$ are the enzymes involved in this reaction sequence. Usually, one considers the system to be in steady state, then the overall flux, J, is equal to any of the individual reaction steps, $v_1, v_2, \ldots$

$$J = v_1 = v_2 = \cdots. \tag{4.2}$$

One can define three types of control parameters of the reaction network. The first is referred to as the flux control coefficient, C_i^J. This type of coefficient is defined as

$$C_i^J = \frac{E_i}{J}\frac{\partial J}{\partial E_i} = \frac{\partial \ln J}{\partial \ln E_i}. \tag{4.3}$$

These coefficients express the response of the steady state reaction flux to a small change in an enzyme concentration. The flux control coefficients are systemic properties of the reaction sequence, for they measure how the system responds to a slight perturbation exerted on a specific enzyme.

The second type of parameter is called the concentration control coefficient, $C_i^{S_j}$, and is defined as

$$C_i^{S_j} = \frac{E_i}{S_j}\frac{\partial S_j}{\partial E_i} = \frac{\partial \ln S_j}{\partial \ln E_i}. \tag{4.4}$$

These coefficients express how a perturbation of the concentration of enzyme E_i affects the concentration of the intermediate S_j. As the concentration of this intermediate is dependent on the behaviour of the whole multienzyme system, this type of control coefficient also reflects a systemic property.

The last type of control coefficient corresponds to a so-called elasticity. Elasticities are definied as

$$\varepsilon_{S_j}^{v_i} = \frac{S_j}{v_i}\frac{\partial v_i}{\partial S_j} = \frac{\partial \ln v_i}{\partial \ln S_j}, \tag{4.5}$$

and express how a perturbation of the concentration of an intermediate S_j affects the reaction rate, v_i, of the ith step. This response is thus clearly not systemic, but rather a local property of the reaction network. By studying the individual response of each enzyme of the reaction sequence to a change in concentration of the reaction intermediates, one can easily determine the elasticities of the enzymes involved in the metabolic network.

4.1.2. The summation theorems

If we disregard for the moment the concentrations of the reaction intermediates, $S_1, S_2, \ldots$, the overall flux of a metabolic sequence is indeed a function of the enzyme concentrations

$$J = f(E_1, E_2, \ldots). \tag{4.6}$$

Moreover, if the enzyme concentrations are changed by small values $\mathrm{d}E_1$, $\mathrm{d}E_2, \ldots$, the overall flux, J, will be changed by a small value $\mathrm{d}J$. This is, indeed, equivalent to stating that J is a state function of enzyme concentrations. One thus has

$$\mathrm{d}J = \sum_{i=1}^{n} \left(\frac{\partial J}{\partial E_i} \right) \mathrm{d}E_i. \tag{4.7}$$

Dividing both sides by J, rearranging and taking into account the definition of flux control coefficients, one finds

$$\frac{\mathrm{d}J}{J} = \sum_{i=1}^{n} C_i^j \frac{\mathrm{d}E_i}{E_i}. \tag{4.8}$$

Now, if all the $\mathrm{d}E_i/E_i$ are the same and equal to α, the relative change of the reaction flux will also be equal to α and therefore

$$\alpha = \sum_{i=1}^{n} C_i^J \alpha, \tag{4.9}$$

which requires that

$$\sum_{i=1}^{n} C_i^J = 1. \tag{4.10}$$

This is the mathematical expression of the summation theorem of flux control coefficients. The summation theorem gives sensible grounds to the view that different enzymes of a given network, and perhaps all the enzymes of this network, partake in the control of the flux.

From a mathematical viewpoint, the summation theorem is nothing but the statement that the flux J is a homogeneous function of degree one in enzyme concentrations, E_1, $E_2, \ldots$. A function $f(x_1, x_2, \ldots)$ is called homogeneous of degree h if

$$f(tx_1, tx_2, \ldots) = t^h f(x_1, x_2, \ldots) \tag{4.11}$$

for all $t \neq 0$. Euler's theorem establishes that a function $f(x_1, x_2, \ldots)$ is homogeneous of degree h in $x_1, x_2, \ldots$, if

$$hf(x_1, x_2, \ldots) = \sum_{j=1}^{n} x_j \frac{\partial f}{\partial x_j}. \tag{4.12}$$

Conversely, any function $g(x_1, x_2, \ldots)$ that obeys the relationship

$$hg(x_1, x_2, \ldots) = \sum_{j=1}^{n} x_j \frac{\partial g}{\partial x_j} \tag{4.13}$$

is homogeneous of degree h in $x_1, x_2, \ldots$. The expression of the sumation theorem can be rewritten as

$$\sum_{i=1}^{n} E_i \left(\frac{\partial J}{\partial E_i} \right) = J, \tag{4.14}$$

and comparison of eqs. (4.13) and (4.14) shows that they are equivalent if $h = 1$. Hence, the summation theorem implies that a flux through a metabolic network is a homogeneous function of degree one in enzyme concentrations.

Equation (4.14) and the conclusion that a reaction flux is a homogeneous function of degree one in enzyme concentrations have important physical implications. Since the flux is assumed to be in steady state, it must be equal to the rate of any of the various steps of the metabolic sequence. Hence, eq. (4.14) becomes

$$\left(\frac{\partial v_i}{\partial E_i} \right) E_i = v_i. \tag{4.15}$$

Therefore, if the steady state flux is a homogeneous function of degree one in the enzyme concentrations, the same statement should apply to any of the reaction steps. Equation (4.15), in turn, implies that

$$t v_i(E_i) = v_i(t E_i). \tag{4.16}$$

In other words, changing the enzyme concentration by a factor t changes the rate by the same factor. This is in fact equivalent to a classical result of steady state kinetics, namely the existence of a linear relationship between enzyme concentration and steady state velocity. The conclusion, however, no longer holds true if the enzymes of the pathway, or some of them at least, associate or dissociate and if the resulting steady state rate depends on the degree of association of these enzymes. These association–dissociation processes may involve identical or different enzymes. In its classical form, Metabolic control theory implicitly postulates that there is no interaction between enzymes. This is a severe restriction indeed to the applicability of the theory and an attempt has recently been made to abandon the postulate of the physical independence of the enzymes of a pathway [20]. This matter will be discussed very briefly later.

Likewise, one may also state that any concentration, S_j, is a function of all the enzyme concentrations

$$S_j = g(E_1, E_2, \ldots). \tag{4.17}$$

Perturbing these enzyme concentrations by different values $\mathrm{d}E_1$, $\mathrm{d}E_2, \ldots$ leads to a change, $\mathrm{d}S_j$, of the concentration S_j, namely,

$$\mathrm{d}S_j = \sum_{i=1}^{n} \left(\frac{\partial S_j}{\partial E_i} \right) \mathrm{d}E_i. \tag{4.18}$$

Dividing both sides of this expression by S_j and taking account of the definition of the substrate control coefficients leads to

$$\frac{\mathrm{d}S_j}{S_j} = \sum_{i=1}^{n} C_i^{S_j} \frac{\mathrm{d}E_i}{E_i}. \tag{4.19}$$

It is obvious that changing all the enzyme concentrations by the same value α will not affect the steady state concentration S_j. Thus one has

$$\sum_{i=1}^{n} C_i^{S} \alpha = 0, \tag{4.20}$$

which requires that

$$\sum_{i=1}^{n} C_i^{S_j} = 0. \tag{4.21}$$

This equation expresses the summation theorem for the substrate control coefficients.

4.1.3. Connectivity between flux control coefficients and elasticities

Let us consider step i of a metabolic network. Its steady state rate, v_i, is a function of the corresponding enzyme and reactant concentrations, $E_i, S_i, \ldots, S_j, \ldots$. Thus one has

$$v_i = f(E_i, S_i, \ldots, S_j, \ldots). \tag{4.22}$$

The total differential of the rate v_i may thus be expressed as

$$\mathrm{d}v_i = \left(\frac{\partial v_i}{\partial E_i}\right)\mathrm{d}E_i + \sum_{j=1}^{m}\left(\frac{\partial v_i}{\partial S_j}\right)\mathrm{d}S_j. \tag{4.23}$$

This expression may be rearranged to

$$\mathrm{d}v_i = \frac{\partial v_i}{\partial E_i}\frac{E_i}{v_i}\frac{\mathrm{d}E_i}{E_i}v_i + \sum_{j=1}^{m}\frac{\partial v_i}{\partial S_j}\frac{S_j}{v_i}\frac{\mathrm{d}S_j}{S_j}v_i. \tag{4.24}$$

If, as already assumed, there is no interaction between enzyme E_i and the other enzymes of the metabolic pathway

$$\frac{\partial v_i}{\partial E_i}\frac{E_i}{v_i} = 1, \tag{4.25}$$

and eq. (4.24) becomes

$$dv_i = \frac{dE_i}{E_i} v_i + \sum_{j=1}^{m} \varepsilon_{S_j}^{v_i} \frac{dS_j}{S_j} v_i \tag{4.26}$$

or

$$\frac{dv_i}{v_i} = \frac{dE_i}{E_i} + \sum_{j=1}^{m} \varepsilon_{S_j}^{v_i} \frac{dS_j}{S_j}. \tag{4.27}$$

This expression offers the rationale for the term "elasticity" given to the ε parameter. It shows that, if the concentration of the enzyme E_i is perturbed, the corresponding rate, v_i, will remain unchanged provided that

$$\frac{dE_i}{E_i} = -\sum_{j=1}^{m} \varepsilon_{S_j}^{v_i} \frac{dS_j}{S_j}. \tag{4.28}$$

This equation represents the condition required to maintain the flux constant, when all the reaction intermediates of the pathway are perturbed by the values $dS_1, dS_2, \ldots$. If the concentration of only one intermediate, S_j, is altered then

$$\frac{dE_i}{E_i} = -\varepsilon_{S_j}^{v_i} \frac{dS_j}{S_j}. \tag{4.29}$$

Therefore, eq. (4.8) may be rewritten as

$$\frac{dJ}{J} = 0 = \sum_{i=1}^{n} C_i^J \frac{dE_i}{E_I} \tag{4.30}$$

or

$$\frac{dS_j}{S_j} \sum_{i=1}^{n} C_i^J \varepsilon_{S_j}^{v_i} = 0. \tag{4.31}$$

As $dS_j/S_j \neq 0$, expression (4.31) will be satisfied if

$$\sum_{i=1}^{n} C_i^J \varepsilon_{S_j}^{v_i} = 0. \tag{4.32}$$

Equation (4.32) represents a connectivity relationship between the flux control coefficients and all the elasticities involving the reaction intermediates S_j. This relationship is absolutely required to maintain the flux unchanged after a perturbation of enzyme concentration. This connectivity property, which applies to a specific intermediate S_j, can be extended to all the reaction intermediates of the network. If the network consists of n enzyme reactions and $n-1$ reaction intermediates, the relationship reads

$$\sum_{j=1}^{n-1}\sum_{i=1}^{n} C_i^J \varepsilon_{S_j}^{v_i} = 0, \tag{4.33}$$

and represents a general connectivity relationship amongst all the flux control coefficients and all the elasticity coefficients pertaining to the $n-1$ reaction intermediates.

4.1.4. Connectivity between substrate control coefficients and elasticities

There is now another type of connectivity that involves the substrate control coefficients and the elasticities. Let us assume that the concentration of the enzyme E_i is perturbed. This perturbation is assumed to reverberate, for instance, to the substrate S_j, and to this substrate only. However, the steady state velocity, v_i, of the corresponding step will not be altered if (eq. (4.29) above)

$$\frac{dE_i}{E_i} = -\varepsilon_{S_j}^{v_i}\frac{dS_j}{S_j}. \tag{4.34}$$

Now, if we consider an unperturbed substrate S_k, in the metabolic sequence, it is a function, h, of the concentrations of all the enzymes partaking in that network. Thus one has

$$S_k = h(E_1, E_2, \ldots), \tag{4.35}$$

and (see eq. (4.19))

$$\frac{dS_k}{S_k} = \sum_{i=1}^{n} C_1^{S_k}\frac{dE_i}{E_i} = 0. \tag{4.36}$$

Substituting dE_i/E_i by its expression derived from eq. (4.29), expression (4.36) reads

$$\sum_{i=1}^{n} C_i^{S_k}\varepsilon_{S_j}^{v_i} = 0. \tag{4.37}$$

Now, if a similar relationship is derived for the perturbed substrate S_j, eq. (4.19) can be rewritten, taking advantage of expression (4.29), as

$$\frac{dS_j}{S_j} = -\sum_{i=1}^{n} C_i^{S_j}\varepsilon_{S_j}^{v_i}\frac{dS_j}{S_j}, \tag{4.38}$$

which is equivalent to

$$\sum_{i=1}^{n} C_i^{S_j} \varepsilon_{S_j}^{v_i} = -1. \tag{4.39}$$

Equations (4.37) and (4.39) can be summarized as

$$\sum_{i=1}^{n} C_i^{S_k} \varepsilon_{S_j}^{v_i} = -\delta_{jk}, \tag{4.40}$$

where δ_{jk} is the Kronecker δ equal to 1 when $j = k$ and 0 when $j \neq k$. This relationship expresses the connectivity between the substrate control coefficients and the elasticities.

4.1.5. Generalized connectivity relationships and the problem of enzyme interactions and information transfer in Metabolic control theory

In its standard formulation, Metabolic control theory assumes that relationship (4.25) holds, which, in turn, implies that the flux is a homogeneous function of the enzyme concentrations. As already outlined (chapter 1), it is increasingly evident that, in a number of biological systems, there exist interactions between enzymes and, possibly, information transfer between them. Since these protein–protein interactions are concentration dependent, the enzyme concentrations cannot be considered independent variables. In order to circumvent this difficulty, expression (4.25) has been replaced by [20]

$$\pi_i^i = \frac{E_i}{v_i}\left(\frac{\partial v_i}{\partial E_i}\right), \tag{4.41}$$

where π_i^i can be different from one. Moreover, it has been claimed that the summation and the connectivity theorems which, in the case of associated enzymes, do not apply to control coefficients C_i^j should apply to the ratios C_i^J/π_i^i. It hardly seems possible, however, that the use of this new parameter, π, be sufficient to reconcile Metabolic control theory with the view that enzymes may interact and exchange information. There are at least two reasons for this. The first is that associated enzymes cannot be considered independent variables that can be varied independently. The second reason is that, when applying the summation theorem to the ratio C_i^J/π_i^i, one is in fact applying the summation theorem to the coefficients

$$C_i^{*J} = \frac{v_i}{J}\left(\frac{\partial J}{\partial v_i}\right), \tag{4.42}$$

but, as the network is in steady state, the overall flux is equal to the velocity of any of the reaction steps ($J = v_i$), so these new coefficient C_i^{*J} are all equal to unity and therefore meaningless.

To have a clear view of the problem under study, it is first necessary to derive and discuss generalized connectivity relationships. For simplicity, these relationships will be derived and discussed for a sequence of three enzyme reactions, assuming either that the three enzymes are physically independent entities, or that two of them are associated.

Let us consider the sequence of three enzyme reactions with two reaction intermediates S_1 and S_2

$$X_0 \xrightarrow[v_1]{E_1} S_1 \xrightarrow[v_2]{E_2} S_2 \xrightarrow[v_3]{E_3} X_f \tag{4.43}$$

If there is no interaction amongst the three enzymes, the rates of the three steps can be expressed as

$$v_1 = f_1(E_1, S_1), \qquad v_2 = f_2(E_2, S_1, S_2), \qquad v_3 = f_3(E_3, S_2). \tag{4.44}$$

The source (X_0) and the sink (X_f), considered reservoirs, do not appear in these expressions. Moreover, since the system is in steady state, one must have

$$J = v_1 = v_2 = v_3. \tag{4.45}$$

The classical theorem of the differentiation of functions of functions, allow us to express the participation of enzyme E_1 to the flux through its control coefficient [9,17]

$$\left(\frac{\partial J}{\partial E_1}\right)\frac{E_1}{J} = \left(\frac{\partial v_1}{\partial E_1}\right)\frac{E_1}{v_1} + \left(\frac{\partial v_1}{\partial S_1}\right)\frac{S_1}{v_1}\left(\frac{\partial S_1}{\partial E_1}\right)\frac{E_1}{S_1}. \tag{4.46}$$

Similarly, the participation of enzymes E_1 and E_2 in the same flux is expressed by

$$\left(\frac{\partial J}{\partial E_2}\right)\frac{E_2}{J} = \left(\frac{\partial v_2}{\partial E_2}\right)\frac{E_2}{v_2} + \left(\frac{\partial v_2}{\partial S_1}\right)\frac{S_1}{v_2}\left(\frac{\partial S_1}{\partial E_2}\right)\frac{E_2}{S_1} + \left(\frac{\partial v_2}{\partial S_2}\right)\frac{S_2}{v_2}\left(\frac{\partial S_2}{\partial E_2}\right)\frac{E_2}{S_2} \tag{4.47}$$

and

$$\left(\frac{\partial J}{\partial E_3}\right)\frac{E_3}{J} = \left(\frac{\partial v_3}{\partial E_3}\right)\frac{E_3}{v_3} + \left(\frac{\partial v_3}{\partial S_2}\right)\frac{S_2}{v_3}\left(\frac{\partial S_2}{\partial E_3}\right)\frac{E_3}{S_2}. \tag{4.48}$$

These three equations are equivalent to

$$\begin{aligned} C_1^J &= 1 + \varepsilon_{S_1}^{v_1} C_1^{S_1}, \\ C_2^J &= 1 + \varepsilon_{S_1}^{v_2} C_2^{S_1} + \varepsilon_{S_2}^{v_2} C_2^{S_2}, \\ C_3^J &= 1 + \varepsilon_{S_2}^{v_3} C_3^{S_2}, \end{aligned} \tag{4.49}$$

which can be rewritten in matrix form as

$$\begin{bmatrix} 1 & -\varepsilon_{S_1}^{v_1} & 0 \\ 1 & -\varepsilon_{S_1}^{v_2} & -\varepsilon_{S_2}^{v_2} \\ 1 & 0 & -\varepsilon_{S_2}^{v_3} \end{bmatrix} \begin{bmatrix} C_1^J & C_2^J & C_3^J \\ C_1^{S_1} & C_2^{S_1} & C_3^{S_1} \\ C_1^{S_2} & C_2^{S_2} & C_3^{S_2} \end{bmatrix} = \begin{bmatrix} 1 & 0 & 0 \\ 0 & 1 & 0 \\ 0 & 0 & 1 \end{bmatrix}. \tag{4.50}$$

As the order of the elasticity and control matrices can be reversed, one also has

$$\begin{bmatrix} C_1^J & C_2^J & C_3^J \\ C_1^{S_1} & C_2^{S_1} & C_3^{S_1} \\ C_1^{S_2} & C_2^{S_2} & C_3^{S_2} \end{bmatrix} \begin{bmatrix} 1 & -\varepsilon_{S_1}^{v_1} & 0 \\ 1 & -\varepsilon_{S_1}^{v_2} & -\varepsilon_{S_2}^{v_2} \\ 1 & 0 & -\varepsilon_{S_2}^{v_3} \end{bmatrix} = \begin{bmatrix} 1 & 0 & 0 \\ 0 & 1 & 0 \\ 0 & 0 & 1 \end{bmatrix}. \tag{4.51}$$

Either of the two matrix relations can be used to derive summation and connectivity relationships that have already been worked out, plus additional connectivity relations. Thus, in the matrix equation (4.51), multiplying the rows of the control matrix by the first column of the elasticity matrix yields the classical summation relationships, namely,

$$\begin{aligned} &C_1^J + C_2^J + C_3^J = 1, \\ &C_1^{S_1} + C_2^{S_1} + C_3^{S_1} = 0, \\ &C_1^{S_2} + C_2^{S_2} + C_3^{S_2} = 0. \end{aligned} \tag{4.52}$$

Similarly, in eq. (4.51), multiplying the first row of the control matrix by the second and third columns of the elasticity matrix, gives the two connectivity relationships between the flux control coefficients and the elasticities, namely,

$$\varepsilon_{S_1}^{v_1} C_1^J + \varepsilon_{S_1}^{v_2} C_2^J = 0, \qquad \varepsilon_{S_2}^{v_2} C_2^J + \varepsilon_{S_2}^{v_3} C_3^J = 0. \tag{4.53}$$

In the same spirit, in the same equation (4.51), multiplying the second and third rows of the control matrix by the second and third columns of the elasticity matrix gives all the connectivity relationships between the substrate control coefficients and the elasticities. Thus, one finds

$$\begin{aligned} &\varepsilon_{S_1}^{v_1} C_1^{S_1} + \varepsilon_{S_1}^{v_2} C_2^{S_1} = -1, \qquad \varepsilon_{S_2}^{v_2} C_2^{S_2} + \varepsilon_{S_2}^{v_3} C_3^{S_2} = -1, \\ &\varepsilon_{S_2}^{v_2} C_2^{S_1} + \varepsilon_{S_2}^{v_3} C_3^{S_1} = 0, \qquad \varepsilon_{S_1}^{v_1} C_1^{S_2} + \varepsilon_{S_1}^{v_2} C_2^{S_2} = 0. \end{aligned} \tag{4.54}$$

Moreover, eq. (4.50) allows one to derive other connectivity relationships for the three enzymes that associate the flux control coefficients, the substrate control coefficients and the elasticities. Thus, in eq. (4.50), multiplying the rows of the elasticity matrix by the first column of the control matrix yields these connectivity relationships for the first enzyme, namely,

$$C_1^J - \varepsilon_{S_1}^{v_1} C_1^{S_1} = 1,$$

$$C_1^J - \varepsilon_{S_1}^{v_2} C_1^{S_1} - \varepsilon_{S_2}^{v_2} C_1^{S_2} = 0,$$
$$C_1^J - \varepsilon_{S_2}^{v_3} C_1^{S_2} = 0. \tag{4.55}$$

For the second enzyme, similar connectivity relationships are obtained by multiplying the rows of the elasticity matrix by the second column of the control matrix, and one finds

$$C_2^J - \varepsilon_{S_1}^{v_1} C_2^{S_1} = 0,$$
$$C_2^J - \varepsilon_{S_1}^{v_2} C_2^{S_1} - \varepsilon_{S_2}^{v_2} C_2^{S_2} = 1,$$
$$C_2^J - \varepsilon_{S_2}^{v_3} C_2^{S_2} = 0. \tag{4.56}$$

For the third enzyme, similar connectivity relations are obtained by multiplication of the rows of the elasticity matrix by the third column of the control matrix. One obtains in that way

$$C_3^J - \varepsilon_{S_1}^{v_1} C_3^{S_1} = 0,$$
$$C_3^J - \varepsilon_{S_1}^{v_2} C_3^{S_1} - \varepsilon_{S_2}^{v_2} C_3^{S_2} = 0,$$
$$C_3^J - \varepsilon_{S_2}^{v_3} C_3^{S_2} = 1. \tag{4.57}$$

The general expression of these connectivity relationships may thus be cast in the following form

$$C_i^J - \sum_{j=1}^{n} \sum_{k=1}^{m} \varepsilon_{S_k}^{v_j} C_i^{S_k} = \delta_{ij}, \tag{4.58}$$

where n is the number of enzymes, m is the number of intermediates and δ_{ij} is the Kronecker delta equal to one if $i = j$ and equal to zero if $i \neq j$.

The matrix relation (4.51) has another virtue. It allows one to derive the expression of the flux control and of the substrate control coefficients in terms of elasticities that can be measured directly with isolated enzymes. As an example, the first summation relation (4.52) and the two connectivity relationships (4.53) can be cast in the following form

$$\begin{bmatrix} C_1^J \\ C_2^J \\ C_3^J \end{bmatrix} = \begin{bmatrix} 1 & 1 & 1 \\ -\varepsilon_{S_1}^{v_1} & -\varepsilon_{S_1}^{v_2} & 0 \\ 0 & -\varepsilon_{S_2}^{v_2} & -\varepsilon_{S_2}^{v_3} \end{bmatrix}^{-1} \begin{bmatrix} 1 \\ 0 \\ 0 \end{bmatrix}, \tag{4.59}$$

and from this system one can derive the expression of the three flux control coefficients as a function of the elasticities. Similar relationships between substrate control coefficients and elasticities can also be derived and will not be considered here.

In the model studied thus far, it was assumed that the flux was a homogeneous function of degree one in the enzyme concentrations and that these enzymes were physically independent entities. Now, if two enzymes, for example E_2 and E_3, are associated as a bienzyme complex, the previous kinetic scheme reads

$$X_0 \xrightarrow[v_1]{E_1} S_1 \xrightarrow[v_2]{E_2E_3} S_2 \xrightarrow[v_3]{E_2E_3} X_f \tag{4.60}$$

The dependence of the three reaction steps on the different variables is

$$v_1 = f_1(E_1, S_1), \qquad v_2 = f_2(C_p, S_1, S_2), \qquad v_3 = f_3(C_p, S_2). \tag{4.61}$$

In these equations C_p is the concentration of the bienzyme complex. For the second step, C_p, S_1 and S_2 are functions of enzyme E_2 concentration. Similarly, for the third step, C_p and S_2 are functions of enzyme E_3 concentration. Therefore, the control coefficients of these three steps can be expressed as

$$\begin{aligned}
\left(\frac{\partial J}{\partial E_1}\right)\frac{E_1}{J} &= \left(\frac{\partial v_1}{\partial E_1}\right)\frac{E_1}{v_1} + \left(\frac{\partial v_1}{\partial S_1}\right)\frac{S_1}{v_1}\left(\frac{\partial S_1}{\partial E_1}\right)\frac{E_1}{S_1},\\
\left(\frac{\partial J}{\partial E_2}\right)\frac{E_2}{J} &= \left(\frac{\partial v_2}{\partial C_p}\right)\frac{C_p}{v_2}\left(\frac{\partial C_p}{\partial E_2}\right)\frac{E_2}{C_p} + \left(\frac{\partial v_2}{\partial S_1}\right)\frac{S_1}{v_2}\left(\frac{\partial S_1}{\partial E_2}\right)\frac{E_2}{S_1}\\
&\quad + \left(\frac{\partial v_2}{\partial S_2}\right)\frac{S_2}{v_2}\left(\frac{\partial S_2}{\partial E_2}\right)\frac{E_2}{S_2},\\
\left(\frac{\partial J}{\partial E_3}\right)\frac{E_3}{J} &= \left(\frac{\partial v_3}{\partial C_p}\right)\frac{C_p}{v_3}\left(\frac{\partial C_p}{\partial E_3}\right)\frac{E_3}{C_p} + \left(\frac{\partial v_3}{\partial S_2}\right)\frac{S_2}{v_3}\left(\frac{\partial S_2}{\partial E_3}\right)\frac{E_3}{S_2}.
\end{aligned} \tag{4.62}$$

Because of the association of enzymes E_2 and E_3, new terms appear in these equations that were absent in the previous model. These terms are the following

$$\begin{aligned}
\left(\frac{\partial v_2}{\partial C_p}\right)\frac{C_p}{v_2} &= \pi_2^{v_2}, & \left(\frac{\partial v_3}{\partial C_p}\right)\frac{C_p}{v_3} &= \pi_3^{v_3},\\
\left(\frac{\partial C_p}{\partial E_2}\right)\frac{E_2}{C_p} &= \tau_2^{v_2}, & \left(\frac{\partial C_p}{\partial E_3}\right)\frac{E_3}{C_p} &= \tau_3^{v_3}.
\end{aligned} \tag{4.63}$$

Because of the association of the enzymes E_2 and E_3, the concentration of the complex C_p varies when the concentration of either enzyme varies. This implies that $\pi_2^{v_2}$ and $\pi_3^{v_3}$ are different from unity. The flux is not a homogeneous function of degree one in enzyme concentrations, and the values of $\tau_2^{v_2}$ and $\tau_3^{v_3}$ are not necessarily equal to one either. Equations (4.62) assume the form

$$\begin{aligned}
C_1^J &= 1 + \varepsilon_{S_1}^{v_1} C_1^{S_1},\\
C_2^J &= \pi_2^{v_2}\tau_2^{v_2} + \varepsilon_{S_1}^{v_2} C_2^{S_1} + \varepsilon_{S_2}^{v_2} C_2^{S_2},\\
C_3^J &= \pi_3^{v_3}\tau_3^{v_3} + \varepsilon_{S_2}^{v_3} C_3^{S_2}.
\end{aligned} \tag{4.64}$$

Setting for simplicity

$$\pi_2^{*v_2} = \pi_2^{v_2} \tau_2^{v_2}, \qquad \pi_3^{*v_3} = \pi_3^{v_3} \tau_3^{v_3}, \tag{4.65}$$

eqs. (4.64) can be cast in matrix form

$$\begin{bmatrix} 1 & -\varepsilon_{S_1}^{v_1} & 0 \\ 1 & -\varepsilon_{S_1}^{v_2} & -\varepsilon_{S_2}^{v_2} \\ 1 & 0 & -\varepsilon_{S_2}^{v_3} \end{bmatrix} \begin{bmatrix} C_1^J & C_2^J & C_3^J \\ C_1^{S_1} & C_2^{S_1} & C_3^{S_1} \\ C_1^{S_2} & C_2^{S_2} & C_3^{S_2} \end{bmatrix} = \begin{bmatrix} 1 & 0 & 0 \\ 0 & \pi_2^* & 0 \\ 0 & 0 & \pi_3^* \end{bmatrix}. \tag{4.66}$$

This equation shows clearly that, depending on the supramolecular organization of the enzymes of the pathway, the connectivity and the summation relationships, as expressed by eqs. (4.50) and (4.51), may not hold because the velocity of each enzyme step and the overall flux are not homogeneous functions of degree one in enzyme concentrations. Under these conditions of supramolecular enzyme association, the applicability of Metabolic control theory to real experimental data may become extremely difficult. The applicability of this theory, in its simple form, to eukaryotic systems is therefore questionable.

4.1.6. Feedback control of a metabolic pathway

The method which has been followed so far to derive connectivity and summation equations [9] can be used to investigate the effect of a feedback loop in an enzyme pathway. Let us assume, for instance, that intermediate S_2 of scheme (4.43) acts as a feedback inhibitor, or activator, of enzyme E_1. The scheme thus becomes

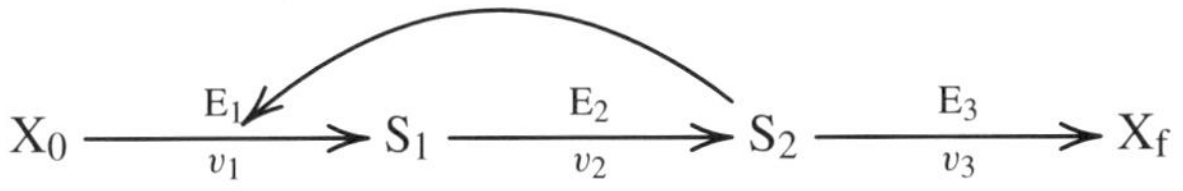

$$X_0 \xrightarrow[v_1]{E_1} S_1 \xrightarrow[v_2]{E_2} S_2 \xrightarrow[v_3]{E_3} X_f \tag{4.67}$$

The first step of the reaction is then a function of E_1, S_1 and S_2

$$v_1 = f_1(E_1, S_1, S_2). \tag{4.68}$$

The corresponding control coefficient is then

$$\left(\frac{\partial J}{\partial E_1}\right)\frac{E_1}{J} = \left(\frac{\partial v_1}{\partial E_1}\right)\frac{E_1}{v_1} + \left(\frac{\partial v_1}{\partial S_1}\right)\frac{S_1}{v_1}\left(\frac{\partial S_1}{\partial E_1}\right)\frac{E_1}{S_1} + \left(\frac{\partial v_1}{\partial S_2}\right)\frac{S_2}{v_1}\left(\frac{\partial S_2}{\partial E_1}\right)\frac{E_1}{S_2}, \tag{4.69}$$

which is equivalent to

$$C_1^J = 1 + \varepsilon_{S_1}^{v_1} C_1^{S_1} + \varepsilon_{S_2}^{v_1} C_1^{S_2}. \tag{4.70}$$

The equations of the flux control coefficients for the second and third processes are still equations (4.49). The system of the three equations may then be cast in matrix form as

$$\begin{bmatrix} 1 & -\varepsilon_{S_1}^{v_1} & -\varepsilon_{S_2}^{v_1} \\ 1 & -\varepsilon_{S_1}^{v_2} & -\varepsilon_{S_2}^{v_2} \\ 1 & 0 & -\varepsilon_{S_2}^{v_3} \end{bmatrix} \begin{bmatrix} C_1^J & C_2^J & C_3^J \\ C_1^{S_1} & C_2^{S_1} & C_3^{S_1} \\ C_1^{S_2} & C_2^{S_2} & C_3^{S_2} \end{bmatrix} = \begin{bmatrix} 1 & 0 & 0 \\ 0 & 1 & 0 \\ 0 & 0 & 1 \end{bmatrix}. \tag{4.71}$$

Comparing this expression with eq. (4.50) reveals that the feedback loop has introduced an additional term $(-\varepsilon_{S_2}^{v_1})$ in the elasticity matrix. The usual summation and connectivity relationships can be derived form this matrix equation.

4.1.7. Control of branched pathways

Let us consider the following model

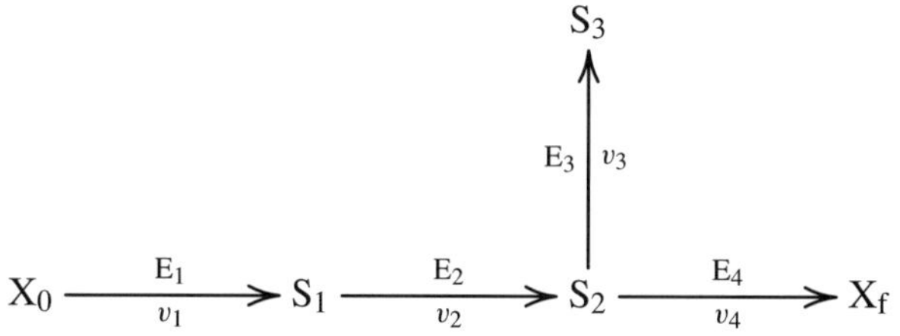

(4.72)

under steady state conditions

$$v_1 = v_2 = J_1 = J_2, \qquad v_3 = J_3, \qquad v_4 = J_4, \tag{4.73}$$

and

$$J_1 = J_3 + J_4. \tag{4.74}$$

Differentiation with respect to any enzyme concentration (for instance the concentration of enzyme E_1) leads to

$$\frac{\partial J_1}{\partial E_1} = \frac{\partial J_3}{\partial E_1} + \frac{\partial J_4}{\partial E_1} = \frac{\partial J_2}{\partial E_1}, \tag{4.75}$$

or to

$$J_1 C_1^{J_1} = J_3 C_1^{J_3} + J_4 C_1^{J_4}. \tag{4.76}$$

Setting

$$J_3^* = J_3/J_1 \quad \text{and} \quad J_4^* = J_4/J_1 \tag{4.77}$$

eq. (4.76) can be rewritten as

$$C_1^{J_1} = J_3^* C_1^{J_3} + J_4^* C_1^{J_4}, \qquad C_2^{J_2} = J_3^* C_2^{J_3} + J_4^* C_2^{J_4}. \tag{4.78}$$

In these expressions the fluxes have been normalized to J_1. The four flux control coefficients can thus be expressed as

$$\begin{aligned}
&\left(\frac{\partial J_1}{\partial E_1}\right)\frac{E_1}{J_1} = \left(\frac{\partial v_1}{\partial E_1}\right)\frac{E_1}{v_1} + \left(\frac{\partial v_1}{\partial S_1}\right)\frac{S_1}{v_1}\left(\frac{\partial S_1}{\partial E_1}\right)\frac{E_1}{S_1},\\
&\left(\frac{\partial J_2}{\partial E_2}\right)\frac{E_2}{J_2} = \left(\frac{\partial v_2}{\partial E_2}\right)\frac{E_2}{v_2} + \left(\frac{\partial v_2}{\partial S_1}\right)\frac{S_1}{v_2}\left(\frac{\partial S_1}{\partial E_2}\right)\frac{E_2}{S_1} + \left(\frac{\partial v_2}{\partial S_2}\right)\frac{S_2}{v_2}\left(\frac{\partial S_2}{\partial E_2}\right)\frac{E_2}{S_2},\\
&\left(\frac{\partial J_3}{\partial E_3}\right)\frac{E_3}{J_3} = \left(\frac{\partial v_3}{\partial E_3}\right)\frac{E_3}{v_3} + \left(\frac{\partial v_3}{\partial S_2}\right)\frac{S_2}{v_3}\left(\frac{\partial S_2}{\partial E_3}\right)\frac{E_3}{S_2},\\
&\left(\frac{\partial J_4}{\partial E_4}\right)\frac{E_4}{J_4} = \left(\frac{\partial v_4}{\partial E_4}\right)\frac{E_4}{v_4} + \left(\frac{\partial v_4}{\partial S_2}\right)\frac{S_2}{v_4}\left(\frac{\partial S_2}{\partial E_4}\right)\frac{E_4}{S_2}.
\end{aligned} \tag{4.79}$$

These equations can be written in more compact form as

$$\begin{aligned}
&C_1^{J_1} = 1 + \varepsilon_{S_1}^{v_1} C_1^{S_1}, \qquad C_2^{J_2} = 1 + \varepsilon_{S_1}^{v_2} C_2^{S_1} + \varepsilon_{S_2}^{v_2} C_2^{S_2},\\
&C_3^{J_3} = 1 + \varepsilon_{S_2}^{v_3} C_3^{S_2}, \qquad C_4^{J_4} = 1 + \varepsilon_{S_2}^{v_4} C_4^{S_2}.
\end{aligned} \tag{4.80}$$

As previously, these equations can be reexpressed in matrix form (not presented here) and this allows the derivation of all the summation and connectivity relationships. Thus, for instance, the summation equations for the flux control coefficients read

$$\begin{aligned}
&J_3^* C_1^{J_3} + J_3^* C_2^{J_3} + C_3^{J_3} = 1, \qquad J_3^* C_1^{J_4} + J_3^* C_2^{J_4} + C_3^{J_4} = 0,\\
&J_4^* C_1^{J_3} + J_4^* C_2^{J_3} + C_3^{J_3} = 0, \qquad J_4^* C_1^{J_4} + J_4^* C_2^{J_4} + C_3^{J_4} = 1.
\end{aligned} \tag{4.81}$$

4.2. Biochemical systems theory

This theoretical model was derived by Savageau and his colleagues [21–27], mostly before the development of the Metabolic control theory, and therefore independently of this model. For some time, these models were thought to be in conflict, but today they appear as alternative and converging approaches to the same basic problem. As these models are convergent, they can be formulated in similar language, which was not the case initially. The way the principles of Biochemical systems theory are presented below is thus somewhat different from that initially followed by Savageau and colleagues.

The basic idea of the Biochemical systems theory is to postulate that any chemical reaction of metabolism can be approximated in an empirical way by a truncated Taylor series in logarithmic space. For instance, if the reaction rate v is function of substrate concentration S, one has

$$\ln v = \ln v_0 + \left(\frac{\partial \ln v}{\partial \ln S}\right)(\ln S - \ln S_0). \tag{4.82}$$

Here v_0 and S_0 represent the values of the rate and of the substrate concentration at a specific operating point. One will notice that $\partial \ln v/\partial \ln S$ is the expression of the elasticity, ε_S^v. Therefore, eq. (4.82) becomes

$$\ln v = \ln v_0 - \varepsilon_S^v(\ln S - \ln S_0). \tag{4.83}$$

Setting

$$\ln \alpha = \ln v_0 - \varepsilon_S^v \ln S_0, \tag{4.84}$$

eq. (4.83) assumes the form

$$\ln v = \ln \alpha + \varepsilon_S^v \ln S \tag{4.85}$$

or

$$v = \alpha S^{\varepsilon_S^v}. \tag{4.86}$$

This empirical approximation has been used in different fields of biology. It is especially known under the name of "allometry" for studying, in a comparative manner, the growth of two organs of the same animal [28–32]. Relations (4.85) and (4.86) can be applied to a sequence of enzyme reactions. For instance, if a step j of a network has rate v_j and depends on a number of reactants R_i, one has

$$v_j = \alpha_j \prod_{i=1}^{m} R_i^{\varepsilon_{S_k}^{v_j}}. \tag{4.87}$$

In the terminology of the Biochemical systems theory, this type of relationship is called a power-law.

Let us consider a sequence of physically independent enzymes

$$\mathrm{X_0} \xrightarrow[v_1]{\alpha_1} \mathrm{S_1} \xrightarrow[v_2]{\alpha_2} \mathrm{S_2} \xrightarrow[v_3]{\alpha_3} \mathrm{X_f} \tag{4.88}$$

If each step depends on the corresponding substrate only (but not on the product), one must have

$$v_j = \alpha_j S_{j-1}^{\varepsilon_{j-1}}, \tag{4.89}$$

where, for simplicity, elasticity has been represented by ε_{j-1}. If alternatively the rate of each step depends on both the substrate and the product, one has

$$v_j = \alpha_j S_{j-1}^{\varepsilon_{j-1}} S_j^{\varepsilon_j}. \tag{4.90}$$

One may then write, for steady state conditions and in the case where eq. (4.86) applies,

$$\begin{aligned}\frac{dS_1}{dt} &= \alpha_1 X_0^{\varepsilon_0} - \alpha_2 S_1^{\varepsilon_1} = 0,\\ \frac{dS_2}{dt} &= \alpha_2 S_1^{\varepsilon_1} - \alpha_3 S_2^{\varepsilon_2} = 0.\end{aligned} \tag{4.91}$$

If eq. (4.89) applies, then the system of consecutive reactions takes the form

$$\begin{aligned}\frac{dS_1}{dt} &= \alpha_1 X_0^{\varepsilon_0^1} S_1^{\varepsilon_1^1} - \alpha_2 S_1^{\varepsilon_1^2} S_2^{\varepsilon_2^2} = 0,\\ \frac{dS_2}{dt} &= \alpha_2 S_1^{\varepsilon_1^2} S_2^{\varepsilon_2^2} - \alpha_3 S_2^{\varepsilon_2^3} = 0.\end{aligned} \tag{4.92}$$

In these expressions, the elasticities

$$\varepsilon_1^1 = \frac{\partial \ln v_1}{\partial \ln S_1} \quad \text{and} \quad \varepsilon_2^2 = \frac{\partial \ln v_2}{\partial \ln S_2} \tag{4.93}$$

have negative values because the rates v_1 and v_2 are inhibited by their respective products S_1 and S_2. Taking the logarithms of eqs. (4.91) or (4.92) yields linear systems. For instance, in the case of eqs. (4.92), one finds

$$\begin{bmatrix} (\varepsilon_1^1 - \varepsilon_1^2) & -\varepsilon_2^2 \\ \varepsilon_1^2 & (\varepsilon_2^2 - \varepsilon_2^3) \end{bmatrix} \begin{bmatrix} \ln S_1 \\ \ln S_2 \end{bmatrix} = \begin{bmatrix} \lambda_{21} - \varepsilon_0^1 \ln X_0 \\ \lambda_{32} \end{bmatrix}. \tag{4.94}$$

In this system,

$$\lambda_{21} = \ln(\alpha_2/\alpha_1) \quad \text{and} \quad \lambda_{32} = \ln(\alpha_3/\alpha_2). \tag{4.95}$$

Moreover, solving this linear system allows one to express the logarithmic concentrations of S_1 and S_2 in terms of the elasticities, of $\ln X_0$ and of λ_{21} and λ_{32}. Thus, the concept of elasticity stems in a logical manner from both Biochemical systems theory and Metabolic control theory. Likewise, one can show that the concept of control coefficient, as well as connectivity reationships, originate from the power-law formalism in the Biochemical systems theory. Thus, one can write

$$\ln v_2 = \ln \alpha_2 + \varepsilon_{S_1}^{v_2} \ln S_1 + \varepsilon_{S_2}^{v_2} \ln S_2. \tag{4.96}$$

Differentiating with respect to $\ln E_2$ yields

$$\frac{\partial \ln v_2}{\partial \ln E_2} = \varepsilon_{S_1}^{v_2} \frac{\partial \ln S_1}{\partial \ln E_2} + \varepsilon_{S_2}^{v_2} \frac{\partial \ln S_2}{\partial \ln E_2}, \tag{4.97}$$

or

$$\frac{\partial \ln v_2}{\partial \ln E_2} = \varepsilon_{S_1}^{v_2} C_2^{S_1} + \varepsilon_{S_2}^{v_2} C_2^{S_2}. \tag{4.98}$$

If the enzyme E_2 is not associated with another one, eq. (4.98) reduces to

$$\varepsilon_{S_1}^{v_2} C_2^{S_1} + \varepsilon_{S_2}^{v_2} C_2^{S_2} = 1, \tag{4.99}$$

which represents a connectivity relationship.

It is thus clear that Metabolic control theory and Biochemical system theory display similarities. The latter is more general because the power-law formalism is an empirical manner to express the relationships between different variables, whatever their nature. On the other hand, the former is perhaps better suited for the direct study of metabolic systems [33].

4.3. An example of the application of Metabolic control theory to a biological problem

To be convincing, any model or theoretical development, must be tested experimentally. Both Metabolic control theory and Biochemical systems theory have now been used in the analysis of experimental systems [32–36]. The aim of the present section is to discuss very briefly the use of Metabolic control theory as a tool for understanding the control of gluconeogenesis in rat liver [33]. Roughly speaking, gluconeogenesis consists of four sets of reactions: first, the reactions that convert lactate to phosphoenolpyruvate; second, the reaction that converts phosphoenolpyruvate back to pyruvate; third, the conversion of phosphoenolpyruvate to glyceraldehyde phosphate and dihydroxyacetone phosphate, and fourth, the conversion of glyceraldehyde phosphate and dihydroxyacetone phosphate to glucose. The gluconeogenetic flux through the four segments of the pathway can be determined. Groen and Westerhoff [33] have also measured the flux control coefficients of the four segments for different values of the flux of glucose. What is striking in their results is the existence of a gradual shift of the control from the first segment, at low flux values, to the third and fourth segments, as the flux is increased. These results clearly show that the control of the flux is not the intrinsic property of one enzyme, but may be distributed over different regions of the pathway, and that this distribution changes as the experimental conditions are varied. To the present author, Metabolic control theory is well suited to the study of many biochemical systems. It requires, however, two conditions that are by no means always fulfilled: the existence of a steady state and a lack of interactions between the enzymes involved in the same metabolic sequence. Although an attempt has been made to modify this theory to take account of a lack of steady state [38], these modifications bring very little, if any, relative to more conventional theories that aim at studying dynamic systems (see chapters 9 and 10). An axiom of Metabolic control theory is to assume that the flux is a homogeneous function of degree one in enzyme concentrations. This leads in turn to the summation theorem. This theorem, however, is no longer valid if the enzymes cannot be considered independent variables and if their activity is not proportional to their concentration. Yet, as will be seen later, there is little doubt that this is precisely the situation that occurs for enzymes integrated in supramolecular edifices. In spite of the efforts that have been made to circumvent the difficulty raised by the summation theorem, it is clear, for reasons that have already been alluded to, that Metabolic control theory cannot be applied if the metabolic flux is not a homogeneous function of degree one in enzyme concentrations.

References

[1] Kacser, H. and Burns, J.A. (1979) Molecular democracy. Who shares the control? Biochem. Soc. Trans. 7, 1149–1160.

[2] Kacser, H. and Burns, J.A. (1972) The control of flux. Symp. Soc. Exp. Biol. 27, 65–104.

[3] Kacser, H. (1983) The control of enzyme systems *in vivo*. Elasticity analysis of the steady state. Biochem. Soc. Trans. 11, 35–40.

[4] Kacser, H. (1987) Control of metabolism. In: P.K. Stumpf and E.E. Conn (Eds.), The Biochemistry of Plants. Academic Press, New York, Vol. 11, pp. 39–67.

[5] Heinrich, R. and Rapoport, T.A. (1974) A linear steady state treatment of enzymatic chains. Critique of the crossover theorem and a general procedure to identify interaction site with an effector. Eur. J. Biochem. 42, 97–105.

[6] Heinrich, R., Rapoport, S.M. and Rapoport, T.A. (1977) Metabolic regulation and mathematical models. Progress Biophys. Mol. Biol. 32, 1–82.

[7] Giersch, C. (1988a) Control analysis of metabolic networks. 1. Homogeneous functions and the summation theorems for control coefficients. Eur. J. Biochem. 174, 509–513.

[8] Giersch, C. (1988b) Control analysis of metabolic networks. 2. Total differentials and general formulation of the connectivity relations. Eur. J. Biochem. 174, 515–519.

[9] Cascante, M., Franco, R. and Canela, E.I. (1989) Use of implicit methods for general sensitivity theory to develop a systematic approach of metabolic control. I. Unbranched pathways. Math. Biosci. 94, 271–288.

[10] Westerhoff, H.V. and Chen, Y.D. (1984) How do enzyme activities control metabolite concentrations? Eur. J. Biochem. 142, 425–430.

[11] Fell, D.A. and Sauro, H.M. (1985) Metabolic control and its analysis. Additional relationships between elasticities and control coefficients. Eur. J. Biochem. 148, 555–561.

[12] Sauro, H.M., Small, J.R. and Fell, D.A. (1987) Metabolic control and its analysis. Extensions to the theory and matrix methods. Eur. J. Biochem. 165, 215–221.

[13] Sen, A.K. (1990) Metabolic control analysis. An application of signal flow graphs. Biochem. J. 269, 141–147.

[14] Shultz, A.R. (1991) Algorithms for the derivation of flux and concentration control coefficients. Biochem. J. 278, 299–304.

[15] Reder, C. (1988) Metabolic control theory: a structural approach. J. Theor. Biol. 135, 175–201.

[16] Cornish-Bowden, A. and Cardenas, M.L. (1990) Control of Metabolic Processes. Plenum Press, New York.

[17] Shultz, A.R. (1994) Enzyme Kinetics. Cambridge University Press, Cambridge.

[18] Cornish-Bowden, A. (1995) Fundamentals of Enzyme Kinetics. Portland Press, London.

[19] Westerhoff, H.V. and Van Dam, K. (1987) Thermodynamics and Control of Biological Free Energy Transduction. Elsevier, Amsterdam.

[20] Kacser, H., Sauro, H.M. and Acerenza, L. (1990) Enzyme–enzyme interactions and control analysis. Eur. J. Biochem. 165, 215–221.

[21] Savageau, M.A. (1969) Biochemical systems analysis. I. Some mathematical properties of the rate law for the component enzymatic reaction. J. Theor. Biol. 25, 365–369.

[22] Savageau, M.A. (1972) The behavior of intact biochemical control systems. Curr. Top. Cell. Regul. 6, 63–130.

[23] Savageau, M.A., Voit, E.O. and Irvine, D.H. (1987a) Biochemical systems theory and metabolic control theory. I. Fundamental similarities and differences. Math. Biosci. 86, 127–145.
[24] Savageau, M.A., Voit, E.O. and Irvine, D.H. (1987b) Biochemical systems theory and metabolic control theory. II. The role of summation and connectivity relationships. Math. Biosci. 86, 147–169.
[25] Voit, E.O. and Savageau, M.A. (1987) Accuracy of alternative representations for integrated biochemical systems. Biochemistry 26, 6869–6880.
[26] Savageau, M.A. (1976) Biochemical Systems Analysis: A Study of Function and Design in Molecular Biology. Addison-Wesley, Reading, MA.
[27] Voit, E.O. (1991) Canonical Nonlinear Modelling. Van Nostrand Reinhold, New York.
[28] Teissier, G. (1937) Les Lois Quantitatives de la Croissance. Act. Sci. Indust. Hermann, Paris.
[29] Teissier, G. (1948) La relation d'allométrie. Sa signification statistique et biologique. Biometrics 4, 14–25.
[30] Teissier, G. (1955a) Allométrie de taille et variabilité chez *Maia squinado*. Arch. Zool. Exp. Gen. 92, 141–151.
[31] Teissier, G. (1955b) Grandeur de référence et allométrie de taille chez *Maia squinado*. C. R. Acad. Sci. Paris 240, 364–368.
[32] Savageau, M.A. (1980) Growth equations: A general equation and a survey of special cases. Math. Biosci. 48, 267–278.
[33] Groen, A.K. and Westerhoff, H.V. (1990) Modern control theories: a consumers' test. In: A. Cornish-Bowden and M.L. Cardenas (Eds.), Control of Metabolic Processes. Plenum Press, New York, pp. 101–118.
[34] Wright, B.E. and Albe, K.R. (1990) A new method for estimating enzyme activity and control coefficients *in vivo*. In: A. Cornish-Bowden and M.L. Cardenas (Eds.), Control of Metabolic Processes. Plenum Press, New York, pp. 317–328.
[35] Berry, M.N., Gregory, R.B., Grivell, A.R., Henly, D.C., Phillips, J.W., Wallace, P.G. and Welch, G.R. (1990) Constraints in the application of control analysis to the study of metabolism in hepatocytes. In: A. Cornish-Bowden and M.L. Cardenas (Eds.), Control of Metabolic Processes. Plenum Press, New York, pp. 343–350.
[36] Stitt, M. (1990) Application of control analysis to photosynthetic sucrose synthesis. In: A. Cornish-Bowden and M.L. Cardenas (Eds.), Control of Metabolic Processes. Plenum Press, New York, pp. 363–376.
[37] Fell, D. (1997) Understanding the Control of Metabolism. Portland Press, London.
[38] Acerenza, L. (1990) Temporal aspects of the control of metabolic processes. In: A. Cornish-Bowden and M.L. Cardenas (Eds.), Control of Metabolic Processes. Plenum Press, New York, pp. 297–302.

CHAPTER 5

Compartmentalization of the living cell and thermodynamics of energy conversion

One of the specific features of eukaryotic cells is the existence, in these living systems, of different compartments limited by membranes. As already mentioned in chapter 1, the membranes are made up of lipid bilayers. Many different enzymes and non-enzymatic proteins are anchored in these membranes and, owing to their interfacial situation, they play a role that would be unimaginable in homogeneous media. How is it thermodynamically feasible, for instance, that ions or molecules can accumulate in a cell "uphill", against an electrochemical gradient? How is it conceivable that an endergonic chemical reaction, which is associated with an increase of free energy, routinely takes place in a cell? Considered in isolation, such events are thermodynamically disfavoured and yet they are commonplace in biological systems. These fundamental processes, which involve energy conversion phenomena can be carried out, without violating thermodynamic principles, thanks to the compartmentalization of the cell. The aim of this chapter is precisely to understand how compartmentalization makes these apparently disfavoured events easy.

5.1. Thermodynamic properties of compartmentalized systems

Let us consider a permeable membrane that separates the available space in two compartments, called cis (′) and trans (″), of equal volumes and different absolute temperatures, T' and T''. Let us assume that each of these two compartments contains various chemical species having mole numbers n_i' and n_i'' and chemical potentials μ_i' and μ_i'' (Fig. 5.1). There must therefore exist a transport of matter and energy from one compartment to the other. As an example, we shall assume that both migrations take place from the cis- to the trans-compartment.

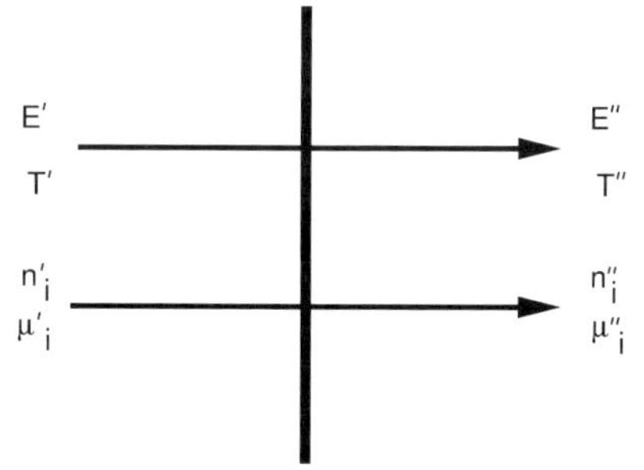

Fig. 5.1. Transfer of energy and matter from a cis- to a trans-compartment. See text.

The Gibbs–Duhem equation for the two compartments are

$$\begin{aligned} \mathrm{d}E' &= T'\,\mathrm{d}S' + \sum_{i=1}^{m} \mu_i'\,\mathrm{d}n_i', \\ \mathrm{d}E'' &= T''\,\mathrm{d}S'' + \sum_{i=1}^{m} \mu_i''\,\mathrm{d}n_i'', \end{aligned} \tag{5.1}$$

where E' and E'' are the internal energies and S' and S'' the entropies of the cis and trans sub-systems. As the entropy of a system is an extensive function one has

$$\mathrm{d}S = \mathrm{d}S' + \mathrm{d}S'', \tag{5.2}$$

where $\mathrm{d}S$ is the differential of the entropy of the whole system. Conservation of energy requires that

$$\mathrm{d}E' = -\mathrm{d}E'', \tag{5.3}$$

and conservation of matter demands that

$$\mathrm{d}n_i' = -\mathrm{d}n_i''. \tag{5.4}$$

Thus, equations (5.1) can be rewritten as

$$\begin{aligned} \mathrm{d}S' &= -\frac{1}{T'}\,\mathrm{d}E'' + \frac{1}{T'} \sum_{i=1}^{m} \mu_i'\,\mathrm{d}n_i'', \\ \mathrm{d}S'' &= \frac{1}{T''}\,\mathrm{d}E'' - \frac{1}{T''} \sum_{i=1}^{m} \mu_i''\,\mathrm{d}n_i''. \end{aligned} \tag{5.5}$$

Therefore, one has

$$\mathrm{d}S = \mathrm{d}S' + \mathrm{d}S'' = \left(\frac{1}{T''} - \frac{1}{T'}\right)\mathrm{d}E'' + \sum_{i=1}^{m} \left(\frac{\mu_i'}{T'} - \frac{\mu_i''}{T''}\right)\mathrm{d}n_i'', \tag{5.6}$$

and the dissipation of entropy of the whole system during the migration of matter and energy is thus

$$\frac{\mathrm{d}S}{\mathrm{d}t} = \left(\frac{1}{T''} - \frac{1}{T'}\right)\frac{\mathrm{d}E''}{\mathrm{d}t} + \sum_{i=1}^{m} \left(\frac{\mu_i'}{T'} - \frac{\mu_i''}{T''}\right)\frac{\mathrm{d}n_i''}{\mathrm{d}t}. \tag{5.7}$$

One may define a flow of energy, J_E, and flows of matter, J_i, as

$$J_E = \frac{\mathrm{d}E''}{\mathrm{d}t}, \qquad J_i = \frac{\mathrm{d}n_i''}{\mathrm{d}t}. \tag{5.8}$$

Similarly the forces, X_E and X_i, that generate these flows are

$$X_E = \frac{1}{T''} - \frac{1}{T'}, \qquad X_i = \frac{\mu_i'}{T'} - \frac{\mu_i''}{T''}. \tag{5.9}$$

Therefore, eq. (5.7) can be rewritten as

$$\frac{\mathrm{d}S}{\mathrm{d}t} = J_E X_E + \sum_{i=1}^{m} J_i X_i \geqslant 0. \tag{5.10}$$

This expression is important, for it shows that storage of energy in a compartment is not forbidden by thermodynamics, provided this storage of energy is accompanied by a migration of matter from one compartment to the other. Alternatively, migration of molecules, or ions, against a concentration, or an electrochemical, gradient is also feasible if it is accompanied by an expenditure of energy. In either case thermodynamics requires that expression (5.10) be fulfilled. As we shall see later, expression (5.10) represents the thermodynamic basis for the synthesis of energy-rich compounds, such as ATP, in mitochondria and chloroplasts, and of active transport of matter across membranes. Moreover, expression (5.10) shows that an exchange of matter across a membrane may also be allowed provided the expression (5.10) is not violated.

It is thus clear, in this thermodynamic context, that storage of energy is possible only because there exists a transport of matter along a concentration gradient, or, alternatively, that active transport against a concentration gradient is possible only because there is an expenditure of energy. Expression (5.10) thus represents the thermodynamic basis for coupling spontaneous, or active, transport of matter with storage, or expenditure, of energy. This expression thus offers the theoretical grounds for the various types of energy coupling systems that may *a priori* be expected to occur in living cells.

On this basis, one can distinguish three different, but related, types of coupling processes: chemiosmotic coupling, osmochemical coupling, and osmosmotic coupling. Let us briefly summarize what all this means. Chemiosmotic coupling is coupling between a spontaneous, exergonic, chemical reaction, or a network of chemical reactions, such as electron transfer in cell respiration, to an active transport of ions (of protons for instance). This process typically occurs on the inner membrane of mitochondria. Electrons are transferred, through an electron transfer chain, to molecular oxygen. Part of the free energy of this exergonic process is used to pump protons located in the mitochondrial matrix against an electrochemical gradient to the intermembrane space of the same mitochondrion (Fig. 5.2). Another example of chemiosmotic coupling is offered by the hydrolysis of ATP coupled to the active transport of ions (calcium or protons for instance). Thus, calcium ATPases, thanks to ATP hydrolysis, pump calcium from the cytoplasm through the endoplasmic reticulum, and proton ATPases pump protons from the cytoplasm through the membrane of lysosomes, thanks to ATP hydrolysis (Fig. 5.2).

By osmochemical coupling is meant coupling between spontaneous transport of an ion and an endergonic, disfavoured chemical reaction. This is precisely the situation that occurs with the ATP synthase of mitochondria, for example. Protons that have been pumped into the intermembrane space of mitochondria, diffuse through the pore of ATP synthase

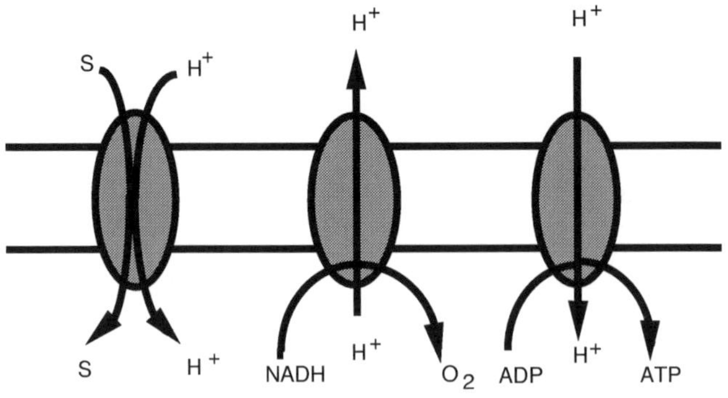

Fig. 5.2. Different types of coupling processes. From left to right: osmosmotic coupling, chemiosmotic coupling, osmochemical coupling. See text.

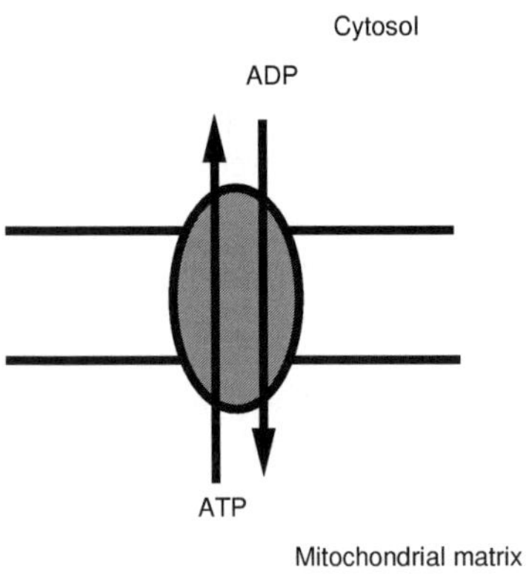

Fig. 5.3. ATP–ADP exchange between a mitochondrion and cytosol.

anchored in the internal mitochondrial membrane. The dissipation of this electrochemical gradient is sufficient to allow ATP synthesis (Fig. 5.2).

In the process of osmosmotic coupling, different ions may be exchanged through a membrane. This is the case, for instance, in the exchange of ATP from the mitochondrion for ADP from the cytosol (Fig. 5.3).

When the transport of matter is, in fact, a transport of neutral molecules, the difference of their concentration in the two compartments generates a free energy difference between these compartments. The chemical potentials of the neutral chemical species L in the two compartments ($''$) and ($'$) are

$$\mu_L'' = \mu_L^\circ + RT \ln[L''] \tag{5.11}$$

and

$$\mu_L' = \mu_L^\circ + RT \ln[L'], \tag{5.12}$$

where μ_L° is, as usual, the standard chemical potential of L. The affinity, A_L, is defined as $A_L = -\Delta\mu_L = -(\mu_L'' - \mu_L')$ and is thus equal to

$$A_L = -\Delta\mu_L = -RT \ln \frac{[L'']}{[L']}. \tag{5.13}$$

In most cases, however, the transport of matter is a transport of ions. There must therefore exist an electric potential difference $\Delta\Psi = \Psi'' - \Psi'$ between the trans and the cis sides of the membrane. If an anion is transported in the direction of a positive $\Delta\Psi$ value, the migration generates electrical work (this implies that the corresponding affinity is positive). If the anion is transported in the direction of a negative $\Delta\Psi$ value, one has to provide the anion with electrical work in order to obtain the migration (and therefore the affinity is negative). The situation is symmetrical for the transport of a cation. Therefore, the affinity component associated with the motion of the charged ligand, $A_{L(e)}$, is

$$A_{L(e)} = -zF\Delta\psi. \tag{5.14}$$

If z' is the valence of the ion, z is assigned a minus sign ($z = -z'$) if the ion is an anion, and a positive sign ($z = z'$) if it is a cation. F is the Faraday constant. The electrochemical potential, $\tilde{\mu}_L'$, of the ion L in the cis compartment is

$$\tilde{\mu}_L' = \mu_L^\circ + RT \ln[L'] + zF\Psi', \tag{5.15}$$

and the electrochemical potential of the same ion in the trans compartment is now

$$\tilde{\mu}_L'' = \mu_L^\circ + RT \ln[L''] + zF\Psi''. \tag{5.16}$$

Therefore, the affinity associated with the migration of the ion L from the cis to the trans side of the membrane is

$$A_L = -\left(\tilde{\mu}_L'' - \tilde{\mu}_L'\right) = -RT \ln \frac{[L'']}{[L']} - zF\left(\Psi'' - \Psi'\right) \tag{5.17}$$

or

$$A_L = -\Delta\tilde{\mu}_L = -RT \ln \frac{[L'']}{[L']} - zF\Delta\Psi, \tag{5.18}$$

where z is defined as previously. If the ligand which is being transported is a proton, eq. (5.18) above becomes

$$\Delta\tilde{\mu}_H = -2.3RT\,\Delta\text{pH} + F\Delta\Psi, \tag{5.19}$$

where

$$\Delta\text{pH} = \text{pH}'' - \text{pH}'. \tag{5.20}$$

Once equilibrium is reached, the affinity A_L is nil and therefore eq. (5.18) reduces to

$$\Delta\Psi = -\frac{RT}{zF}\ln\frac{[L'']}{[L']}. \tag{5.21}$$

$\Delta\Psi$ defines the so-called Nernst potential.

Owing to the existence of this potential, there is an electric field, E, across the membrane, defined as

$$E = -\frac{\partial\Psi}{\partial x}, \tag{5.22}$$

where x is the distance. In the case of an ion subjected to this electric field, the corresponding flow is

$$J = -zq\frac{D}{k_B T}[L]\frac{\partial\Psi}{\partial x}. \tag{5.23}$$

D is the diffusion coefficient, q is the charge of the electron, z is the positive or negative value of the valence, as already defined, and k_B is the Boltzmann constant. The expression for the flow can be rewritten as

$$J = -D[L]\frac{zNq}{Nk_B T}\frac{\partial\Psi}{\partial x}, \tag{5.24}$$

where N is the Avogadro number. Since $Nq = F$ and $Nk_B = R$, this expression can be reexpressed as

$$J = -D[L]\frac{zF}{RT}\frac{\partial\Psi}{\partial x}. \tag{5.25}$$

An ion may indeed be under the combined influence of the concentration gradient $\partial[L]/\partial x$ and the electric field. The mathematical expression for the flow is then

$$J = -D\frac{\partial[L]}{\partial x} - D[L]\frac{zF}{RT}\frac{\partial\Psi}{\partial x}. \tag{5.26}$$

This is known as the Nernst–Planck equation.

To integrate this equation, one assumes that the potential varies linearly within the membrane. Hence one has

$$\frac{\partial\Psi}{\partial x} = \frac{\Psi'' - \Psi'}{m} = \frac{\Delta\Psi}{m}, \tag{5.27}$$

where m is the membrane thickness. In fact the Nernst–Planck equation should apply to all ionic species present in the medium, so that

$$J_i = -D_i\frac{\partial[L_i]}{\partial x} - D_i[L_i]\frac{z_i F}{RT}\frac{\Delta\Psi}{m}. \tag{5.28}$$

Setting for simplicity

$$\alpha_i = \frac{D_i}{m}\frac{z_i F}{RT}\Delta\Psi, \tag{5.29}$$

eq. (5.28) can be rearranged to

$$\left\{(J_i/\alpha_i) + [\mathrm{L}_i]\right\} \mathrm{d}x = -\frac{D_i}{\alpha_i}\,\mathrm{d}[\mathrm{L}_i]. \tag{5.30}$$

Now, for the boundary conditions $x = 0$ and $x = m$, one must have

$$\begin{aligned}[\mathrm{L}_i] &= \pi_i[\mathrm{L}'] \quad \text{for } x = 0,\\ [\mathrm{L}_i] &= \pi_i[\mathrm{L}''] \quad \text{for } x = m,\end{aligned} \tag{5.31}$$

where π_i is the partition coefficient between the bulk concentration of L_i and the corresponding concentrations of this ligand in the membrane on the cis (L_i') and on the trans (L_i'') sides. Equation (5.30) can then be integrated and one has

$$\int_0^m \mathrm{d}x = -\frac{D_i}{\alpha_i}\int_{\pi_i[\mathrm{L}_i']}^{\pi_i[\mathrm{L}_i'']} \frac{\mathrm{d}[\mathrm{L}_i]}{J_i/\alpha_i + [\mathrm{L}_i]}. \tag{5.32}$$

Integration yields

$$m = -\frac{D_i}{\alpha_i}\ln\frac{J_i + \alpha_i\pi_i[\mathrm{L}_i'']}{J_i + \alpha_i\pi_i[\mathrm{L}_i']}, \tag{5.33}$$

or

$$\frac{J_i + \alpha_i\pi_i[\mathrm{L}_i'']}{J_i + \alpha_i\pi_i[\mathrm{L}_i']} = \exp\left(-\frac{\alpha_i m}{D_i}\right), \tag{5.34}$$

which can be rearranged to

$$J_i = \frac{[\mathrm{L}_i']\exp(-\alpha_i m/D_i) - [\mathrm{L}_i'']}{1 - \exp(-\alpha_i m/D_i)}\,\alpha_i\pi_i. \tag{5.35}$$

Since

$$\frac{\alpha_i m}{D_i} = \frac{z_i F}{RT}\Delta\Psi, \tag{5.36}$$

eq. (5.35) can be reexpressed as

$$J_i = \frac{D_i\pi_i}{m}\frac{z_i F\Delta\Psi}{RT}\frac{[\mathrm{L}_i']\exp(-z_i F\Delta\Psi/RT) - [\mathrm{L}_i'']}{1 - \exp(-z_i F\Delta\Psi/RT)}. \tag{5.37}$$

The ratio $D_i\pi_i/m$ defines the permeability, P_i, of the membrane to the ion L_i, thus

$$J_i = P_i \frac{z_i F\Delta\Psi}{RT} \frac{[L_i'] \exp(-z_i F\Delta\Psi/RT) - [L_i'']}{1 - \exp(-z_i F\Delta\Psi/RT)}. \tag{5.38}$$

This equation defines the expression of the net flow of an ion across a membrane. If one considers different cations and anions, one can derive from expression (5.38) the so-called Goldman–Hodgkin–Katz equation which expresses, for near-equilibrium conditions, the $\Delta\Psi$ value of a membrane and the respective contributions of different ions to this potential. In the case of monovalent anions, B_i^-, and cations, A_i^+, eq. (5.38) shows that the net flows will be zero when

$$\sum_{i=1}^{m} P_i^- [B_i^{-\prime}] \exp(F\Delta\Psi/RT) = \sum_{i=1}^{m} P_i^- [B_i^{-\prime\prime}] \tag{5.39}$$

and

$$\sum_{i=1}^{m} P_i^+ [A_i^{+\prime}] \exp(-F\Delta\Psi/RT) = \sum_{i=1}^{m} P_i^+ [A_i^{+\prime\prime}]. \tag{5.40}$$

From these equations one has

$$\exp(F\Delta\Psi/RT) = \frac{\sum_{i=1}^{m} P_i^- [B_i^{-\prime\prime}]}{\sum_{i=1}^{m} P_i^- [B_i^{-\prime}]} = \frac{\sum_{i=1}^{m} P_i^+ [A_i^{+\prime}]}{\sum_{i=1}^{m} P_i^+ [A_i^{+\prime\prime}]}, \tag{5.41}$$

or

$$\Delta\Psi = \frac{RT}{F} \ln \frac{\sum_{i=1}^{m} P_i^- [B_i^{-\prime\prime}] + \sum_{i=1}^{m} P_i^+ [A_i^{+\prime}]}{\sum_{i=1}^{m} P_i^- [B_i^{-\prime}] + \sum_{i=1}^{m} P_i^+ [A_i^{+\prime\prime}]}. \tag{5.42}$$

This expression is the Goldman–Hodgkin–Katz equation and expresses the contributions of the various ionic species to the membrane potential. The ions that contribute significantly to the definition of this potential should have relatively large permeability coefficients [1–5].

5.2. Brief description of molecular events involved in energy coupling

The fact that a given event is energetically feasible is not sufficient *per se* to allow one to conclude that the event indeed takes place *in vivo*. In order for this event to occur, the cell, or the organelle, has to possess the required molecular machinery. The aim of this section is precisely to present an overview of the molecular processes involved in energy coupling.

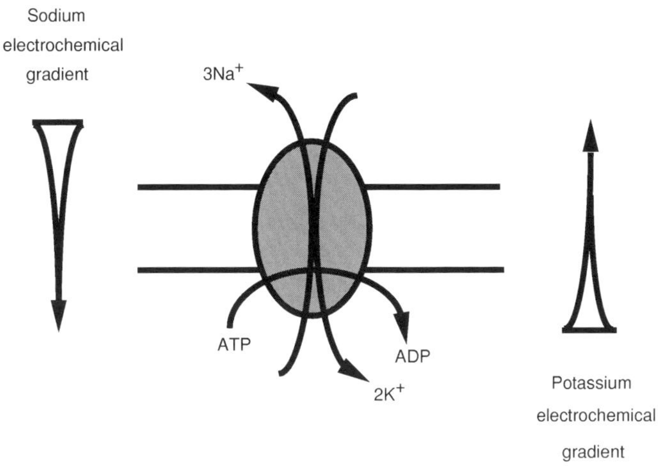

Fig. 5.4. Sodium–potassium ATPase from animal cells. Sodium and potassium are exchanged against electrochemical gradients. See text.

5.2.1. Carriers and channels

The cell membranes allow the direct transport, by simple diffusion, of many neutral molecules. They must also allow passage of many polar molecules such as inorganic ions, sugars, aminoacids, etc. There must therefore exist membrane proteins that allow this passage. These proteins are in fact multipass transmembrane proteins, i.e. their polypeptide chain traverses the lipid bilayer many times. There exist two major classes of transport proteins, called carrier and channel proteins.

5.2.1.1. Carriers

Although ions can, in principle, traverse lipid bilayers by simple diffusion, this process is so slow that it cannot play any part in biological events. Moreover, many ions may be transported "uphill", i.e. against an electrochemical gradient. Such processes of active transport require the participation of protein carriers that can bind the ligand to be transported on the cis side of the membrane and, thanks to a conformation change, release it on the trans side of the same membrane. This can be achieved through the consumption of free energy brought about by the hydrolysis of an ATP molecule. Carriers carry out precisely this function. A formal, speculative scheme of active ligand transport carried out by a transmembrane protein is shown in Fig. 5.4.

Because it is very difficult, for obvious reasons, to obtain three-dimensional crystals of transmembrane carrier proteins, our knowledge of the detailed structure of these proteins is limited and often indirect. It is, however, possible to gain some information on these proteins by the use of recombinant DNA technology. Cloning and sequencing the genes of these proteins provides the biochemist with their primary structure. This allows us, in turn, to know the number and the length of hydrophobic α-helices and leads to sensible speculation as to how the polypetide chain traverses the lipid bilayer. Most membrane carriers known today, in different organisms, have related structures. Although protein carriers are required for active transport, they may also carry out facilitated diffusion, i.e. transport processes that do not require energy consumption. Because carrier-mediated transport involves the preliminary binding of the ligand to a site of the carrier and then its release from that site, the kinetics of ligand transport is somewhat comparable to classical steady state enzyme kinetics. This matter will be studied in some detail in the next section of this chapter. These carriers display a certain specificity towards the ligand to be transported. Some of them transport one ion at a time. They are called *uniports*. Others simultaneously transport two ions in the same direction. They are referred to as *symports*. Last, many carriers may transport two ligands, but in opposite directions, and are thus termed *antiports*.

Protein carriers that carry out active transport are often ATPases, i.e. they associate "uphill" ligand transport with ATP hydrolysis. These ATPases thus act as pumps. There are two main classes of ATPases: the so-called P-type and V-type ATPases. P-type ATPases carry out their own phosphorylation, together with the transport of different cations and possibly protons. V-type ATPases allow the transport of protons only. The best known, among P-type ATPases, is probably the NA^+–K^+-ATPase from the plasma membrane of animal cells. This carrier acts as an antiport. It allows the exit of 3 Na^+ (against an electrochemical gradient) coupled with the entry of 2 K^+ (also against an electrochemical gradient) and the hydrolysis of ATP. The carrier thus acts as an electrogenic pump and, owing to its activity, the membrane becomes polarized. The phosphorylation site of this ATPase, as for the other P-type ATPases, is an aspartyl residue. The enzyme is made up of a large multipass catalytic chain of 110 kDa molecular mass as well as of a glycosylated chain of 55 kDa. A cardiotoxic drug, ouabain, binds to the carrier, competitively with K^+, and thus inhibits its entry. This carrier plays a major part in maintaining a low level of Na^+ in the cell, as well as in controlling its osmotic balance (Fig. 5.4). Another example of P-type ATPase is the Ca^{2+}-ATPase from the endoplasmic and sarcoplasmic reticulum, which is responsible for the ATP-coupled transport of Ca^{2+} in the lumen of the reticulum.

The so-called V-type ATPases couple the hydrolysis of ATP with the active transport of protons. They are located in the membranes of various organelles (vacuoles, lysosomes, ...). They lead to the acidification of the lumen of these organelles. These proton carriers are extremely complex multimolecular edifices with a molecular mass about 500 kDa [6–16].

5.2.1.2. Channels

Channel proteins form hydrophobic pores in the membrane. They are involved in the transport of small inorganic ions. Unlike carrier-mediated transport, channel-mediated transport is not coupled to the hydrolysis of ATP. This transport, however, is much faster than any carrier-mediated transport. The passage of an ion through a channel can only take place "downhill", along an electrochemical gradient.

Channels display a selectivity with respect to ions to be transported. This is probably a consequence of the topology of the narrow hydrophobic pore. Moreover, these channels may exist in open and closed conformations (Fig. 5.5). Channel opening may be brought about by a change of voltage across the membrane (voltage-gated channels), the binding of a specific ligand such as a neurotransmitter or an ion (ligand-gated channels), or a mechanical stress (mechanically-gated channels).

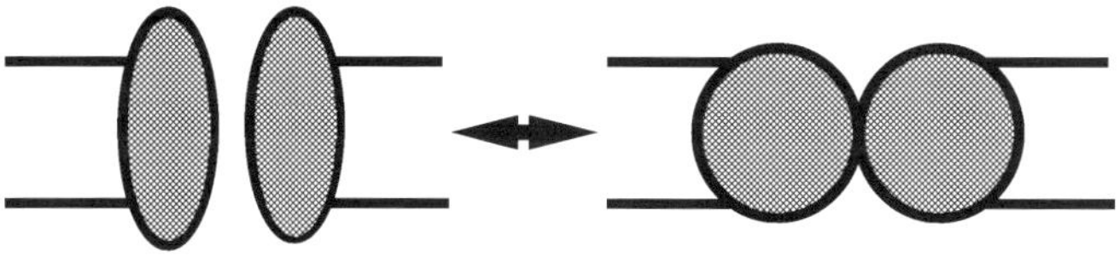

Fig. 5.5. Opening and closing of a channel. See text.

Fig. 5.6. Release of neurotransmitter molecules in a synaptic cleft. Neurotransmitter molecules, located in vesicles of the presynaptic cell, are released in the synaptic cleft and diffuse to the transmitter-gated channels of the post-synaptic cell, inducing in turn their opening.

A membrane potential arises when a difference of the distribution of ions appears on the two sides of the membrane. This difference may be due, for instance, to the active pumping of a cation outside the cell. In the case of the action potential of electrically excitable cells, the emergence of this potential is triggered by the depolarization of the plasma membrane. This causes the voltage-gated ion channels to open and slight amounts of sodium to enter the cell through them, "downhill" the electrochemical gradient. This depolarizes the membrane further, causing other channels to open and bringing about the entry of new sodium ions in the cell. This local depolarization is sufficient to depolarize neighbouring regions of the membrane. The action potential then propagates as a travelling wave. In order to avoid further leakage of sodium, the cell displays a process of temporary inactivation of the open channels and, after several milliseconds, the channel closes again.

It is now possible to record currents flowing through individual channels. This can be done by the technique of patch-clamp recording. A patch of membrane is sealed at the tip of a micropipette, thus allowing one to record currents passing through a small number of channels, possibly only one. It has been shown with this technique that the voltage-gated ion channels open or close in an all-or-nothing fashion. The times of opening and closing are random, but when a channel is open it is completely open and allows about 1000 ions to pass freely per millisecond.

The transmission of an action potential from neuron to neuron at the level of a synapse is an extremely interesting phenomenon which displays highly controlled dynamics. A synapse is a specialized site of contact betwen two neurons. The presynaptic cell is separated from the post-synaptic cell by a narrow synaptic cleft (Fig. 5.6). A change of the electric potential of the presynaptic cell triggers the release of neurotransmitter molecules located in vesicles. This release is effected by exocytosis and the neurotransmitter molecules diffuse to transmitter-gated channels of the post-synaptic target cell. After the neurotransmitter has been secreted, it is rapidly removed from the synaptic cleft by specific enzymes.

Some neurotransmitter molecules, however, bind to their specific channels, making them open and thus bringing about ionic leakage and localized depolarization of the target cell which propagates (Fig. 5.6) [17–25].

5.2.2. Energy storage in mitochondria and chloroplasts

In eukaryotic cells, most of the energy stemming from the degradation of foodstuff is stored in mitochondria in the form of ATP molecules. In green plants, light energy is stored in chloroplasts as chemical energy in the form of ATP and NADPH. The molecular processes that preside over this energy conversion and storage are briefly described below.

5.2.2.1. Mitochondria

As already outlined (chapter 1), mitochondria are surrounded by two membranes. The inner membrane is convoluted, forming infoldings referred to as cristae. These cristae very much increase the surface of the inner membrane. The inner and outer membranes are separated by an intermembrane space. The central region of the mitochondrion, called the matrix, is crammed with enzymes and multienzyme complexes that are involved in different metabolic processes and in particular the Krebs cycle.

The functioning of this cycle results in the generation of carbon dioxide and, above all, of NADH. Oxidation of NADH and succinate, one of the intermediates of the Krebs cycle, takes place on the inner membrane of the mitochondrion. The oxidation of these two metabolites requires the participation of an electron transfer chain, the supramolecular organization of which, in the inner mitochondrial membrane, is remarkably invariant among eukaryotic organisms. The oxidation of NADH requires the sequential participation of three multiprotein complexes that are embedded in the inner membrane and, in fact, traverse this membrane (Fig. 5.7). Succinate oxidation requires the participation of an additional complex.

The three complexes involved in the oxidation of NADH are the NADH dehydrogenase complex, the cytochrome b–c_1 complex and the cytochrome c oxidase complex. If it is

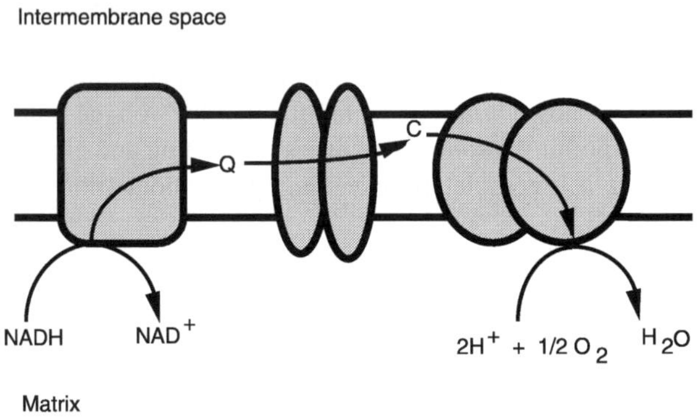

Fig. 5.7. Simplified scheme of the electron transfer chain of mitochondria. Q and C mean ubiquinone and cytochrome C, respectively. See text.

succinate that is beeing oxidized, the additional complex is called the succinate dehydrogenase complex. In either case, the electrons taken up by NADH or succinate are finally transferred to molecular oxygen which is then converted to water. The succinate dehydrogenase complex has a molecular mass of about 200 kDa, is made up of about 8 polypeptide chains and contains one cytochrome b, one flavine adenine dinucleotide (FAD) and 4 iron–sulfur centres. The NADH dehydrogenase complex is the largest complex. Its molecular mass is about 800 kDa, it is made up of about 25 polypeptide chains and passes electrons to a flavin and about 6 iron–sulfur centres. The cytochrome b–c_1 complex is probably a dimer of about 500 kDa. Each monomer apparently consists of 8 polypeptide chains and contains one cytochrome b, one cytochrome c and one iron–sulfur centre. The last complex is the cytochrome oxidase complex and is the best known of these complexes. It is a dimer of about 300 kDa. Each monomer is made up of at least 10 polypeptide chains and contains two cytochrome a and two copper proteins. These complexes are mutually connected by two relatively soluble electron transfer proteins, namely a ubiquinone and a cytochrome c.

The three complexes involved in the electron transfer from NADH to oxygen are also involved in the active pumping of protons through the inner membrane to the intermembrane space. As the lipid bilayer is almost totally impermeable to protons, these protons are transported through the three protein complexes (Fig. 5.8). There are *a priori* two extreme possible mechanisms for these proton transfers called the oxidoreduction loop and the conformational coupling process (Fig. 5.9).

In the oxidoreduction loop mechanism it is postulated that, within a multiprotein complex, electrons and protons follow opposite pathways. Moreover protons transferred across the membrane originate from the electron donor molecule NADH. The inescapable consequence of this scheme is that the ratio between electrons and protons transferred must be equal to unity. There is little doubt that, in many cases, this ratio may be different from one. Hence another possibility of explaining proton transfer is to consider that the reduction of a protein from the relevant complex induces a conformation change and a reorientation of this molecule, thus resulting in an increase of the pK of at least one group of this protein molecule. This will result in the capture of a proton taken up from the cis side of the membrane (the matrix space). When this reduced protein releases its electron it regains its initial conformation and delivers the captured proton but, owing to its reorientation, this proton release will take place on the trans side (the transmembrane space) of the mem-

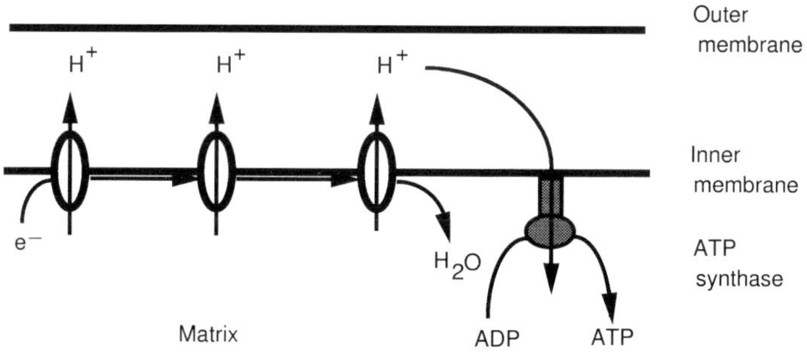

Fig. 5.8. Electron transfer coupled to proton transfer and ATP synthesis. See text.

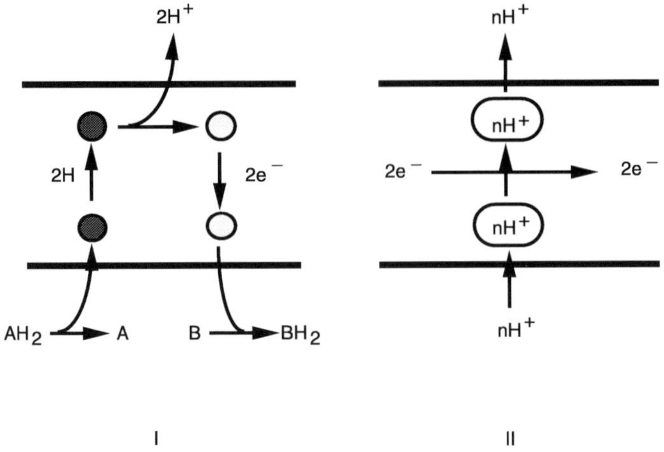

Fig. 5.9. Oxidoreduction loop and conformational coupling. I – Oxidoreduction loop. II – Conformational coupling. See text.

brane. Indeed, this transfer process may involve several protons, and there is *a priori* no fixed stoichiometry between the number of electrons and the number of protons transferred. It seems that cytochrome oxidase follows precisely this type of mechanism. This protein accepts electrons from cytochrome c. The electrons are then transferred to heme a and to copper, both located on the external side of the membrane. The electrons then pass on to heme a_3 and to another copper, this time located on the matrix side of the membrane. It is, apparently in this region of the supramolecular edifice that proton pumping takes place.

The fact that protons are transported in the intermembrane space means that this region of the mitochondrion will become more acidic than the matrix space. Hence there exists, between the intermembrane space and the matrix, a $\Delta\Psi$ and a ΔpH, both of which being, as already mentioned, components of an electrochemical proton gradient. These two components can be determined experimentally. $\Delta\Psi$ can be measured by using the Goldman–Hodgkin–Katz equation, after blocking the electron transfer chain. Its value is 150 mV (negative in the matrix). ΔpH is about one pH unit (alkaline in the matrix). The resulting $\Delta\tilde{\mu}_H$ is therefore of the order of 20 kJ. As the free energy change associated with the transfer of two electrons from NADH to oxygen is about 22 kJ, it is worth pointing out that most of the free energy released by the oxidation of NADH is stored in the electrochemical gradient that spans the inner mitochondrial membrane. As will be discussed later, this conclusion represented a major breakthrough in our understanding of the mechanism of energy conversion for it clearly shows that the classical reductionist approach that aims at understanding biological phenomena at the level of individual macromolecules may be biased. In the present case it is the property of a supramolecular *system* that has to be understood.

It is obvious, however, that the energy originating from the oxidation of NADH cannot be permanently stored in the form of an electrochemical gradient [26–36]. As a matter of fact, this gradient tends to dissipate spontaneously. Dissipation occurs by spontaneous transfer of protons across the pore of a supramolecular edifice anchored in the inner membrane and referred to as ATP-synthase. The energy released by the migration of protons through the pore of this edifice is, in part, stored as ATP molecules. Since ATP-synthase is probably one of the most important multimolecular edifices in biology, its mode of action will be considered more specifically in another section of this chapter (section 5.2.2.3).

5.2.2.2. Chloroplasts

Chloroplasts of green plants are surrounded by two membranes. Unlike mitochondria, the inner membrane does not display infoldings forming cristae. In the central region of the chloroplast, called stroma, are located flattened disclike sacs referred to as thylakoids. The energy conversion process of photosynthesis takes place in the thylakoids. Two photosystems, termed I and II, are anchored in the thylakoids. Both are complex supramolecular edifices containing proteins, chlorophyll and other pigments. The pigments, together with chlorophyll molecules, play the role of an antenna that collects photons and directs them to a specific pair of chlorophyll molecules, which play the main role in two photochemical acts. An electron from a chlorophyll belonging to photosystem II (called P 680) is excited and immediately passed on to an electron transfer chain in close contact with this chlorophyll molecule, which then becomes oxidized. The electrons expelled from the chlorophyll are immediately replaced by electrons taken up from water molecules and this results in the splitting of water and the release of molecular oxygen and protons. These electrons are passed to a plastoquinone, then to a cytochrome b_6–f complex, which is similar to the b_6–f complex of mitochondria. This complex acts as a proton pump, resulting in a decrease of the local pH in the lumen of the thylakoids. As with mitochondria, part of the energy available from the oxidation process is stored as an electrochemical gradient between the thylakoid lumen and the stroma. The electrons are then passed to a plastocyanin and to photosystem I. Chlorophyll from this photosystem (called P 700) undergoes light excitation and passes electrons to a plastoquinone and to a ferredoxin. A NADP reductase may then catalyse the reduction of NADP to NADPH + H^+. A very simplified scheme of the electron transfer process is shown in Fig. 5.10. This process is the one that takes place in green plants but is different from that occurring in photosynthetic bacteria (with the exception of Cyanobacteria).

As with mitochondria, this gradient spontaneously dissipates, for the protons stored in the lumen of thylakoids tend to travel through the pore of the ATP-synthase anchored in the thylakoid membrane, with their globular head directed towards the stroma [37–43]. As with mitochondria, dissipation of the gradient leads to ATP synthesis through a process which is apparently identical to that taking place in mitochondria.

5.2.2.3. ATP-synthase

ATP-synthase from mitochondria and chloroplasts looks like a spherical knob protruding from the inner mitochondrial membrane towards the matrix, or from the thylakoid membrane towards the stroma. This knob has a diameter of about 90–100 Å. When exposed to low ionic strength media, the knobs dissociate from the mitochondria, or from the thylakoid membrane, and are able to hydrolyze ATP. This soluble part of the ATP-synthase is called F_1. It contains different types of subunits, referred to as α_3, β_3, γ, δ and ε. The F_1 part of ATP-synthase is anchored in the membrane by a relatively narrow stalk (about

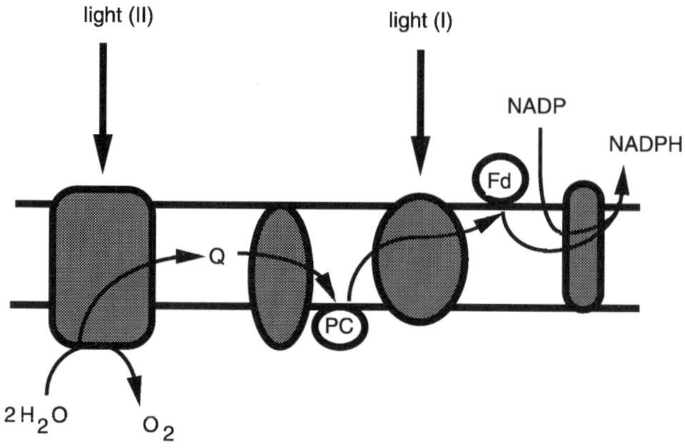

Fig. 5.10. Simplified scheme of electron transfer in chloroplasts. Q, PC and Fd mean plastoquinone, plastocyanin and ferredoxin, respectively. The whole process is triggered by two light excitations. See text.

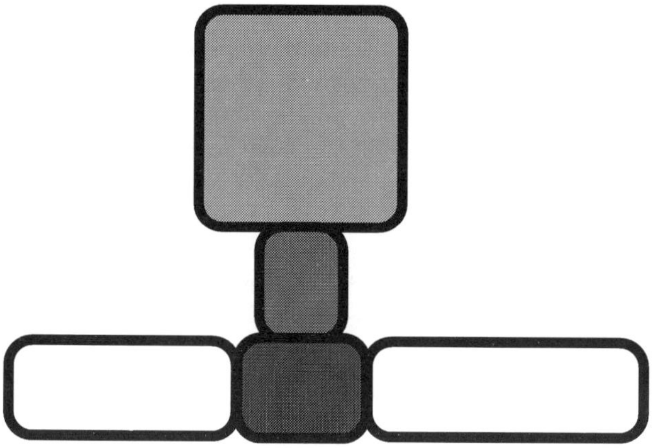

Fig. 5.11. Simplified scheme of ATP-synthase. The globular part, called F_1, is anchored in the membrane by the F_0 region. This region is surrounded by proton carriers.

45 Å in diameter). The part of the supramolecular edifice that is buried in the membrane, is called F_0. F_0 is the element of the multimolecular edifice which is permeable to protons. The stalk region is made up of polypeptide chains that belong to both F_1 and F_0. Whereas ATP hydrolysis can be catalysed by mixtures of α- and β-subunits, ATP synthesis requires the presence of the complete edifice. The three-dimensional structure of F_1 ATP-synthase has been solved by crystallography [44] and is schematically pictured in Fig 5.11. Last but not least, three catalytic sites, located on the β-subunits, may convert ADP plus phosphate into ATP. α-subunits bear nucleotide binding sites, but are unable to carry out catalysis.

The mode of action of ATP-synthase has proved intriguing. The binding of ADP and of inorganic phosphate are both exergonic, spontaneous, processes ($\Delta G° = -37$ kJ/mole for

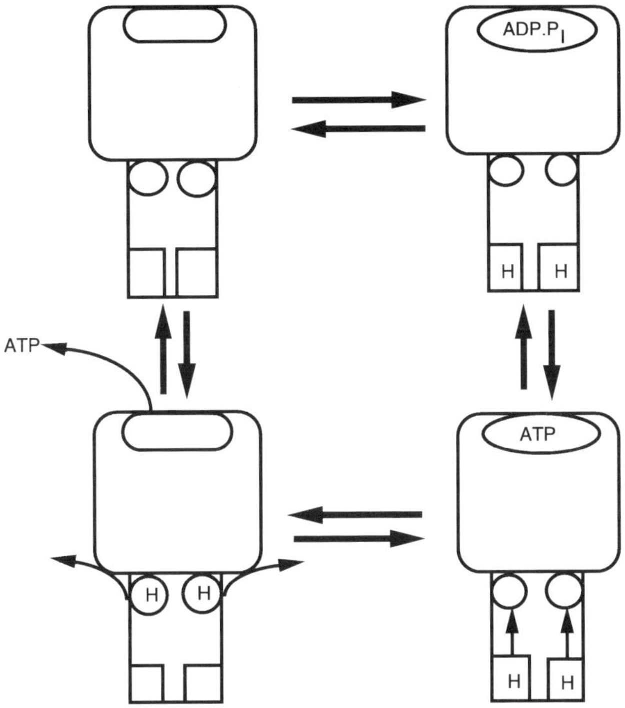

Fig. 5.12. The steps of ATP synthesis. ADP and inorganic phosphate bind to a specific site of the F_1 region and protons bind to carriers surrouding the F_0 region. ADP and phosphate are converted to ATP that cannot be released from the corresponding site. Proton migration induces a conformation change of the ATP binding site that allows ATP release.

ADP binding and $\Delta G° = -18$ kJ/mole for phosphate binding). Surprisingly, the catalytic step for the phosphorylation of ADP is also exergonic ($\Delta G° = -1.7$ kJ/mole). It is the release of ATP from F_1 which is strongly endergonic ($\Delta G° = 69$ kJ/mole). This means that it is the release of ATP which is promoted by the transfer of protons through F_0. The whole process is schematized as shown in Fig 5.12. ADP and inorganic phosphate bind to F_1, then protons are bound to low affinity sites on F_0. This binding process is possible because the value of $\tilde{\mu}_H$ is large in the compartment opposite F_1 (intermembrane space of mitochondria, or lumen of thylakoids). After the conversion of ADP into ATP, which induces a conformation change of F_1, the protons migrate through F_0 and bind to high affinity sites in the region of F_0 which is close to the face of the membrane bearing F_1. Since the value of $\tilde{\mu}_H$ is low in this region of space, the protons tend to dissociate from the corresponding sites and the F_1 region of ATP-synthase reaches its initial conformation, thus resulting in a release of ATP (Fig. 5.12).

Moreover, there exists between the three subunits some sort of connection that results in their circular and alternative functioning. If the enzyme carries out ATP synthesis, a first β-subunit (SU_1) is able to bind ADP and phosphate and to catalyse their conversion

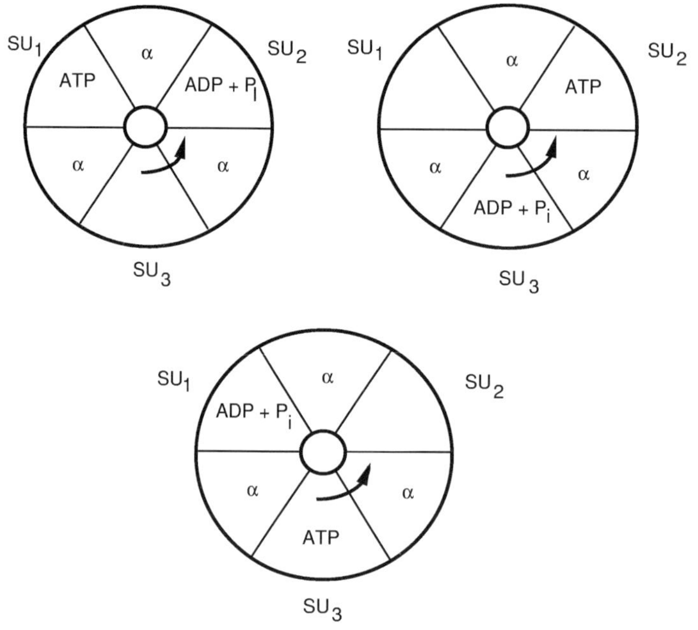

Fig. 5.13. Rotational catalysis by ATP-synthase. See text.

into ATP. A second β-subunit (SU_2) is able to bind ADP and phosphate weakly but does not catalyse their conversion to ATP. The last β-subunit (SU_3) does not bind anything. As soon as ATP has been synthesized on SU_1, it is released and SU_1 is, for a while, unable to bind anything. SU_2 can now catalyse the synthesis of ATP and SU_3 can bind ADP and phosphate but does not catalyse ATP synthesis. Thus SU_1 has become SU_3, SU_2 has been transformed into SU_1, and SU_3 has become equivalent to SU_2 (Fig. 5.13) [2,45,46].

Therefore, the system behaves as if the activity of the three subunits were rotating clockwise. Boyer [47] assumed, on the basis of indirect arguments, that this apparent rotation is in fact real and that the central part of the ATP-synthase (the γ-subunit) rotates counterclockwise relative to the α- and β-subunits. This contention has recently been proved valid by Noji et al. [48], who attached a fluorescent actin filament to the γ-subunit of an ATP-synthase crystal and saw, under the microscope, a counterclockwise rotation of γ relative to the $\alpha\beta$-hexamer. This circular motion occurs in the presence of ATP and, in these conditions, the enzyme behaves as an ATPase. As a matter of fact, the γ-subunit behaves like a rotor turning in a stator barrel made up of the six $\alpha\beta$-subunits. Hence ATP-synthase may be considered a motor protein.

ATP synthase can carry out not only ATP synthesis but also, in the opposite direction, ATP hydrolysis. The balance between these two antagonistic functions is determined by the steepness of the electrochemical potential gradient across the inner mitochondrial, or thylakoid, membrane as well as by the concentrations of ATP, ADP and phosphate in the cell. One must realize that *in vivo* the ratio [ADP][P_i]/[ATP] (where P_i is the orthophosphate) is such that minus the affinity ($-A_S$) of ATP hydrolysis is usually -46 to -54 kJ/mole whereas the ΔG° of the same reaction is only -30 kJ/mole. This implies that, *in vivo*, the above ratio is far from the equilibrium constant of ATP hydrolysis. We shall see later, in the last section of this chapter, how the respective values of the affinity (A_S) of ATP hydrolysis and of the ΔG° of the same reaction play a central part in the control of both active ion transport and storage of free energy.

5.3. Compartmentalization of the living cell and the kinetics and thermodynamics of coupled scalar and vectorial processes

5.3.1. The model

Let us now consider a complex vectorial process that takes place on a biomembrane and is associated with a scalar chemical reaction

$$\mathrm{S} + \mathrm{W} \rightleftarrows \mathrm{P} + \mathrm{Q} \tag{5.43}$$

which is spontaneously shifted towards the breakdown of S into P and Q (hydrolysis exerted by water W). This reaction could be, for instance, the spontaneous, exergonic, hydrolysis of ATP into ADP and phosphate (P and Q), and it is assumed to take place on one side of the membrane. This scalar chemical process may, possibly, be coupled to the vectorial transport of a ligand (a proton for example) across the membrane. The transfer, as already noted, usually requires the participation of a carrier, X, which may exist under several conformations. One conformation is able to bind the ligand on the cis side of the membrane whereas another binds the same ligand on the trans side of the same membrane. Thanks to the supramolecular organization of the membrane, some of these conformation changes may be coupled to the polypeptide chain responsible for the corresponding catalytic process. This situation is depicted in Fig 5.14.

In this model, X′ and X″ are the two conformations of the carrier that can bind the ligand on either side of the membrane. $\widetilde{K}_1'$ and $\widetilde{K}_1''$ are the apparent binding constants of the ligand on the cis (′) and on the trans (″) sides of the membrane. These apparent binding constants are in fact the product of the real binding constant and of the local ligand concentration, namely

$$\widetilde{K}_1' = K_1'[\mathrm{L}'], \qquad \widetilde{K}_1'' = K_1''[\mathrm{L}'']. \tag{5.44}$$

K_1' and K_1'' are the real binding constants, [L′] and [L″] the local concentrations on either side of the membrane. $\widetilde{K}_2' = K_2'$ is the real equilibrium constant of an isomerization of

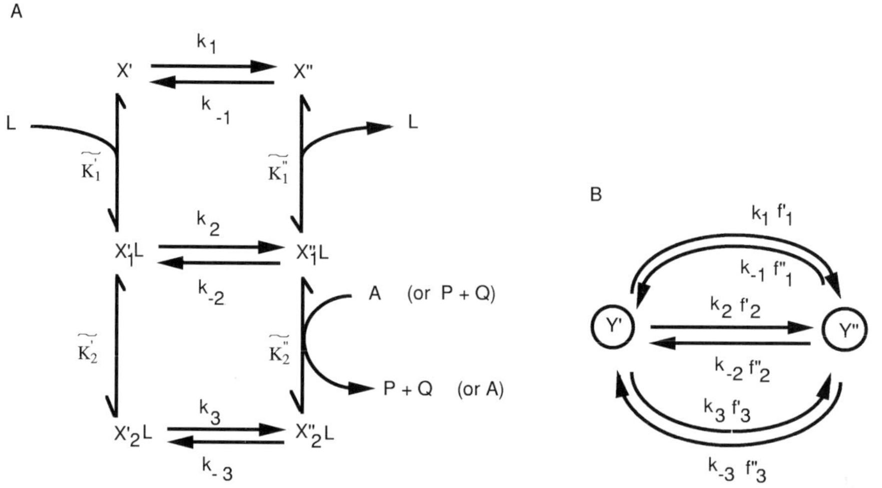

Fig. 5.14. A phenomenological kinetic scheme that explains ATP synthesis coupled to proton transfer and active transport coupled to ATP hydrolysis. A – Kinetic scheme. B – Node contraction of the kinetic scheme.

the carrier. $\widetilde{K}_2''$ is the overall apparent transconformation constant of the carrier which is coupled to the chemical reaction. Thus one may have

$$\begin{array}{ccccc} LX_1'' & \underset{}{\overset{K_{21}''S}{\rightleftharpoons}} & SLX_1'' & \underset{}{\overset{K_{22}''W}{\rightleftharpoons}} & QPLX_2'' \\ & & & & P \Big\updownarrow K_{23}'' \\ & & & & QLX_2'' \underset{Q}{\overset{K_{24}''}{\rightleftharpoons}} LX_2'' \end{array} \tag{5.45}$$

or

$$\begin{array}{ccccc} LX_1'' & \underset{}{\overset{K_{21}''Q}{\rightleftharpoons}} & QLX_1'' & \underset{}{\overset{K_{22}''P}{\rightleftharpoons}} & PQLX_1'' \\ & & & & W \Big\updownarrow K_{23}'' \\ & & & & SLX_2'' \underset{S}{\overset{K_{24}''}{\rightleftharpoons}} LX_2'' \end{array} \tag{5.46}$$

depending on whether the process spends (model (5.45)), or stores (model (5.46)) free energy. LX_1'' and LX_2'' are two conformations of the carrier associated with the enzyme, in the membrane. The apparent equilibrium constant, $\widetilde{K}''_2$, is thus

$$\widetilde{K}''_2 = K''_{21}K''_{22}K''_{23}K''_{24}\frac{[S]}{[P][Q]} = K''_2\frac{[S]}{[P][Q]} \tag{5.47}$$

for model (5.45) and

$$\widetilde{K}''_2 = K''_{21}K''_{22}K''_{23}K''_{24}\frac{[P][Q]}{[S]} = K''_2\frac{[P][Q]}{[S]} \tag{5.48}$$

for model (5.46). In eqs. (5.47) and (5.48) water concentrations have been left out for simplicity, i.e. this concentration has been included in the expressions of K''_{22} and K''_{23}. This notation will be systematically used henceforth. It is worth stressing that the ratio [P][Q]/[S] is not the equilibrium constant of hydrolysis of S (the equilibrium constant of ATP hydrolysis for instance), but a concentration ratio that contributes to the definition of the equilibrium constant, K''_2, of the equilibrium

$$LX''_1 + S \rightleftharpoons LX''_2 + P + Q \tag{5.49}$$

namely,

$$K''_2 = \frac{[P][Q]}{[S]}\frac{[LX''_2]}{[LX''_1]}. \tag{5.50}$$

In the case of ATP hydrolysis by free, isolated ATPase, the ratio [P][Q]/[S] is indeed the equilibrium constant of ATP hydrolysis and therefore has a large value (see section 5.2.2.3). The same ratio, in eq. (5.50) above may not have a large value because part of the free energy of ATP hydrolysis may have been used to force the carrier to adopt the energetically unfavourable LX''_2 conformation in the membrane. In other words, eq. (5.50) represents the simple mathematical expression of an energy conversion process, where the free energy of a chemical process is converted into conformational energy. Substitution for K''_2 in expression (5.47) with its value obtained from eq. (5.50) leads to

$$\widetilde{K}''_2 = \frac{[LX''_2]}{[LX''_1]} \tag{5.51}$$

and shows that $\widetilde{K}''_2$ is indeed dimensionless. A similar conclusion can be derived if K''_2 is defined as the equilibrium constant of the process depicted in model (5.46).

Moreover, the complex kinetic scheme of Fig. 5.14 is asumed to display a time hierarchy. That is the conformation changes of the carrier that "close" or "open" a site for the ligand on either side of the membrane are assumed to be slow relative to all the other steps. This means that the fast steps may be considered to be in quasi-equilibrium, whereas the other dynamic events are not. This assumption allows a node contraction of the graph, as shown in Fig. 5.14(B) [49]. Although the node contraction was not made intially, it greatly simplifies the mathematical treatment of the coupled vectorial–scalar system. In

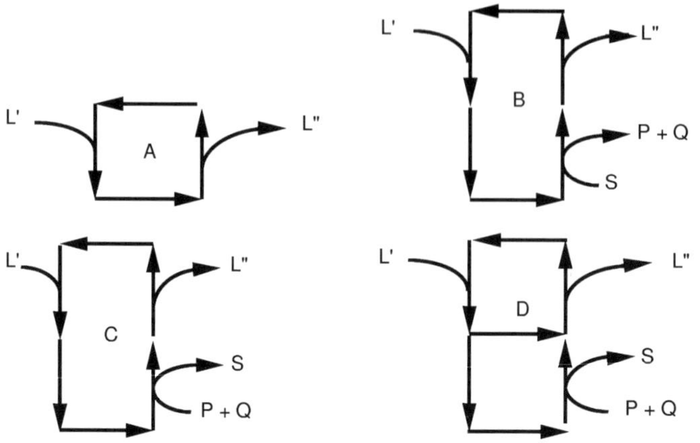

Fig. 5.15. Different types of events predicted by the kinetic scheme of Fig. 5.14. A – Facilitated diffusion. B – Active transport coupled to ATP hydrolysis. C – ATP synthesis coupled to spontaneous ion transport. D – Facilitated diffusion associated with ATP synthesis.

the contracted scheme, Y' and Y'' represent the sums of species concentrations that are in quasi-equilibrium, namely,

$$\begin{aligned} Y' &= [\mathrm{X}'] + [\mathrm{LX}'_1] + [\mathrm{LX}'_2], \\ Y'' &= [\mathrm{X}''] + [\mathrm{LX}''_1] + [\mathrm{LX}''_2]. \end{aligned} \tag{5.52}$$

The f_s that appear in the contracted scheme are called the fractionation factors [49] and are defined as

$$\begin{aligned} f'_1 &= \frac{[\mathrm{X}']}{[\mathrm{X}'] + [\mathrm{LX}'_1] + [\mathrm{LX}'_2]} = \frac{1}{1 + \widetilde{K}'_1 + \widetilde{K}'_1 \widetilde{K}'_2}, \\ f'_2 &= \frac{[\mathrm{LX}'_1]}{[\mathrm{X}'] + [\mathrm{LX}'_1] + [\mathrm{LX}'_2]} = \frac{\widetilde{K}'_1}{1 + \widetilde{K}'_1 + \widetilde{K}'_1 \widetilde{K}'_2}, \\ f'_3 &= \frac{[\mathrm{LX}'_2]}{[\mathrm{X}'] + [\mathrm{LX}'_1] + [\mathrm{LX}'_2]} = \frac{\widetilde{K}'_1 \widetilde{K}'_2}{1 + \widetilde{K}'_1 + \widetilde{K}'_1 \widetilde{K}'_2}, \\ f''_1 &= \frac{[\mathrm{X}'']}{[\mathrm{X}''] + [\mathrm{LX}''_1] + [\mathrm{LX}''_2]} = \frac{1}{1 + \widetilde{K}''_1 + \widetilde{K}''_1 \widetilde{K}''_2}, \\ f''_2 &= \frac{[\mathrm{LX}''_1]}{[\mathrm{X}''] + [\mathrm{LX}''_1] + [\mathrm{LX}''_2]} = \frac{\widetilde{K}''_1}{1 + \widetilde{K}''_1 + \widetilde{K}''_1 \widetilde{K}''_2}, \\ f''_3 &= \frac{[\mathrm{LX}''_2]}{[\mathrm{X}''] + [\mathrm{LX}''_1] + [\mathrm{LX}''_2]} = \frac{\widetilde{K}''_1 \widetilde{K}''_2}{1 + \widetilde{K}''_1 + \widetilde{K}''_1 \widetilde{K}''_2}. \end{aligned} \tag{5.53}$$

The general model of Fig 5.14 predicts several possible types of events. Some of them are depicted in Fig. 5.15. By convention, they involve the transport of a ligand from the cis

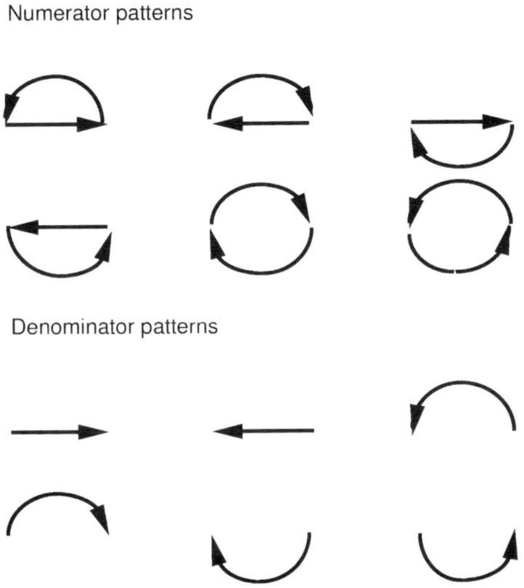

Fig. 5.16. Cyclical and non-cyclical patterns of the kinetic scheme of Fig. 5.14.

(′) to the trans (″) compartment. The first event is the transport of the ligand by a carrier thanks to facilitated diffusion, without recourse to any coupling with a chemical reaction (Fig. 5.15(A)). The second event is the transport of a ligand against an electrochemical gradient, coupled to the hydrolysis of molecule S (ATP, for instance). Active ion transport coupled to ATPase activity falls precisely into this category (Fig. 5.15(B)). The third type of event is the transport of a ligand along an electrochemical gradient which drives an endergonic reaction to completion (synthesis of ATP, for instance) owing to the tight coupling that exists between the scalar and vectorial processes (Fig. 5.15(C)). The last type of event (Fig. 5.15(D)) represents the partial coupling between facilitated diffusion of a ligand and an endergonic chemical reaction. The two last models can be viewed as a formal analysis of ATP-synthase activity. These topics have been studied in details in refs. [1] and [50].

5.3.2. The steady state equations of coupled scalar-vectorial processes

After node contraction, as shown in Fig. 5.14(B), the subsequent kinetic scheme can be treated with the standard methods used to derive steady state rate equations (see chapter 2). The patterns that appear in the numerator and the denominator of the reaction rate are shown in Fig. 5.16. Not all these patterns, however, appear in the equations that describe the events of Fig. 5.15. We shall thus derive these equations below.

The relevant kinetic scheme of facilitated diffusion of ligand L is that of Fig 5.15(A). Moreover, the expression for the f' and f'' is simpler than in expressions (5.53). One has

$$
\begin{aligned}
f_1' &= \frac{1}{1+\widetilde{K}_1'}, \qquad & f_1'' &= \frac{1}{1+\widetilde{K}_1''}, \\
f_2' &= \frac{\widetilde{K}_1'}{1+\widetilde{K}_1'}, \qquad & f_1'' &= \frac{\widetilde{K}_1''}{1+\widetilde{K}_1''}.
\end{aligned} \tag{5.54}
$$

The relevant steady state equation is

$$J = \frac{k_2 f_2' k_{-1} f_1'' - k_{-2} f_2'' k_1 f_1'}{k_2 f_2' + k_{-2} f_2'' + k_{-1} f_1'' + k_1 f_1'}, \tag{5.55}$$

which can be rewritten in explicit form as

$$J = \frac{k_{-1}k_2 K_1'[\mathrm{L}'] - k_1 k_{-2} K_1''[\mathrm{L}'']}{k_{-1} + k_1 + (k_{-1} + k_2)K_1'[\mathrm{L}'] + (k_1 + k_{-2})K_1''[\mathrm{L}''] + (k_2 + k_{-2})K_1'[\mathrm{L}']K_1''[\mathrm{L}'']}. \tag{5.56}$$

J represents the net flow, i.e. the difference between the influx, J^+, and the outflux, J^-. If the ligand L is absent on the trans side of the membrane, eq. (5.56) reduces to

$$J = J^+ = \frac{k_{-1}k_2 K_1'[\mathrm{L}']}{k_1 + k_{-1} + (k_{-1} + k_2)K_1'[\mathrm{L}']}, \tag{5.57}$$

which is the expression of a rectangular hyperbola.

The second situation, that of Fig. 5.15(B), will be considered now. It describes the active transport of ligand L against an electrochemical gradient thanks to the expenditure of ATP. The relevant equation of the net flow is now

$$J = \frac{k_{-1} f_1'' k_3 f_3' - k_1 f_1' k_{-3} f_3''}{k_{-1} f_1'' + k_1 f_1' + k_{-3} f_3'' + k_3 f_3'}, \tag{5.58}$$

and, taking advantage of the expression of the fractionation factors f' and f'' and of the expression of $\widetilde{K}_2''$, which is

$$\widetilde{K}_2'' = K_2'' \frac{[\mathrm{P}][\mathrm{Q}]}{[\mathrm{S}]}, \tag{5.59}$$

this equation becomes

$$\begin{aligned} J = {} & \left(k_{-1}k_3 K_2' K_1'[\mathrm{L}'] - k_1 k_{-3} K_1''[\mathrm{L}''] \frac{[\mathrm{P}][\mathrm{Q}]}{[\mathrm{S}]}\right) \\ & \times \left\{ (k_1 + k_3 K_2' K_1'[\mathrm{L}']) \left\{ 1 + K_1''[\mathrm{L}''] \left(1 + K_2'' \frac{[\mathrm{P}][\mathrm{Q}]}{[\mathrm{S}]} \right) \right\} \right. \\ & \left. + \left(k_{-1} + k_{-3} K_1''[\mathrm{L}''] K_2'' \frac{[\mathrm{P}][\mathrm{Q}]}{[\mathrm{S}]} \right) \{ 1 + K_1'[\mathrm{L}'](1 + K_2') \} \right\}^{-1}. \end{aligned} \tag{5.60}$$

Since $[\mathrm{L}'] < [\mathrm{L}'']$, the net flow will be oriented from the cis to the trans side if the ratio $[\mathrm{P}][\mathrm{Q}]/[\mathrm{S}]$ is small enough. We shall come back later to the thermodynamic significance of this statement, and of eqs. (5.58) and (5.60).

The third interesting situation, that depicted in Fig. 5.15(C), will be considered now. The corresponding model describes the synthesis of an energy-rich compound thanks to the transfer of a ligand "downhill" an electrochemical gradient. The mathematical formulation of the net flow, oriented from the cis to the trans compartment, is still eq. (5.58) but the expression of $\widetilde{K}_2''$ is now

$$\widetilde{K}_2'' = K_2'' \frac{[\mathrm{S}]}{[\mathrm{P}][\mathrm{Q}]}. \tag{5.61}$$

Therefore, the expression of the net flow is

$$\begin{aligned} J = & \left(k_{-1}k_3K_2'K_1'[\mathrm{L}'] - k_1k_{-3}K_1''[\mathrm{L}'']K_2'' \frac{[\mathrm{S}]}{[\mathrm{P}][\mathrm{Q}]}\right) \\ & \times \left[\left(k_1 + k_3K_2'K_1'[\mathrm{L}']\right)\left\{1 + K_1''[\mathrm{L}'']\left(1 + K_2''\frac{[\mathrm{S}]}{[\mathrm{P}][\mathrm{Q}]}\right)\right\} \right. \\ & \left. + \left(k_{-1} + k_{-3}K_1''[\mathrm{L}'']K_2''\frac{[\mathrm{S}]}{[\mathrm{P}][\mathrm{Q}]}\right)\left\{1 + K_1'[\mathrm{L}'](1 + K_2')\right\} \right]^{-1}. \end{aligned} \tag{5.62}$$

In the present situation $[\mathrm{L}'] > [\mathrm{L}'']$, and the net flow will now be oriented from the cis to the trans side of the membrane if the ratio $[\mathrm{S}]/[\mathrm{P}][\mathrm{Q}]$ is not too large. This point will be discussed later.

The last situation, described in Fig. 5.15(D), will be considered now. This scheme postulates that there is partial coupling between ligand transport along an electrochemical gradient and an endergonic chemical reaction, such as the synthesis of ATP. The relevant rate equation of this net flow is then

$$J = \frac{k_{-1}f_1''(k_2f_2' + k_3f_3') - k_1f_1'(k_{-2}f_2'' + k_{-3}f_3'')}{k_2f_2' + k_{-2}f_2'' + k_{-1}f_1'' + k_1f_1' + k_{-3}f_3'' + k_3f_3'}. \tag{5.63}$$

The expression of the fractionation factors f' and f'' is already known (eq. (5.53)) and $\widetilde{K}_2''$ is defined, in the present case, as

$$\widetilde{K}_2'' = K_2'' \frac{[\mathrm{S}]}{[\mathrm{P}][\mathrm{Q}]}. \tag{5.64}$$

This allows to write the net flow of the ligand, from the cis to the trans side of the membrane, as

$$\begin{aligned} J = & \left[k_{-1}K_1'[\mathrm{L}'](k_2 + k_3K_2') - k_1K_1''[\mathrm{L}'']\left(k_{-2} + k_{-3}K_2''\frac{[\mathrm{S}]}{[\mathrm{P}][\mathrm{Q}]}\right)\right] \\ & \times \left[\left\{1 + K_1''[\mathrm{L}'']\left(1 + K_2''\frac{[\mathrm{S}]}{[\mathrm{P}][\mathrm{Q}]}\right)\right\}\left\{k_1 + K_1'[\mathrm{L}'](k_2 + k_3K_2')\right\} \right. \end{aligned}$$

$$+\{1+K_1'[\mathrm{L}'](1+K_2')\}\left\{k_{-1}+K_1''[\mathrm{L}'']\left(k_{-2}+k_{-3}K_2''\frac{[\mathrm{S}]}{[\mathrm{P}][\mathrm{Q}]}\right)\right\}\Bigg]^{-1}. \quad (5.65)$$

This equation makes it obvious that the net flow is controlled both by facilitated diffusion of the ligand across the membrane and by the chemical reaction, but it tells us nothing about the relative influence of these two factors.

5.3.3. Thermodynamics of coupling betwen scalar and vectorial processes

It was pointed out by Hill [50] that the equations that express the net flow as a function of the ligand concentration, in the cis and trans compartments, can also be used to study the thermodynamics of coupling between scalar and vectorial processes. When the transport is not coupled to any chemical reaction, eq. (5.55) can be rewritten as

$$J=k_{-2}f_2''k_1f_1'\left(\frac{k_2f_2'k_{-1}f_1''}{k_{-2}f_2''k_1f_1'}-1\right)\left[k_2f_2'+k_{-2}f_2''+k_{-1}f_1''+k_1f_1'\right]^{-1}. \quad (5.66)$$

Substituting for the fractionation factors f' and f'' their mathematical expression leads to

$$\frac{k_2f_2'k_{-1}f_1''}{k_{-2}f_2''k_1f_1'}=\frac{K_2K_1'}{K_1K_1''}\frac{[\mathrm{L}']}{[\mathrm{L}'']}, \quad (5.67)$$

where K_1, K_1', K_1'' and K_2 are the real equilibrium constants already defined. Moreover, thermodynamics imposes that

$$K_2K_1'=K_1K_1''. \quad (5.68)$$

Thus eq. (5.67) reduces to

$$\frac{k_2f_2'k_{-1}f_1''}{k_{-2}f_2''k_1f_1'}=\frac{[\mathrm{L}']}{[\mathrm{L}'']}. \quad (5.69)$$

As the electrochemical potential of the ligand L is

$$\tilde{\mu}_\mathrm{L}=\tilde{\mu}_\mathrm{L}^\circ+RT\ln[\mathrm{L}]+zF\Psi, \quad (5.70)$$

where F is the Faraday, Ψ is the electrostatic potential and z is the valence of the ligand, defined as negative for an anion and positive for a cation, one can express the ligand concentration ratio $[\mathrm{L}']/[\mathrm{L}'']$ in terms of the difference of electrochemical potentials across the membrane. If one defines $\Delta\tilde{\mu}_\mathrm{L}$ and $\Delta\Psi$ as

$$\Delta\tilde{\mu}_\mathrm{L}=\tilde{\mu}_\mathrm{L}''-\tilde{\mu}_\mathrm{L}', \qquad \Delta\Psi=\Psi''-\Psi', \quad (5.71)$$

and takes advantage of eqs. (5.70) and (5.71), it can easily be shown that

$$\frac{[\mathrm{L}']}{[\mathrm{L}'']} = \exp\left\{-\frac{\Delta\tilde{\mu}_{\mathrm{L}} - zF\Delta\Psi}{RT}\right\}. \tag{5.72}$$

Equation (5.66) shows that ligand L will be transported from the cis to the trans compartment if, and only if,

$$\frac{k_2 f_2' k_{-1} f_1''}{k_{-2} f_2'' k_1 f_1'} = \frac{[\mathrm{L}']}{[\mathrm{L}'']} > 1, \tag{5.73}$$

and this, in turn, implies that

$$-\left(\Delta\tilde{\mu}_{\mathrm{L}} - zF\Delta\Psi\right) > 0. \tag{5.74}$$

This represents the thermodynamic condition that defines the existence of a net flow of the ligand oriented from the cis to the trans compartment. Inserting expressions (5.71), (5.72) and (5.73) into eq. (5.66) leads to

$$J = J^{-}\left[\exp\left\{-\frac{\Delta\tilde{\mu}_{\mathrm{L}} - zF\Delta\Psi}{RT}\right\} - 1\right], \tag{5.75}$$

where J^{-} is, as previously, the flow in the backward direction. Thus, from a thermodynamic viewpoint, the efficiency of the transport of ligand L depends solely on the respective values of the electrochemical and electrostatic potential differences across the membrane. If the value of the exponential term in eqs. (5.72) and (5.75) is such that

$$\exp\left\{-\frac{\Delta\tilde{\mu}_{\mathrm{L}} - zF\Delta\Psi}{RT}\right\} \approx 1 - \frac{\Delta\tilde{\mu}_{\mathrm{L}} - zF\Delta\Psi}{RT}, \tag{5.76}$$

then eq. (5.75) reduces to

$$J = J^{-}\frac{zF\Delta\Psi - \Delta\tilde{\mu}_{\mathrm{L}}}{RT}, \tag{5.77}$$

which is a typical flow-force relationship.

This reasoning can indeed be extended to situations where vectorial and scalar processes are coupled (Fig. 5.15(B)). Thus, one has for eq. (5.58)

$$J = k_1 f_1' k_{-3} f_3'' \left(\frac{k_{-1} f_1'' k_3 f_3'}{k_1 f_1' k_{-3} f_3''} - 1\right)\left[k_{-1} f_1'' + k_1 f_1' + k_{-3} f_3'' + k_3 f_3'\right]^{-1}. \tag{5.78}$$

Replacing f' and f'' by their expressions leads to

$$\frac{k_{-1} f_1'' k_3 f_3'}{k_1 f_1' k_{-3} f_3''} = \frac{K_3 K_2' K_1'}{K_1 K_1'' K_2''} \frac{[\mathrm{L}']}{[\mathrm{L}'']} \frac{[\mathrm{S}]}{[\mathrm{P}][\mathrm{Q}]}. \tag{5.79}$$

It is important, at this stage, to come back to the significance of the ratio [S]/[P][Q] as well as to that of its reciprocal, which appears in eq. (5.60) above. If one considers the breakdown of S into P and Q (for instance, the hydrolysis of ATP into ADP and phosphate), this reaction has a standard free energy difference, ΔG_S°, which is negative. The affinity of the corresponding hydrolytic reaction is thus

$$A_S = -\left(\Delta G_S^\circ + RT \ln \frac{[P][Q]}{[S]}\right). \tag{5.80}$$

If the ratio

$$\rho = \frac{[P][Q]}{[S]} \tag{5.81}$$

is chosen to have the same value it had in the expression of the equilibrium constant K_2'' (eq. (5.50)), then the corresponding affinity A_S is of necessity positive. From expressions (5.80) and (5.81) one finds

$$\rho = \exp\left\{-\frac{\Delta G_S^\circ + A_S}{RT}\right\}. \tag{5.82}$$

Therefore, [S]/[P][Q], which appears in eq. (5.62), must be equal to

$$\frac{[S]}{[P][Q]} = \frac{1}{\rho} = \exp\left\{\frac{\Delta G_S^\circ + A_S}{RT}\right\}. \tag{5.83}$$

Expressions (5.82) and (5.83) show that

$$\Delta G_S^\circ + A_S < 0 \quad \text{if } \rho > 1, \tag{5.84}$$

and

$$\Delta G_S^\circ + A_S > 0 \quad \text{if } \rho < 1. \tag{5.85}$$

In eq. (5.79) thermodynamics demands that

$$K_2 K_2' K_1' = K_1 K_1'' K_2''. \tag{5.86}$$

Moreover, the ratio [L′]/[L″] is still given by expression (5.72). Therefore, one has for eq. (5.58)

$$\frac{k_{-1} f_1'' k_3 f_3'}{k_1 f_1' k_{-3} f_3''} = \exp\left[\frac{\Delta G_S^\circ + A_S - (\Delta\tilde{\mu}_L - zF\Delta\Psi)}{RT}\right]. \tag{5.87}$$

Equation (5.58) shows that the ligand will be transported from the cis to the trans compartment if

$$\frac{k_{-1} f_1'' k_3 f_3'}{k_1 f_1' k_{-3} f_3''} > 1, \tag{5.88}$$

that is if

$$\left(\Delta G_S^\circ + A_S\right) - \left(\Delta\tilde{\mu}_L - zF\Delta\Psi\right) > 0. \tag{5.89}$$

Of particular interest is the situation where the transport is an active process which takes place against an electrochemical gradient. Then

$$zF\Delta\Psi - \Delta\tilde{\mu}_L < 0. \tag{5.90}$$

Active transport requires that

$$\Delta G_S^\circ + A_S > 0 \tag{5.91}$$

in such a way that condition (5.89) be fulfilled.

In the case of an endergonic chemical reaction driven by ligand transport (Fig. 5.15(C)), eq. (5.58) still applies but expression (5.79) has to be replaced by

$$\frac{k_{-1} f_1'' k_3 f_3'}{k_1 f_1' k_{-3} f_3''} = \frac{K_3 K_2' K_1'}{K_1 K_1'' K_2''} \frac{[\mathrm{L}']}{[\mathrm{L}'']} \frac{[\mathrm{P}][\mathrm{Q}]}{[\mathrm{S}]}, \tag{5.92}$$

which can be rewritten as

$$\frac{k_{-1} f_1'' k_3 f_3'}{k_1 f_1' k_{-3} f_3''} = \exp\left[\frac{-(\Delta G_S^\circ + A_S) - (\Delta\tilde{\mu}_L - zF\Delta\Psi)}{RT}\right]. \tag{5.93}$$

As previously, to obtain net transport of ligand from the cis to the trans compartment, thermodynamics demands that

$$-\left(\Delta G_S^\circ + A_S\right) - \left(\Delta\tilde{\mu}_L - zF\Delta\Psi\right) > 0. \tag{5.94}$$

The larger the exponent of eq. (5.93), the faster the flow is expected to be. Large values of this exponent will be obtained if $\Delta G_S^\circ + A_s < 0$, i.e. if $\rho > 1$, which implies that little ATP is being synthesized. Alternatively, smaller values of this exponent will be obained if $\Delta G_S^\circ + A_S > 0$, i.e. if $\rho > 1$, which indicates that larger quantities of ATP are synthesized. As expected, there is an antagonism between the rate of net flow and ATP synthesis. The energy, which has been stored in the membrane, can be used either to accelerate the net flow, or to help synthesize ATP. The membrane, however, may compromise between these important functions.

Taking advantage of eqs. (5.87) and (5.93), expression (5.58) can be rewritten as

$$J = J^{-}\left[\exp\left\{\frac{\Delta G_{\mathrm{S}}^{\circ} + A_{\mathrm{S}} - \Delta\tilde{\mu}_{\mathrm{L}} + zF\Delta\Psi}{RT}\right\} - 1\right] \tag{5.95}$$

for active transport of ligand (Fig. 5.15(B)), or as

$$J = J^{-}\left[\exp\left\{-\frac{\Delta G_{\mathrm{S}}^{\circ} + A_{\mathrm{S}} + \Delta\tilde{\mu}_{\mathrm{L}} - zF\Delta\Psi}{RT}\right\} - 1\right] \tag{5.96}$$

for ATP synthesis driven by a flow of ligand (Fig. 5.15(C)). In these equations, J^{-} is still the flow in the backward direction. Close to equilibrium, these expressions are equivalent to linear flow-force relationships.

The last situation discussed in Fig. 5.15(D) is partial coupling between the synthesis of S (ATP) and spontaneous ligand transfer from the cis to the trans side of a membrane. The relevant equation of the net flow (5.63) can be rewritten as

$$\begin{aligned} J &= k_1 f_1'\left(k_{-2} f_2'' + k_{-3} f_3''\right)\left\{\frac{k_{-1} f_1''(k_2 f_2' + k_3 f_3')}{k_1 f_1'(k_{-2} f_2'' + k_{-3} f_3''} - 1\right\} \\ &\quad \times \left[k_2 f_2' + k_{-2} f_2'' + k_{-1} f_1'' + k_1 f_1' + k_{-3} f_3'' + k_3 f_3'\right]^{-1}. \end{aligned} \tag{5.97}$$

Within the contracted kinetic scheme (Fig. 5.15(B)), thermodynamics requires that

$$\frac{k_2 f_2'}{k_{-2} f_2''} = \frac{k_3 f_3'}{k_{-3} f_3''}. \tag{5.98}$$

Therefore, one has

$$\frac{k_{-1} f_1''(k_2 f_2' + k_3 f_3')}{k_1 f_1'(k_{-2} f_2'' + k_{-3} f_3'')} = \frac{k_{-1} f_1'' k_2 f_2'}{k_1 f_1' k_{-2} f_2''} = \frac{k_{-1} f_1'' k_3 f_3'}{k_1 f_1' k_{-3} f_3''}, \tag{5.99}$$

and

$$\frac{k_{-1} f_1'' k_2 f_2'}{k_1 f_1' k_{-2} f_2''} = \frac{K_1' K_2'}{K_1 K_1''}\frac{[\mathrm{L}']}{[\mathrm{L}'']} = \frac{[\mathrm{L}']}{[\mathrm{L}'']}, \tag{5.100}$$

$$\frac{k_{-1} f_1'' k_3 f_3'}{k_1 f_1' k_{-3} f_3''} = \frac{K_1' K_2 K_2'}{K_1 K_1'' K_2''}\frac{[\mathrm{L}']}{[\mathrm{L}'']}\frac{[\mathrm{P}][\mathrm{Q}]}{[\mathrm{S}]} = \frac{[\mathrm{L}']}{[\mathrm{L}'']}\frac{[\mathrm{P}][\mathrm{Q}]}{[\mathrm{S}]}. \tag{5.101}$$

Comparison of eqs. (5.99), (5.100) and (5.101) shows that

$$\frac{[\mathrm{P}][\mathrm{Q}]}{[\mathrm{S}]} = \rho = 1, \tag{5.102}$$

and this result implies, in turn, that

$$A_{\mathrm{S}} = -\Delta G_{\mathrm{S}}^{\circ}. \tag{5.103}$$

In other words, in the model presented in Fig. 5.15(D), the partial coupling between ATP synthesis and ligand transport demands that the affinity of ATP hydrolysis be equal in size, but opposite in sign, to its standard hydrolysis free energy. This type of coupling will not affect the rate of ligand transport but will define a ratio between ADP and ATP concentrations which is different from that imposed by thermodynamic equilibrium. As the ratio $\rho = 1$, this partial coupling between scalar and vectorial processes results in a significant increase of ATP relative to the concentration one would have expected if the system were in equilibrium.

We have already outlined that, in mitochondria and chloroplasts, the ratio $\rho = [\text{ADP}][\text{P}_i]/[\text{ATP}]$ is smaller than the equilibrium constant of ATP hydrolysis. In the light of the above thermodynamic considerations, one may wonder whether there is a functional advantage in this situation. In order to couple ATP hydrolysis to active transport against an electrochemical gradient, we have seen that

$$\begin{aligned} &(\Delta G_{\text{S}}^{\circ} + A_{\text{S}}) - (\Delta\tilde{\mu}_{\text{L}} - zF\Delta\Psi) > 0, \\ &zF\Delta\Psi - \Delta\tilde{\mu}_{\text{L}} < 0, \qquad \Delta G_{\text{S}}^{\circ} < 0. \end{aligned} \tag{5.104}$$

Hence, in order to have the first of these expressions fulfilled, the affinity A_{S} must, of necessity, be positive and large. This requires, in turn, that ρ be small and therefore that the concentration of ATP be relatively large with respect to that of ADP and phosphate. If ρ were equal to the equilibrium constant of ATP hydrolysis, the affinity of this reaction would have been equal to zero and active transport would have been impossible.

Alternatively, if we consider the transport of an ion "downhill" its electrochemical gradient, and coupled to the phosphorylation of ADP, this requires that

$$\begin{aligned} &-(\Delta G_{\text{S}}^{\circ} + A_{\text{S}}) - (\Delta\tilde{\mu}_{\text{L}} - zF\Delta\Psi) > 0, \\ &zF\Delta\Psi - \Delta\tilde{\mu}_{\text{L}} > 0, \qquad \Delta G_{\text{S}}^{\circ} > 0. \end{aligned} \tag{5.105}$$

Hence the first condition above will be more easily fulfilled if the affinity of phosphorylation of ADP is negative. This will be the case if the reaction is not in equilibrium and if there is excess ATP over its equilibrium concentration. There is, therefore, an obvious functional advantage for a mitochondrion, or a chloroplast, to have ATP, ADP and phosphate in a concentration ratio $\rho = [\text{ADP}][\text{P}_i]/[\text{ATP}]$ smaller than the standard equilibrium constant of ATP hydrolysis.

The results and the ideas presented in this chapter show how compartmentalization of the living cell makes thermodynamically feasible and efficient the synthesis of ATP coupled to ion transport, "downhill" an electrochemical gradient. It was long assumed that ATP synthesis in the cell required the existence of a high energy chemical species. In spite of many efforts, this chemical species has never been isolated, for it probably does not exist. The discovery by Mitchell [31] that the free energy released from electron transfer processes can be stored as an electrochemical gradient which is, in turn, used for ATP synthesis, probably represents a paradigmatic change in the history of ideas of modern biology. As a matter of fact, it clearly shows that a major biological process, such as free energy storage in mitochondria and chloroplasts, cannot be understood from the sole properties of isolated macromolecules but requires understanding of the thermodynamics of coupled scalar chemical and vectorial transport processes.

References

[1] Westerhoff, H.V. and Van Dam, K. (1987) Thermodynamics and Control of Biological Free Energy Transduction. Elsevier, Amsterdam.

[2] Schechter, E. (1990) Biochimie et Biophysique des Membranes. Masson, Paris.

[3] Hill, T.L. (1968) Thermodynamics for Chemists and Biologists. Addison-Wesley, Reading, MA.

[4] Heinz, E. (1978) Mechanics and Energetics of Biological Transport. Springer-Verlag, Berlin.

[5] Heinz, E. (1981) Electrical Potentials in Biological Membrane Transport. Springer-Verlag, Berlin.

[6] Martonosi, A.N. (1985) The Enzymes of Biological Membrane. Vol. 3, Membrane Transport, 2nd edition. Plenum Press, New York.

[7] Stein, W.D. (1990) Channels, Carriers and Pumps: An Introduction to Membrane Transport. Academic Press, San Diego, CA.

[8] Finkelstein, A. (1984) Water movement through membrane channels. Curr. Top. Membr. Transport 21, 295–308.

[9] Walter, A. and Gutknecht, J. (1986) Permeability of small nonelectrolytes through lipid bilayer membranes. J. Membr. Biol. 90, 207–217.

[10] Tanford, C. (1989) Mechanism of free energy coupling in active transport. Annu. Rev. Biochem. 52, 379–409.

[11] Kaback, H.R. (1989) Molecular biology of active transport: from membrane to molecule, to mechanism. Harvey Lect. 83, 77–105.

[12] Stein, W.D. (1986) Transport and Diffusion Across Cell Membranes. Academic Press, Orlando, FL.

[13] Horisberger, J.D., Memas V., Kraehenbühl, J.P. and Rossier, B.C. (1991) Structure–function relationships of Na–K ATPase. Annu. Rev. Physiol. 53, 565–584.

[14] Mercer, R.W. (1993) Structure of the Na–K ATPase. Int. Rev. Cytol. 137 C, 139–168.

[15] Carafoli, E. (1991) Calcium pump of the plasma membrane. Physiol. Rev. 71, 129–153.

[16] Jencks, W.P. (1989) How does a calcium pump pump calcium? J. Biol. Chem. 264, 18855–18858.

[17] Alberts, B., Bray, D., Lewis, J., Raff, M., Roberts, K. and Watson, J. (1994) The Molecular Biology of the Cell, 3rd edition. Garland, New York and London.

[18] Jessel, T.M. and Kandel, E.R. (1993) Synaptic transmission: a bi-directional and self-modifiable form of cell-cell communication. Cell 72, 1–30.

[19] Kandel, E.R., Schartz, J.H. and Jessel, T.M. (1991) Principles of Neural Science, 3rd edition. Elsevier, Amsterdam.

[20] Unwin, N. (1989) The structure of ion channels in membranes of excitable cells. Neuron 3, 665–676.

[21] Baker, P.F., Hodgkin, A.L. and Shaw, T.L. (1962) The effects of changes in internal ionic concentration on the electrical properties of perfused giant axons. J. Physiol. 164, 355–374.

[22] Hodgkin, A.L. and Keynes, R.D. (1955) Active transport of cations in giant axons from *Sepia* and *Loligo*. J. Physiol. 128, 26–60.

[23] Hodgkin, A.L. and Huxley, A.F. (1952) A quantitative description of membrane current and its application to conduction and excitation in nerve. J. Physiol. 117, 500–544.

[24] Katz, B. (1966) Nerve, Muscle and Synapse. McGraw-Hill, New York.

[25] Neher, E. and Sakmann, B. (1992) The patch clamp technique. Sci. Amer., March, 503–512.

[26] Ernster, L. and Schatz, G. (1981) Mitochondria: a historical review. J. Cell Biol. 91, 227–255.

[27] Bereiter-Hahn, J. (1990) Behavior of mitochondria in the living cell. Int. Rev. Cytol. 122, 1–63.

[28] Nicholls, D.G. (1982) Bioenergetics: An Introduction to the Chemiosmotic Theory. Academic Press, New York.

[29] De Pierre, J.W. and Ernster, L. (1977) Enzyme topology of intracellular membranes. Annu. Rev. Biochem. 46, 201–262.

[30] Srere, P.A. (1982) The structure of the mitochondrial inner membrane-matrix compartment. Trends Biochem. Sci. 7, 375–378.
[31] Mitchell, P. (1961) Coupling of phosphorylation to electron and hydrogen transfer by a chemi-osmotic type of mechanism. Nature 191, 144–148.
[32] Racker, E. (1980) From Pasteur to Mitchell: a hundred years of bioenergetics. Fed. Proc. 39, 210–215.
[33] Hatefi, Y. (1985) The mitochondrial electron transport and oxidative phosphorylation system. Annu. Rev. Biochem. 54, 1015–1070.
[34] Durand, R., Briand, Y., Touraille, S. and Alziari, S. (1981) Molecular approaches to phosphate transport in mitochondria. Trends Biochem. Sci. 6, 211–214.
[35] Hinckle, P.C. and McCarty, R.E. (1978) How cells make ATP. Sci. Amer., March 104–123.
[36] Klingenberg, M. (1979) The ADP, ATP shutle of the mitochondrion. Trends Biochem. Sci. 4, 249–252.
[37] Bogorad, L. (1981) Chloroplasts. J. Cell Biol. 91, 256–270.
[38] Clayton, R.K. (1980) Photosynthesis: Physical Mechanisms and Chemical Patterns. Cambridge University Press, Cambridge.
[39] Hoober, J.K. (1984) Chloroplasts. Plenum Press, New York.
[40] Barber, J. (1987) Photosynthetic reaction centres: a common link. Trends Biochem. Sci. 12, 321–326.
[41] Zuber, H. (1986) Structure of the light-harvesting antenna complexes of photosynthetic bacteria, cyanobacteria and red algae. Trends Biochem. Sci. 11, 414–419.
[42] Blankenship, R.E. and Prince, R.C. (1985) Excited state redox potentials and the Z scheme of photosynthesis. Trends Biochem. Sci. 10, 382–383.
[43] Jagendorf, A.T. (1967) Acid–base transitions and phosphorylation by chloroplasts. Fed. Proc. 26, 1361–1369.
[44] Abrahams, J.P., Leslie, A.G.W., Lutter, R. and Walker, J.E. (1994) Structure at 2.8 Å resolution of F_1-ATPase from bovine heart mitochondria. Nature 370, 621–628.
[45] Grubmeyer, C., Cross, R.L. and Penefsky, H.S. (1982) Mechanism of ATP hydrolysis by beef heart mitochondrial ATPase. Rate constants for elementary steps in catalysis at a single site. J. Biol. Chem. 257, 12092–12100.
[46] Senior, A.E. (1988) ATP synthesis by oxidative phospshorylation. Physiol. Rev. 68, 177–231.
[47] Boyer, P.D. (1997) The ATP synthase. A splendid molecular machine. Annu. Rev. Biochem. 66, 717–749.
[48] Noji, H., Yasuda, R., Yoshida, M. and Kinosita, K. (1977) Direct observation of the rotation of F_1-ATPase. Nature 386, 299–302.
[49] Cha, S. (1968) A simple method for derivation of rate equations under the rapid equilibrium assumption or combined assumption of equilibrium and steady state. J. Biol. Chem. 243, 820–825.
[50] Hill, T.L. (1977) Free Energy Transduction in Biology. Academic Press, New York.

CHAPTER 6

Molecular crowding, transfer of information and channelling of molecules within supramolecular edifices

Protein molecules in the living cell, and in particular enzymes, have often been viewed, explicitly or implicitly, as acting in free, dilute, solutions. Even if this assumption is useful, there is no doubt that it is an oversimplification, at least for eukaryotic cells. As a matter of fact, molecular crowding in these cells has two distinct, but related effects, both of which can alter enzyme behaviour. First, molecular crowding generates significant differences between concentrations and activities, and this will change the kinetic properties of enzymes. Second, and more importantly, molecular crowding favours the formation of supramolecular complexes within the cell. In eukaryotic cells, as already outlined, many enzymes are associated with other enzymes in multienzyme complexes, with membranes, with the cell wall, etc., and one may thus wonder whether these associations play a role in the expression of the enzyme function *in situ*. Classical molecular biology has made familiar, to all biologists, the view that information is stored in a structural gene and can be translated into a polypeptide chain that, under definite conditions of pH, ionic strength, etc., folds into a globular functional enzyme [1]. Although a gene codes for an aminoacid sequence and nothing else, this sequence directs the proper folding of the polypeptide chain. One may thus conclude that, under these experimental conditions, the functional properties are, in a way, stored in the corresponding structural gene. This reasoning is indeed staightforward if the enzyme is not associated with other proteins or with cell organelles. But one may wonder whether the functional properties of an enzyme can still be considered as fully defined by the structure of the corresponding structural gene, if this enzyme is associated with another cell component. The enzyme's conformation might be completely changed through this association, thus leading to a dramatic alteration of its function. This altered biological function could not be considered fully coded in the corresponding structural gene. It would also be the expression of a multimolecular association, which is itself the result of the action of several structural genes.

It is of interest to stress that this problem is distinct from that of a simple modulation of enzyme activity by different agents (pH, ligands, etc.) or that of postranslational effects such as glycosylation, phosphorylation, etc. The issue at stake is to know whether the function of a protein is, in part, a consequence of the supramolecular organization of the living cell. In that case, one would expect an instruction, or imprinting, to be transferred fom protein to protein, or from a cell organelle to a protein.

Many enzyme complexes catalyse consecutive reactions of the same metabolic sequence. One can therefore speculate that a selective pressure has been exerted to allow channelling of reaction intermediates from active site to active site within the same multienzyme complex, without diffusing in the cell milieu. If this process is taking place *in vivo* one would expect the efficiency of the metabolic pathway to increase relative to that of a sequence of physically independent enzymes. This chapter is devoted to the discussion of precisely these problems.

6.1. Molecular crowding

The chemical potential of a reagent usually increases when another neutral compound is added to the reaction medium. This is due to an exclusion of the reagent from part of the solvent [2–7]. Therefore an experimentally determined value of an equilibrium constant, or of a rate constant, apparently varies as the concentration of the neutral compound increases. Thus, for instance, if equilibrium is reached amongst different reagents

$$\mu_1 A_1 + \mu_2 A_2 + \cdots \leftrightarrow \nu_1 B_1 + \nu_2 B_2 + \cdots \tag{6.1}$$

the equilibrium constant will be

$$K = \frac{\gamma_{B_1}^{\nu_1} \gamma_{B_2}^{\nu_2} \cdots}{\gamma_{A_1}^{\mu_1} \gamma_{A_2}^{\mu_2} \cdots}. \tag{6.2}$$

The factor $(\gamma_{B_1}^{\nu_1} \gamma_{B_2}^{\nu_2})/(\gamma_{A_1}^{\nu_1} \gamma_{A_2}^{\nu_2})$ is close to unity under the usual conditions of dilution but, in a concentrated medium, this factor may be significantly different from unity. Thus, owing to the exclusion effect that takes place under conditions of molecular crowding, an enzyme should not behave the same way in dilute and in concentrated conditions. Let us consider, as an example, the ideally simple one-substrate reaction

$$\mathrm{E} + \mathrm{S} \longleftrightarrow \mathrm{ES} \longrightarrow \mathrm{E} + \mathrm{P}. \tag{6.3}$$

Assuming that the K_m is nearly identical to the dissociation constant of the ES complex, one has

$$K_m = \frac{[\mathrm{E}][\mathrm{S}]}{[\mathrm{ES}]} \frac{\gamma_E \gamma_S}{\gamma_{ES}}, \tag{6.4}$$

and the corresponding rate equation assumes the form

$$v = \frac{V_m}{1 + \dfrac{K_m}{[\mathrm{S}]} \dfrac{\gamma_{ES}}{\gamma_E \gamma_S}}, \tag{6.5}$$

where V_m is indeed the maximum reaction rate. Hence one has, in reciprocal form

$$\frac{1}{v} = \frac{K_m}{V_m} \frac{\gamma_{ES}}{\gamma_E \gamma_S} \frac{1}{[\mathrm{S}]} + \frac{1}{V_m}, \tag{6.6}$$

and the slope of the reciprocal plot is altered by the ratio $\gamma_{ES}/\gamma_E \gamma_S$ whereas the maximum rate remains unchanged. Thus, molecular crowding tends to alter the apparent K_m of reaction [4]. In addition to these simple effects, molecular crowding tends to favour the formation of supramolecular edifices, in particular the formation of multienzyme complexes.

6.2. Statistical mechanics of ligand binding to supramolecular edifices

Let us consider a multisubunit protein bearing n identical binding sites for a given ligand. Moreover, let us assume that this protein is embedded in a supramolecular edifice composed of different proteins unable to bind the ligand. The multiple equilibria can be expressed as shown in Fig. 6.1.

The corresponding isotherm, $\bar{\nu}$ or $\overline{Y}$, of the ligand L on protein P within the supramolecular edifice is defined as

$$\bar{\nu} = n\overline{Y} = \frac{[P_1] + 2[P_2] + \cdots + n[P_n]}{[P_0] + [P_1] + \cdots + [P_n]}, \tag{6.7}$$

where $[P_0]$, $[P_1]$, ... represent the protein concentration that has bound no ligand, one ligand molecule and so forth. Setting for simplicity

$$\Psi_1 = K_1, \qquad \Psi_2 = K_1 K_2, \qquad \Psi_n = K_1 K_2 \ldots K_n, \tag{6.8}$$

where K_i are the macroscopic ligand binding constants, eq. (6.7) assumes the form

$$\bar{\nu} = n\overline{Y} = \frac{\sum_{i=1}^{n} i\Psi_i [L]^i}{1 + \sum_{i=1}^{n} \Psi_i [L]^i}, \tag{6.9}$$

which is the classical Adair equation [8,9] for a multienzyme complex. The macroscopic binding constants [9–12] are defined as

$$K_i = \frac{[P_{i,T}]}{[P_{i-1,T}][L]}, \tag{6.10}$$

where $[P_{i,T}]$ and $[P_{i-1,T}]$ are the *total* concentrations of protein molecules that have bound i and $i-1$ ligand molecules to *any combination* of sites. If the sites are all equivalent, i.e. if there is an equal probability of binding the ligand to all the sites, one can define a microscopic binding constant, K_i', equal to

$$K_i' = \frac{[P_i]}{[P_{i-1}][L]}, \tag{6.11}$$

where $[P_i]$ and $[P_{i-1}]$ represent now the concentrations of protein molecules that have bound i and $i-1$ ligand molecules on i and $i-1$ *specific* sites. One thus has

$$[P_{i,T}] = \binom{n}{i}[P_i] = \frac{n!}{i!(n-i)!}[P_i], \tag{6.12}$$

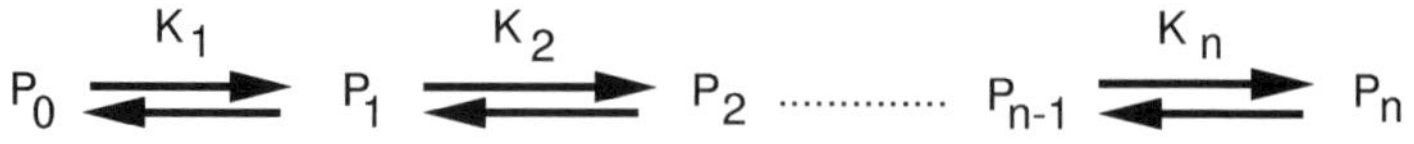

Fig. 6.1. Multiple equilibria for a multienzyme complex.

and the relationship between the macroscopic and the microscopic binding constants is

$$K_i = \left\{ \binom{n}{i} \Big/ \binom{n}{i-1} \right\} K_i' = \frac{n-i+1}{i} K_i'. \tag{6.13}$$

Hence the product Ψ_i of the i macroscopic binding constants takes the form

$$\Psi_i = K_1 K_2 \ldots K_i = \binom{n}{i} K_1' K_2' \ldots K_i', \tag{6.14}$$

and the equation of the binding isotherm can be rewritten as

$$\bar{\nu} = \frac{\sum_{i=1}^{n} i \binom{n}{i} K_1' K_2' \ldots K_i' [\mathrm{L}]^i}{1 + \sum_{i=1}^{n} \binom{n}{i} K_1' K_2' \ldots K_i' [\mathrm{L}]^i}. \tag{6.15}$$

In this form, it becomes obvious that if all the microscopic binding constants are identical and equal to K, the binding isotherm reduces to

$$\bar{\nu} = \frac{nK[\mathrm{L}]}{1 + K[\mathrm{L}]}. \tag{6.16}$$

If the binding constants are different, this indicates the existence of co-operativity between identical sites of the protein.

It is important, at this stage, to understand, on a thermodynamic basis, what the origin is of the co-operativity of a protein within a supramolecular edifice. This can be done on the basis of reasoning which relies on a principle that may be referred to as the "structural kinetic principle", and which can be used for both kinetic and binding data [14–25]. This principle was first applied to free oligomeric enzymes [14–25], and then to the thermodynamic study of oligomeric and monomeric enzymes embedded in supramolecular edifices [26–28]. If ΔG_i° is the standard free energy of ligand binding to the ith site of the protein within the supramolecular edifice and $\Delta G^{\circ *}$ the standard free energy on an ideally isolated site, the relationship between these two thermodynamic parameters is

$$\Delta G_i^\circ = \Delta G^{\circ *} + \sum \Delta G_\alpha^{\mathrm{int}} + \sum \Delta G_\sigma^{\mathrm{int}}. \tag{6.17}$$

$\sum \Delta G_\alpha^{\mathrm{int}}$ expresses how the spatial arrangement of identical and different polypeptide chains may affect this binding. Similarly, $\sum \Delta G_\sigma^{\mathrm{int}}$ shows how the mutual constraints between identical, or different, chains within the supramolecular edifice may alter the binding. It is obvious from eqs. (6.15) and (6.17) that a necessary, although not sufficient, condition for co-operativity in ligand binding is

$$\sum \Delta G_{\alpha}^{\text{int}} + \sum \Delta G_{\sigma}^{\text{int}} \neq 0. \tag{6.18}$$

Each of the two energy components, $\sum \Delta G_{\alpha}^{\text{int}}$ and $\sum \Delta G_{\sigma}^{\text{int}}$, can be split into two sub-components, namely

$$\begin{aligned} \sum \Delta G_{\alpha}^{\text{int}} &= \sum \Delta G_{\alpha\text{I}}^{\text{int}} + \sum \Delta G_{\alpha\text{H}}^{\text{int}}, \\ \sum \Delta G_{\sigma}^{\text{int}} &= \sum \Delta G_{\sigma\text{I}}^{\text{int}} + \sum \Delta G_{\sigma\text{H}}^{\text{int}}, \end{aligned} \tag{6.19}$$

where the subscripts 'I' and 'H' refer to isologous (I) and heterologous (H) interactions between identical (I) or different (H) polypeptide chains. Setting

$$\begin{aligned} \overline{U}^{\alpha\text{H}\sigma\text{H}} &= \sum \Delta G_{\alpha\text{H}}^{\text{int}} + \sum \Delta G_{\sigma\text{H}}^{\text{int}}, \\ \overline{U}^{\alpha\text{I}\sigma\text{I}} &= \sum \Delta G_{\alpha\text{I}}^{\text{int}} + \sum \Delta G_{\sigma\text{I}}^{\text{int}}, \end{aligned} \tag{6.20}$$

the expression for the microscopic constant K_i' assumes the form

$$K_i' = K^* \exp\left\{-\frac{\overline{U}^{\alpha\text{I}\sigma\text{I}} + \overline{U}^{\alpha\text{H}\sigma\text{H}}}{RT}\right\}. \tag{6.21}$$

This equation shows that the binding constant of a ligand on a polymeric enzyme embedded in a supramolecular edifice is directly related to the binding of the ligand on an ideally isolated site, to the interaction energy, $\overline{U}_i^{\alpha\text{I}\sigma\text{I}}$, between identical (isologous) chains and to the interaction energy, $\overline{U}_i^{\alpha\text{H}\sigma\text{H}}$, between different (heterologous) chains. This equation leaves open the possibility that $\overline{U}_i^{\alpha\text{I}\sigma\text{I}} = 0$ and $\overline{U}_i^{\alpha\text{H}\sigma\text{H}} \neq 0$. In that case the isolated polymeric enzyme displays no co-operativity but, when embedded in a multienzyme complex, it may possibly become co-operative. Thus co-operativity may be an intrinsic property of the polymeric enzyme ($\overline{U}_i^{\alpha\text{I}\sigma\text{I}} \neq 0$), or it may be a property imposed on the enzyme by the other proteins of the same multimolecular edifice. We shall come back to this important matter later.

It is now of interest to know whether the binding isotherm expressed by eq. (6.15) has a physical significance, i.e. whether it can be expressed in terms of statistical physics. If one considers the distribution of differently liganded supramolecular complexes $P_0, P_1, \ldots$ (Fig. 6.1) over all the allowed energy levels, the probability of drawing a complex that has bound i ligand molecules is

$$p_i = B\,\text{e}^{-E_i/k_\text{B}T}, \tag{6.22}$$

where B is a constant which is defined below. Similarly, the probability of drawing an enzyme complex that has bound no ligand is

$$p_0 = B\,\mathrm{e}^{-E_0/k_\mathrm{B}T}. \tag{6.23}$$

Therefore, the significance of constant B is

$$B = \frac{p_0}{\mathrm{e}^{-E_0/k_\mathrm{B}T}}. \tag{6.24}$$

As already seen in chapter 2, $E_0, \ldots, E_i, \ldots$ are the energy levels of differently liganded complexes, k_B is the Boltzmann constant and T is the absolute temperature. Thus, the Boltzmann equation (6.23) can be rewritten as

$$N_i = N_0\,\mathrm{e}^{-(E_i-E_0)/k_\mathrm{B}T} = N_0\,\mathrm{e}^{-N(E_i-E_0)/RT}, \tag{6.25}$$

where N_i and N_0 are the numbers of enzymes complexes that ocur in states i and 0, and N is the Avogadro number ($N = R/k_\mathrm{B}$). From eq. (6.25)

$$N(E_i - E_0) = -RT\ln\left(\frac{N_i}{N_0}\right). \tag{6.26}$$

The ratio N_i/N_0 stems from the multiple equilibria associated with the binding of i ligand molecules to P_0. Thus one has

$$\mathrm{e}^{-N(E_i-E_0)/RT} = K_1K_2\ldots K_i[\mathrm{L}]^i = \binom{n}{i}K_1'K_2'\ldots K_i'[\mathrm{L}]^i, \tag{6.27}$$

and the denominator, Π, of the binding isotherm (6.15) is a partition function, namely

$$\Pi = 1 + \sum_{i=1}^{n}\binom{n}{i}K_1'K_2'\ldots K_i'[\mathrm{L}]^i = 1 + \sum_{i=1}^{n}\mathrm{e}^{-N(E_i-E_0)/RT}. \tag{6.28}$$

Moreover,

$$\mathrm{d}\ln\Pi = \mathrm{d}\ln\left\{1+\sum\Psi_i[\mathrm{L}]^i\right\} = \frac{\mathrm{d}\left\{1+\sum\Psi_i[\mathrm{L}]^i\right\}}{1+\sum\Psi_i[\mathrm{L}]^i}, \tag{6.29}$$

and

$$\mathrm{d}\sum\Psi_i[\mathrm{L}]^i = \sum i\Psi_i[\mathrm{L}]^{i-1}\,\mathrm{d}[\mathrm{L}]. \tag{6.30}$$

Therefore,

$$\mathrm{d}\ln\Pi = \frac{\sum i\Psi_i[\mathrm{L}]^i}{1+\sum\Psi_i[\mathrm{L}]^i}\,\mathrm{d}\ln[\mathrm{L}], \tag{6.31}$$

which is equivalent to

$$\bar{\nu} = n\overline{Y} = \frac{\partial \ln \Pi}{\partial \ln[\mathrm{L}]}. \tag{6.32}$$

This equation, derived by Wyman [10,13], shows that the binding isotherm of a ligand to a protein can be fully defined by the corresponding partition function.

It is now of interest to come back to the idea that co-operativity may be imposed upon a polymeric enzyme by its association with other proteins and to understand the physical basis of this phenomenon. As the aim of this section is not to review the co-operativity of oligomeric enzymes, but rather to understand how different proteins, physically associated with a non co-operative oligomeric enzyme, may generate such co-operativity we shall assume that

$$\overline{U}_0^{\alpha_\mathrm{I}\sigma_\mathrm{I}} = \overline{U}_i^{\alpha_\mathrm{I}\sigma_\mathrm{I}} = 0 \quad (i = 1, \ldots, n), \tag{6.33}$$

which means that the isolated oligomeric enzyme is not co-operative. When this enzyme is embedded in a multi-protein complex and binds a ligand, the corresponding partition function takes the form

$$\Pi = 1 + \sum_{i=1}^{n} \binom{n}{i} K^{*i}[\mathrm{L}]^i \exp\left\{-\frac{\overline{U}_1^{\alpha_\mathrm{H}\sigma_\mathrm{H}} + \overline{U}_2^{\alpha_\mathrm{H}\sigma_\mathrm{H}} + \cdots + \overline{U}_i^{\alpha_\mathrm{H}\sigma_\mathrm{H}}}{RT}\right\} \tag{6.34}$$

and the binding isotherm will display co-operativity. This co-operativity is the consequence of protein–protein interactions within the supramolecular edifice. However, if

$$\overline{U}_j^{\sigma_\mathrm{H}\sigma_\mathrm{H}} = \lambda \quad (j = 1, \ldots, i), \tag{6.35}$$

where λ is a constant, then one has

$$\overline{U}_1^{\alpha_\mathrm{H}\sigma_\mathrm{H}} + \overline{U}_2^{\alpha_\mathrm{H}\sigma_\mathrm{H}} + \cdots + \overline{U}_i^{\alpha_\mathrm{H}\sigma_\mathrm{H}} = i\lambda. \tag{6.36}$$

Put in more mathematical terms this means that $\overline{U}^{\alpha_\mathrm{H}\sigma_\mathrm{H}}$ does not vary along the reaction co-ordinate, that is

$$\frac{\partial \overline{U}^{\alpha_\mathrm{H}\sigma_\mathrm{H}}}{\partial \xi} = 0, \tag{6.37}$$

where ξ is the advancement of the reaction. Under these conditions, the partition function (6.34) reduces to

$$\Pi = 1 + \sum_{i=1}^{n} \binom{n}{i} \mathrm{e}^{-i\lambda/RT} K^{*i}[\mathrm{L}]^i = \left(1 + \mathrm{e}^{-\lambda/RT} K^{*}[\mathrm{L}]\right)^n, \tag{6.38}$$

and the resulting binding isotherm is a hyperbola, namely

$$\bar{\nu} = \frac{n\,\mathrm{e}^{-\lambda/RT} K^*[\mathrm{L}]}{1 + \mathrm{e}^{-\lambda/RT} K^*[\mathrm{L}]}. \tag{6.39}$$

When expression (6.36) or (6.37) applies, there is no co-operativity, but the ligand binding constant has been altered by a factor $\mathrm{e}^{-\lambda/RT}$. Thus, the association of an oligomeric enzyme with other proteins may alter its ligand binding properties and may even induce the emergence of ligand binding co-operativity.

6.3. Statistical mechanics and catalysis within supramolecular edifices

Let us consider an oligomeric enzyme embedded in a supramolecular edifice and catalysing a chemical reaction, the relevant kinetic scheme is shown in Fig. 6.2. The apparent microscopic affinity constants of the substrate for the enzyme are

$$\overline{K}_i = \frac{k_i}{k_{-i} + k'_i} \quad (i = 1, \ldots, n), \tag{6.40}$$

where k'_i are the catalytic constants. The corresponding steady state rate equation takes the form

$$\frac{v}{[\mathrm{E}]_0} = \frac{\sum_{i=1}^{n} i\binom{n}{i} k'_i \overline{K}_1 \overline{K}_2 \ldots \overline{K}_i [\mathrm{S}]^i}{1 + \sum_{i=1}^{n} \binom{n}{i} \overline{K}_1 \overline{K}_2 \ldots \overline{K}_i [\mathrm{S}]^i}, \tag{6.41}$$

where $[\mathrm{E}]_0$ is, as usual, the total active enzyme concentration and v the steady state rate.

If we consider the free energies of activation associated with the rate constants of the same process carried out by an ideally isolated active site of the oligomeric enzyme, or by the oligomeric enzyme embedded in the multi-protein complex, these free energies of activation can be set up so as to constitute a thermodynamic box [26], as shown in Fig. 6.3.

If the interactions between identical and different polypeptide chains affect the free energy of activation of the process in the supramolecular edifice, these interactions will stabilize, or destabilize, the ground and the transition states (Fig. 6.3). As the first principle of thermodynamics applies within this "box", one has

$$\Delta G^{\neq} = \Delta G^{\neq *} + U_{\gamma}^{\alpha\sigma} - U_{\tau}^{\alpha\sigma}. \tag{6.42}$$

In this expression, $\Delta G^{\neq *}$ represents the intrinsic free energy of activation, i.e. what the free energy would be if the active sites were not in interaction. $\Delta G^{\neq}$ is the free energy of activation of the same process taking place in the supramolecular complex. $U_{\gamma}^{\alpha\sigma}$ and $U_{\tau}^{\alpha\sigma}$ are the stabilization–destabilization energies of the ground (γ) and of the transition (τ) states. As above (see preceding section), these stabilization–destabilization energies

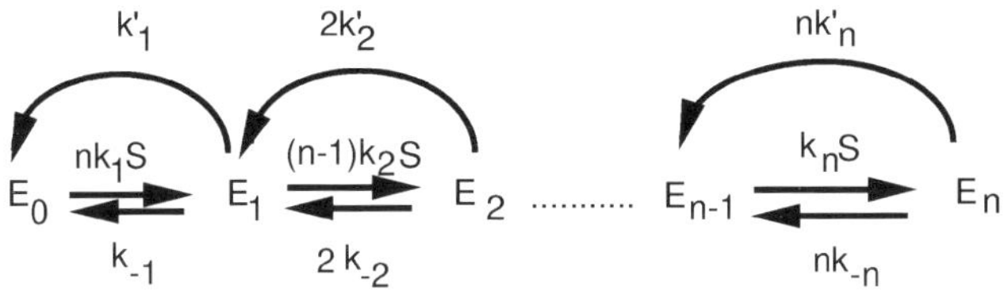

Fig. 6.2. Catalytic process involving a multienzyme complex.

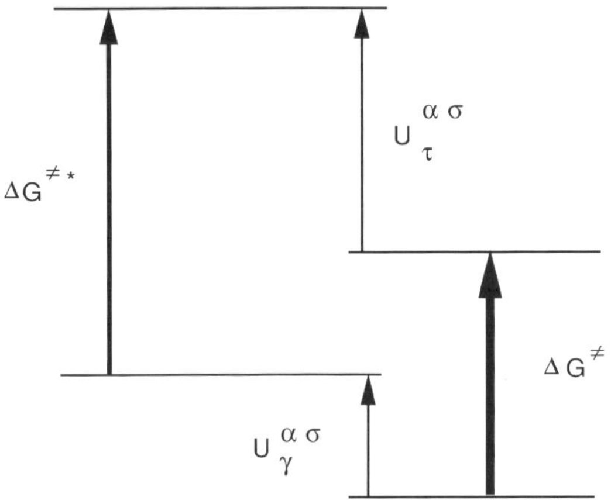

Fig. 6.3. Stabilization–destabilization of a chemical process by polypeptide chain interactions. In this scheme, association of different polypeptide chains stabilize the ground and the transition states of the reaction. However, as the stabilization energy is smaller for the ground than for the transition state, the chemical process will be faster if catalysed by the multienzyme complex rather than by the free enzyme.

are due to the arrangement (α) and to the conformational constraints (σ) between identical and different polypeptide chains, both in the ground and in the transition states [26]. Thus one has

$$U_{\gamma}^{\alpha\sigma} = U_{\gamma}^{\alpha_{\rm I}\sigma_{\rm I}} + U_{\gamma}^{\alpha_{\rm H}\sigma_{\rm H}} \tag{6.43}$$

and

$$U_{\tau}^{\alpha\sigma} = U_{\tau}^{\alpha_{\rm I}\sigma_{\rm I}} + U_{\tau}^{\alpha_{\rm H}\sigma_{\rm H}}. \tag{6.44}$$

These equations mean that, within the complex, both isologous (I) and heterologous (H) interactions contribute to the stabilization–destabilization energies of the ground and of the transition states. Moreover, one has

$$U_{\gamma}^{\alpha_{\rm I}\sigma_{\rm I}} = U_{\gamma}^{\alpha_{\rm I}} + U_{\gamma}^{\sigma_{\rm I}}, \qquad U_{\gamma}^{\alpha_{\rm H}\sigma_{\rm H}} = U_{\gamma}^{\alpha_{\rm H}} + U_{\gamma}^{\sigma_{\rm H}}, \tag{6.45}$$

and similar relationships apply to the transition states.

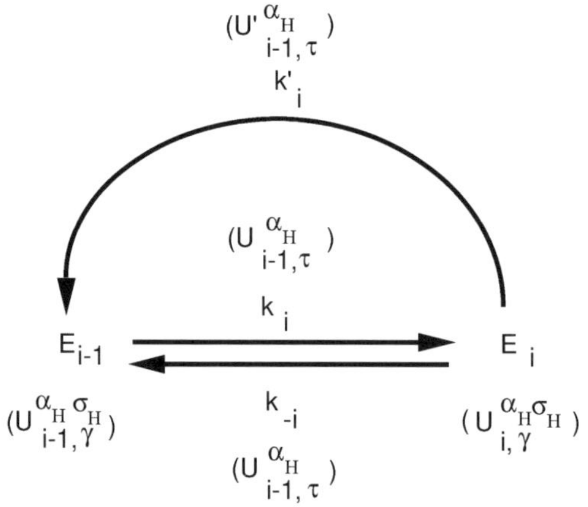

Fig. 6.4. Stabilization–destabilization energies of an enzyme step within a multienzyme complex. See text.

One can define an intrinsic rate constant k^* as [23,26]

$$k^* = \frac{k_B T}{h} \exp\left\{-\frac{\Delta G^{\neq *}}{RT}\right\} \tag{6.46}$$

where k_B and h are the Boltzmann and the Planck constants, T is the absolute temperature, and R is the gas constant. This expression defines what the rate constant would be if the association between identical and different polypeptide chains did not affect the rate process. The rate constant, k, associated with the free energy of activation $\Delta G^{\neq}$ may thus be expressed as

$$k = k^* \exp\left\{-\frac{U_{\gamma}^{\alpha_I \sigma_I} - U_{\tau}^{\alpha_I \sigma_I} + U_{\gamma}^{\alpha_H \sigma_H} - U_{\tau}^{\alpha_H \sigma_H}}{RT}\right\}. \tag{6.47}$$

If we assume that the isolated oligomeric enzyme does not display any co-operativity *per se*, the subunits behave as a collection of independent catalytic units, then

$$U_{\gamma}^{\alpha_I \sigma_I} = U_{\tau}^{\sigma_I \sigma_I} = 0, \tag{6.48}$$

and the expression of the rate constant reduces to

$$k = k^* \exp\left\{-\frac{U_{\gamma}^{\alpha_H \sigma_H} - U_{\tau}^{\alpha_H \sigma_H}}{RT}\right\}. \tag{6.49}$$

Let us consider a step in the reaction sequence of the model of Fig. 6.4. Equation (6.50) allows one to determine whether an oligomeric enzyme, intrinsically devoid of kinetic co-operativity but embedded in a supramolecular edifice, may acquire this kinetic co-operativity owing to the heterologous interactions it displays with different proteins.

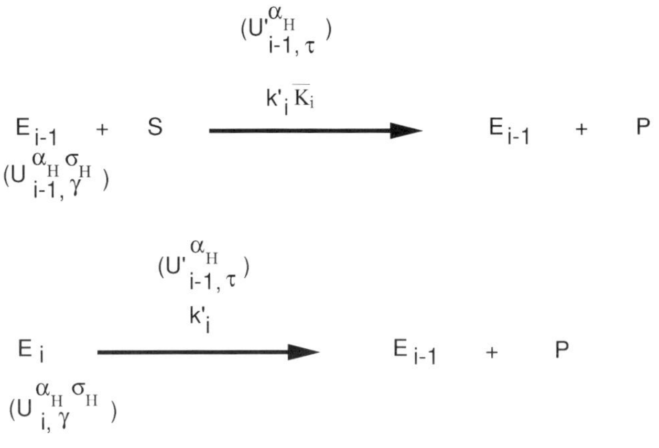

Fig. 6.5. Equivalence of the stabilization–destabilization energies for a substrate desorption step and for a catalytic step in a multienzyme complex. See text.

Taking account of eq. (6.51), one can write

$$k_i = k_s^* \exp\left\{-\frac{U_{i-1,\gamma}^{\alpha_H\sigma_H} - U_{i-1,\tau}^{\alpha_H\sigma_H}}{RT}\right\}, \qquad k_{-i} = k_{-s}^* \exp\left\{-\frac{U_{i,\gamma}^{\alpha_H\sigma_H} - U_{i-1,\tau}^{\alpha_H\sigma_H}}{RT}\right\},$$
$$k_i' = k^* \exp\left\{-\frac{U_{i,\gamma}^{\alpha_H\sigma_H} - U_{i-1,\tau}^{\prime\alpha_H\sigma_H}}{RT}\right\}. \tag{6.50}$$

In these expressions, k_s^*, k_{-s}^* and k^* are the relevant intrinsic rate constants. $U_{i-1,\tau}^{\alpha_H\sigma_H}$ and $U_{i-1,\tau}^{\prime\alpha_H\sigma_H}$ represent the stabilization–destabilization energies of the transition states for substrate release and catalysis. Simple inspection of these equations shows that the relationship between the $\overline{U}^{\alpha_H\sigma_H}$ of the previous section and the $U_\gamma^{\alpha_H\sigma_H}$ is

$$\overline{U}_i^{\alpha_H\sigma_H} = U_{i-1,\gamma}^{\alpha_H\sigma_H} - U_{i,\gamma}^{\alpha_H\sigma_H}. \tag{6.51}$$

To find out a relationship between the apparent substrate binding constants $k_i/(k_{-i} + k_i')$ and $k_s^*/(k_{-s}^* + k^*)$, one has to discover a relationship between the stabilization–destabilization energies $U_{i-1,\tau}^{\alpha_H\sigma_H}$ and $U_{i-1,\tau}^{\prime\alpha_H\sigma_H}$. For this purpose, let us consider the second order and first order processes shown in Fig. 6.5. One will notice that they have the same stabilization–destabilization energies for their transition states. Hence one has

$$k_i'\overline{K}_i = k^*\overline{K}^* \exp\left\{-\frac{U_{i-1,\gamma}^{\alpha_H\sigma_H} - U_{i-1,\tau}^{\prime\alpha_H\sigma_H}}{RT}\right\} \tag{6.52}$$

and

$$k_i' = k^* \exp\left\{-\frac{U_{i,\gamma}^{\alpha\text{H}\sigma\text{H}} - U_{i-1,\tau}^{\prime\alpha\text{H}\sigma\text{H}}}{RT}\right\}. \tag{6.53}$$

From these equations it follows that

$$\overline{K}_i = \overline{K}^* \exp\left\{-\frac{U_{i-1,\gamma}^{\alpha\text{H}\sigma\text{H}} - U_{i,\gamma}^{\alpha\text{H}\sigma\text{H}}}{RT}\right\}. \tag{6.54}$$

Comparison of eqs. (6.50) and (6.54) leads to

$$U_{i-1,\tau}^{\alpha\text{H}\sigma\text{H}} = U_{i-1,\tau}^{\prime\alpha\text{H}\sigma\text{H}}. \tag{6.55}$$

This relationship has two important implications and demonstrates the validity of a principle that traces back to Pauling [29] and has been used in enzymology ever since (see chapter 2). Expression (6.55) means that the stabilization–destabilization energy exerted by heterologous interactions is the same whether the enzyme in the supramolecular edifice has bound a transition state, or another one. This can be achieved, only if one assumes that the conformational constraints, which exist between the enzyme and the other proteins of the complex, are transiently relieved at the top of the free energy barrier. Hence the interactions between the enzyme and the other proteins of the supramolecular edifice should be the same as those existing in the ground state of a supramolecular complex, lacking heterologous conformational constraints. This implies that eq. (6.55) is equivalent to

$$U_{i-1,\tau}^{\alpha\text{H}\sigma\text{H}} = U_{i-1,\tau}^{\prime\alpha\text{H}\sigma\text{H}} = U_{i-1,\gamma}^{\alpha\text{H}}. \tag{6.56}$$

It is now possible to express how heterologous interactions between different proteins may control the chemical reaction. One has

$$\overline{K}_1\overline{K}_2\ldots\overline{K}_i = \overline{K}^{*i} \exp\left\{-\frac{U_{0,\gamma}^{\alpha\text{H}\sigma\text{H}} - U_{i,\gamma}^{\alpha\text{H}\sigma\text{H}}}{RT}\right\} \tag{6.57}$$

and

$$k_i'\overline{K}_1\overline{K}_2\ldots\overline{K}_i = k^*\overline{K}^{*i} \exp\left\{-\frac{U_{0,\gamma}^{\alpha\text{H}\sigma\text{H}} - U_{i-1,\gamma}^{\alpha\text{H}}}{RT}\right\}. \tag{6.58}$$

The denominator of the rate eq. (6.41) assumes the form

$$D = 1 + \sum_{i=1}^{n} \binom{n}{i} \overline{K}^{*i} [\text{S}]^i \exp\left\{-\frac{U_{0,\gamma}^{\alpha\text{H}\sigma\text{H}} - U_{i,\gamma}^{\alpha\text{H}\sigma\text{H}}}{RT}\right\}, \tag{6.59}$$

which is the expression of a partition function of degree n under steady state conditions. The numerator of the corresponding rate equation is

$$N = \sum_{i=1}^{n} i \binom{n}{i} k^* \overline{K}^{*i} [\mathrm{S}]^i \exp\left\{ -\frac{U_{0,\gamma}^{\alpha\mathrm{H}\sigma\mathrm{H}} - U_{i-1,\gamma}^{\alpha\mathrm{H}}}{RT} \right\}. \tag{6.60}$$

This expression is not *sensu stricto* a partition function, but can be shown to be related to a partition function. It can be rearranged to

$$N = k^* \overline{K}^* [\mathrm{S}] \exp\left(-\frac{U_{0,\gamma}^{\sigma\mathrm{H}}}{RT} \right) n \Pi_\tau (n-1), \tag{6.61}$$

where $n\Pi_\tau(n-1)$ is equal to

$$n\Pi_\tau(n-1) = \sum_{i=1}^{n} i \binom{n}{i} \overline{K}^{*i-1} [\mathrm{S}]^{i-1} \exp\left\{ -\frac{U_{0,\gamma}^{\alpha\mathrm{H}} - U_{i-1,\gamma}^{\alpha\mathrm{H}}}{RT} \right\}. \tag{6.62}$$

This expression can be rearranged to

$$n\Pi_\tau(n-1) = n\left[1 + \sum_{i=1}^{n} \frac{i+1}{n} \binom{n}{i+1} \overline{K}^{*i} [\mathrm{S}]^i \exp\left\{ -\frac{U_{0,\gamma}^{\alpha\mathrm{H}} - U_{i,\gamma}^{\alpha\mathrm{H}}}{RT} \right\} \right]. \tag{6.63}$$

Moreover, as

$$\frac{i+1}{n} \binom{n}{i+1} = \binom{n-1}{i}, \tag{6.64}$$

one has

$$\Pi_\tau(n-1) = 1 + \sum_{i=1}^{n} \binom{n-1}{i} \overline{K}^{*i} [\mathrm{S}]^i \exp\left\{ -\frac{U_{0,\gamma}^{\alpha\mathrm{H}} - U_{i,\gamma}^{\alpha\mathrm{H}}}{RT} \right\}, \tag{6.65}$$

which is a partition function of degree $n-1$ for

$$\binom{n-1}{i} = 0, \tag{6.66}$$

if $i = n$. Therefore the expression for the rate takes a new thermodynamic formulation, namely [26],

$$\frac{v}{[\mathrm{E}]_0} = n k^* \overline{K}^* [\mathrm{S}] \exp\left(-\frac{U_{0,\gamma}^{\sigma\mathrm{H}}}{RT} \right) \frac{\Pi_\tau(n-1)}{\Pi_\gamma(n)}. \tag{6.67}$$

$\Pi_\gamma(n)$ is a partition function for the n ground states and $\Pi_\tau(n-1)$ is a partition function for the $n-1$ transition states. Whereas the first function describes the distribution of the ground states over all the available energy levels, the seond function pictures the distribution of the transition states over the corresponding energy levels.

The heterologous interactions that take place between an oligomeric non co-operative enzyme and other proteins may indeed generate co-operativity, as eq. (6.67) shows, but is by no means compulsory. If the oligomeric enzyme intrinsically follows Michaelis–Menten kinetics, the interactions between the enzyme and the other proteins may also change the kinetic parameters of the enzyme within the supramolecular complex. Let us consider, for instance, the case where

$$\frac{\partial U_\gamma^{\alpha\mathrm{H}\sigma\mathrm{H}}}{\partial \xi} = 0 \tag{6.68}$$

and

$$\frac{\partial U_\gamma^{\alpha\mathrm{H}}}{\partial \xi} = 0. \tag{6.69}$$

This implies that

$$U_{0,\gamma}^{\alpha\mathrm{H}\sigma\mathrm{H}} - U_{i,\gamma}^{\alpha\mathrm{H}\sigma\mathrm{H}} = 0 \tag{6.70}$$

and

$$U_{0,\gamma}^{\sigma\mathrm{H}} - U_{i,\gamma}^{\alpha\mathrm{H}} = 0. \tag{6.71}$$

The two corresponding partition functions assume the form

$$\Pi_\gamma(n) = \left(1 + \overline{K}^*[\mathrm{S}]\right)^n, \qquad \Pi_\tau(n-1) = \left(1 + \overline{K}^*[\mathrm{S}]\right)^{n-1}. \tag{6.72}$$

The steady state rate equation reduces to

$$\frac{V}{[\mathrm{E}]_0} = \frac{nk^* \exp(-U_{0,\gamma}^{\sigma\mathrm{H}}/RT)\overline{K}^*[\mathrm{S}]}{1 + \overline{K}^*[\mathrm{S}]}, \tag{6.73}$$

and the catalytic rate constant is multiplied by a factor of $\exp(-U_{0,\gamma}^{\sigma\mathrm{H}}/RT)$ but without any change of the apparent binding constant.

If, alternatively,

$$\frac{\partial U_\gamma^{\alpha\mathrm{H}\sigma\mathrm{H}}}{\partial \xi} = \frac{\partial U_\gamma^{\alpha\mathrm{H}}}{\partial \xi} = \lambda, \tag{6.74}$$

where λ is, as previously, a constant. Then

$$U_{0,\gamma}^{\alpha\mathrm{H}\sigma\mathrm{H}} - U_{i,\gamma}^{\alpha\mathrm{H}\sigma\mathrm{H}} = U_{0,\gamma}^{\alpha\mathrm{H}} - U_{i,\gamma}^{\alpha\mathrm{H}} = i\lambda \quad (i = 1, \ldots, n), \tag{6.75}$$

and the two partition functions, $\Pi_\gamma(n)$ and $\Pi_\tau(n-1)$, take the form

$$\Pi_\gamma(n) = \{1 + \exp(-\lambda/RT)\overline{K}^*[\mathrm{S}]\}^n \tag{6.76}$$

and

$$\Pi_\tau(n-1) = \{1 + \exp(-\lambda/RT)\overline{K}^*[\mathrm{S}]\}^{n-1}. \tag{6.77}$$

The resulting rate equation is thus

$$\frac{v}{[\mathrm{E}]_0} = \frac{nk^* \exp(-U_{0,\gamma}^{\sigma\mathrm{H}}/RT)\overline{K}^*[\mathrm{S}]}{1 + \exp(-\lambda/RT)\overline{K}^*[\mathrm{S}]}, \tag{6.78}$$

and the apparent binding and catalytic constants are

$$\overline{K} = \overline{K}^* \exp\left(-\frac{\lambda}{RT}\right), \qquad k_\mathrm{c} = k^* \exp\left\{-\frac{U_{0,\gamma}^{\sigma\mathrm{H}} - \lambda}{RT}\right\}. \tag{6.79}$$

The theoretical results discussed in the present and in the previous sections show that proteins which have no catalytic function *per se*, or are lacking their substrates, can dramatically alter the activity of another enzyme if they are all embedded in the same supramolecular edifice. This effect is expected to be exerted thanks to a stabilization–destabilization of the ground and transition states of the active enzyme. If these theoretical expectations can be observed experimentally, i.e. if the same enzyme can be found to occur as a free molecular entity and as a component of a multimolecular edifice, with different catalytic properties in the two situations, one will have to admit that a piece of information, or rather an instruction, has been transferred from protein to protein within the supramolecular edifice. It is therefore important to show experimentally that an instruction can indeed be transferred from protein to protein without contradicting thermodynamic principles and information theory.

6.4. Statistical mechanics of imprinting effects

If this information transfer were an experimental reality and not only a sensible assumption, one would logically expect the existence of imprinting effects exerted between proteins. Thus upon dissociating a multienzyme complex by dilution, one would expect that some free enzymes, originating from the breakage of the complex, may retain for a while the conformation and the function they had within the complex. These enzymes would then bear, for a certain time period, an imprint of the other enzymes that were initially associated with them. Some of the free enzymes released upon the dissociation of the complex would thus be in a metastable state that would ultimately relapse to a more stable conformation. This concept of imprinting is reminiscent of that of enzyme memory [30] already discussed in chapter 2. We shall see later that imprinting may indeed exist between two (or several) enzymes or between an enzyme and a membrane, and so this is not an arbitrary assumption.

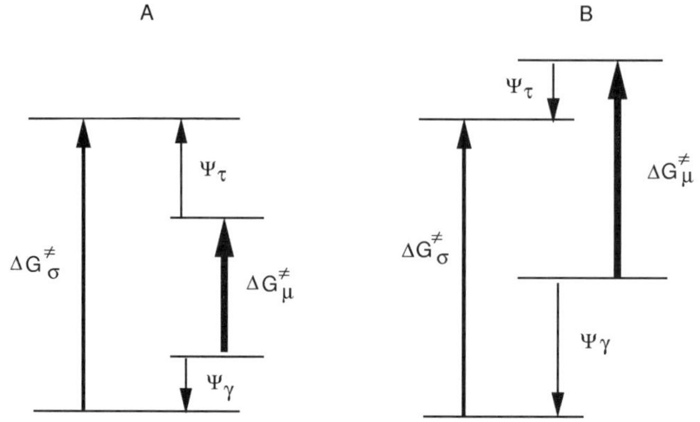

Fig. 6.6. Imprinting of an instruction in an enzyme. As explained in the text, the difference $\Psi_\gamma - \Psi_\tau$ must be negative. The situations shown in A and B are compatible with this requirement.

The idea that the folding and the function of a polymer may be determned not only by its sequence but also by its transient association with other proteins has received support from experiments coming from a different field. Thus metacrylic acid and ethylene glycol metacrylate can be copolymerized in the presence of theophylline. If the latter is then removed the copolymer is still able to bind theophylline. However when the same copolymer is synthesized in the absence of theophylline, it is unable to bind this ligand. The binding of theophylline to the copolymer is explained by assuming that this macromolecule has kept an imprinting of the ligand [31]. The studies of Klibanov and colleagues [32,33,35] and Yennawar et al. [34] on enzymes in organic solvents have led to similar conclusions. If subtilisin, for instance, is mixed with a competitive inhibitor of the enzyme reaction and if the mixture is lyophylised, dissolved in an organic solvent and the inhibitor washed out, the enzyme is about 100-fold more active than it would normally be if it had not been in contact with the inhibitor. These results are consistent with the concepts of imprinting and of enzyme memory already discussed (chapter 2).

To evaluate the importance of these imprinting effects for an enzyme process, one can follow the same reasoning as before. If $\Delta G_\sigma^{\neq}$ is the free energy of activation of a given chemical process carried out by the stable enzyme form and $\Delta G_\mu^{\neq}$ is the corresponding free energy of the same process catalysed by the metastable form, one has

$$\Delta G_\mu^{\neq} = \Delta G_\sigma^{\neq} + \Psi_\gamma(t) - \Psi_\tau(t). \tag{6.80}$$

$\Psi_\gamma(t)$ and $\Psi_\tau(t)$ represent the stabilization–destabilization energies of the ground and the transition states of the metastable enzyme form relative to the stable enzyme state. As previously, eq. (6.80) stems from the first principle of thermodynamics. Indeed, $\Psi_\gamma(t)$ and $\Psi_\tau(t)$ vary as a function of time and the difference $\Psi_\gamma(t) - \Psi_\tau(t)$ approaches zero as time increases. However, if the relaxation of the metastable state to the stable state is very slow relative to the time required for the kinetic measurements, these variables can be considered constants and can be represented, for simplicity, as Ψ_γ and Ψ_τ. If the activity of

$$E_\mu + S \xrightarrow{\substack{(\Psi_{0,\tau}) \\ k'_\mu \; \overline{K}_\mu}} E_\mu + P$$
$(\Psi_{0,\gamma})$

$$E_\mu S \xrightarrow{\substack{(\Psi_{0,\tau}) \\ k'_\mu}} E_\mu + P$$
$(\Psi_{1,\gamma})$

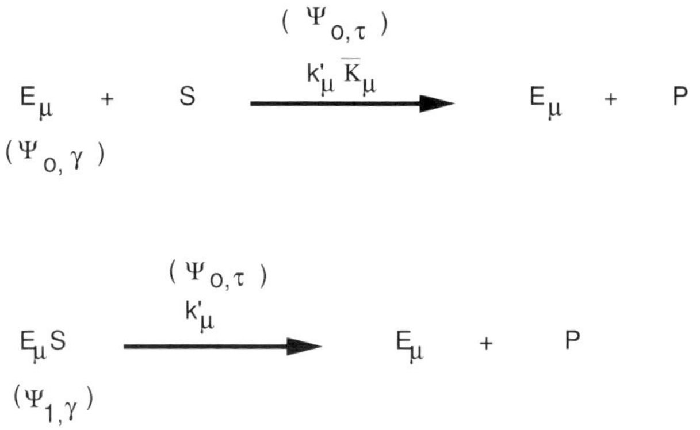

Fig. 6.7. Equivalence of the stabilization–destabilization energies for a substrate desorption step and for a catalytic step in an imprinted enzyme. See text.

the imprinted metastable enzyme form is larger than that of the stable form, $\Delta G_\sigma^{\neq} > \Delta G_\mu^{\neq}$ and therefore $\Psi_\gamma - \Psi_\tau < 0$. This is the situation illustrated in Fig. 6.6.

If one considers the hydrolysis of a substrate S into products P and Q, catalysed by a metastable form of an enzyme E (Fig. 6.7), the overall second-order rate constant can be expressed as a function of the corresponding second order rate constant that describes the activity of the enzyme in its stable state. One thus has

$$k'_\mu \overline{K}_\mu = k'_\sigma \overline{K}_\sigma \exp\left\{-\frac{\Psi_{0,\gamma} - \Psi_{0,\tau}}{RT}\right\}, \tag{6.81}$$

where $\overline{K}_\mu$ and $\overline{K}_\sigma$ are the apparent binding constants of the substrate to the metastable (μ) and stable (σ) enzyme forms and k'_μ and k'_σ the corresponding catalytic rate constants. $\Psi_{0,\gamma}$ and $\Psi_{0,\tau}$ represent the stabilization–destabilization energies of the ground and of the transition states, respectively. This is illustrated in Fig. 6.7. If we consider now the catalytic process itself, the relevant stabilization–destabilization energies are $\Psi_{1,\gamma}$ and $\Psi_{0,\tau}$ (Fig. 6.7) and the catalytic constant of the process carried out by the metastable state can be expressed as

$$k'_\mu = k'_\sigma \exp\left\{-\frac{\Psi_{1,\gamma} - \Psi_{0,\tau}}{RT}\right\}. \tag{6.82}$$

From equations (6.81) and (6.82) ones deduces that

$$\overline{K}_\mu = \overline{K}_\sigma \exp\left\{-\frac{\Psi_{0,\gamma} - \Psi_{1,\gamma}}{RT}\right\}. \tag{6.83}$$

This equation has an interesting thermodynamic implication. Since any simple enzyme must be able to bind several transition states, for instance $S^{\neq}$ and $X^{\neq}$, there may exist some relationship between the binding energies of these transition states to the enzyme.

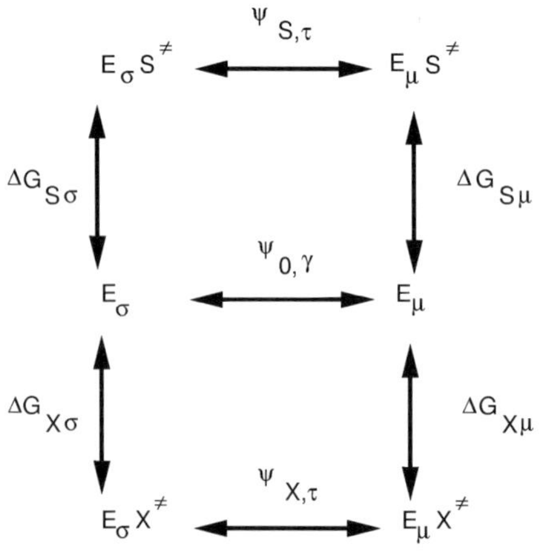

Fig. 6.8. The Ψ parameters and their equivalence to equilibrium constants. The free enzyme is assumed to occur under a stable, E_σ, and a metastable state, E_μ. Either of these states may bind the transition states $S^{\neq}$ and $X^{\neq}$.

This putative relationship can be illustrated by the thermodynamic box of Fig. 6.8. The first principle of thermodynamics implies that

$$\Psi_{S,\tau} + \Delta G_{S\mu} - \Delta G_{X\mu} = \Psi_{X,\tau} + \Delta G_{X\sigma} - \Delta G_{S\sigma}. \tag{6.84}$$

In this expression, $\Psi_{S,\tau}$ and $\Psi_{X,\tau}$ are the stabilization–destabilization energies of the metastable state, relative to the stable state, when the enzyme has bound either transition state $S^{\neq}(\Psi_{S,\tau})$ or transition state $X^{\neq}(\Psi_{X,\tau})$. $\Delta G_{S\mu}$, $\Delta G_{X\mu}$, $\Delta G_{X\sigma}$ and $\Delta G_{S\sigma}$ are the binding free energies of the transition states $S^{\neq}$ and $X^{\neq}$ to the metastable (μ) and to the stable (σ) forms of the enzyme. For the ideally simple enzyme process

$$E + S \underset{k_{-1}}{\overset{k_1}{\rightleftarrows}} ES \xrightarrow{k'_1} E + P + Q$$

the reaction should overcome two energy barriers associated with the two transition states $S^{\neq}$ and $X^{\neq}$. The expression of the three rate constants is (see Fig. 6.7)

$$k_{1,\mu} = k_{1,\sigma} \exp\left\{-\frac{\Psi_{0,\gamma} - \Psi_{S,\tau}}{RT}\right\},$$

$$k_{-1,\mu} = k'_{-1,\sigma} \exp\left\{-\frac{\Psi_{1,\gamma} - \Psi_{S,\tau}}{RT}\right\},$$

$$k'_{1,\mu} = k'_{1,\sigma} \exp\left\{-\frac{\Psi_{1,\gamma} - \Psi_{X,\tau}}{RT}\right\}. \tag{6.85}$$

It then becomes obvious that eq. (6.83) can be derived if, and only if,

$$\Psi_{S,\tau} = \Psi_{X,\tau} = \Psi_{0,\tau} \tag{6.86}$$

and this, in turn, leads to

$$\Delta G_{s\mu} + \Delta G_{s\sigma} = \Delta G_{x\mu} + \Delta G_{x\sigma}. \tag{6.87}$$

Thus, the sum of the binding energies of the transition state $S^{\neq}$ to the two enzyme forms must be equal to the sum of the binding energies of the transition state $X^{\neq}$ to these two enzyme states.

After the dissociation of a multienzyme complex, it is possible to measure experimentally, for an enzyme that follows Michaelis–Menten kinetics and is in a metastable state, the catalytic constant k'_{μ}, the apparent substrate binding constant $\overline{K}_{\mu}$, the second-order constant $k'_{\mu}\overline{K}_{\mu}$ and the corresponding constants in the stable state. It should therefore be possible to measure $\Psi_{0,\gamma} - \Psi_{1,\gamma}$, $\Psi_{1,\gamma} - \Psi_{0,\tau}$ and $\Psi_{0,\gamma} - \Psi_{0,\tau}$. These values express how the instruction and the energy transferred from enzyme to enzyme is used to alter the functional properties of one of the enzymes.

6.5. *Statistical mechanics of instruction transfer within supramolecular edifices*

Classical molecular biology is shaped by the central dogma which states that all the information of a polypeptide chain is stored in the corresponding structural gene, and that this information is transferred from DNA to protein but never in the opposite direction [1]. Moreover this axiom of molecular biology has been associated with another one, namely that information transfer can take place from DNA to DNA, or from RNA to DNA, but never from protein to protein [1]. Here the term "information" is taken to mean the sequence of non-overlapping triplets in nucleic acids, or the sequence of aminoacids in polypeptides. As outlined by Yokey [36], the statement that information transfer from DNA to protein is not a specific principle, or axiom, of molecular biology, but a property of any code in which a source alphabet is larger than the destination alphabet. This is a consequence of the degeneracy of a code, whatever its nature. This makes it possible, from a DNA sequence, to predict the aminoacid sequence of a polypeptide, whereas the converse is indeed impossible. The fundamental property of all the codes mentioned above does not forbid the transfer of information from DNA to DNA, from RNA to RNA, from RNA to DNA or even from protein to protein [36]. Nevertheless Crick [1] considers that the transfer of information from protein to protein would "shake the whole intellectual basis of molecular biology". This is indeed true insofar as the concept of information is exclusively identified with the primary sequence of nucleic acids and proteins, but it is not compulsory if the term "information" is given a broader meaning more closely related to the functions of the proteins. It is a time-honoured idea of genetics to consider that the biological function of a protein is, in a way, encoded in the corresponding gene. The studies that aim at discovering, for instance, the genes that are responsible for many diseases are meanigful only if a biological function (or one of its disorders) is encoded in a given gene. Strictly speaking, however, a gene does not code for a function, but for an aminoacid sequence. Therefore, in order to be able to associate a function with a gene, one must be able to associate first the function of a protein in the cell with a linear sequence of aminoacids. This goal has been reached in the work of Anfinsen [37,38] and many other authors [39–41]. When a protein is denatured, by urea for instance, and the urea is then washed out, the polypeptide chains fold, under definite experimental conditions (pH, ionic strength, etc.), and an active protein can then be obtained. Thus, in some specific experimental conditions, the function appears to be fully defined by the aminoacid sequence, which is itself determined by the structure of the corresponding structural gene. This is the simple picture, which is offered by classical molecular biology, of the relationship between a functional protein and the corresponding sructural gene. There seems to be no room for protein-to-protein information transfer. Thus the theoretical considerations of the previous sections do not fit well with these general considerations about information transfer in the cell. It is therefore of interest to spend some time on this concept of information transfer in relation to the functional properties of proteins.

Let us consider for instance a polypeptide chain of N aminoacids and let $n_1, n_2, \ldots$ be the number of aminoacids belonging to the different species present in the sequence (for instance glycine, leucine, ...). The number of complexions, Ω_S, associated with this polypeptide chain is defined as

$$\Omega_S = \frac{N!}{n_0!n_1!\ldots}. \tag{6.88}$$

Thus Ω_S increases with the total number, N, of aminoacids and with the number of different aminoacids present in the polypeptide chain. Hence the information content of this aminoacid sequence is defined as [36,42]

$$I_S = \log_2 \Omega_S. \tag{6.89}$$

As already outlined, this way of defining information does not fit so well with what we know on functional proteins, for many enzymes (isoenzymes) may have different sequences, and, therefore, different information content, yet still have the same activity. It thus appears mandatory to define the information content in a different manner, more directly related to the function of the protein. This can be done through an analysis of the distribution of the population of enzyme molecules over all possible quantum states. As a matter of fact, a folded polypeptide chain in solution is not a rigid entity, as seen in crystals, but displays internal motions and adopts different states associated with different energies. This approach relies on the well-known idea that, in order to overcome the energy barriers of the transition states, the enzyme must have stored a certain amount of free energy. The enzyme can then give the substrate(s) and the product(s) the amount of energy required for the conversion of the reagents into the corresponding transition states. The number of complexions, Ω_E, defined on this basis, would still be expressed by eq. (6.88), but now N is the total number of enzyme molecules and $n_0, n_1, \ldots$ are the the numbers of enzyme molecules having the energy levels $E_0, E_1, \ldots$. As previously, the information content of this population of molecules is

$$I_E = \log_2 \Omega_E. \tag{6.90}$$

There is an obvious similarity between this information content and the entropy, S, of the population of molecules, for this entropy has the form

$$S = k_B \ln \Omega_E, \tag{6.91}$$

where k_B is, as previously, the Boltzmann constant.

During the association process of an enzyme with another protein or with a cell organelle which, for simplicity, may be considered a "rigid body", the enzyme loses translational and rotational degrees of freedom and the mobility of the polypeptide chain in the vicinity of the zone of contact between the two partners decreases as well. As the number of quantum states of the bound enzyme is of necessity smaller than that of the corresponding free enzyme, the number of complexions, Ω_E, becomes smaller as well. This means that both the entropy, S, and the information content, I_E, have smaller values for the bound than for the free enzyme. One can calculate the variation of entropy upon binding the enzyme to a "rigid body". One has

$$\Delta S = S^b - S^f, \tag{6.92}$$

where the superscripts 'b' and 'f' stand for "bound" and "free", and this expression is indeed equivalent to

$$\Delta S = k_B \ln \frac{\Omega_E^b}{\Omega_E^f}. \tag{6.93}$$

Since

$$\Omega_E^f > \Omega_E^b, \tag{6.94}$$

ΔS is of necessity negative.

The value of the partition function Q also reflects the change of entropy associated with the binding of the enzyme to the "rigid body". Let us consider the partition function of an enzyme

$$Q = \sum_{i=0}^{n} e^{-E_i/k_B T}. \tag{6.95}$$

Q will assume large values if the population of enzyme molecules is distributed over many quantum states and if the ratio $E_i/k_B T$ is small. Alternatively, Q will have smaller values if the population of enzyme molecules is spread over a smaller number of quantum states and if the ratio $E_i/k_B T$ is larger. As the enzyme is assumed to be bound to a "rigid body", the number of energy states of the bound enzyme is less than what it would be if the enzyme were free. This means that the decline of the corresponding Boltzmann law is faster for the bound than for the free enzyme. Therefore the ratio $E_i/k_B T$ is larger for the former than for the latter. The conclusion of this reasoning is that the partition function is smaller for the bound than for the free enzyme. Hence one must have

$$Q^{\mathrm{f}} > Q^{\mathrm{b}}. \tag{6.96}$$

The expression for the entropy change can also be expressed with respect to the partition functions Q^{f} and Q^{b}. Since

$$S = k_{\mathrm{B}} \ln Q + k_{\mathrm{B}} T \left(\frac{\partial \ln Q}{\partial T} \right)_V, \tag{6.97}$$

expression (6.93) can be rewritten as

$$\Delta S = k_{\mathrm{B}} \ln \frac{Q^{\mathrm{b}}}{Q^{\mathrm{f}}} + k_{\mathrm{B}} T \left\{ \left(\frac{\partial \ln Q^{\mathrm{b}}}{\partial T} \right)_V - \left(\frac{\partial \ln Q^{\mathrm{f}}}{\partial T} \right)_V \right\}. \tag{6.98}$$

The first term of this equation is of necessity negative because $Q^{\mathrm{f}} > Q^{\mathrm{b}}$. Moreover,

$$\left(\frac{\partial \ln Q}{\partial T} \right)_V > 0 \tag{6.99}$$

for on increasing the temperature, the number of available quantum states rises and so does the corresponding partition function Q. Moreover, this increase in the number of quantum states is expected to be larger for the free than for the bound enzyme. Therefore,

$$\left(\frac{\partial \ln Q^{\mathrm{f}}}{\partial T} \right)_V > \left(\frac{\partial \ln Q^{\mathrm{b}}}{\partial T} \right)_V \tag{6.100}$$

and $\Delta S < 0$, consistent with the predictions of eq. (6.93).

Moreover, one must have

$$\left(\frac{\partial \ln Q}{\partial V} \right)_T < 0. \tag{6.101}$$

This relationship becomes obvious if one recalls that the expression for the pressure is

$$p = k_{\mathrm{B}} T \left(\frac{\partial \ln Q}{\partial V} \right)_T \tag{6.102}$$

and it decreases as V increases. Statistical thermodynamics also predicts what the change of free energy stored in the enzyme will be if the enzyme binds to a relatively rigid structure of the cell. The Gibbs free energy can be expressed in terms of the partition function as

$$G = -k_{\mathrm{B}} T \left\{ \ln Q - V \left(\frac{\partial \ln Q}{\partial V} \right)_T \right\}. \tag{6.103}$$

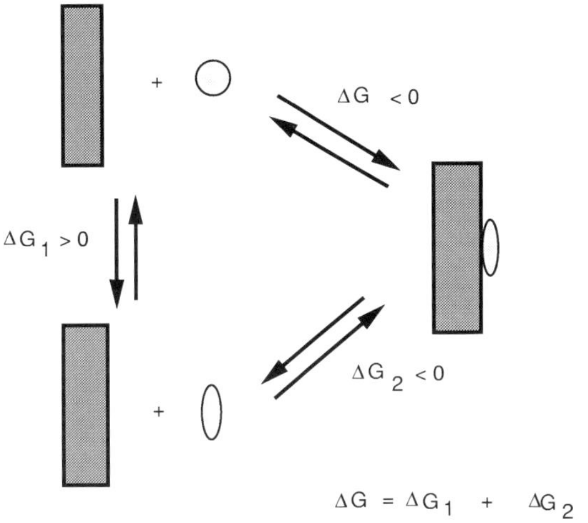

Fig. 6.9. Thermodynamic scheme showing that the binding of a protein to a rigid body results in the storage of energy in the protein. See text.

Therefore, the conformational free energy that can be stored in the enzyme when it binds to a "rigid body" is

$$\Delta G = -k_{\mathrm{B}}T\left\{\ln\frac{Q^{\mathrm{b}}}{Q^{\mathrm{f}}} - V\left[\left(\frac{\partial \ln Q^{\mathrm{b}}}{\partial V}\right)_T - \left(\frac{\partial \ln Q^{\mathrm{f}}}{\partial V}\right)_T\right]\right\}. \tag{6.104}$$

We have already seen that when the volume V increases, the corresponding partition function Q decreases. The reason for this is that the increase in volume is associated with an increase in the values of $E_i/k_{\mathrm{B}}T$. This increase is larger for the bound than for the free enzyme. Therefore one must expect that

$$\left(\frac{\partial \ln Q^{\mathrm{b}}}{\partial V}\right)_T > \left(\frac{\partial \ln Q^{\mathrm{f}}}{\partial V}\right)_T, \tag{6.105}$$

and, from eq. (6.105), $\Delta G > 0$. Therefore the binding of a protein to a "rigid body" should be accompanied by a storage of conformational free energy in this protein. It should be stressed that this positive ΔG value does not represent the free energy of binding of the protein to the "rigid body", but the conformational energy stored in this molecule. The thermodynamic scheme of Fig. 6.9 may help clarify this point. In this scheme, it is assumed that the free energy of binding of the enzyme to a cell organelle (the "rigid body") is negative ($\Delta G < 0$). This energy can be split into two ideal components: the energy required to change the conformation of the enzyme in order to bind to a "rigid body", which is positive ($\Delta G_1 > 0$); and the binding free energy of this modified conformation to the same "rigid body", which is negative ($\Delta G_2 < 0$). The sum of these two contributions is negative. The

ΔG of eq. (6.104) is the ΔG_1 component only, and thus represents the energy stored in the enzyme molecule.

It is now of interest to come back to the information content of an enzyme that binds to another one, or to a "rigid body". Its entropy decreases, and so does its information content, I_E. This means that the number of messages that may be conveyed by the enzyme decreases in going from the free to the bound state. Discarding a number of messages, is indeed equivalent to make a "choice" amongst the available information contained in the population of free protein molecules. It therefore corresponds to an "instruction" received by the bound enzyme. The main conclusion of this section is that an enzyme which binds to another protein, or to a cell structure, receives an instruction and stores a certain amount of free energy. From a thermodynamic viewpoint, this means that the enzyme may, possibly, use this instruction and this energy to perform a catalytic function it could not perform if it were in the free state.

6.6. *Instruction, chaperones and prion proteins*

The previous sections of this chapter were mainly theoretical. Their aim was primarily to know whether it is thermodynamically feasible that an instruction be transferred from protein to protein. If this is so, it would mean that the information content, I_E, of a protein in a supramolecular edifice has its origin not only in the corresponding structural gene, but also in the interactions that take place between this protein and other constituents of the eukaryotic cell. In order to be able to test this idea experimentally, a protein must be present in the cell in a free and in a bound state, in such a way that the properties of these two states of the protein can be compared. If the properties are different, it means that the organizational complexity of the eukaryotic cell plays a certain part in the biological function of a given protein.

It is important to stress that the notion of control of enzyme activity through cell complexity is quite different from the classical notion of enzyme regulation. In the former case, it is really the structural and functional organization of the cell that controls enzyme function, whereas in the latter case, this function is modulated by covalent modification of the enzyme, or by the reversible binding of a ligand to the protein. In the first case, we are dealing with a supramolecular *system* whereas in the second case we are interested in the irreversible, or reversible, alteration of an intrinsic property of *a macromolecule*. Although different in their essence, these two fields may somewhat overlap. The aim of this section is certainly not to offer an extensive review on chaperones and prion proteins, but rather to discuss, on the basis of our knowledge of these two important types of macromolecules, whether the association of different polypeptide chains can result in an instruction transfer between proteins.

6.6.1. *Chaperones*

After the work of Anfinsen [37–39], it was considered that *in vitro*, in dilute solutions, a denatured protein spontaneously folds to its native and active conformation. This conclusion has been confirmed many times with different enzymes [40,41], and implies that the active conformation of a protein is fully determined by its primary structure. In other words, it means that the information content of a given protein is already present in the corresponding structural gene. By the same token, it was also postulated that the folding processes that take place *in vitro* also operate *in vivo,* without additional complications.

It was later discovered, however, that the correct folding of polypeptide chains in the cell requires the participation of specialized proteins called chaperones [43–47]. There are at least two conceptual reasons that allow one to understand that the correct folding of a polypeptide chain may require the participation of these chaperone proteins. First, in eukaryotic cells, newly synthesized polypeptide chains are transported within the cell from compartment to compartment, from cytoplasm to mitochondria for instance, and it is unthinkable that transport takes place if the protein is in its final globular state; second, owing to the molecular crowding of the living cell, the probability of cross-linking between hydrophobic regions of different polypeptide chains becomes very high, thus leading to aggregation. To avoid such aggregation processes, and to allow the transport of polypeptide chains through membranes, one can easily imagine that a specific device be required. Chaperones represent this specific device. One may thus wonder whether these chaperones do not give the nascent polypeptide chain additional information that is not present in the corresponding structural gene. To shed some light on this matter, it is important to recall briefly some of the properties of these macromolecules. Class I chaperones bind to nascent polypeptide chains, thus preventing their spontaneous folding and allowing their transport through membranes [44,45]. Once the unfolded chain has been transported across the membrane of a cell organelle (mitochondrion or chloroplast, for example), a class II chaperone directs the correct folding of the polypeptide chain. Other chaperones, sometimes called chaperonins, are large supramolecular edifices with a cleft that may accomodate a polypeptide chain in the molten globule state [48]. The molten globule is a state in the protein folding process in which the polypeptide chain is incompletely folded, more open and less ordered than the active protein. This sequestration of a polypeptide chain is believed to prevent molecular crowding and aggregation in a sort of "isolation chamber". There is also a relationship between heat shocks and chaperones. *In vivo*, under the stress conditions of a heat shock, many proteins of the cell tend to unfold. This process is prevented by a stimulation of chaperone synthesis which results in the correct refolding of aggregated polypeptides [49–52].

On the basis of this very brief survey of the properties of chaperones, one may conclude that these proteins prevent the aggregation of hydrophobic regions of different polypeptide chains, or avoid molecular crowding, by temporary sequestration of polypeptide chains during the folding process. On this basis alone, there is no reason to believe that chaperones give the folding polypeptide chain additional information not already present in the primary structure of the chain. In addition to these two functions, however, chaperones may break bonds of incorrectly folded polypeptide chains, thus giving these misfolded chains a new chance to fold correctly. It is difficult, in this case, to decide whether or not the chaperone has given the misfolded polypeptide chain additional information that was not already encoded in the chain.

The situation is perhaps even more complex with what is known today as intramolecular chaperones. Some proteins require a specific region of their sequence to fold properly. During the folding process, this region of the polypeptide is removed by an autolytic, or an exogeneous proteolytic, cleavage. The fragment removed is referred to as an intramolecular chaperone, for it allows the proper folding of the polypeptide chain. Subtilisin [53–58], α-lytic protease [59], aqualysin [60] and carboxypeptidase Y [61] are examples of this type of proteins. At low concentrations, and after removal of the chaperone region, subtilisin and α-lytic protease fold only up to the molten globule state [62–64]. Addition of the intramolecular chaperone domain allows proper refolding and the appearance of enzyme activity [63,64]. These results have been interpreted as indicating that the chaperone allows the polypeptide to ovecome a free energy barrier, which takes place in a late stage of the folding process [63,64]. The sequence of these intramolecular chaperones can be altered by mutation. After cleavage of this region, one may expect the active enzyme to be present in two different conformations depending on whether or not the structure of the chaperone region was altered by the mutation. This is precisely what appears to occur with subtilisin, which, after excision of the chaperone region, can exist in two different conformations with different activities. These two forms of the enzyme have exactly the same sequence. It is thus difficult at the moment to decide whether an instruction is transferred from a chaperone to a polypeptide chain.

6.6.2. Prions

Although it is difficult to know whether there is an information transfer from a chaperone to a nascent polypeptide, there is little doubt that this transfer indeed takes place between prion proteins. Scrapie and other diseases have been shown to be due to a protein, called a prion protein. This protein does not appear to be associated with any nucleic acid, and yet the disease propagates [65–68]. Surprisingly, however, the protein has been shown to be present in healthy organisms [66]. Thus the origin and propagation of the disease were for a long time very difficult to understand [67]. In healthy organisms the prion protein is called PrPc. Its structure is known and its gene has been isolated and cloned [65]. One of the most surprising results is that the primary structure of this protein and of its corresponding gene are unaffected in animals with the scrapie disease. Many different results have led to the conclusion that it is the conformation of the protein which has been affected [68]. The new conformation of the prion protein is called PrPsc. Propagation of the disease is believed to be due to the association of the "normal" (PrPC) and "pathogenic" (PrPsc) forms of the prion protein, whereby the latter (PrPsc) forces the former (PrPc) to change its conformation and adopt the PrPsc structure [66,69]. Although this interpretation led to a controversy, there is little doubt today as to its validity. Thus the "pathogenic" prion protein, PrPsc, gives a "non-pathogenic" protein, PrPc, the "instruction" to become pathogenic. The apparently irreversible conversion of PrPc into PrPsc has been interpreted has an example of a bistability phenomenon [70,71]. When these ideas were presented for the first time, they were considered incompatible with the most widely accepted ideas of classical molecular biology and with the interpretation usually given to the central dogma. When the information transfer between prion proteins received ample confirmation, this was considered the exception that confirms the rule. Nevertheless, the considerations of statistical thermodynamics of the previous sections show that it is perfectly feasible, from a physical viewpoint, that an instruction, or piece of information, be transferred fom protein to protein. One of the ideas of this chapter is that it the rule rather than the exception.

6.7. *Multienzyme complexes, instruction and energy transfer*

Many multienzyme complexes catalyse consecutive reactions, but some carry out chemical processes that do not occur sequentially. If there is a functional advantage in the physical association of enzymes that catalyse non-consecutive reactions, this advantage would be expected to be an increase of their catalytic activity, or an alteration of their properties. In order to test experimentally the instruction and energy transfer which may occur between different proteins embedded in the same supramolecular edifice, the best system is a bienzyme complex that does not catalyse consecutive reactions. Three systems will be considered below: the plasminogen–streptokinase system; the phosphoribulokinase–glyceraldehyde phosphate dehydrogenase system; and the RAS–GAP system.

6.7.1. *The plasminogen–streptokinase system*

Inactive plasminogen can be converted into active plasmin through a proteolytic attack by a serine protease, urokinase. However, a polypeptide of 414 aminoacids, called streptokinase, since it can be isolated from *Streptococcus* strains, can also induce plasminogen activation. Streptokinase, however, has no intrinsic proteolytic activity but binds to plasminogen and forms a 1 : 1 molar complex with it [72–75]. The productive binding of streptokinase to plasminogen must induce a conformation change of this protein which becomes catalytically active without undergoing any proteolytic cleavage. Recombinant streptokinases may, or may not, bind to plasminogen and they can be tested for their ability to induce the activation of the protein [76]. The recombinant streptokinase fragments spanning aminoacid residues 1–352, 120–352, 244–414 and 244–352 bind to plasminogen and competitively inhibit streptokinase binding, but none of them is able to induce the conformation change required for catalytic activity [76,78]. The emergence of a catalytic activity of plasminogen after it has bound streptokinase implies that this polypeptide has given the enzyme an instruction via the 1 : 1 molar complex.

6.7.2. *The phosphoribulokinase–glyceraldehyde phosphate dehydrogenase system*

Several multienzyme complexes are present in the chloroplast and it is not clear, at present, whether this type of organization plays a role in the process of photosynthetic metabolism [79–87]. The results presented below concern a bienzyme complex from *Chlamydomonas reinhardtii*. As the thermodynamic considerations already discussed have been applied to this bienzyme complex, the structure and the functional properties of the system will be discussed in some detail.

6.7.2.1. *Properties of the complex*

This complex is made up of two molecules of glyceraldehyde phosphate dehydrogenase and two molecules of phosphoribulokinase. The molecular mass of this complex is about 460 kDa [88]. The two enzymes catalyse non-consecutive reactions of the Benson–Calvin cycle in the chloroplast. This cycle, associated with photosynthesis, is responsible for the fixation and reduction of carbon dioxide. Under non-denaturing conditions, the complex can be labelled with antibodies raised against isolated phosphoribulokinase and glyceraldehyde phosphate dehydrogenase. As antibodies raised against phosphoribulokinase do not recognize isolated glyceraldehyde phosphate dehydrogenase, and conversely antibodies raised against glyceraldehyde phosphate dehydrogenase do not recognize isolated phosphoribulokinase, this represents the best direct demonstration of the existence of the complex.

Chloroplast phosphoribulokinase and glyceraldehyde phosphate dehydrogenase can also be obtained in a free state. In this state, phosphoribulokinase does not display any significant activity if it is oxidized [88,89]. However, when reduced with dithiothreitol or reduced thioredoxin, reduced phosphoribulokinase becomes extremely active. This complex spon-

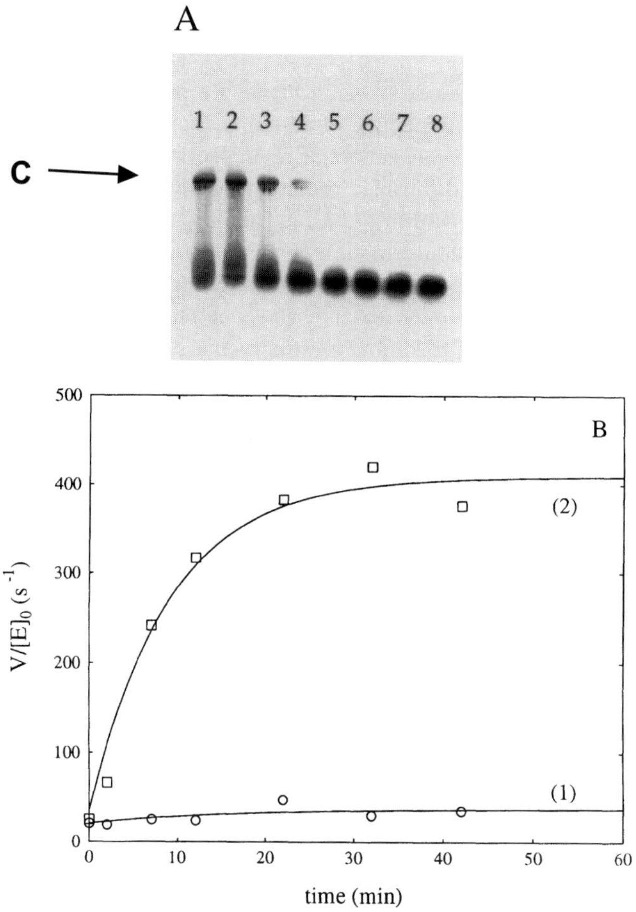

Fig. 6.10. Activation of phosphoribulokinase by dithiothreitol. A – Dissociation of the bienzyme complex (C) in the presence of dithiothreitol, and as a function of time. B – Activation of phosphoribulokinase during dissociation of the complex (curve 2). Curve 1 shows the activity of the complex in the absence of dithiothreitol. From ref. [88].

taneously dissociates in the presence of these reducing agents. In parallel with the dissociation of the complex, phosphoribulokinase becomes more and more active (Fig. 6.10). The dissociation process is reversible. Under oxidizing conditions, isolated phosphoribulokinase and glyceraldheyde phosphate dehydrogenase associate to form a complex which is indistinguishable from the native complex extracted from *Chlamydomonas* cells.

The gene coding for phosphoribulokinase has been isolated, cloned and expressed in *E. coli*. The recombinant protein, with the native glyceraldehyde phosphate dehydrogenase from chloroplasts, can form a complex indistinguishable from that isolated from *Chlamydomonas* cells [90]. A phosphoribulokinase has been isolated from the 12–2 B mutant of *Chlamydomonas*. In this organism, arginine 64 is replaced by cysteine, and the bienzyme complex does not exist. This result suggests that arginine 64 is involved, directly or indirectly, in the formation and the stability of the complex. This conclusion may be reinforced by site-directed mutagenesis. Arginine 64 has been replaced by alanine [A 64], lysine [K 64] or glutamic acid [E 64]. Whereas the [A 64] and [E 64] mutants are unable to form a complex with glyceraldehyde phosphate dehydrogenase, the [K 64] mutant does, but only slightly.

6.7.2.2. Phosphoribulokinase activity in the bienzyme complex: instruction transfer and imprinting effects

Although isolated oxidized phosphoribulokinase is nearly totally devoid of activity, this enzyme becomes active when bound to glyceraldehyde phosphate dehydrogenase. This activity may be detected even under oxidizing conditions, in the presence of glutathione. If a solution of free phosphoribulokinase is mixed with a solution of glyceraldehyde phosphate dehydrogenase, the complex is formed and, at the same time, the activity of phosphoribu-

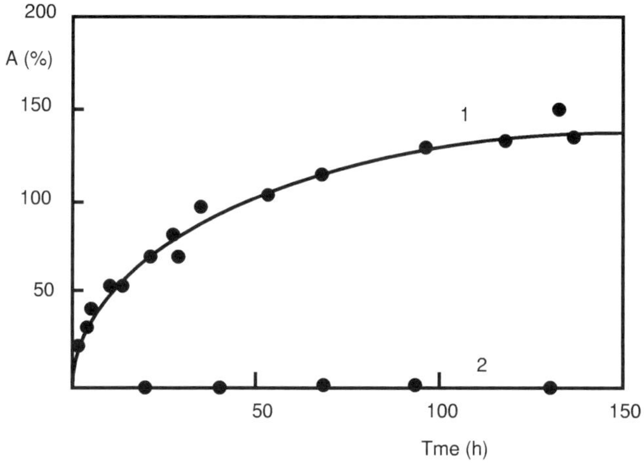

Fig. 6.11. Appearance of phosphoribulokinase activity in the presence of glyceraldehyde phosphate dehydrogenase. 1 – Appearance of phosphoribulokinase activity induced by glyceraldehyde phosphate dehydrogenase. 2 – Lack of activity in the presence of serum albumin.

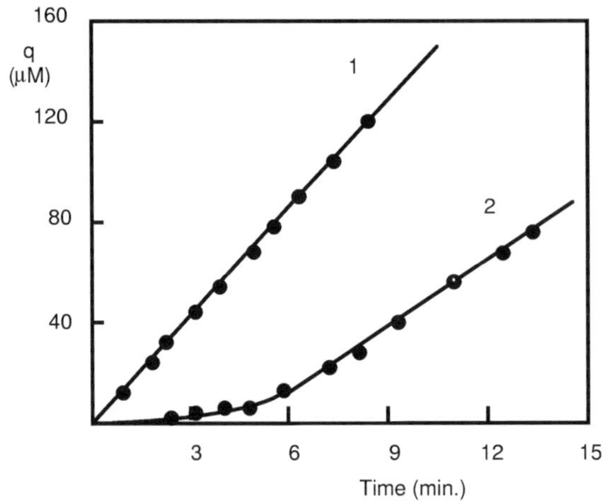

Fig. 6.12. Progress curve of phosphoribulokinase reaction in the presence of the bienzyme complex. 1 – The complex is preincubated for 15 minutes in a suitable medium but in the absence of the substrates. The reaction is initiated by adding the substrates. No lag is observed. 2 – The progress curve is monitored without any preincubation by adding the complex to a reaction mixture containing the substrates. Adapted from ref. [89].

lokinase rises (Fig. 6.11). It is thus clear that the expression of phosphoribulokinase activity requires the presence of glyceraldehyde phosphate dehydrogenase [89].

The progress curve of the reaction, in the presence of the bienzyme complex, displays a lag at the beginning of the reaction. This lag correlates with the dissociation of the complex on dilution. If the complex is incubated first for about 15 minutes, in the absence of the substrates of the reaction, and the substrates are then added to the reaction mixture, the progress curve is linear for a rather long period and does not display any lag (Fig. 6.12). The recombinant phosphoribulokinase associated with glyceraldehyde phosphate dehydrogenase displays the same behaviour and so does the [K 64] mutant, although its activity is lower than that of the wild type enzyme, native or recombinant (Fig. 6.12).

A simple model that represents the dissociation and change of activity of the complex is shown in Fig. 6.13. It takes account of the ordered binding of substrates, ribulose bisphosphate being bound first and ATP afterwards. To fit the rate data correctly, the model must postulate that both the free phosphoribulokinase and the bienzyme complex are active. The active oxidized phosphoribulokinase form, which is released on dissociation of the complex, is not stable and slowly relapses to an inactive stable enzyme form. The active enzyme form released can be considered a metastable state of the phosphoribulokinase which differs from the stable state by its spectrofluorimetric features [90].

Thus, although there is a single structural gene coding for chloroplast phosphoribulokinase, three enzyme forms have been isolated that have indeed the same sequence but very different functional properties, as exemplified by their k_{cat} and K_m values (Table 6.1). From these results, it is thus clear that glyceraldehyde phosphate dehydrogenase has given phosphoribulokinase the instruction and the energy that allow it to carry out a catalytic process it would be unable to perform in a free state. The free metastable enzyme form,

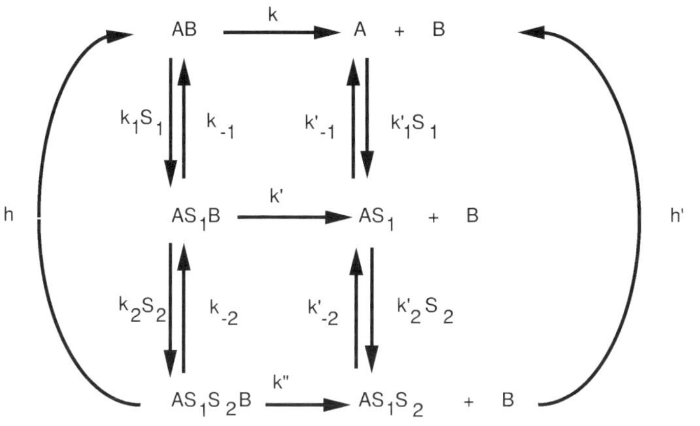

Fig. 6.13. Simple theoretical model for the dissociation of the bienzyme complex during enzyme reaction. AB is the complex, S_1 and S_2 are the two substrates. This model allows to fit the progress curves of the reaction. From ref. [89].

Table 6.1
Kinetic parameters of the three forms of phosphoribulokinase

State of oxidized phosphoribulokinase	Km (Ru5P) (μM)	Km (ATP) (μM)	k_{cat} s^{-1}/site
As part of the complex	30	46	3.25
Free "metastable" state	59	48	56
Free "stable" state	115	89	0.06

which is released after dissociation of the complex, is more active than the same enzyme embedded in the complex. This result strongly suggests that the free phosphoribulokinase keeps the imprinting of glyceraldehyde phosphate dehydrogenase after dissociation of the complex for some time. Moreover the internal mobility of the free metastable phosphoribulokinase is probably larger than that of the same enzyme embedded in the complex. Taken together, these two effects may explain the higher activity of the free metastable form of phosphoribulokinase relative to that of the same enzyme within the bienzyme complex.

6.7.2.3. Thermodynamics of instruction transfer and imprinting effects

Free stable phosphoribulokinase is a dimer which does not display any interaction between its subunits and which follows classical Michaelis–Menten kinetics relative to either substrate. Thus the corresponding steady state rate, when one substrate is saturating is

$$\frac{v}{2k^*[\mathrm{E}]_0} = \frac{\overline{K}^*[\mathrm{S}]}{1 + \overline{K}^*[\mathrm{S}]} \tag{6.106}$$

Table 6.2
Thermodynamic parameters associated with phosphoribulokinase reaction in the bienzyme complex

	Ru5P	ATP
λ (KJ/mol)	−3	−1.4
U_γ^σ (KJ/mol)	−12.7	−10.8
$\overline{\Delta G^*}$ (kJ/mol)	−21.7	−22.3
$\overline{\Delta G}$ (kJ/mol)	−24.8	−23.8
$\Delta G^{\neq *}$ (kJ/mol)	−77.3	−77.3
$\Delta G^{\neq}$ (kJ/mol)	−67.6	−67.3

$\overline{\Delta G}$ and $\overline{\Delta G^*}$ are the apparent binding free energies of the substrates to phosphoribulokinase in the complex and in the free state, respectively.

where $[\mathrm{E}]_0$ is, as ever, the total enzyme concentration, k^* is the catalytic constant and $\overline{K}^*$ is the apparent affinity constant for the non-saturating substrate. When the same enzyme is embedded in a supramolecular complex, the corresponding rate equation is now

$$\frac{v}{4k^*[\mathrm{E}]_0} = \frac{\overline{K}^*[\mathrm{S}]\exp(-U_\gamma^{\sigma\mathrm{H}}/RT)}{1+\overline{K}^*[\mathrm{S}]\exp(-\lambda/RT)}. \tag{6.107}$$

The number 4, which appears in the denominator of the left-hand side member of this equation, refers to the fact that there are four subunits of phosphoribulokinase in the bienzyme complex (two dimeric enzyme molecules). Comparison of eqs. (6.106) and (6.107) shows that the spatial organization of the supramolecular edifice and conformational constraints between the two enzyme species have altered the properties of phosphoribulokinase. More specifically, $U_\gamma^{\sigma\mathrm{H}}$ represents how the constraints exerted by different polypeptide chains within the complex tend to stabilize, or to destabilize, the ground state of the enzyme that has not yet bound any substrate. By the same token, λ represents the variation, along the reaction co-ordinate, of the stabilization–destabilization exerted on the enzyme by the other protein of the complex. As the enzyme in the complex follows Michaelis–Menten kinetics, λ must be a constant.

From the kinetic data one can determine: two values of λ and $U_\lambda^{\sigma\mathrm{H}}$ (for either ribulose 5-phosphate or ATP saturating); two apparent substrate binding energies, $\overline{\Delta G^*}$, for the free enzyme; two apparent substrate binding energies, $\overline{\Delta G}$, for the enzyme embedded in the complex; the free energy of catalysis, $\Delta G^{\neq *}$, for the free enzyme; the free energy of catalysis, $\Delta G^{\neq}$, for the enzyme in the complex. The numerical values are reported in Table 6.2.

The variation of the stabilization–destabilization along the reaction co-ordinate, as expressed by λ, favours the binding of ribulose 5-phosphate by a value of about 3 kJ/mole and the binding of ATP by about 1.5 kJ/mole. These values are indeed modest compared to that observed on the catalytic constant (Table 6.2). The value of the free energy of activation associated with this constant is decreased by a value of about 10 kJ/mole. Thus, within the complex, glyceraldehyde phosphate dehydrogenase has given phosphoribuloki-

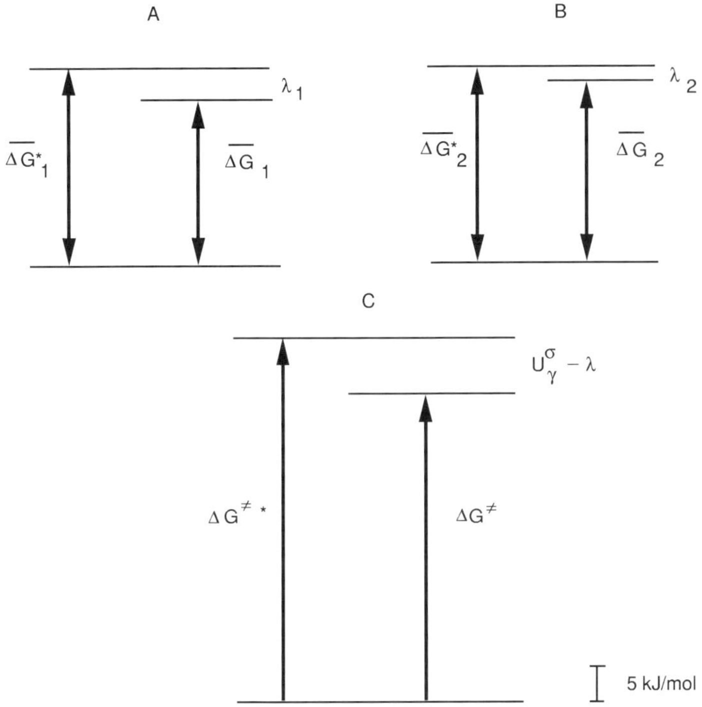

Fig. 6.14. Instruction transfer within the bienzyme complex. Instruction transfer facilitates ribulose bisphosphate (A) and ATP (B) binding to the phosphoribulokinase. It also decreases the height of the energy barrier associated with catalysis (C). Adapted from ref. [91].

nase an instruction and an energy that have mainly been used to decrease the height of the energy barrier associated with catalysis. Thus, the values of λ and $U_{\gamma}^{\sigma \mathrm{H}} - \lambda$, which appear in equations (6.73), represent how the energy transferred from glyceraldehyde phosphate dehydrogenase to phosphoribulokinase is used by the latter enzyme to alter the binding of the substrates and the efficiency of catalysis. This situation can be illustrated by the thermodynamic boxes of Fig. 6.14.

It has already been outlined that, when the bienzyme complex breaks down on dilution, the phosphoribulokinase thus released is in a metastable active state which slowly relapses to the inactive, stable conformation. Both enzyme forms display Michaelis–Menten kinetics. From these data, one can derive the apparent binding constants ($\overline{K}_{\sigma}$ and $\overline{K}_{\mu}$) of the two substrates to the stable (σ) and metastable (μ) enzyme forms. Similarly, one can also determine the two catalytic constants (k'_{σ} and k'_{μ}) pertaining to these stable (σ) and metastable (μ) forms, as well as the apparent second-order constants ($k'_{\sigma}\overline{K}_{\sigma}$ and $k'_{\mu}\overline{K}_{\mu}$) for the two substrates [91]. The corresponding values are collected in Table 6.3.

From these parameters, one can derive the energy values $\Psi_{0,\gamma} - \Psi_{1,\gamma}$, $\Psi_{1,\gamma} - \Psi_{0,\tau}$ and $\Psi_{0,\gamma} - \Psi_{0,\tau}$, which are reported in the same table. These energy values show how the imprinting exerted by gyceraldehyde phosphate dehydrogenase on phosphoribulokinase has

Table 6.3
Thermodynamic parameters associated with the imprinting of glyceraldehyde phosphate dehydrogenase on phosphoribulokinase

		Ru5P	ATP
Transferred energy used to alter $\overline{K}$	$\Psi_{0\gamma} - \Psi_{1\gamma}$ (KJ/mol)	−1.6	−1.4
Transferred energy used to alter k	$\Psi_{1\gamma} - \Psi_{0\tau}$ (KJ/mol)	−16.3	−16.3
Transferred energy used to alter $k\overline{K}$	$\Psi_{0\gamma} - \Psi_{0\tau}$ (kJ/mol)	−17.8	−17.8

k is the catalytic constant and $\overline{K}$ is the apparent substrate binding constant.

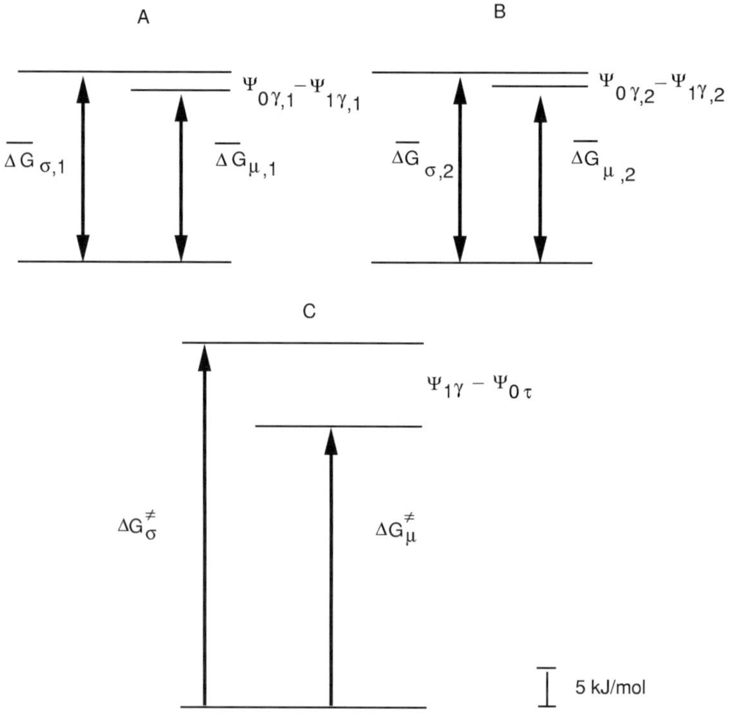

Fig. 6.15. Imprinting on isolated phosphoribulokinase. Imprinting facilitates slightly ribulose bisphosphate (A) and ATP binding to the enzyme. It decreases significantly the height of the energy barrier associated with catalysis (C).

altered the energy parameters of this enzyme, namely: the apparent binding energy of the two substrates; the free energy of activation of catalysis; the free energy associated with the second-order rate constant. The imprinting effect exerted on phosphoribulokinase facilitates the binding of either substrate by about the same value (1.5 kJ/mole), but decreases the height of the energy barrier associated with catalysis by about 16 kJ/mole. These conclusions are formulated in the thermodynamic boxes of Fig. 6.15.

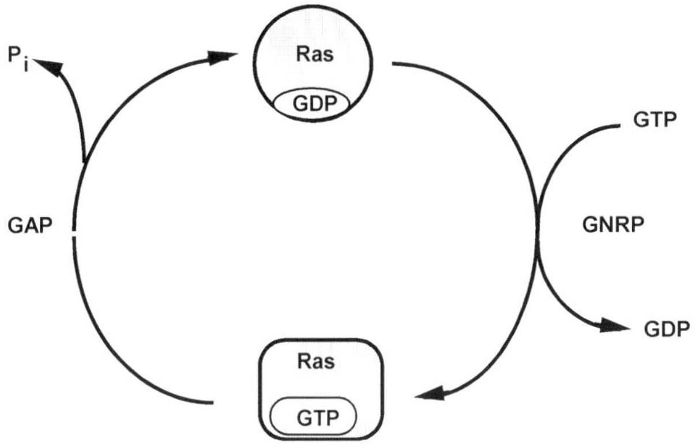

Fig. 6.16. The Ras system. The two forms of Ras are interconverted in the presence of guanine nucleotide releasing proteins (GNRP) and GTPase activating proteins (GAP).

6.7.3. The Ras–Gap complex

Guanosine triphosphate binding proteins (Ras) partake in the transfer of signals from the cell surface to the nucleus and stimulate cell proliferation and differentiation. These proteins play the part of a switch in signal transduction, and oscillate between an inactive GDP-bound and an active GTP-bound conformation [92–94]. The conversion of the inactive to the active form requires the phosphorylation of the bound GDP, and this phosphoryl exchange is stimulated by guanine nucleotide releasing proteins (GNRP). Alternatively, the conversion of the active to the inactive form of the protein involves the hydrolysis of bound GTP and the release of orthophosphate (Fig. 6.16). In the absence of any external factor, however, this process is extremely slow and so that active Ras protein predominates.

The hydrolysis of GTP is enhanced in the presence of a variety of proteins, including guanine-nucleotide exchange factors (GEF) and GTPase-activating proteins (GAP). The enhancement may be considerable. Thus, at saturating concentrations of GAP, the catalytic constant of GTP hydrolysis by Ras is enhanced by a factor of 10^5 [95,96]. The steady state ratio between these two protein forms is no doubt important for the living cell. In principle, any alteration of the structure of the Ras protein that results in an increase of its activity could promote cancer by inducing the cell to proliferate out of control of the appropriate extracellular signals. The genes that code for Ras proteins are called proto-oncogenes. If they have been altered by a mutation, they are called oncogenes. Point mutations, which cause the replacement of a single aminoacid residue, are sufficient to turn ras proto-oncogenes into oncogenes. All these aminoacid substitutions result in a decrease in Ras GTPase activity and shifts the steady state ratio towards the GTP-bound protein form.

The control of GTP hydrolysis by Ras and GAP is no doubt an important problem of cell biochemistry. Different hypotheses have been formulated to explain the GAP activation of GTP hydrolysis carrried out by Ras [95–98]. The most likely is the so-called arginine finger hypothesis [99–101] which has now received firm experimental support. The central idea of this proposed mechanism is that GAP reaches the active site of Ras and, through positive charges, stabilizes the transition state of the reaction [102]. The catalytic domain of GAP has been solved by X-ray crystallography and, indeed, shows the presence of several arginine residues in the putative active site of this protein, in agreement with the arginine finger hypothesis [103]. In the course of single turnover stopped-flow experiments performed with GAP-active neurofibromin and Ras [101,104] a sudden rise followed by a decline of fluorescence is observed. This is interpreted as fast binding of neurofibromin to Ras followed by hydrolysis of bound GTP and finally the dissociation of neurofibromin from Ras. If the arginine residues of the neurofibromin finger are replaced by another uncharged residue one observes, in single turnover experiments, a rise in fluorescence which is not followed by a decay. This means that the modified neurofibromin binds to the Ras protein, but no significant GTP hydrolysis takes place. If, on the other hand, the arginine residues of the neurofibromin finger are replaced by lysine, significant catalytic activity is restored [101]. Hence, in this system, the hydrolysis of GTP requires a complex of two proteins, Ras and GAP. The part played by GAP is apparently to stabilize the transition state by neutralization of γ-phosphate oxygens [101,102].

6.8. Proteins at the lipid–water interface and instruction transfer to proteins

The theoretical considerations of section 4 above suggest that the binding of an enzyme to the lipid–water interface may give this enzyme an instruction and an energy that can be used to promote catalytic activity. This enhancement of catalytic activity has been observed, for instance, with protein kinase C and pancreatic lipase. The main properties of these two systems will be briefly considered now.

6.8.1. Protein kinase C

Protein kinases C, like the other protein kinases, are involved in signal transduction. They can phosphorylate glycogen synthase and phosphorylase kinase. Protein kinases are found in the cytoplasm of the cell but are also translocated and bound to lipid membranes. The structure of the conventional protein kinase C is rather complex and reveals four domains [105–111]: the C_1 domain bears a cysteine-rich motif involved in ester binding, which is preceded by an aminoacid sequence called the pseudo-substrate; the C_2 domain contains the recognition site for acidic lipids and calcium ions [105,106]; the C_3 and C_4 domains bear the ATP- and substrate-binding sites of the kinase [106]. The so-called pseudo-substrate sequence is located in the immediate vicinity of the active site of the C_4 domain. It prevents the substrate from entering the site, and the enzyme is therefore inactive. Maturation of the enzyme takes place in the cytoplasm and involves phosphorylation of three residues of the C_4 domain, namely threonine 500, threonine 641 and serine 660. The translocation of the protein to the membrane leads to the removal of the pseudo-substrate from the active site and thereby to the activation of the enzyme [112,113]. The binding of protein kinase C to the membrane requires in fact the binding of membrane diacylglycerol and phosphatidylserine to the C_1 and C_2 domains of the enzyme, respectively. This process is strengthened in the presence of calcium [114].

6.8.2. *Pancreatic lipase*

Pancreatic lipase is almost totally inactive in the presence of soluble substrates of the enzyme. However, when the concentration of these substrates is increased, they become insoluble and form micelles and the lipase becomes highly active. This is the process of interfacial activation discovered by Desnuelle and co-workers [115]. This process has been explained by the existence of a conformation change of the enzyme that allows one to distinguish true lipases from esterases [116–118]. The reality of this conformation change has been confirmed by X-ray crystallography [119–121]. The active site possesses a Ser-His-Asp/Glu catalytic triad which is reminiscent of serine proteases [122]. The triad, however, is not directly accessible to the substrate for it is covered by surface loops that constitute a "lid" [123–125]. On interfacial activation these surface loops may show large interfacial rearrangements. Specific experimental conditions may favour the occurrence of "open" or "closed" conformations of lipases. Thus, polyethylene glycol promotes the crystallization of the "closed" form whereas alchohols and detergents lead to the crystallization of the "open" state. In the "open" conformation, some hydrophobic regions become exposed to the solvent, whereas other previously exposed regions become buried in the protein.

6.9. *Information transfer between proteins and enzyme regulation*

Communication between proteins relies on the existence of conformation changes of these proteins. But these conformation changes are also the basis of enzyme regulation. It is thus sensible to raise the question of the similarities and differences that may exist between enzyme regulation and information transfer between proteins.

Enzyme regulation stems from the propagation of information within an enzyme molecule. This propagation originates from different events that may affect the conformation of the active site and therefore enzyme activity. The events may be due to a local alteration of the structure of the polypeptide chain, for instance phosphorylation, glycosylation, etc., or to the reversible binding of a ligand to the protein. Thus, phosphorylation or dephosphorylation of some specific aminoacid residues of the enzyme may result in a conformation change that is propagated to the active site and results in a dramatic change of catalytic activity. Moreover, if an enzyme bears several catalytically competent sites, the occupancy of one of them by a substrate molecule may modulate the activity of another site. Similarly, in addition to the catalytic sites, there may exist other sites which are not involved in the catalytic process but which can bind an effector (activator or inhibitor) of the reaction. In either case, this means that information is propagated from site to site within the enzyme molecule. This type of event, which takes place within the same enzyme molecule, may also play a major role in the communication between different molecules.

In many cases, the function of the protein is, in a way, encoded in the corresponding structural gene and the expression of this function is modulated by the molecular mechanisms of enzyme regulation. In the processes that have been considered in this chapter, the expression of a biological function results, in fact, from the information contained in different structural genes coding for the proteins of the same molecular edifice which interact with each other.

Thus, although enzyme regulation and communication between proteins use the same molecular devices of conformational changes, they are markedly different. These differences are summarized below. In enzyme regulation, the information transfer is intramolecular. In the process of enzyme communication, it is intermolecular. In enzyme regulation that does not involve covalent modification of the protein, the conformation changes often modulate a reaction rate that is already significant in the absence of the effector. For most experimental evidence of enzyme communication thus far obtained, the response of the active enzyme is of the all-or-none type and does not correspond to a modulation of an existing activity. This means that the enzyme is active in the complex and nearly totally inactive when it is free. Last but not least, in conventional enzyme regulation, enzyme activity is considered to be solely dependent on the corresponding structural gene. If information transfer between proteins takes place, several structural genes are required to specify the function of an enzyme within a supramolecular edifice. The distinction between these two types of processes, however, is not as sharp as one might have expected, because certain enzymes, such as aspartate transcarbamylase, are made up of different polypeptide chains and can, in a way, be considered supramolecular edifices.

6.10. *Channelling of reaction intermediates within multienzyme complexes*

We have considered so far an information transfer from protein to protein within enzyme complexes. But there may also exist a transfer of molecules from active site to active site within these complexes. The possibility of this transfer was suggested by the observation that many multienzyme complexes catalyse consecutive reactions of the same metabolic sequence. If the association of different enzymes represent a functional advantage, it would be the direct transfer of reaction intermediates from site to site within the same molecular edifice. Channelling should thus avoid dilution of the reaction intermediates in the cell milieu and should increase the efficiency of the corresponding metabolic pathway [126–134].

The increase in efficiency can be expressed in different ways. The best is probably offered by the measurement of the transient of the overall metabolic flux. If the flux is in steady state and is perturbed, it returns back to its initial steady state. The time required for returning to the initial state is the transient time, or the transient, τ, of the overall process. In a metabolic system, in which all the enzymes are physically independent, the difference between the actual progress curve of the last product and the theoretical progress curve pertaining to the last enzyme of the reaction sequence is due to the accumulation of reaction intermediates. Although this will not be demonstrated here, it is intuitively obvious that the overall sequence must be [135–139].

$$\tau = \sum_i ([\widetilde{\mathrm{I}}_i]/J) = \sum_i \tau_i, \tag{6.108}$$

where $[\widetilde{\mathrm{I}}_i]$ is the steady state concentration of the intermediate I_i, J is the overall flux and τ_i is the transient of the ith reaction. This equation has several interesting implications:

first, the transient τ increases with the number of enzyme steps of the pathway; second, the transient time increases with the steady state concentration of the intermediates I; third, since this concentration depends on the volume of the corresponding cell compartment, the τ value may be expected to be larger for large cells than for smaller ones. Equation (6.108) is indeed valid if there is no channelling, but only a diffusion of the reaction intermediates from active site to active site of physically independent enzymes.

It is thus essential to obtain direct experimental evidence for the existence of channelling of reaction intermediates. This goal has ben reached in a number of cases. Direct evidence for channelling has beeen obtained by different techniques: crosslinking in a matrix of different enzymes that catalyse consecutive reactions [140]; fusion of genes coding for consecutive enzyme reactions [141]; resolution of the structure and of the structure changes of a native multienzyme complex [142]. Using the first and the second techniques, one can observe an increase in the performances of the associated enzymes, compared to the reaction flux measured when these enzymes are in a free state. Although the last approach does not always offer the possibility of measuring the functional improvement brought about by channelling, it allows the possibility to "see" the channel. For this reason, we shall present very briefly the wealth of informations that have been obtained on the bienzyme complex tryptophan synthase of enteric bacteria [143–149]. The complex catalyses the overall reaction

$$\text{IGP} + \text{Serine} \longleftrightarrow \text{GAP} + \text{Tryptophan}$$

where IGP is indole glycerol phosphate and GAP glyceraldehyde phosphate. This overall reaction is in fact the sum of two elementary processes, namely

$$\text{IGP} \longleftrightarrow \text{Indole} + \text{GAP}$$

and

$$\text{Indole} + \text{Serine} \longleftrightarrow \text{Tryptophan}$$

The first process is catalysed by the α-enzyme and the second by the β-enzyme. The latter enzyme requires pyridoxal, which is attached to the protein through a Schiff base linkage with an ε-amino group. The overall complex has an $\alpha_2\beta_2$ structure. The β-reaction occurs in two stages. During the first stage, L-serine reacts with enzyme-bound pyridoxal phosphate to form a rather stable α-aminoacrylate intermediate. Then, in the second stage, indole reacts with this intermediate to form L-tryptophan. Indole, which is formed as a product of the first reaction, is channelled to the β-site.

The interactions of the α and β polypeptide chains lead to interesting effects that have already been discussed earlier, on a theoretical basis, in this chapter. The isolated α-enzyme exists as a monomer, α, and the isolated β-enzyme as a dimer, β_2. Temperature-jump studies have shown that α exists in, at least, three conformations, $\alpha, \alpha', \alpha''$, and β_2 in two

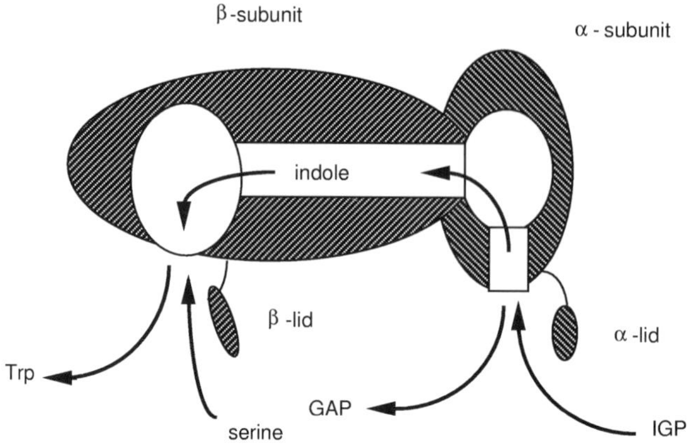

Fig. 6.17. The $\alpha\beta$ complex of tryptophan synthase. Indole is channelled between α and β subunits as described in the text.

states, β_2 and β_2'. However, when the bienzyme complex is formed by their association

$$2(\alpha \longleftrightarrow \alpha' \longleftrightarrow \alpha'') + (\beta_2 \longleftrightarrow \beta_2') \longleftrightarrow \alpha_2\beta_2$$

the $\alpha_2\beta_2$ complex apparently displays only one conformation [150,151]. This result illustrates the theoretical idea, already discussed, that the association of two proteins must result in a decrease of their information content.

An $\alpha\beta$ subunit pair of tryptophan synthase is very schematically pictured in Fig. 6.17. Indole glycerol phosphate enters the α-site, is cleaved and the resulting glyceradehyde leaves the site while indole is channelled through a tunnel of 25 Å-long to the β-site. Serine enters the β-site, forms, with bound pyridoxal phosphate, an α-aminoacrylate intermediate that is transformed, after a sequence of chemical reactions that will be not described here, into tryptophan and water that both leave the β-site. It is hardly conceivable that the coupling of the two sites is solely explained by the free diffusion of indole through a 25 Å-long tunnel. There must exist some devices that allow the co-ordination of the functioning of the two sites. They have been studied in detail [142]. The co-ordination of the activity of the two sites relies on the existence of polypeptide loops, or "lids", of the two enzymes that change their conformation and alternatively "close" or "open" the two sites. Thus, as soon as indole glycerol phosphate is bound to the α-site, the corresponding polypeptide loop "closes" this site thereby preventing the escape of the ligand to the external milieu. The only possibility left for indole glycerol phosphate is to travel via the tunnel from the α- to the β-site. Moreover, the "lid" temporarily closes the β-site after the binding of serine, thus preventing its release to the external milieu [142]. It is therefore obvious that this co-operation between the two sites can only increase the performance of the bienzyme complex. This is indeed what is observed experimentally, for the reaction of L-serine with the β-site activates the α-site 25- or 30-fold [142].

6.11. *The different types of communication within multienzyme complexes*

There may therefore exist different types of communication in multienzyme complexes. First, an instruction may be transferred, within the complex, from protein to protein. This means that the functional properties of an individual enzyme are not necessarily the same if they are studied *in vivo*, or in the isolated state, *in vitro*. Hence the classical views we have on metabolism may be biased. Thus, for instance, phosphofructokinase is considered the main site of regulation of glycolysis on the basis of its properties *in vitro*. In free solution, the enzyme is allosteric and is inhibited by ATP and citrate, which can be considered the end products of the glycolytic flux. This classical view, which is found in most textbooks, appears, however, to be seriously flawed for, in the living cell, most of the phosphofructokinase molecules are in a bound state and do not appear to be allosteric. Moreover, in most eukaryotic cells, for instance in muscle cells, ATP and citrate levels are so high that the enzyme should be inhibited if its properties were the same *in vivo* and *in vitro* [152].

Second, the instruction transfer may lead to the dissociation of the complex, thus producing isolated proteins that temporarily have properties, for instance catalytic properties, that are different from those they have in their stable, isolated state. After dissociation of the complex, one of the enzymes appears in an active metastable state that slowly relapses to its stable inactive state. This means that the transiently active enzyme displays the imprinting of another protein. These two types of communication can take place whether or not the enzymes of the multienzyme complex catalyse consecutive reactions.

Last but not least, metabolites can be channelled within a multienzyme complex. At least in a number of cases which are particularly well documented, this process requires both the transfer of a reaction intermediate from site to site and induced conformation changes that prevent the intermediate from diffusing out of the complex. This situation thus involves both a transfer of information and a transfer of molecules from site to site.

It is striking that these properties, which certainly play an important role in cell functions, are a consequence of the complexity of the supramolecular edifices that exist in eukaryotic cells and even, to a lesser extent, in prokaryotic organisms.

References

[1] Crick, F. (1970) Central dogma of molecular biology. Nature, 227, 561–563.

[2] Laurent, T.C. (1963) The interaction between polysaccarides and other macromolecules. 5. The solubility of proteins in the presence of dextran. Biochem. J. 89, 253–257.

[3] Laurent, T.C. and Ogston, A.G. (1963) The interaction between polysaccharides and other macromolecules. 4. The osmotic pressure of mixtures of serum albumin and hyaluronic acid. Biochem. J. 89, 249–253.

[4] Laurent, T.C. (1971) Enzyme reactions in polymer media. Eur. J. Biochem. 21, 498–506.

[5] Minton, A.P. (1981) Excluded volume as a determinant of macromolecular structure and reactivity. Biopolymers 20, 2093–2120.

[6] Zimmerman, S.B. and Minton, A.P. (1993) Macromolecular crowding: biochemical, biophysical and physiological significance. Annu. Rev. Biophys. Biomol. Struct. 22, 27–65.

[7] Minton, A.P. (1998) Molecular crowding: analysis of effects of high concentrations of inert cosolutes on biochemical equilibria and rates in terms of volume exclusion. In: Methods Enzymol., Vol. 295 (Part B), Academic Press, New York, pp. 127–149.

[8] Adair, G.S. (1925) The hemoglobin system: VI. The oxygen dissociation curve of hemoglobin. J. Biol. Chem. 63, 529–545.
[9] Levitzki, A. (1978) Quantitative Aspects of Allosteric Mechanisms. Springer-Verlag, Berlin.
[10] Cantor, C.R. and Schimmel, P.R. (1980) Biophysical Chemistry. Part III. The Behavior of Biological Macromolecules. Freeman, San Francisco, CA.
[11] Tanford, C. (1961) Physical Chemistry of Macromolecules. Wiley, New York.
[12] Wyman, J. (1964) Linked functions and reciprocal effects in haemoglobin: A second look. Adv. Prot. Chem. 19, 223–286.
[13] Wyman, J. (1967) Allosteric linkage. J. Amer. Chem. Soc. 89, 2202–2218.
[14] Ricard, J., Mouttet, C. and Nari, J. (1974) Subunit interactions in enzyme catalysis. I. Kinetic models for one-substrate polymeric enzymes. Eur. J. Biochem. 41, 479–497.
[15] Nari, J., Mouttet, C., Fouchier, F. and Ricard, J. (1974) Subunit interactions in enzyme catalysis. II. Kinetic analysis of subunit interactions in the enzyme L-phenylalanine ammonia lyase. Eur. J. Biochem. 41, 419–515.
[16] Ricard, J. and Noat, G. (1984) Subunit interactions in enzyme transition states. Antagonism between substrate binding and reaction rate. J. Theor. Biol. 111, 737–753.
[17] Ricard, J. and Noat, G. (1985) Subunit coupling and kinetic co-operativity of polymeric enzymes. Amplification, attenuation and inversion effects. J. Theor. Biol. 117, 633–649.
[18] Ricard, J. and Cornish-Bowden, A. (1987) Co-operative and allosteric enzymes: 20 years on. Eur. J. Biochem. 166, 255–272.
[19] Ricard, J., Giudici-Orticoni, M.T. and Buc, J. (1990) Thermodynamics of information transfer between subunits in oligomeric enzymes and kinetic cooperativity. 1. Thermodynamics of subunit interactions, partition functions and enzyme reaction rate. Eur. J. Biochem. 194, 463–473.
[20] Giudici-Orticoni, M.T., Buc, J. and Ricard, J. (1990) Thermodynamics of information transfer between subunits in oligomeric enzymes and kinetic cooperativity. 2. Thermodynamics of kinetic cooperativity. Eur. J. Biochem. 194, 475–481.
[21] Giudici-Orticoni, M.T., Buc, J., Bidaud, M. and Ricard, J. (1990) Thermodynamics of information transfer between subunits in oligomeric enzymes and kinetic cooperativity. 3. Information transfer between the subunits of chloroplast fructose bisphosphatase. Eur. J. Biochem. 194, 483–490.
[22] Ricard, J. (1985) Organized polymeric enzyme systems: catalytic properties. In: G.R. Welch (Ed.), Organized Multienzyme Systems. Catalytic properties. Academic Press, New York, pp. 177–240.
[23] Ricard, J. (1989) Concepts and models of enzyme cooperativity. In: G. Hervé (Ed.), Allosteric Enzymes. CRC Press, Boca Raton, FL, pp. 1–25.
[24] Neet, K.E. (1995) Cooperativity in enzyme function: equilibrium and kinetic aspects. In: D.L. Purich (Ed.), Methods Enzymol., Vol. 249 (Part D). Academic Press, New York, pp. 519–567.
[25] Shultz, A.R. (1994) Enzyme Kinetics. From Diastase to Multi-Enzyme Systems. Cambridge University Press, Cambridge.
[26] Ricard, J., Giudici-Orticoni, M.T. and Gontéro, B. (1994) The modulation of enzyme reaction rates within multi-enzyme complexes. I. Statistical thermodynamics and information transfer through multi-enzyme complexes. Eur. J. Biochem. 226, 993–998.
[27] Gontéro, B., Giudici-Orticoni, M.T. and Ricard J. (1994) The modulation of enzyme reaction rates within multi-enzyme complexes. II. Information transfer within a chloroplast multienzyme complex containing ribulose bisphosphate carboxylase oxygenase. Eur. J. Biochem. 226, 998–1006.
[28] Welch, G.R., Somogyi, B. and Damjanovich, S. (1982) The role of protein fluctuations in enzyme action: a review. Progress Biophys. Mol. Biol. 39, 109–146.
[29] Pauling, L. (1948) Nature of forces between large molecules of biological interest. Nature 161, 707–709.
[30] Ricard, J., Meunier, J.C. and Buc, J. (1974) Regulatory behavior of monomeric enzymes. I. The mnemonical enzyme concept. Eur. J. Biochem. 49, 195–208.
[31] Vlatakis, G., Andersson, L.I., Muller, R. and Mosbach, K. (1993) Drug assay using antibody mimics made by molecular imprinting. Nature 361, 645–647.
[32] Russel, A.J. and Klibanov, A.M. (1998) Inhibition induced enzyme activation in organic solvents. J. Biol. Chem. 263, 11624–11626.
[33] Klibanov, A.M. (1995) Enzyme memory. What is remembered and why? Nature 374, 596.

[34] Yennawar, M.P., Yennawar, N.M. and Farber, G.K. (1995) A structural explanation for enzyme memory in nonaqueous solvents. J. Am. Chem. Soc. 117, 577–585.
[35] Dabulis, K. and Klibanov, A.M. (1993) Dramatitc enhancement of enzyme activity in organic solvents by lyoprotectants. Biotechnol. Bioeng. 41, 566–571.
[36] Yokey, H.P. (1992) Information Theory and Molecular Biology. Cambridge University Press, Cambridge.
[37] Anfinsen, C.B., Haber, E., Sela, M. and White, F.H. (1961) The kinetics of formation of native ribonuclease during oxidation of the reduced polypeptide chain. Proc. Natl. Acad. Sci. USA 47, 1309–1314.
[38] Anfinsen, C.B. (1973) Principles that govern the folding of protein chains. Science 181, 223–230.
[39] Anfinsen, C.B. and Scheraga, H.A. (1975) Experimental and theoretical aspects of protein folding. Advances Prot. Chem. 29, 205–300.
[40] Creighton, T.E. (1983) Proteins. Freeman, New York.
[41] Yon, J. (1997) Protein folding: concepts and perspectives. Cell. Mol. Life Sci. 53, 557–567.
[42] Shannon, C.E. and Weaver, W. (1949) The Mathematical Theory of Communication. University of Illinois Press, Urbana, IL.
[43] Deshaies, R.J., Koch, B.D., Xerner-Washburne, M., Craig, E.A. and Schekman, R. (1988) A subfamily of stress proteins facilitates translation of secretory and mitochondrial precursor polypeptides. Nature 332, 800–805.
[44] Chirico, W.J., Waters, M.G. and Blobel, G. (1988) 70 K heat shock related proteins stimulate protein translocation into microsomes. Nature 332, 805–810.
[45] Ellis, R.J. (1991) Molecular chaperones. Annu. Rev. Biochem. 60, 321–347.
[46] Langer, T. and Neupert, W. (1996) Chaperonin-mediated and assembly of proteins in mitochondria. In: R.J. Ellis (Ed.), The Chaperonins. Academic Press, San Diego, pp. 91–106.
[47] Ellis, R.J. and Hemmingsen, S.M. (1989) Molecular chaperones: proteins essential for the biogenesis of some macromolecular structures. Trends Biochem. Sci. 14, 339–342.
[48] Martin, J., Langer, T., Boteva, R., Schrammel, A., Horwich, A.L. and Hartl, F.U. (1991) Chaperonin-mediated protein folding at the surface of GroEL through a "molten globule" like intermediate. Nature 352, 36–42.
[49] Pelham, H.R.B. (1986) Speculations on the functions of the major heat shock and glucose-regulated proteins. Cell 46, 959–966.
[50] Nguyen, V.T., Morange, M. and Bensaude, O. (1989) Protein denaturation during heat shock and related stress. *Escherichia coli* beta-galactosidase and *Photinus pyralis* luciferase inactivation in mouse cells. J. Biol. Chem. 264, 10487–10492.
[51] Pinto, M., Morange, M. and Bensaude, O. (1991) Denaturation of proteins during heat shock. In vivo recovery of solubility and activity of reporter enzymes. J. Biol. Chem. 266, 13941–13946.
[52] Ellis, R.J. (1996) The Chaperonins. Academic Press, San Diego, CA.
[53] Ikemura, H., Takagi, H. and Inouye, M. (1988) Requirement of a pro-sequence for the production of active subtilisin E in *Escherichia coli*. J. Biol. Chem. 262, 7859–7864.
[54] Shinde, U., Li, Y., Chatterjee, S. and Inouye, M. (1993) Folding pathway mediated by an intramolecular chaperone. Proc. Natl. Acad. Sci. USA 90, 6924–6928.
[55] Zhu, X.L., Ohta, Y., Jordan, F. and Inouye, M. (1989) Pro-sequence of subtilisin can guide the refolding of denatured subtilisin in an intermolecular process. Nature 339, 483–484.
[56] Eder, J., Rheinnecker, M. and Fersht, A.R. (1993) Folding of subtilisin BPN: role of the prosequence. J. Mol. Biol. 233, 293–304.
[57] Strausberg, S., Alexander, P., Wang, L., Schwarz, F. and Bryan, P. (1993) Catalysis of a protein folding reaction: thermodynamic and kinetic analysis of subtilisin BPN interactions with its propeptide fragment. Biochemistry 32, 8312–8319.
[58] Shinde, U.P., Liu, J.J. and Inouye, M. (1997) Protein memory through altered folding mediated by intramolecular chaperones. Nature 389, 520–522.
[59] Silen, J.L. and Agard, D.A. (1989) The α-lytic protease pro-region does not require a physical linkage to activate the protease domain in vivo. Nature 341, 462–464.
[60] Lee, Y.C., Ohta, T. and Matsuzawa, H. (1992) A non-covalent NH_2-terminal pro-region aids the production of active aqualysin I (a thermophylic protease) without the COOH-terminal pro-sequence in *Escherichia coli*. FEMS Microbiol. Lett. 71, 72–77.

[61] Winther, J.R., Sorensen, P. and Kielland-Grandt, M.C. (1994) Refolding of carboxypeptidase Y folding intermediate in vitro by low affinity binding of the pro-region. J. Biol. Chem. 268, 22007–220013.
[62] Eder, J., Rheinnecker, M. and Fersht, A.R. (1993) Folding of subtilisin BPN: characterization of a folding intermediate. Biochemistry 32, 18–26.
[63] Shinde, U.P. and Inouye, M. (1995) Folding pathway mediated by an intramolecular chaperone: characterization of the structural changes in pro-subtilisin E coincident with autoprocessing. J. Mol. Biol. 252, 25–30.
[64] Baker, D., Sohl, Y. and Agard, D.A. (1992) A protein-folding reaction under kinetic control. Nature 356, 263–265.
[65] Prusiner, S.B. (1989) Scrapie prions. Annu. Rev. Microbiol. 43, 347–374.
[66] Prusiner, S.B. (1991) Molecular biology of prion diseases. Science 252, 1515–1522.
[67] Prusiner, S.B. (1997) Prion disease and the BSE crisis. Science 278, 245–251.
[68] Bolton, D.C., McKinley, M.P. and Prusiner, S.B. (1982) Identification of a protein that purifies with the scrapie prion. Science 218, 1309–1311.
[69] Gabizon, R., McKinley, M.P., Groth, D. and Prusiner, S.B. (1988) Immunoaffinity purification and neutralization of scrapie prion infectivity. Proc. Natl. Acad. Sci. USA 85, 6617–6622.
[70] Kacser, H. and Rankin Small, J. (1996) How many phenotypes from one genotype? The case of prion disease. J. Theor. Biol. 182, 209–218.
[71] Laurent, M. (1996) Prion diseases and the "protein only" hypothesis: a theoretical dynamic study. Biochem. J. 318, 35–39.
[72] McClintock, D.K. and Bell, P.M. (1971) The mechanism of activation of human plasminogen by streptokinase. Biochem. Biophys. Res. Com. 43, 694–702.
[73] Reddy, K.N. and Markus, G. (1972) Mechanism of activation of human plasminogen by streptokinase. Presence of active center in streptokinase–plasminogen complex. J. Biol. Chem. 247, 1683–1691.
[74] Schlick, L.A. and Castellino, F.Y. (1973) Interaction of streptokinase and rabbit plasminogen. Biochemistry 12, 4315–4321.
[75] Summaria, L., Wohl, R.C., Boreisha, I.G. and Robbins, K.C. (1982) A virgin enzyme derived from human plasminogen. Specific cleavage of the arginyl-560-valyl peptide bond in the diisopropoxyphosphinyl virgin enzyme by plasminogen activators. Biochemistry 21, 2056–2059.
[76] Buck, F.F., Hummel, B.C.N. and De Renzo, E.C. (1968) Interaction of streptokinase and human plasminogen. V. Studies on the nature and mechanism of formation of the enzymatic site of the activator complex. J. Biol. Chem. 243, 3648–3654.
[77] Groskopf, W.R., Summaria, L. and Robbins, K.C. (1969) Studies on the active center of human plasmin. The serine and histidine residues. J. Biol. Chem. 244, 359–365.
[78] Reed, G.L., Lin, L.F., Parkami-Seren, B. and Kussie, P. (1965) Identification of a plasminogen binding region in streptokinase that is necessary for the creation of a functional streptokinase–plasminogen activation complex. Biochemistry 34, 10266–10271.
[79] Muller, B. (1972) A labile CO_2-fixing complex in spinach chloroplasts. Z. Naturforsch. 27b, 925–932.
[80] Gontéro, B., Cardenas, M.L. and Ricard, J. (1988) A functional five-enzyme complex of chloroplasts involved in the Calvin cycle. Eur. J. Biochem. 173, 437–446.
[81] Rault, M., Giudici-Orticoni, M.T., Gontéro, B. and Ricard, J. (1993) Structural and functional properties of a multi-enzyme complex of spinach chloroplasts. I. Stoichiometry of the polypeptide chains. Eur. J. Biochem. 217, 1065–1073.
[82] Gontéro, B., Mulliert, G., Rault, M., Giudici-Orticoni, M.T. and Ricard, J. (1993) Structural and functional properties of a multi-enzyme complex of spinach chloroplasts. II. Modulation of the kinetic properties of enzymes in the aggregated state. Eur. J. Biochem. 217, 1075–1082.
[83] Nicholson, S., Easterby, J.S. and Powls, R. (1987) Properties of a multimeric protein complex from chloroplasts possessing potential activities of NADPH-dependent glyceraldehyde phosphate dehydrogenase and phosphoribulokinase. Eur. J. Biochem. 162, 423–431.
[84] Clasper, S., Easterby, J.S. and Powls, R. (1991) Properties of two high-molecular mass forms of GraPDH from spinach leaf, one of which also possesses latent phoshoribulokinase activity. Eur. J. Biochem. 202, 1239–1246.
[85] Süss, K.H., Arkona, C., Manteuffel, R. and Adler, K. (1993) Calvin cycle multienzyme complexes are bound to chloroplast thylakoid membranes of higher plants in situ. Proc. Natl. Acad. Sci. USA 90, 5514–5518.

[86] Sainis, J.K., Merriam, K. and Harris, G. (1989) The association of D-ribulose-1,5-bisphosphate carboxylase/oxygenase with phosphoribulokinase. Plant Physiol. 89, 368–374.

[87] Anderson, L.E., Goldhaber-Gordon, I.M., Li, D., Tang, X., Xsiang, M. and Prakash, N. (1995) Enzyme–enzyme interactions in the chloroplast: glyceraldehyde phosphate dehydrogenase, triose phosphate isomerase and aldolase. Planta 196, 245–255.

[88] Avilan, L., Gontéro, B., Lebreton, S. and Ricard, J. (1997) Memory and impronting effects in multi-enzyme complexes. I. Isolation, dissociation and reassociation of a phosphoribulokinase–glyceraldehyde phosphate dehydrogenase from *Chlamydomonas reinhardtii* chloroplasts. Eur. J. Biochem. 246, 78–84.

[89] Lebreton, S., Gontéro, B., Avilan, L. and Ricard, J. (1997) Memory and imprinting effects in multienzyme complexes. II. Kinetics of the bi-enzyme complex from *Chlamydomonas reinhardtii* and hysteretic activation of chloroplast oxidized phosphoribulokinase. Eur. J. Biochem. 246, 85–91.

[90] Avilan, L., Gontéro, B., Lebreton, S. and Ricard, J. (1997) Information transfer in multienzyme complexes. II. The role of Arg 64 of *Chlamydomonas reinhardtii* phosphoribulokinase in the information transfer between glyceraldehyde phosphate dehydrogenase and phosphoribulokinase. Eur. J. Biochem. 250, 296–302.

[91] Lebreton, S., Gontéro, B., Avilan, L. and Ricard, J. (1997) Information transfer in multienzyme complexes. I. Thermodynamics of conformational constraints and memory effects in the bienzyme glyceraldehyde phosphate dehydrogenase–phosphoribulokinase of *Chlamydomonas renhardtii* chloroplasts. Eur. J. Biochem. 250, 286–295.

[92] Bourne, H.R., Sanders, D.A. and McCormick, F. (1990) The GTPase superfamily: a conserved switch for diverse cell functions. Nature 348, 125–132.

[93] Bourne, H.R., Sanders, D.A. and McCormick, F. (1991) The GTPase superfamily: conserved structure and molecular mechanism. Nature 349, 117–127.

[94] Lowy, D.R. and Willumsen, B.M. (1993) Function and regulation of Ras. Annu. Rev. Biochem. 62, 851–891.

[95] Gideon, P., John, J., Freech, M., Lautwein, A., Clark, R. and Scheffer, J.E. (1992) Mutational and kinetic analysis of the GTPase-activating protein (GAP)-p21 interaction: the C-terminal domain of GAP is not sufficient for full activity. Mol. Cell. Biol. 12, 2050–2056.

[96] Eccleston, J.F., Moore, K.J.M., Morgan, L., Sjinner, R.H. and Lowe, P.N. (1993) Kinetics of interaction between normal and proline 12 Ras and the GTPase-activating proteins, p120-GAP and neurofibromin. J. Biol. Chem. 268, 27012–27019.

[97] Neal, S.A., Eccleston, J.F. and Webb, M.R. (1990) Hydrolysis of GTP by p21NRAS, the NRAS protooncogene product is accompanied by a conformational change in the wild-type protein: use of a single fluorescent probe at the catalytic site. Proc. Natl. Acad. Sci. USA 87, 3562–3565.

[98] Rensland, H., Lautwein, A., Wittinghofer, A. and Goody, R.S. (1991) Is there a rate-limiting step before GTP cleavage by H-ras p21? Biochemistry 36, 11181–11185.

[99] Schweins, T., Geyer, M., Scheffzek, K., Warshel, A., Kalbitzer, M.R. and Wittinghofer, A. (1995) Substrate assisted catalysis as a mechanism for GTP hydrolysis of p21 ras and other GTP-binding proteins. Nature Struct. Biol. 2, 36–44.

[100] Wittinghofer, A., Scheffzek, K. and Ahmadian, M.R. (1997) The interaction of Ras with GTPase-activating proteins. FEBS Letters 410, 63–67.

[101] Ahmadian, M.R., Stege, P., Scheffzek, K. and Wittinghofer, A. (1997) Conformation of the arginine-finger hypothesis for the GAP-stimulated GTP-hydrolysis reaction of Ras. Nature Struct. Biol. 4, 686–689.

[102] Bourne, H.R. (1997) The arginine-finger hypothesis strikes again. Nature 389, 673–674.

[103] Scheffzek, K., Lautwein, A., Kabsch, W., Ahmadian, M.R. and Wittinghofer, A. (1996) Crystal structure of GTPase-activating domain of human p120 GAP and implications for the interactions with Ras. Nature 384, 591–596.

[104] Ahmadian, M.R., Hoffmann, U., Goody, R.S. and Wittinghofer, A. (1997) Individual rate constants for the interaction of Ras proteins with GTPase-activating proteins determined by fluorescence spectroscopy. Biochemistry 36, 4535–4541.

[105] Newton, A.C. (1995) Seeing two domains. Curr. Biol. 5, 973–976.

[106] House, C. and Kemp, B.E. (1987) Protein kinase C contains a pseudo prototype in its regulatory domain. Science 238, 1726–1728.

[107] Bell, R.M. and Burns, D.J. (1991) Lipid activation of protein kinase C. J. Biol. Chem. 266, 4661–4664.

[108] Borner, C., Filipuzzi, I., Wartmann, M., Eppenberger, U. and Fabbro, D. (1989) Biosynthesis and post-translational modifications of protein kinase C in human breast cancer cells. J. Biol. Chem. 264, 13902–13909.

[109] Dutil, E.M., Keranen, L.M., De Paoli-Roach, A.A. and Newton, A.C. (1994) In vivo regulation of protein kinase C by transphosphorylation followed by autophosphorylation. J. Biol. Chem. 269, 29359–29362.

[110] Cazaubon, S., Bornancin, F. and Parker, P.J. (1994) Threonine-487 is a critical site for permissive activation of protein kinase C alpha. Biochem. J. 301, 443–448.

[111] Orr, J.W. and Newton, A.C. (1994) Requirement for negative charge on "activation loop" of protein kinase C. J. Biol. Chem. 269, 27715–27718.

[112] Orr, J.W. and Newton, A.C. (1994) Intrapeptide regulation of protein kinase C. J. Biol. Chem. 269, 8383–8387.

[113] Orr, J.W., Keranen, L.M. and Newton, A.C. (1992) Reversible exposure of the pseudosubstrate domain of protein kinase C by phosphatidylserine and diacylglycerol. J. Biol. Chem. 267, 15263–15266.

[114] Newton, A.C. (1995) Protein kinase C: structure, function and regulation. J. Biol. Chem. 270, 28495–28498.

[115] Sarda, L. and Desnuelle, P. (1958) Action de la lipase pancréatique sur les esters en émulsion. Biochim. Biophys. Acta 30, 513–521.

[116] Borgström, B. and Brukman, M.L. (1984) Lipases. Elsevier, Amsterdam.

[117] Cygler, M. and Schrag, J.D. (1997) Structure as basis for understanding interfacial properties of lipases. Methods Enzymol. 284, 3–27. Academic Press, New York.

[118] Ferrato, F., Carriere, F., Sarda, L. and Verger, R. (1997) A critical reevaluation of the phenomenon of interfacial activation. Methods Enzymol. 286, 327–347. Academic Press, New York.

[119] Brady, L., Brzozowski, A.M., Derewenda, Z.S., Dodson, E., Dodson, G., Tolley, S., et al. (1990) A serine protease triad forms the catalytic centre of a triacylglycerol lipase. Nature 343, 767–770.

[120] Winkler, F.K., D'Arcy, A. and Hunziker, W. (1990) Structure of human pancreatic lipase. Nature 343, 771–774.

[121] Schrag, J.D., Li, T.G., Wu, S. and Cygler, M. (1991) Ser-His-Glu triad forms the catalytic site of the lipase from *Geotrichum candidum*. Nature 351, 761–764.

[122] Blow, D.M. (1990) Enzymology: more of the catalytic triad. Nature 343, 694–695.

[123] Van Tilburgh, H., Egloff, M.P., Martinez, C., Rugani, N., Verger, R. and Cambillau, C. (1993) Interfacial activation of the lipase–colipase complex by mixed micelles revealed by X-ray crystallography. Nature 362, 814–820.

[124] Brzozowski, A.M., Derewenda, U.D., Derewenda, Z.S., Dodson, G.G., Lawson, D.M., Turkenburg, J.P., et al. (1991) A model of interfacial activation in lipases from the structure of a fungal lipase–inhibitor complex. Nature 351, 491–494.

[125] Derewenda, U., Brzozowski, A.M., Lawson, D.M. and Derewenda, U.D. (1992) Catalysis at the interface: the anatomy of a conformational change in triglyceride lipase. Biochemistry 31, 1532–1541.

[126] Srere, P.A. (1980) The infrastructure of the mitochondrial matrix. Trends Biochem. Sci. 5, 120–121.

[127] Srere, P.A. (1987) Complexes of sequential metabolic enzymes. Annu. Rev. Biochem. 56, 89–124.

[128] Srere, P.A. (1967) Enzyme concentrations in tissues. Science 158, 936–937.

[129] Srere, P.A. (1972) Is there an organization of the Krebs cycle enzymes in the mitochondrial matrix? In: M.A. Mehlman, R.W. Hanson (Eds.), Energy Metabolism and the Regulation of Metabolic Processes in Mitochondria. Academic Press, New York, pp. 79–91.

[130] Srere, P.A. (1985) Organization of proteins within the mitochondrion. In: G.R. Welch (Ed.), Organized Multienzyme Systems: Catalytic Properties. Academic Press, New York, pp. 1–61.

[131] Srere, P.A. (1985) The metabolon. Trends Biochem. Sci. 10, 109–110.

[132] Srere, P.A. and Ovadi, J. (1990) Enzyme–enzyme interactions and their metablic role. FEBS Letters 268, 360–368.

[133] Ovadi, J. (1991) Physiological significance of metabolite channelling. J. Theor. Biol. 152, 1–22.

[134] Mendes, P., Kell, D.B. and Westerhoff, H.V. (1996) Why and when channelling can decrease pool size at constant net flux in a simple dynamic channel. Biochim. Biophys. Acta 1289, 175–186.

[135] Easterby, J.S. (1981) A generalized theory of the transition time for sequential enzyme reactions. Biochem J. 199, 155–161.

[136] Easterby, J.S. (1980) Temporal analysis of the transition between steady states. In: A. Cornish-Bowden and M.L. Cardenas (Eds.), Control of Metabolic Processes. Plenum Press, New York, pp. 281–290.

[137] Easterby, J.S. (1989) The analysis of metabolite channelling in multienzyme complexes and multifunctional proteins. Biochem. J. 264, 89–124.

[138] Keleti, T. and Ovadi, J. (1988) Control of metabolism by dynamic macromolecular interactions. Curr. Top. Cell. Regul. 29, 1–33.

[139] Keleti, T. (1984) Channelling in enzyme complexes. In: J. Ricard and A. Cornish-Bowden (Eds.), Dynamics of Biochemical Systems. Plenum Press, New York, pp. 103–114.

[140] Siegbahn, N., Mosbach, K. and Welch, G.R. (1985) Models of organized multienzyme systems: use in microenvironmental characterization and in practical application. In: G.R. Welch (Ed.), Organized Multienzyme systems: catalytic properties. Academic Press, New York, pp. 271–301.

[141] Shatalin, K., Lebreton, S., Rault-Leonardon, M., Vélot, C. and Srere, P.A. (1999) Electrostatic channelling in a fusion protein of porcine citrate synthase and porcine mitochondrial malate dehydrogenase. Biochemistry 38, 881–889.

[142] Pan, P., Woehl, E. and Dunn, M.F. (1997) Protein architecture, dynamics and allostery in tryptophan synthase channelling. Trends Biochem. Sci. 22, 22–27.

[143] Hyde, C.C., et al. (1988) Three-dimensional structure of the tryptophan synthase $\alpha_2\beta_2$ multienzyme complex from *Salmonella thyphimurium*. J. Biol. Chem. 263, 15857–15871.

[144] Rhee, S., et al. (1996) Exchange of K^+ and Cs^+ for Na^+ induces local and long-range changes in the three-dimensional structure of the tryptophan synthase $\alpha_2\beta_2$ complex. Biochemistry 35, 4211–4221.

[145] Dunn, M.F., et al. (1990) The tryptophan synthase bienzyme complex transfers indole between the α- and β-sites via a 25–30 Å long tunnel. Biochemistry 29, 8598–8607.

[146] Lane, A.N. and Kirschner, K. (1991) Mechanism of the physiological reaction catalyzed by tryptophan synthase from *Escherichia coli*. Biochemistry 30, 479–484.

[147] Anderson, K.S., Miles, E.W. and Johnson, K.A. (1991) Serine modulates substrate channelling in tryptophan synthase. J. Biol. Chem. 266, 8020–8033.

[148] Houben, K.F. and Dunn, M.F. (1990) Allosteric effects acting over distance of 20–25 Å in the *Escherichia coli* tryptophan synthase increase ligand affinity and cause redistribution of covalent intermediates. Biochemistry 29, 2421–2429.

[149] Bzoric, P.S., Ngo, K. and Dunn, M.F. (1992) Allosteric interactions coordinate catalytic activity between successive metabolic enzymes in tryptophan synthase bienzyme complex. Biochemistry 31, 3821–3829.

[150] Faeder, E.J. and Hammes, G.G. (1970) Kinetic studies of tryptophan synthase. Interaction of substrates with the B subunit. Biochemistry 9, 4043–4049.

[151] Hammes, G.G. (1982) Enzyme Catalysis and Regulation. Academic Press, New York.

[152] Srere, P.A. (1994) Complexities of metabolic regulation. Trends Biochem. Sci. 19, 519–520.

CHAPTER 7

Cell complexity, electrostatic partitioning of ions and bound enzyme reactions

As outlined previously, the living cell is an extremely complex system. This system is made up of supramolecular edifices of macromolecules that often behave as polyelectrolytes. Understanding the kinetics, or the dynamics, of an enzyme reaction within this complex medium requires the physical analysis of the interplay between the dynamics of the enzyme reaction and the electrostatic partitioning of mobile ions between the inside and the outside of the polyelectrolyte matrix. Many cell organelles, membranes, cell walls, ribosomes, etc. behave as insoluble polyelectrolytes. The subcellular system best suited for studying how electrostatic interactions, between fixed and mobile charges, may affect enzyme reactions, is probably the plant cell wall. We shall describe later the structure of this organelle, but it is sufficient, for the time being, to mention that the plant cell wall behaves as a polyanion, owing to the fixed negative charges of polygalacturonic acids. Given the fact that many enzymes are located within this matrix, understanding how these enzymes work is an important problem, for both physical biochemistry and cell biology.

7.1. *Enzyme reactions in a homogeneous polyelectrolyte matrix*

Within a polyelectrolyte matrix, an enzyme reaction which involves ions must be sensitive to the electrostatic partitioning of these ions between the inside and outside of the matrix, whether these ions modulate, or are the substrates, of the reaction. It is therefore of interest to describe briefly the physical principles that govern the electrostatic partitioning of small, mobile, ions by an insoluble polyelectrolyte matrix. In this section we shall assume that the enzyme molecules and the fixed charges are randomly distributed within the matrix.

7.1.1. *Electrostatic partitioning of mobile ions by charged matrices*

Let us consider an insoluble polyelectrolyte matrix permeable to a mobile ion, the electrochemical potential of this ion in the matrix is

$$\tilde{\mu}_i = \mu^\circ + RT\ln(\gamma_i c_i) + zF\Psi_i, \tag{7.1}$$

where $\tilde{\mu}_i$ is this potential, γ_i and c_i are the activity coefficient and concentration of the ion inside the matrix, Ψ_i is the corresponding electrostatic potential, z is the valence of the ion multiplied by $+1$ or -1 depending on it is a cation or an anion and μ° is the standard chemical potential of this ion. R and T have their usual significance. The subscript 'i' refers to the "inside" of the matrix. Similarly the electrochemichal potential of this ion outside the matrix is

$$\tilde{\mu}_o = \mu^\circ + RT\ln\gamma_o, c_o + zF\Psi_o. \tag{7.2}$$

The symbols have the same significance as above except that the subscript 'o' refers to the "outside" of the matrix. At thermodynamic equilibrium

$$\tilde{\mu}_{\mathrm{o}} - \tilde{\mu}_{\mathrm{i}} = 0 = RT \ln \frac{\gamma_{\mathrm{o}} c_{\mathrm{o}}}{\gamma_{\mathrm{i}} c_{\mathrm{i}}} + zF(\Psi_{\mathrm{o}} - \Psi_{\mathrm{i}}), \tag{7.3}$$

which can be rearranged to

$$\frac{1}{z} \ln \frac{\gamma_{\mathrm{i}} c_{\mathrm{i}}}{\gamma_{\mathrm{o}} c_{\mathrm{o}}} = \frac{F \Delta \Psi}{RT} \tag{7.4}$$

with

$$\Delta \Psi = \Psi_{\mathrm{o}} - \Psi_{\mathrm{i}}. \tag{7.5}$$

Expression (7.4) is equivalent to the Nernst equation already encoutered in chapter 5. If a cation $A^{z'_A+}$ and an anion $B^{z'_B-}$ (with $z'_A = z_A$ and $z'_B = -z_B$) are present in these two phases, one has

$$\frac{1}{z'_A} \ln \frac{\gamma_{\mathrm{i}}^{\mathrm{A}} c_{\mathrm{i}}^{\mathrm{A}}}{\gamma_{\mathrm{o}}^{\mathrm{A}} c_{\mathrm{o}}^{\mathrm{A}}} = \frac{1}{z'_B} \ln \frac{\gamma_{\mathrm{o}}^{\mathrm{B}} c_{\mathrm{o}}^{\mathrm{B}}}{\gamma_{\mathrm{i}}^{\mathrm{B}} c_{\mathrm{i}}^{\mathrm{B}}} = \frac{F \Delta \Psi}{RT}, \tag{7.6}$$

where $\gamma^{\mathrm{A}}, \gamma^{\mathrm{B}}, c^{\mathrm{A}}, c^{\mathrm{B}}$ represent the activity coefficients and the concentrations of the two ions, respectively. Defining an electrostatic partition coefficient Π_{e} as

$$\Pi_{\mathrm{e}} = \exp\left(\frac{F \Delta \Psi}{RT}\right), \tag{7.7}$$

eq. (7.6) becomes

$$\left(\frac{\gamma_{\mathrm{i}}^{\mathrm{A}} c_{\mathrm{i}}^{\mathrm{A}}}{\gamma_{\mathrm{o}}^{\mathrm{A}} c_{\mathrm{o}}^{\mathrm{A}}}\right)^{1/z'_A} = \left(\frac{\gamma_{\mathrm{o}}^{\mathrm{B}} c_{\mathrm{o}}^{\mathrm{B}}}{\gamma_{\mathrm{i}}^{\mathrm{B}} c_{\mathrm{i}}^{\mathrm{B}}}\right)^{1/z'_B} = \Pi_{\mathrm{e}}, \tag{7.8}$$

which is the well-known Donnan equation.

If, inside and outside the polyelectrolyte matrix, there are different anions of different valences, $B^-, \ldots, B^{z'-}$, and different cations of the same valence, A^+, one has, if the activity coefficients are close to unity

$$\Pi_{\mathrm{e}} = \frac{\sum B_{\mathrm{o}}^-}{\sum B_{\mathrm{i}}^-} = \cdots = \left(\frac{\sum B_{\mathrm{o}}^{z'-}}{\sum B_{\mathrm{i}}^{z'-}}\right)^{1/z'} = \frac{\sum A_{\mathrm{i}}^+}{\sum A_{\mathrm{o}}^+}. \tag{7.9}$$

Moreover, electroneutrality should hold inside and outside the matrix. Therefore, one has

$$\begin{aligned} &\sum B_{\mathrm{o}}^- + \cdots + z' \sum B_{\mathrm{o}}^{z'-} = \sum A_{\mathrm{o}}^+, \\ &\sum B_{\mathrm{i}}^- + \cdots + z' \sum B_{\mathrm{i}}^{z'-} \pm \Delta^{\pm} = \sum A_{\mathrm{i}}^+, \end{aligned} \tag{7.10}$$

where $\Delta^{\pm}$ represents the density of fixed, negative or positive, charges of a polyanion, or of a polycation. The second equation of (7.10) can be rewritten as

$$\frac{\sum B_o^-}{\Pi_e} + \cdots + \frac{z' \sum B_o^{z'-}}{\Pi_e^{z'}} \pm \Delta^{\pm} = \Pi_e \sum A_o^+, \tag{7.11}$$

and taking into account the first expression of (7.10), eq. (7.11) can be rearranged to

$$\Pi_e^{z'+1} \pm \frac{\Delta^{\pm}}{\sum B_o^- + \cdots + z' \sum B_o^{z'-}} \Pi_e^{z'} - \cdots - \frac{z' \sum B_o^{z'-}}{\sum B_o^- + \cdots + z' \sum B_o^{z'-}} = 0. \tag{7.12}$$

Solving this equation allows one to obtain the expression for the electrostatic partition coefficient Π_e as a function of the anion concentrations and of the fixed charge density of the polyelectrolyte matrix. Let us consider, as an example, the simple situation where all the anions are monoanions and where the matrix is a polyanion. Then one has

$$\Pi_e = \frac{\sum B_o^-}{\sum B_i^-} = \frac{\sum A_i^+}{\sum A_o^+}. \tag{7.13}$$

Moreover, the two electroneutrality equations assume the form

$$\begin{aligned} &\sum B_o^- = \sum A_o^+, \\ &\sum B_i^- + \Delta^- = \sum A_i^+, \end{aligned} \tag{7.14}$$

and taking advantage of the expressions of the partition coefficient and of the electroneutrality in the bulk phase, the second equation of (7.14) can be rewritten as

$$\Pi_e^2 - \frac{\Delta^-}{\sum B_o^-} \Pi_e - 1 = 0. \tag{7.15}$$

For a polycation matrix, the relevant electroneutrality equation would be

$$\Pi_e^2 + \frac{\Delta^+}{\sum B_o^-} \Pi_e - 1 = 0. \tag{7.16}$$

The expression of Π_e is then, for a polyanion

$$\Pi_e = \frac{\Delta^-}{2 \sum B_o^-} + \frac{1}{2} \sqrt{\left(\frac{\Delta^-}{\sum B_o^-} \right)^2 + 4}, \tag{7.17}$$

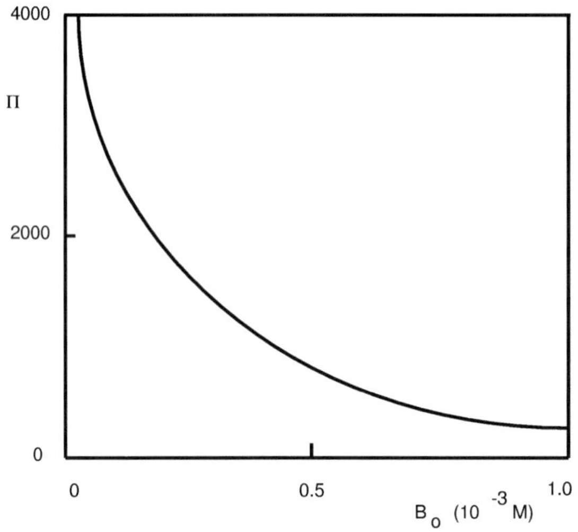

Fig. 7.1. Variation of the electrostatic partition coefficient as a function of a bulk anion concentration. See text. Adapted from ref. [7].

and for a polycation

$$\Pi_e = \frac{1}{2}\sqrt{\left(\frac{\Delta^+}{\sum B_o^-}\right)^2 + 4} - \frac{\Delta^+}{2\sum B_o^-}. \tag{7.18}$$

Since many biological polyelectrolytes, such as membranes, plant cell walls, etc., are negatively charged, eq. (7.17) is of particular interest. For the ideally simple situation that involves an insoluble polyanion, a monovalent anion, B^-, and cation, A^+, the variation of the electrostatic partition coefficient is expressed as a function of the bulk anion concentration, B_o^- (Fig. 7.1).

The value of this partition coefficient declines sharply for low values of B_o^- and approaches an asymptotic value of one as $B_o^- \to \infty$. For a suspension of plant cells in a liquid medium, one would expect the local fixed charge density Δ^- in the cell wall, to be much larger than the concentrations of the anions in the external medium. One then has

$$\Delta^- \gg \sum B_o^-, \tag{7.19}$$

and the expression for the electrostatic partition coefficient reduces to

$$\Pi_e = \frac{\Delta^-}{\sum B_o^-}. \tag{7.20}$$

As we shall see later, this relationship is important, for it allows one to study how the complexity of charge distribution in a polyanionic matrix affects the behaviour of a polyelectrolyte-bound enzyme.

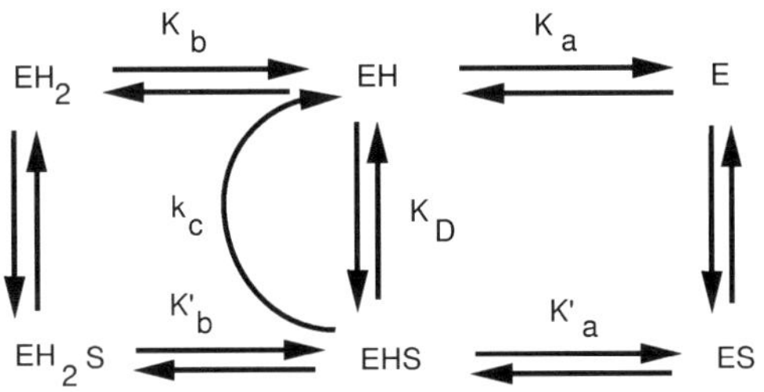

Fig. 7.2. The various ionization states of an enzyme and of an enzyme–substrate complex. See text.

7.1.2. *pH effects of polyelectrolyte-bound enzymes*

A first effect expected for polyelectrolye-bound enzymes should be exerted on the pH profile of the reaction. It is a well-known fact, for enzyme kineticists, that most enzyme reactions in free solution are sensitive to pH. For a simple enzyme reaction following Michaelis–Menten kinetics, the pH effect is ascribed to the existence of different states of protonation of both the free enzyme and the enzyme–substrate complex, as shown in Fig. 7.2.

If this system is in quasi-equilibrum, in such a way that the Michaelis constant of the reaction can be approximated by the dissociation constant of the substrate from the EHS complex (Fig. 7.2) the relevant rate equation assumes the form

$$v = \frac{k_o^* E_T S_i}{K_D^* + S_i}, \tag{7.21}$$

where k_c^* is the apparent catalytic constant, K_D^* is the apparent dissociation constant, S_i is the local substrate concentration within the matrix and E_T is the total enzyme "concentration" within this polyelectrolyte matrix. From the analysis of the kinetic model of Fig. 7.2, it can easily be shown that

$$k_c^* = k_c \left[1 + \frac{K_a'}{H_i} + \frac{H_i}{K_b'} \right]^{-1},$$

$$K_D^* = K_D \left[1 + \frac{K_a}{H_i} + \frac{H_i}{K_b} \right] \left[1 + \frac{K_a'}{H_i} + \frac{H_i}{K_b'} \right]^{-1}. \tag{7.22}$$

In these equations, k_c and K_D are the real catalytic and dissociation constants and K_a, K_a', K_b and K_b' are acid and base ionization constants of the enzyme and of the enzyme–substrate complexes (Fig. 7.2). H_i is the local proton concentration in the polyelectrolyte matrix. We shall discuss below, for a polyanion, two possible cases: the substrate is uncharged; the substrate is a monoanion.

7.1.2.1. *The substrate is uncharged*

If the substrate is a neutral molecule, its local concentration inside the matrix, S_i, is just equal to the corresponding bulk concentration, S_o. From the definition of the electrostatic partition coefficient, Π_e, one must have

$$H_i = \Pi_e H_o. \tag{7.23}$$

Simple inspection of eq. (7.22) shows that two parameters, namely, k_c^* and k_c^*/K_D^*, play an important role in the study of pH effects on enzyme reactions. The first gives information on the pKs of the strategic groups of the enzyme–substrate complex. The second offers information on the strategic groups of the free enzyme.

Taking advantage of eq. (7.23), the expression for k_c^* and k_c^*/K_D^* for a polyelectrolyte-bound enzyme, assumes the form

$$k_c^* = k_c\left[1 + \frac{K_a'}{\Pi_e H_o} + \frac{H_o \Pi_e}{K_b'}\right]^{-1},$$
$$\frac{k_c^*}{K_D^*} = \frac{k_c}{K_D}\left[1 + \frac{K_a}{\Pi_e H_o} + \frac{H_o \Pi_e}{K_b}\right]^{-1}. \tag{7.24}$$

Plots of $\log k_c^*$ and $\log(k_c^*/K_D^*)$ as a function of pH are qualitatively similar to those one would obtain for a free enzyme in solution, but the plots obtained for the bound enzyme are shifted relative to those for the free enzyme. From the first equation (7.24), the linear region of the ascending part of the pH curve can be expressed as

$$\log k_c^* = \log k_c - \mathrm{pK}_b' - \log \Pi_e + \mathrm{pH}. \tag{7.25}$$

The horizontal part of the same curve is

$$\log k_c^* = \log k_c, \tag{7.26}$$

and the descending branch is

$$\log k_c^* = \log k_c + \log \Pi_e + \mathrm{pK}_a' - \mathrm{pH}. \tag{7.27}$$

Therefore, the abscissae of the two pKs are

$$\mathrm{pH} = \mathrm{pK}_b' + \log \Pi_e, \qquad \mathrm{pH} = \mathrm{pK}_a' + \log \Pi_e. \tag{7.28}$$

These equations show that, with respect to a free enzyme in solution, the pH profile of a polyelectrolyte-bound enzyme is shifted by a value of $\log \Pi_e$ (Fig. 7.3). Similar reasoning can be applied to the plot of $\log(k_c^*/K_D^*)$ *versus* pH and one finds, in the same way, that

$$\mathrm{pH} = \mathrm{pK}_b + \log \Pi_e, \qquad \mathrm{pH} = \mathrm{pK}_a + \log \Pi_e. \tag{7.29}$$

Here again, the pH profile of $\log(k_c^*/K_D^*)$ *versus* pH is shifted by $\log \Pi_e$ relative to that of the free enzyme (Fig. 7.3).

If the polyelectrolyte is a polyanion, $\Pi_e > 1$ and the pH profile of the bound enzyme is shifted towards high pH values. This is precisely the situation shown in Fig. 7.3. If,

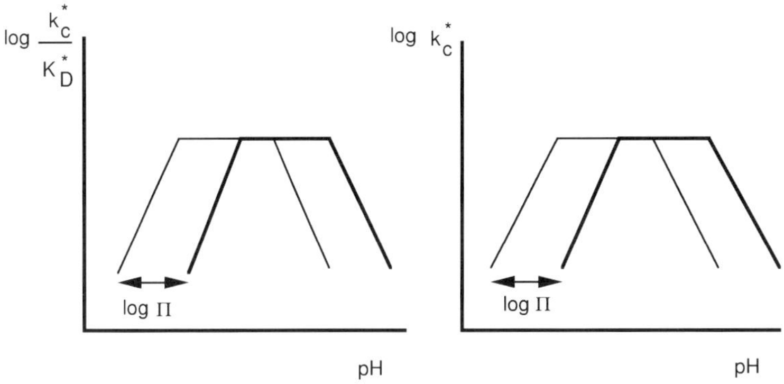

Fig. 7.3. Shift of the pH profile of a polyelectrolyte-bound enzyme. The substrate is assumed to be uncharged. The polyelectrolyte is a polyanion and the pH profile is shifted towards high values. From ref. [1].

alternatively, the polyelectrolyte is a polycation, $\Pi_e < 1$ and the pH profile of the bound enzyme is shifted towards low pH values. An estimation of the values of the electrostatic partition coefficients Π_e can thus be obtained from the shifts of the pH curves [1]. These conclusions are not just theoretical predictions, for they have been confirmed from the experimental study of enzymes bound to biological membranes [2], or to non-biological polyelectrolytes [3]. The pH dependence of bacterial cell wall lysozyme may be considered as an example of these views. As expected, increasing the ionic strength brings about a decrease of Π_e values. Therefore the pH profiles are progressively shifted from high to low pH values as the ionic strength is increased.

7.1.2.2. The substrate of the enzyme is an anion

If the substrate itself is a monoanion, the rate equation (7.21) assumes the form

$$\frac{v}{E_T} = \frac{k_o^* S_o/\Pi_e}{K_D^* + S_o/\Pi_e} = \frac{k_c^* S_o}{K_D^* \Pi_e + S_o}. \tag{7.30}$$

For a fixed bulk substrate concentration, the product $K_D^* \Pi_e$ is equivalent to a dissociation constant. Let us call $\widetilde{K}_D^*$ this apparent dissociation constant. The expressons for k_c^* and $k_c^*/\widetilde{K}_D^*$ are thus

$$\begin{aligned} k_c^* &= k_c \left[1 + \frac{K_a'}{H_o \Pi_e} + \frac{H_o \Pi_e}{K_b'} \right]^{-1}, \\ \frac{k_c^*}{\widetilde{K}_D^*} &= \frac{k_c}{K_D} \left[\Pi_e + \frac{K_a}{H_o} + \frac{\Pi_e^2 H_o}{K_b} \right]^{-1}. \end{aligned} \tag{7.31}$$

A plot of $\log k_c^*$ *versus* pH will thus be unaltered, whether the substrate is neutral or charged. But a plot of $\log(k_c^*/\widetilde{K}_D^*)$ *versus* pH, will be different for a charged and for a neutral substrate. For a charged substrate (second equation of (7.31)), the linear region, on the acid side of the plot, obeys the equation

$$\log\left(\frac{k_c^*}{\widetilde{K}_D^*}\right) = \log\left(\frac{k_c}{K_D}\right) + \mathrm{pH} - \mathrm{pK_b} - 2\log\Pi_e. \tag{7.32}$$

For the optimum pH region, one has

$$\log\left(\frac{k_c^*}{\widetilde{K}_D^*}\right) = \log\left(\frac{k_C}{K_D}\right) - \log\Pi_e, \tag{7.33}$$

and on the basic side, the linear part of the curve follows the equation

$$\log\left(\frac{k_c^*}{\widetilde{K}_D^*}\right) = \log\left(\frac{k_c}{K_D}\right) + \mathrm{pK_a} - \mathrm{pH}. \tag{7.34}$$

Thus the presence of the negative charge borne by he substrate does not alter the plot of $\log k_c^*$ against pH, but it completely changes the curve of $\log(k_c^*/\widetilde{K}_D^*)$ plotted as a function of pH. The acid side of the curve is shifted by $2\log\Pi_e$ (instead of $\log\Pi_e$ for the uncharged substrate). Moreover the intermediate region of the curve, near the optimum, is shifted by $\log\Pi_e$. Last but not least, the descending branch of the plot does not experience any change relative to the situation observed with the enzyme in free solution. If the polyelectrolyte is a polyanion, the acid branch of the curve is moved towards high pHs values and the intermediate plateau is moved downwards. If, alternatively, the polyelectrolyte is a polycation, the acid branch of the plot is shifted towards low pH values and the intermediate plateau is moved upwards [1] (Fig. 7.4).

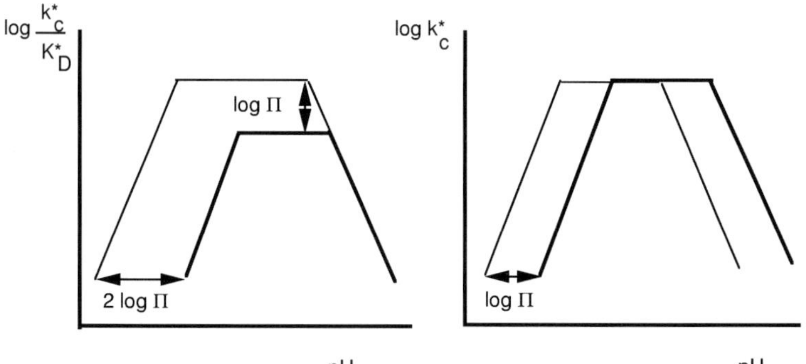

Fig. 7.4. Shift of the pH profile of a polyelectrolyte-bound enzyme. The same conditions as for Fig. 7.3 but the substrate is negatively charged. From ref. [1].

7.1.3. Apparent kinetic co-operativity of a polyelectrolyte-bound enzyme

We have considered so far the situation in which the enzyme molecules are randomly distributed in a homogeneous polyelectrolyte matrix, and where the interactions betweeen the fixed charges of this matrix and protons can alter the pH profile of the bound enzyme. But, if the substrate itself is an ion, as considered in the previous section, it should also experience electrostatic interactions with the fixed charges of the polyelectrolyte. One may therefore wonder whether these electrostatic interactions do not affect the kinetics of the bound enzyme. This is precisely the aim of the present section of this book.

If, as previously assumed, enzyme molecules and fixed charges are randomly distributed within the matrix, and if the enzyme follows Michaelis–Menten kinetics and acts on a charged substrate, then

$$v = \frac{V_m S_i}{K_m + S_i}, \tag{7.35}$$

where S_i is indeed the local concentration of the charged substrate. K_m and V_m have their usual significance. Taking advantage of the definition of the electrostatic partition coefficient, eq. (7.35) can be rewritten as [4–6]

$$v = \frac{V_m S_o}{K_m \Pi_e^{z'} + S_o}, \tag{7.36}$$

where S_o is now the bulk substrate concentration and z' is the valence. When the substrate concentration S_o is varied, Π_e varies as well, and in a certain range of substrate concentration the variation of Π_e may be extremely large (Fig. 7.1). Therefore, equation (7.36) is no longer the equation of a rectangular hyperbola. The departure from Michaelis–Menten kinetics will mimic positive co-operativity if the substrate is repelled by the fixed charges of the matrix, and negative co-operativity if it is attracted by the fixed charges [4,6,7]. A simple and interesting situation is expected to occur if the fixed charge density is much larger than the substrate concentration. This is precisely the situation likely to occur with biological membranes and cell walls. Then one must have

$$\Pi_e = \frac{\Delta^-}{S_o} \tag{7.37}$$

and eq. (7.36) becomes

$$v = \frac{V_m S_o^2}{K_m \Delta^- + S_o^2}. \tag{7.38}$$

In this form it becomes obvious that one should obtain a sigmoidal curve when plotting v *versus* S_o. The sigmoidicity depends on the value of the fixed charge density. For various reasons, which will appear in the next section, it is often advantageous to define a new variable, σ_o, as

$$\sigma_o = \frac{S_o^2}{K_m}, \tag{7.39}$$

which still has the dimension of a concentration. Then eq. (7.38) can be reexpressed as

$$v = \frac{V_m \sigma_o}{\Delta^- + \sigma_o}. \tag{7.40}$$

A plot of v *versus* this new variable σ_o should yield a rectangular hyperbola.

Another interesting consequence of this reasoning is that the reaction rate should become extremely sensitive to variations of the ionic strength when the enzyme is embedded in a polyelectrolyte. This sensitivity should exist, even if the free enzyme *per se* is nearly totally insensitive to changes in ionic strength. In the present case, increasing the ionic strength should result in an apparent activation of the bound enzyme and in a decrease, or even suppression, of the apparent positive co-operativity. If the substrate, instead of being repelled from the matrix, is attracted by the fixed charges, one would expect that, on increasing the ionic strength, apparent inhibition of the bound enzyme will take place [5, 6]. Since many enzymes in the cell are bound to charged subcellular structures, one would expect these bound enzymes to be extremely sensitive to changes in ionic strength, which then appear as a modulator of enzyme activity.

7.2. *Enzyme reactions in a complex heterogeneous polyelectrolyte matrix*

It has been assumed so far that both the fixed charges of the polyanion and the enzyme molecules are randomly distributed in the matrix. According to the definition of complexity already given, the structure of this system can be considered as "simple". One may expect, however, that for most biological polyelectrolytes, membranes and cell walls for instance, the fixed charges and enzyme molecules are more or less organized in space. This organization may in fact be fuzzy and, according to our definition in chapter 1, the system should be considered as "complex". One may thus wonder whether this complexity will affect the reaction rate. If the answer to this question is positive, it would imply the response of a bound enzyme, to a given concentration of a charged substrate, is not solely controlled by the local enzyme and fixed charge density, but also by the complexity of their spatial distribution.

7.2.1. *Can the fuzzy organization of a polyelectrolyte affect a bound enzyme reaction?*

The importance of the fuzzy organization of fixed charges and enzyme molecules in an insoluble polyanion can be intuitively illustrated by an ideal example. Let us first assume that the fixed charges and enzyme molecules are randomly distributed in a unit volume. The resulting steady state velocity (if the enzyme follows Michaelis–Menten kinetics and if the substrate is a monoanion) will be

$$v = \frac{VS_o^2}{K\Delta^- + S_o^2} = \frac{V\sigma_o}{\Delta^- + \sigma_o}. \tag{7.41}$$

Here, for simplicity, V and K represent V_m and K_m, respectively. If one now assumes that the same number of fixed charges and enzyme molecules are distributed in the same unit volume as two clusters of enzyme molecules and fixed charges, the resulting reaction velocity will be

$$v = \frac{V_1 S_o^2}{K\Delta_1^- + S_o^2} + \frac{V_2 S_o^2}{K\Delta_2^- + S_o^2} = \frac{V_1\sigma_o}{\Delta_1^- + \sigma_o} + \frac{V_2\sigma_o}{\Delta_2^- + \sigma_o}, \tag{7.42}$$

where V_1 and V_2, Δ_1^- and Δ_2^- are the V_m and the fixed charge densities associated with the two clusters. Hence V_1 and V_2 are proportional to the enzyme densities in the two clusters. Whereas equation (7.41) is that of a rectangular hyperbola, if v is plotted against σ_o, eq. (7.42) is not and displays apparent kinetic co-operativity relative to σ_o, if $\Delta_1^- \neq \Delta_2^-$. This means that, when plotting $1/v$ *versus* $1/\sigma_o$, the plots are concave downwards. The situation is thus formally analogous to that already encountered in classical enzyme kinetics, where the same enzyme acts on two different substrates, with different K_m values.

One can derive the extreme Hill coefficient (see chapter 2) of this apparent co-operativity relative to σ_o and $v/(V_1 + V_2)$. One finds

$$h_{\text{ext},\sigma} = \frac{2}{1 + \left\{\dfrac{\Delta_1^-\Delta_2^- - \Gamma/2}{\Delta_1^-\Delta_2^-}\right\}^{1/2}}. \tag{7.43}$$

The Γ coefficient (see chapter 2) can be expressed as

$$\Gamma = -2V_1V_2\frac{(\Delta_1^- - \Delta_2^-)^2}{(V_1 + V_2)^2}. \tag{7.44}$$

These equations allow three conclusions: first, as expected, the co-operativity relative to σ_o is always negative, i.e. the extreme Hill coefficient is always smaller than unity; second, this co-operativity with respect to σ_o is the consequence of the spatial organization of the charges and of the enzyme molecules in the two clusters; third, both the fixed charge and enzyme densities contribute to the extent of this co-operativity. It is now important to determine how the complexity of this spatial organization affects the reaction rate catalysed by the bound enzyme.

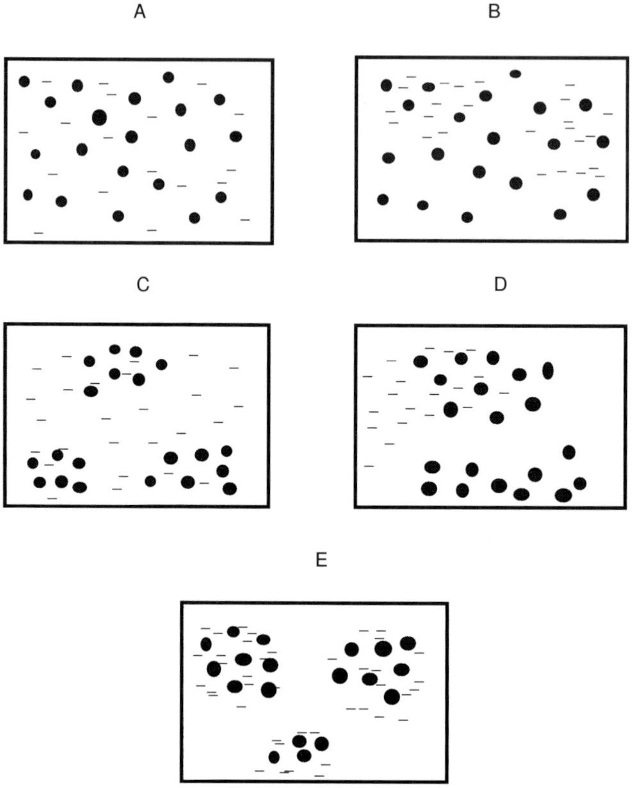

Fig. 7.5. Different types of spatial organization of fixed negative charges and enzyme molecules on a surface. A – Fixed charges and enzyme molecules are randomly distibuted. B – Fixed charges are clustered but enzyme molecules are randomly distributed. C – Enzyme molecules are clustered but fixed charges are randomly distributed. D – Fixed charges and enzyme molecules are clustered and the clusters partly overlap. E – Fixed charges and enzyme molecules are clustered and perfectly overlap. From ref. [7].

7.2.2. Statistical formulation of a fuzzy organization of fixed charges and bound enzyme molecules in a polyanionic matrix

The fixed charges and the enzyme molecules may be distributed, in a polyanionic matrix, according to different types of organization (Fig. 7.5). Both the fixed charges and the enzyme molecules may be distributed randomly (Fig. 7.5(A)). This is the situation that has already been discussed. The second possibility is that shown in Fig. 7.5(B). The enzyme molecules are randomly distributed but the fixed charges are clustered. The third possibility is the converse of the previous one. The enzyme molecules are clustered and the fixed charges are randomly distributed (Fig. 7.5(C)). The fourth possibility requires that both enzyme molecules and fixed charges be clustered, but the clusters of charges and the clusters of enzyme molecules do not overlap perfectly (Fig. 7.5(D)). Last but not least, the clusters of charges and of enzyme molecules are perfectly superimposed (Fig. 7.5(E)). As the last situation is relatively simple, we shall use it as a model that will allow us to understand how spatial organization can affect enzyme activity in a polyelectrolyte.

One can express the charge density of a cluster i, Δ_i, with respect to the mean charge density, $\langle \Delta \rangle$, of the population of clusters. One thus has

$$\Delta_i = \langle \Delta \rangle + \delta_i \quad (i = 1, \ldots, n), \tag{7.45}$$

where δ_i is the deviation, positive or negative, of the actual charge density relative to the mean. Similarly, the maximum reaction rate (proportional to enzyme density) of enzyme cluster j, V_j, may be expressed with respect to the corresponding mean value $\langle V \rangle$. One has thus

$$V_j = \langle V \rangle + \varepsilon_j \quad (j = 1, \ldots, n), \tag{7.46}$$

where ε_j is the deviation (positive or negative) of the actual enzyme density relative to that of the mean $\langle V \rangle$. Expressing, in quantitative terms, the degree of spatial organization of fixed charges and enzyme molecules requires prior analysis of the frequency distribution of both charge and enzyme densities in the clusters. These distributions, like any distribution, can be characterized by their moments [8,9]. The moments may be monovariate, that is they take account of one random variable only (δ_i or ε_j) and are defined as

$$\begin{aligned}
\sum_{i=1}^{n} f_i \delta_i &= N\mu_1(\delta) = 0, \\
\sum_{i=1}^{n} f_i \delta_i^2 &= N\mu_2(\delta) = N\,\mathrm{var}(\delta), \\
\sum_{i=1}^{n} f_i \delta_i^3 &= N\mu_3(\delta), \\
&\vdots
\end{aligned} \tag{7.47}$$

and

$$\begin{aligned}
\sum_{j=1}^{n} f_j \varepsilon_j &= N\mu_1(\varepsilon) = 0, \\
\sum_{j=1}^{n} f_j \varepsilon_j^2 &= N\mu_2(\varepsilon) = N\,\mathrm{var}(\varepsilon), \\
\sum_{j=1}^{n} f_j \varepsilon_j^3 &= N\mu_3(\varepsilon), \\
&\vdots
\end{aligned} \tag{7.48}$$

In expressions (7.47) and (7.48), the monovariate moments are $\mu(\delta)$ and $\mu(\varepsilon)$. var(δ) and var(ε) are the variances of the two random variables. But the moments may also be bivariate, i.e. they associate charge and enzyme densities, for each cluster is characterized by its own charge and enzyme densities. These bivariate moments assume the form

$$\begin{aligned}
\sum_{i=1}^{n}\sum_{j=1}^{n} f_{ij}\delta_i\varepsilon_j &= N\mu_{11}(\delta,\varepsilon) = N\operatorname{cov}(\delta,\varepsilon),\\
\sum_{i=1}^{n}\sum_{j=1}^{n} f_{ij}\delta_i^2\varepsilon_j &= N\mu_{21}(\delta,\varepsilon),\\
\sum_{i=1}^{n}\sum_{j=1}^{n} f_{ij}\delta_i^3\varepsilon_j &= N\mu_{31}(\delta,\varepsilon),\\
&\ \ \vdots
\end{aligned} \tag{7.49}$$

The $\mu(\delta,\varepsilon)$ represent the bivariate moments. The best known of these moments is the covariance cov(δ,ε). The first order monovariate moments $\mu_1(\delta)$ and $\mu_1(\varepsilon)$ are the means of δ and ε. They are in fact equal to zero. The second order monovariate moments $\mu_2(\delta)$ and $\mu_2(\varepsilon)$ are the variances already referred to. The first bivariate moment $\mu_{11}(\delta,\varepsilon)$ is the covariance of the bivariate distribution.

Let us now return to the various types of fuzzy organization of fixed charges and bound enzyme molecules of Fig. 7.5, and the question of how the monovariate and bivariate moments match these differents types of organization. In the case of the first type (Fig. 7.5(A)), there is no organization. Hence there is no spatial macroscopic heterogeneity in the matrix. Therefore, in any macroscopic region of space, $\mu(\delta)$ and $\mu(\varepsilon)$ are equal to zero and, similarly, the bivariate moments $\mu(\delta,\varepsilon)$ are also equal to zero. In the second type of organization (Fig. 7.5(B)), the charges are spatially organized but the enzyme molecules are not. Thus one must expect some moments $\mu(\delta)$ to be different from zero, whereas all the moments $\mu(\varepsilon)$ must be nil. As a consequence, all the bivariate moments $\mu(\delta,\varepsilon)$ should be equal to zero. In the third type of organization (Fig. 7.5(C)), the enzyme molecules are spatially organized, but the fixed charges are not. Hence some of the moments $\mu(\varepsilon)$ should be different from zero, whereas all the moments $\mu(\delta)$ should be nil, and again all the bivariate moments $\mu(\delta,\varepsilon)$ should also be equal to zero. In the D and E types of organization one should expect, in all generality, that some moments, $\mu(\delta)$, $\mu(\varepsilon)$ and $\mu(\delta,\varepsilon)$, be different from zero. However, for the last type of organization, one must expect the clusters of fixed charges and enzyme densities to be exactly superimposed. Moreover, these clusters may all have the same charge density and the same enzyme density. For this type of highly organized system, one will then expect that all the monovariate and bivariate moments be equal to zero. An interesting conclusion of this reasoning is thus that, for fully disorganized and highly organized systems, the monovariate and bivariate moments of charge and enzyme densities are all equal to zero. It is only when the spatial organization of fixed charges and enzyme molecules become complex and fuzzy that some of these moments are different from zero. The question which will now be addressed is to know how these moments, which express the fuzziness of the organization, may affect an enzyme reaction rate in a polyelectrolyte matrix.

7.2.3. Apparent co-operativity generated by the complexity of the polyelectrolyte matrix

We have shown, in the previous section, that the polyelectrolyte has a complex structure when the monovariate and the bivariate moments associated with the distributions of fixed charge and enzyme densities are not all equal to zero. This implies in turn that the enzyme reaction rate should encompass the expression of these moments. If, as assumed previously, the substrate is a monovalent anion, and if the fixed charges and enzyme molecules are clustered, as shown in Fig. 7.5(E), the overall reaction rate of the enzyme process is

$$v = \sum_{i=1}^{n}\sum_{j=1}^{n} \frac{f_{ij} V_j S_o}{K\Pi_i + S_o}, \tag{7.50}$$

where Π_i and V_j are the partition coefficient and maximum rate (proportional to the enzyme densities) of the various clusters. Given the fact that the substrate is a monoanion, this expression can be rewritten as

$$v = \sum_{i=1}^{n}\sum_{j=1}^{n} \frac{f_{ij} V_j S_o^2}{K\Delta_i + S_o^2}, \tag{7.51}$$

and if we take advantage of the new variable σ_o, the expression becomes

$$v = \sum_{i=1}^{n}\sum_{j=1}^{n} \frac{f_{ij} V_j \sigma_o}{\Delta_i + \sigma_o}. \tag{7.52}$$

It is then obvious that the non-Michaelian character of this expression, with respect to σ_o, results from a spatial organization of the fixed charges. Let

$$v_{ij} = (\langle V\rangle + \varepsilon_j)\frac{\sigma_o}{\langle\Delta\rangle + \delta_i + \sigma_o}. \tag{7.53}$$

Expanding this expression in Taylor series with respect to the variable δ_i yields

$$\begin{aligned} v_{ij} &= (\langle V\rangle + \varepsilon_j)\left\{\frac{\sigma_o}{\langle\Delta\rangle + \sigma_o} - \frac{\delta_i\sigma_o}{(\langle\Delta\rangle + \sigma_o)^2}\right. \\ &\qquad \left. + \frac{\delta_i^2\sigma_o}{(\langle\Delta\rangle + \sigma_o)^3} - \frac{\delta_i^3\sigma_o}{(\langle\Delta\rangle + \sigma_o)^4} + \cdots\right\} \end{aligned} \tag{7.54}$$

and, defining the dimensionless variables σ_o^*, δ_i^* and ε_j^* as

$$\sigma_o^* = \frac{\sigma_o}{\langle\Delta\rangle}, \qquad \delta_i^* = \frac{\delta_i}{\langle\Delta\rangle}, \qquad \varepsilon_j^* = \frac{\varepsilon_j}{\langle V\rangle}, \tag{7.55}$$

eq. (7.54) takes the form

$$v_{ij} = \langle V\rangle(1+\varepsilon_j^*)\left\{\frac{\sigma_o^*}{1+\sigma_o^*} - \frac{\delta_i^*\sigma_o^*}{(1+\sigma_o^*)^2} + \frac{\delta_i^{*2}\sigma_o^*}{(1+\sigma_o^*)^3} - \frac{\delta_i^{*3}\sigma_o^*}{(1+\sigma_o^*)^4} + \cdots\right\}. \tag{7.56}$$

At this stage, nothing can be stated as to the convergence, or divergence, of this series. The overall rate equation is obtained after summing up the elementary rates for the set of N clusters. One thus has

$$\begin{aligned} v &= \sum_{i=1}^{n}\sum_{j=1}^{n}\langle V\rangle f_{ij}(1+\varepsilon_j^*) \\ &\quad\times\left\{\frac{\sigma_o^*}{1+\sigma_o^*} - \frac{\delta_i^*\sigma_o^*}{(1+\sigma_o^*)^2} + \frac{\delta_i^{*2}\sigma_o^*}{(1+\sigma_o^*)^3} - \frac{\delta_i^{*3}\sigma_o^*}{(1+\sigma_o^*)^4} + \cdots\right\}. \end{aligned} \tag{7.57}$$

Introducing into this equation the expression of the moments of the dimensionless variables δ_i^* and ε_j^*, namely,

$$\begin{aligned} &\sum_{i=1}^{n}\sum_{j=1}^{n} f_{ij}\delta_i^* = \sum_{i=1}^{n} f_i\delta_i^* = N\mu_1(\delta^*), \\ &\sum_{i=1}^{n}\sum_{j=1}^{n} f_{ij}\varepsilon_j^* = \sum_{j=1}^{n} f_j\varepsilon_j^* = N\mu_1(\varepsilon^*), \\ &\sum_{i=1}^{n}\sum_{j=1}^{n} f_{ij}\delta_i^*\varepsilon_j^* = N\mu_{11}(\delta^*,\varepsilon^*), \\ &\qquad\vdots \end{aligned} \tag{7.58}$$

and recalling that

$$\sum_{i=1}^{n}\sum_{j=1}^{n} f_{ij} = N, \tag{7.59}$$

eq. (7.57) can be rewritten as

$$\frac{v}{N\langle V\rangle} = \frac{\sigma_o^*}{1+\sigma_o^*}\left\{1+\sum_{r=1}^{m}(-1)^r\frac{\mu_r(\delta^*)+\mu_{r1}(\delta^*,\varepsilon^*)}{(1+\sigma_o^*)^r}\right\}. \tag{7.60}$$

In this form, it becomes obvious that both the monovariate moments $\mu_r(\delta^*)$ and the bivariate moments $\mu_{r1}(\delta^*,\varepsilon^*)$ contribute to the expression of the rate, whereas the monovariate moments $\mu_r(\varepsilon^*)$ do not.

This equation, however, does not allow to evaluate quantitatively the importance of the various moments on the overall reaction velocity, because we do not know whether this series, in eqs. (7.57) and (7.60), is convergent or not, and because nothing is known, in eq. (7.60), about the moments that are nil or negligible. This problem, however, can be solved if we know the nature of the distribution functions of charge and enzyme densities. As most, if not all, statistical distribution functions (hypergeometric, binomial, ...) approach the normal distribution (Laplace–Gauss) when N increases, it is quite sensible to assume that both δ^* and ε^* are normally distributed. The assumption of normality has several consequences: first, the dimensionless variables δ^* and ε^* are of necessity smaller than unity, which in turn implies that the series, in eq. (7.57), is convergent; second, the moments that are different from zero have values that are also smaller than unity, and these values decrease as the order of the moments increases, for instance $\mu_2(\delta^*) > \mu_4(\delta^*)$; third, all the monovariate moments of odd degree are nil and some bivariate moments, such as $\mu_{21}(\delta^*, \varepsilon^*)$, are also nil. Taken together, these conditions allow one to simplify the rate equation. Thus, in practice, if the distributions of charge and enzyme densities are normal, the rate equation can be approximately given by

$$\frac{v}{N\langle V\rangle} = \frac{\sigma_o^*}{1+\sigma_o^*}\left\{1 - \frac{\operatorname{cov}(\delta^*, \varepsilon^*)}{1+\sigma_o^*} + \frac{\operatorname{var}(\delta^*)}{(1+\sigma_o^*)^2}\right\}. \tag{7.61}$$

Hence, if there is no fuzzy organization of fixed charges and enzyme molecules

$$\operatorname{cov}(\delta^*, \varepsilon^*) = 0, \qquad \operatorname{var}(\delta^*) = 0, \tag{7.62}$$

and, therefore, eq. (7.61) reduces to

$$\frac{v}{N\langle V\rangle} = \frac{\sigma_o^*}{1+\sigma_o^*}. \tag{7.63}$$

Equation (7.61) can be rewritten as

$$\frac{v}{N\langle V\rangle} = \frac{\sigma_o^*}{1+\sigma_o^*} + \Xi(\sigma_o^*), \tag{7.64}$$

where

$$\Xi(\sigma_o^*) = -\frac{\sigma_o^*}{(1+\sigma_o^*)^2}\operatorname{cov}(\delta^*, \varepsilon^*) + \frac{\sigma_o^*}{(1+\sigma_o^*)^3}\operatorname{var}(\delta^*). \tag{7.65}$$

The function $\Xi(\sigma_o^*)$ expresses how the fuzzy organization may enhance, or decrease, the reaction rate. The term in $\operatorname{var}(\delta^*)$ results in an enhancement of the rate and the term in $\operatorname{cov}(\delta^*, \varepsilon^*)$ generates a decrease of the same rate. For a classical enzymologist, it may appear somewhat puzzling that a certain degree in the complexity of spatial organization can generate an enhancement of the enzyme reaction rate. Besides this increase, or decrease, of the reaction rate, the fuzzy organization of fixed charges and enzyme molecules may generate co-operativity relative to σ_o^* (Fig. 7.6). The sign and the extent of this co-operativity can

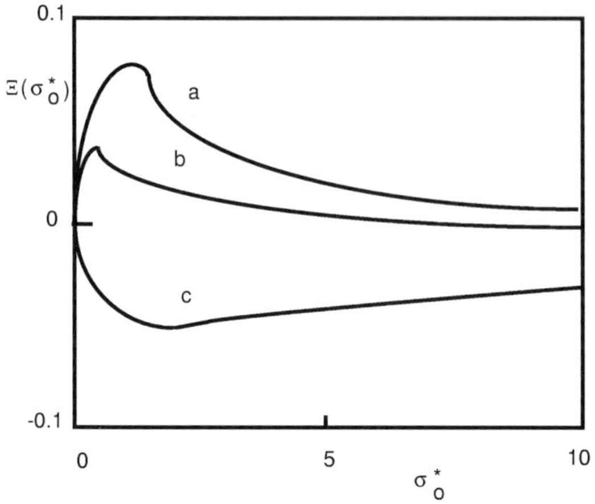

Fig. 7.6. Enhancement and inhibition of enzyme activity through spatial arrangement of fixed charges and enzyme molecules. When the function $\Xi(\sigma_0^*)$ assumes positive values the spatial organization produces an increase of the reaction rate. When $\Xi(\sigma_0^*)$ takes negative values the spatial organization produces a decrease of the reaction rate. Adapted from ref. [7].

be quantitatively determined from the Hill function (see chapter 2). This function, $h(\sigma_o^*)$, may be found and after some algebra to be given by

$$h(\sigma_o^*) = 1 + \Omega(\sigma_o^*), \tag{7.66}$$

where

$$\Omega(\sigma_o^*) = \frac{\zeta_3 \sigma_o^{*3} + \zeta_2 \sigma_o^{*2} + \zeta_1 \sigma_o^*}{\zeta_4' \sigma_o^{*4} + \zeta_3' \sigma_o^{*3} + \zeta_2' \sigma_o^{*2} + \zeta_1' \sigma_o^* + \zeta_o'}. \tag{7.67}$$

The coefficients ζ and ζ' can be expressed in terms of var(δ^*) and cov$(\delta^*, \varepsilon^*)$ and can all be shown to be negative. Hence $\Omega(\sigma_o^*) < 0$ and the kinetic co-operativity with respect to σ_o^* can only be negative.

We are now in a position to determine the various expressions of the rate equation of an enzyme bound to a polyelectrolyte system which can display different degrees of organization (Fig. 7.5(A)–(E)). If the system is disorganized (Fig. 7.5(A)), i.e. if the fixed charges and the enzyme molecules are randomly distributed within the polyanionic matrix, the relevant rate equation reduces to

$$v = \frac{V S_o^2}{K\Delta + S_o^2} \tag{7.68}$$

for

$$\mu_2(\delta^*) = \mathrm{var}(\delta^*) = 0, \qquad \mu_{12}(\delta^*, \varepsilon^*) = \mathrm{cov}(\delta^*, \varepsilon^*) = 0. \tag{7.69}$$

If the charges are clustered, but the enzyme molecules are randomly distributed in the matrix (Fig. 7.5(B)), some of the substrate ions will not be submitted to electrostatic repulsion effects, whereas other substrate ions will experience this repulsion. The relevant rate equation will then be

$$v = \frac{V_1 S_o}{K + S_o} + \frac{N \langle V_2 \rangle S_o^2}{K \langle \Delta \rangle + S_o^2} \left\{ 1 + \frac{K^2 \langle \Delta \rangle^2 \operatorname{var}(\delta^*)}{(K \langle \Delta \rangle + S_o^2)^2} \right\} \tag{7.70}$$

for

$$\begin{aligned} &\mu_2(\delta^*) = \operatorname{var}(\delta^*) \neq 0, \\ &\mu_{12}(\delta^*, \varepsilon^*) = \operatorname{cov}(\delta^*, \varepsilon^*) = 0. \end{aligned} \tag{7.71}$$

In eq. (7.70), V_1 and $\langle V_2 \rangle$ are the enzyme clusters that do not overlap (V_1), and that overlap ($\langle V_2 \rangle$) the charge clusters. If the opposite situation occurs, i.e. if the fixed charges are randomly distributed but the enzyme molecules are clustered, then all the substrate ions will be equally subjected to electrostatic repulsion. The relevant rate equation will be eq. (7.68), because the conditions (7.69) apply. The situations described in Fig. 7.5(A) and C are thus indistinguishable. If now the charges and the enzyme molecules are clustered, and if the clusters of charges partly overlap the clusters of enzyme molecules (Fig. 7.5(D)), some of the substrate ions will be submitted to electrostatic repulsion whereas others will be insensitive to this repulsion. The relevant rate equation will then be

$$v = \frac{V_1 S_o}{K + S_o} + \frac{N \langle V_2 \rangle S_o^2}{K \langle \Delta \rangle + S_o^2} \left\{ 1 - \frac{K \langle \Delta \rangle \operatorname{cov}(\delta^*, \varepsilon^*)}{K \langle \Delta \rangle + S_o^2} + \frac{K^2 \langle \Delta \rangle^2 \operatorname{var}(\delta^*)}{(K \langle \Delta \rangle + S_o^2)^2} \right\}. \tag{7.72}$$

In the present case, one must have

$$\begin{aligned} &\mu_2(\delta^*) = \operatorname{var}(\delta^*) \neq 0, \\ &\mu_{12}(\delta^*, \varepsilon^*) = \operatorname{cov}(\delta^*, \varepsilon^*) \neq 0. \end{aligned} \tag{7.73}$$

Finally, if the clusters of fixed charges are exactly superimposed on the clusters of enzyme molecules (Fig. 7.5(E)), the first term in eq. (7.72) is absent, and one has

$$v = \frac{N \langle V \rangle S_o^2}{K \langle \Delta \rangle + S_o^2} \left\{ 1 - \frac{K \langle \Delta \rangle \operatorname{cov}(\delta^*, \varepsilon^*)}{K \langle \Delta \rangle + S_o^2} + \frac{K^2 \langle \Delta \rangle^2 \operatorname{var}(\delta^*)}{(K \langle \Delta \rangle + S_o^2)^2} \right\}. \tag{7.74}$$

In all generality, conditions (7.73) should apply. If the organization of the system is very strict however, in such a way that the charge densities are exactly the same for all clusters, and the enzyme densities are also the same in the clusters, eq. (7.68) still applies, because conditions (7.69) become valid.

A general conclusion may be drawn from these results. Neither an absolute lack of organization, nor a strict organization, of fixed charges and enzyme molecules will reveal the potential wealth of the catalytic behaviour of a polyelectrolyte-bound enzyme. It is the fuzzy organization of these fixed charges and enzyme molecules that possesses the largest potential for the emergence of new types of behaviour. These new types of behaviour are an enhancement, or an inhibition, of the reaction rate brought about by the complexity of the spatial organization, as well as a rather special type of co-operativity. This co-operativity is the consequence of an electrostatic repulsion of the substrate by the polyanionic matrix, superimposed on the specific effect exerted by the complexity of the spatial organization of the polyelectrolyte system. An example of this complex behaviour is shown in Fig. 7.8. Indeed all these effects, which are predicted by the above theoretical developments, should depend on the local ion concentration. When the ionic strength is increased, the amplitude of these effects should fall off, and ultimately the bound enzyme should behave exactly as if it were in free solution.

7.3. An example of enzyme behaviour in a complex biological system: the kinetics of an enzyme bound to plant cell walls

Plant cell walls represent a typical example of a biological polyelectrolyte. Moreover, many enzymes are embedded in the supramolecular structure of this organelle. As some of these enzymes are involved in the building up and extension of the cell wall, it is logical to try to explain cell wall-bound enzyme kinetics, as well as cell wall extension, by a development of the ideas considered above.

7.3.1. Brief overview of the structure and dynamics of primary cell wall

Young plant cells are surrounded by a primary cell wall. This relatively rigid cell envelope is a complex supramolecular edifice basically made up of interconnected cellulose microfibrils. Cellulose is a linear glucan made up of glucopyranose units linked by $\beta(1 \rightarrow 4)$ bonds. Glucose units are in a "chair" conformation and are oriented in such a way that their hydroxyl groups occupy alternate positions in space. This allows the association, through hydrogen bonds, of several glucan chains which form a cellulose microfibril. Besides cellulose, the primary plant cell wall contains hemicelluloses and pectic compounds. Hemicelluloses are polymers made up of one or several types of sugar residue(s). Thus, for instance, xyloglucans are made up of glucose, xylose, fucose and galacturonic acid. Xyloglucans play an important role in the supramolecular structure of the cell wall because, through hydrogen bonds, they associate different cellulose microfibrils. Cell walls also contain another type of polymer called pectins. Pectins are acidic polysaccharides because they contain polygalacturonic acid. Under "physiological" pH conditions these acidic groups are ionized. Thus, the presence of pectins in the cell wall should give this organelle its polyanionic character. Some of the acidic groups are methylated. Thus the methylation-demethylation of pectins should control the net charge of the wall [10,11]. The acidic groups of pectins are usually stacked as segments of the polysaccharide chain called "blocks". One can directly demonstrate that, under most experimental conditions, cell wall fragments indeed behave as an insoluble polyanion. If cell wall fragments are submitted to changes of ionic strength in the bulk phase, protons are expelled from the cell wall and can be titrated in the bulk phase [12] (Fig. 7.7).

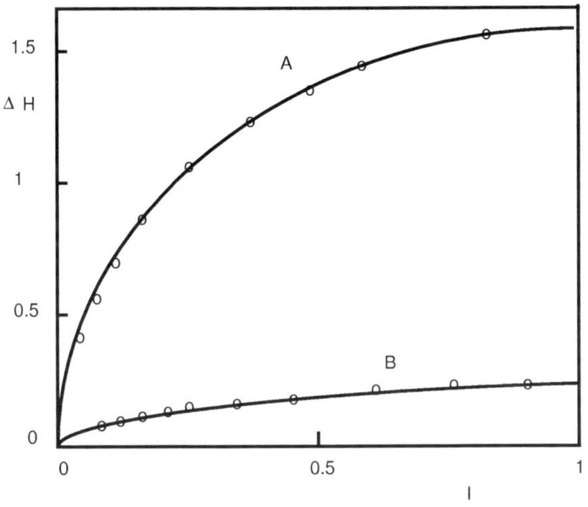

Fig. 7.7. Proton extrusion from plant cell wall fragments. When ionic strength of the bulk phase is increased, cell wall fragments release protons that can be titrated. Curve A: the initial ionic strength is small. Curve B: the initial ionic strength is larger. Adapted from ref. [1].

In addition to polysaccharides, plant cell wall contains a number of proteins. A protein, apparently devoid of enzymatic activity and called extensin, is specific to plant cell walls. It is extremely rich in hydroxyproline and is covalently bound to oligosaccharide chains. Besides this non-enzymatic protein, the cell wall contains different enzymes that are often ionically bound to the organelle. This is the case for an acid phosphatase, several peroxidases, pectin methyl esterase, glycosyl transferases, etc. As we shall see later, some of these enzymes are involved in plant cell wall extension. The plant cell wall is not an inert organelle. It can respond to various stimuli, to pathogenic agents (viruses, fungi, ...) and displays a strictly controlled extension process. Although the detailed mechanism of the extension process is still not completely understood, its main molecular events are known and are in fact the immediate consequence of the supramolecular structure of the wall. Under the influence of turgor pressure, and under acidic pH conditions, some hydrogen bonds, which associate cellulose microfibrils with xyloglucans, break down. This results in the unzipping of xyloglucans from cellulose microfibrils, and in the sliding of these microfibrils, under the influence of turgor pressure exerted by the vacuole. This creep might be a purely physical process but it might also be enzymatic [13]. Only limited regions of the wall undergo this loosening process. This event, however, is not sufficient *per se* to generate a significant increase of the cell wall and must be followed by the splitting of $\beta(1 \rightarrow 4)$ bonds of xyloglucans that interconnect cellulose microfibrils. This splitting results in an enhanced loosening of the wall and in a significant extension of this organelle. After the cxtension has taken place, new $\beta(1 \rightarrow 4)$ bonds of xyloglucan chains are formed, again crosslinking cellulose microfibrils. The breaking and making of $\beta(1 \rightarrow 4)$ bonds of xyloglucan chains is an enzymatic process carried out by an endotransglycosylase. This process should not take place over all the cell surface but

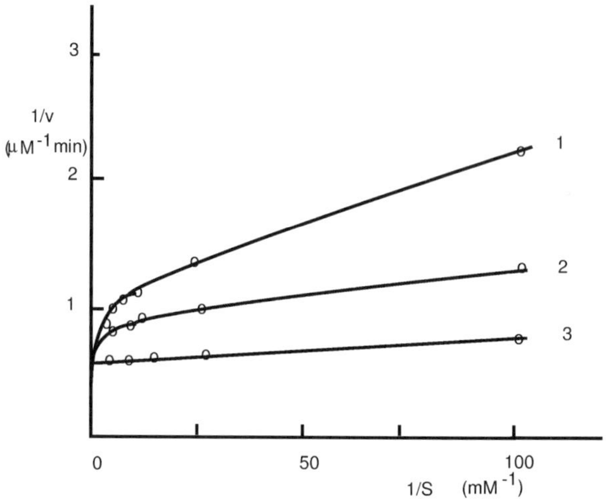

Fig. 7.8. Apparent co-operativity of cell wall-bound acid phosphatase under different conditions. 1 – Cell wall fragments are free from calcium. 2 and 3 – Cell wall fragments contain small amounts (2) or large amounts (3) of calcium. Adapted from ref. [12].

in many different micro-regions of the wall. Cell wall extension should be accompanied by the extrusion through the cell membrane, and the incorporation into the wall, of new material, mainly polysaccharide material. Pectins are incorporated as neutral, methylated, molecules that, aferwards, undergo partial demethylation thanks to pectin methylesterases [13–19].

7.3.2. Kinetics of a cell wall bound enzyme

As already mentioned, many enzymes are ionically bound to plant cell walls. These enzymes can be isolated in free solution simply by increasing the ionic strength. The free enzymes thus obtained, can be purified and their kinetic behaviour can be compared to that observed for the bound enzymes. It is then possible to study how the association of an enzyme with a biological polyelectrolyte apparently alters its kinetic properties. This kind of study has been performed with different enzymes, in particular with an acid phosphatase from sycamore and soybean cell walls. Cells from these plants can be cultivated *in vitro* in liquid medium, under sterile conditions. The cells then divide and extend. An acid phosphatase is ionically bound to the cell walls [20]. It can be isolated and purified to homogeneity. It is a monomeric glycoprotein of about 100 kDa molecular mass. Isolated in free solution, it displays classical Michaelis–Menten kinetics. At a rather "high" ionic strength, but under conditions in which the enzyme is still bound to cell wall fragments or to the surface of unbroken cells, its behaviour is identical to that of the free isolated enzyme. At lower ionic strength, complex mixed co-operativity is observed (Fig. 7.8) but it is progressively suppressed as the ionic strength is increased.

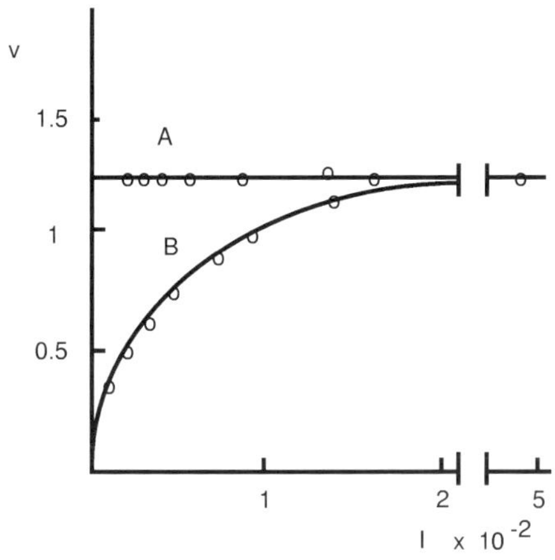

Fig. 7.9. Action of ionic strength on free and cell wall-bound acid phosphatase. Curve A – Acid phosphatase is free. Curve B – Acid phosphatase bound to cell wall fragments. Adapted from ref. [7].

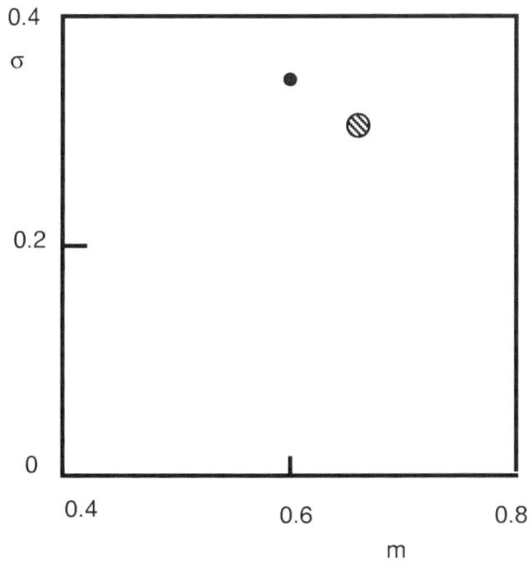

Fig. 7.10. Quantitative estimation of the spatial organization of enzyme molecules in cell wall fragments. In the (σ, m) plane the "large" and the "small" circles represent a perfectly random distribution and the real distribution of enzyme molecules, respectively. Adapted from ref. [7].

One can show that these effects are exerted through electrostatic repulsion of the substrate (an organic phosphate) by the fixed negative charges of the cell wall. The steady state reaction velocity of the isolated enzyme, in free solution and at constant substrate concentration, is measured as a function of the ionic strength. To a large extent, the rate is nearly independent of the ionic strength (Fig. 7.9). However, if the same experiment is performed with the cell wall-bound enzyme, one observes an apparent activation of the enzyme when the ionic strength is increased (Fig. 7.9)

These results show that there is an electrostatic repulsion of the substrate by the fixed charges of the walls, and that this repulsion is progressively abolished when increasing the ionic strength. If, however, the fixed charges and the enzyme molecules were homogeneously distributed in the wall, one would have expected the apparent co-operativity, depicted in Fig. 7.8, to be positive, which is indeed not the case. As a matter of fact, the kinetic results of Fig. 7.10 are best fitted by eq. (7.70) although the value of var(δ^*) in this equation is small, not very significantly different from zero. These results suggest that the enzyme molecules and the fixed charges are probably subject to fuzzy organization in clusters that only partly overlap (see sections 7.2.2 and 7.2.3).

The view that enzyme molecules may be clustered can be directly confirmed by image analysis. The degree of organization that may exist within a set of points in a plane can be quantitatively estimated through a mathematical technique called the minimal spanning tree [21–23]. A minimal spanning tree is a connected graph obtained by joining each point to its nearest neighbour. The frequency distribution of the edge lengths of all the possible minimal spanning trees pertaining to a large set of points follows a Laplace–Gauss distribution. Its mean, m, and its standard deviation, σ, can be scaled so as to allow comparison of the spatial distribution of different sets of points. If the points are randomly distributed, they take fixed scaled m and σ values, namely $m = 0.662$ and $\sigma = 0.311$. In the (m, σ) plane, any deviation from these values, which expresses absolute randomness, implies that some sort of spatial organization, for instance clustering, exists in the distribution of points (Fig. 7.10). This type of technique can be applied to any kind of material points and to any experimental situation. The phosphatase molecules can be visualized by different techniques on electron micrographs of sycamore, or soybean cell walls [22]. In either case, one observes deviations from pure randomness, indicating that the enzyme molecules tend to be clustered in the wall (Fig. 7.10).

Owing to the presence of fixed negative charges, cations, and in particular calcium, tend to be attracted within cell walls. In fact it is well known that calcium inhibits cell extension. Moreover, one should expect that cations, by decreasing the $\Delta\Psi$ value of the wall, may alter the behaviour of cell wall-bound enzymes. This is precisely what is occurring. If cell walls are deprived of their endogenous calcium by acid treatment, the reciprocal plots of bound acid phosphatase are concave downwards and exhibit strong negative co-operativity (Fig. 7.8). If cell wall fragments are loaded with calcium, the enzyme activity increases and the co-operativity declines. For relatively "high" calcium concentrations, the bound enzyme displays no co-operativity at all and its kinetic behaviour becomes identical to what it would have if it were in free solution (Fig. 7.8). Moreover calcium, at the concentrations used in this type of experiment, has strictly no effect on isolated phosphatase in free solution [12]. These results show that calcium binding to the fixed negative charges of the wall induces a decline of the corresponding $\Delta\Psi$ value, thus resulting in a decrease of the electrostatic repulsion of the substrate, which is reflected in the apparent kinetic behaviour of the phosphatase.

7.3.3. The two-state model of the primary cell wall and the process of cell elongation

As already outlined, extension and building up of the cell wall involve the sliding of cellulose microfibrils and the incorporation into the wall of neutral polysaccharide molecules. An endotransglycosylase certainly takes part in the process of microfibril sliding. However, when pectins, which are incorporated in the cell envelope during expansion of the cell volume, are methylated, the fixed charge density of the wall should decline. If this is the case, the electrostatic repulsion effects exerted by the cell envelope on mobile anions should fade away during growth process. However, this is not the case. There must therefore exist a biochemical process that maintains the fixed charge density at an approximately constant level. This process is the demethylation of pectins that have been incorporated into the wall and this demethylation reaction is exerted thanks to a pectin methyl esterase [10,11]. This enzyme should thus be responsible for the building up of the $\Delta\Psi$ of the cell wall.

7.3.3.1. The two-state model

If one considers a micro-domain of fixed volume, located in an expanding region of the wall, this micro-domain may occur in either of the two following states: a state X_1 where most of the acid groups of the pectins are ionized, and a state X_2M where most of these groups are methylated. The conversion $X_1 \rightarrow X_2M$ requires the creep of cellulose microfibrils, leading to an extension of the wall that includes the micro-domain. This process is controlled by an endotransglycosylase and is accompanied by the incorporation, in the extending region of the wall, of neutral precursors, in particular methylated pectins. The reverse process, $X_2M \rightarrow X_1$, requires the demethylation of methylated pectins by the enzyme pectin methyl esterase. The reversible conversion of X_1 and X_2M can then be schematically represented by a monocyclic cascade, as shown in Fig. 7.11.

Since a high endotransglycosylase activity is associated with the presence of carboxylate groups in the environment of the enzyme, this means that the optimum pH of this enzyme should be acidic. Alternatively, high pectin methyl esterase activity takes place when pectins are methylated. Therefore the optimum pH of this enzyme should be neutral, or close to neutrality. We shall see later that these predictions of the model are met experimentally. The view that a micro domain may exist in only two states is in fact a simplification, but this simplification is both sensible and useful. If we consider now an extending region of the wall, this region contains many micro-domains, some of which are in the "ionized" state and others in the "methylated" state. When the cell wall is extending, the system may possibly reach a steady state. This means that the density of micro-domains located in the expanding region remains constant whether these micro-domains are in a "methylated", or in an "ionized" state (Fig. 7.11).

Let us consider now, in more detail, the two processes $X_1 \rightarrow X_2M$ and $X_2M \rightarrow X_1$. The first process requires the participation of both an endotransglycosylase and micro-domains in state X_1 because the enzyme requires low pH conditions to be active. The region of the

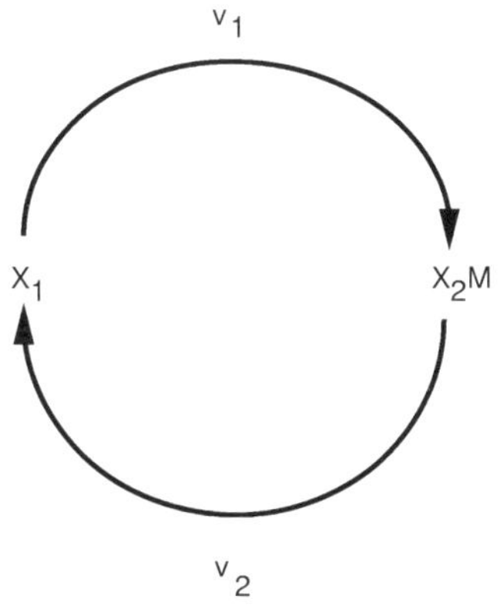

Fig. 7.11. The two-state model as a monocyclic cascade. See text.

cell wall then extends and neutral compounds are incorporated into the micro-domains that become X_2M. From a formal point of view, this process is equivalent to the conversion of X_1 into X_2M. The second process describes the demethylation of X_2M micro-domains by pectin methyl esterase, and the X_1 micro-domains are thus regenerated. If, in the expanding region of the wall, the system is in steady state, the total density of micro-domains should remain constant and one has thus

$$d(\mathrm{X}) = d(\mathrm{X}_1) + d(\mathrm{X}_2\mathrm{M}), \tag{7.75}$$

where $d(\mathrm{X})$ is the constant density of micro-domains in the expanding region, $d(\mathrm{X}_1)$ and $d(\mathrm{X}_2\mathrm{M})$ the densities of micro-domains in states X_1 and X_2M, respectively. The present model, as shown in Fig. 7.11, is based on several simplifying assumptions which are not necessarily met. First, a micro-domain is assumed to exist in only two states. Second, the system displays a steady state, which is far from certain. Third, the system is considered "closed" from a thermodynamic viewpoint. In chapter 9, we shall come back to the validity of these assumptions. Their usefulness, however, is in the number of predictions that can be experimentally tested.

From eq. (7.75), one can write, under steady state conditions

$$\frac{d(\mathrm{X}_2\mathrm{M})}{d(\mathrm{X})} = 1 - \frac{d(\mathrm{X}_1)}{d(\mathrm{X})}. \tag{7.76}$$

Moreover, one has

$$d(\mathrm{X}_1) = \frac{\delta_1}{z'_1}, \qquad d(\mathrm{X}_2\mathrm{M}) = \frac{\delta_2}{z'_2}, \tag{7.77}$$

where δ_1 and δ_2 are the fixed charge densities of the X_1 and X_2M micro-domains. As a X_1 micro-domain is a negatively charged entity, its apparent "valence", that is its number of carboxylate groups, is z'_1. Similarly, z'_2 is the apparent "valence" of X_2M. One must have indeed $z'_1 \gg z'_2$. Therefore, eq. (7.75) can be rewritten as

$$d(\mathrm{X}) = \frac{\delta_1}{z'_1} + \frac{\delta_2}{z'_2} = \frac{z'_2\delta_1 + z'_1\delta_2}{z'_1 z'_2} \tag{7.78}$$

and

$$\frac{d(\mathrm{X}_2\mathrm{M})}{d(\mathrm{X})} = \frac{z'_1\delta_2}{z'_2\delta_1 + z'_1\delta_2} = \frac{\delta_2}{\delta_2 + \frac{z'_2}{z'_1}\delta_1} = \delta_2^*. \tag{7.79}$$

δ_2^* is a normalized charge density. One could define the other charge density, δ_1^*, as

$$\frac{d(\mathrm{X}_1)}{d(\mathrm{X})} = \frac{\delta_1}{\delta_1 + \frac{z'_1}{z'_2}\delta_2} = \delta_1^*, \tag{7.80}$$

and one may note that

$$\delta_1^* + \delta_2^* = 1, \tag{7.81}$$

as expected for normalized variables. If one assumes, for simplicity, that, under steady state, the two reaction rates follow hyperbolic kinetics, then

$$v_1 = \frac{\widetilde{V}_1 d(\mathrm{X}_1)}{K_1 + d(\mathrm{X}_1)}, \qquad v_2 = \frac{\widetilde{V}_2 d(\mathrm{X}_2\mathrm{M})}{K_2 + d(\mathrm{X}_2\mathrm{M})}, \tag{7.82}$$

where $\widetilde{V}_1$ and $\widetilde{V}_2$ are the apparent maximum rates, K_1 and K_2 the corresponding apparent "Michaelis" constants.

Within the frame of this simplified model, wall-loosening enzymes such as the endo-transglycosylase, tend to produce a decrease of the fixed charge density because they generate the loosening and extension of the wall, associated with the incorporation of neutral precursors. Alternatively, pectin methyl esterase must be involved in the building up of the electrostatic potential of the wall, for this enzyme generates the fixed charges of the X_1 micro-domains. In order to understand how pectin methyl esterase, in the complex system, is controlled by the local pH, one must first express eq. (7.82) as a function of the normalized charge density, δ_1^*, generated by this enzyme. One must also study, in the overall steady state equation, how pH changes affect the maximum rates, or the ratios of these rates to the corresponding "Michaelis" constants. The two equations (7.82) can then be rewritten as

$$v_1 = \frac{\widetilde{V}_1 \delta_1^*}{K_1^* + \delta_1^*}, \qquad v_2 = \frac{\widetilde{V}_2 (1 - \delta_1^*)}{K_2^* + (1 - \delta_1^*)}, \tag{7.83}$$

where $K_1^* = K_1/d(\mathrm{X})$ and $K_2^* = K_2/d(\mathrm{X})$. In these equations, it is implicitely assumed that only the maximum rates are sensitive to pH, not the "Michaelis" constants. But it may well happen that both the maximum rates and the "Michaelis" constants are sensitive to pH changes. Let us consider these two possibilities in succession. Since the overall system is assumed to be in steady state, the two rates must be equal. Thus

$$\frac{\widetilde{V}_1 \delta_1^*}{K_1^* + \delta_1^*} = \frac{\widetilde{V}_2 (1 - \delta_1^*)}{K_2^* + (1 - \delta_1^*)}. \tag{7.84}$$

As there is no reason, from this time onwards, to distinguish between the two normalized charge densities, δ_1^* and δ_2^*, because δ_2^* is probably negligible with respect to δ_1^*, the charge density δ_1^* will be referred to as δ_s^*, and represents the charge density generated by pectin methyl esterase in the overall system. Equation (7.84) can be cast into either of two particularly useful forms, namely

$$\left(\frac{\widetilde{V}_2}{\widetilde{V}_1} - 1\right)\delta_s^{*2} - \left\{\left(\frac{\widetilde{V}_2}{\widetilde{V}_1} - 1\right) - K_1^*\left(\frac{\widetilde{V}_2}{\widetilde{V}_1} + \frac{K_2^*}{K_1^*}\right)\right\}\delta_s^* - \frac{\widetilde{V}_2}{\widetilde{V}_1} K_1^* = 0 \tag{7.85}$$

and

$$\left(\frac{\widetilde{V}_2}{\widetilde{V}_1}-1\right)\delta_s^{*2}-\left\{\left(\frac{\widetilde{V}_2}{\widetilde{V}_1}-1\right)-\widetilde{V}_2\left(\frac{\widetilde{K}_1^*}{\widetilde{V}_1}+\frac{\widetilde{K}_2^*}{\widetilde{V}_2}\right)\right\}\delta_s^*-\frac{\widetilde{K}_1^*}{\widetilde{V}_1}\widetilde{V}_2=0. \tag{7.86}$$

In the first formulation (7.85), $\widetilde{V}_1$ and $\widetilde{V}_2$ are assumed to be pH-dependent, but not K_1^* and K_2^*. In the second formulation (7.86) both $\widetilde{V}_1$, $\widetilde{V}_2$ and $\widetilde{K}_1^*/\widetilde{V}_1$, $\widetilde{K}_2^*/\widetilde{V}_2$ are pH-dependent (see section 7.1.2). As outlined previously, the plot of $\log(\widetilde{V})$ *versus* pH allows one to estimate the pKs of the enzyme–substrate complex whereas a plot of $\log(\widetilde{V}/\widetilde{K})$ *versus* pH allows the estimation of the pKs of the enzyme itself.

Wall loosening and growth enzymes (endotransglycosylase for instance) are very active under acid pH conditions. This means that $\widetilde{V}_1$ (if it is the enzyme–substrate complex alone that undergoes ionization) or $\widetilde{V}_1$ and $\widetilde{V}_1/\widetilde{K}_1^*$ (if both the enzyme–substrate complex and the enzyme undergo ionization) are activated by protons in the acid pH range. Thus one should have

$$\widetilde{V}_1=V_1\frac{H_i}{K_a'+H_i},\qquad \frac{\widetilde{V}_1}{\widetilde{K}_1^*}=\frac{V_1}{K_1^*}\frac{H_i}{K_a+H_i}, \tag{7.87}$$

where K_a and K_a' are the acid ionization constants of the enzyme and of the enzyme–substrate complex. H_i is the local proton concentration, V_1 and K_1^* are the real maximum rate and the normalized "Michaelis" constant, respectively. Whereas most cell wall enzymes have an acidic optimum pH, pectin methyl esterase, which is involved in the building up of the cell wall $\Delta\Psi$, has a neutral, or even slightly alkaline, optimum pH. In the acid pH-range, the enzyme is therefore inhibited by protons. The expressions of $\widetilde{V}_2$ and $\widetilde{V}_2/\widetilde{K}_2^*$ in this acid pH-range is thus

$$\widetilde{V}_2=V_2\frac{K_b'}{K_b'+H_i},\qquad \frac{\widetilde{V}_2}{\widetilde{K}_2^*}=\frac{V_2}{K_2^*}\frac{K_b}{K_b+H_i}. \tag{7.88}$$

K_b and K_b' are the base ionization constants of the enzyme and of the enzyme–substrate complex, respectively. V_2 and K_2^* are the real V_{max} of pectin methyl esterase and its normalized Michaelis constant. The pH-dependence of $\widetilde{V}_2/\widetilde{V}_1$, $\widetilde{V}_2\{(\widetilde{K}_1^*/\widetilde{V}_1)+(\widetilde{K}_2^*/\widetilde{V}_2)\}$ and $(\widetilde{K}_1^*/\widetilde{V}_1)\widetilde{V}_2$ that appear in eqs. (7.85) and (7.86) is easily derived. One finds

$$\begin{aligned}
&\frac{\widetilde{V}_2}{\widetilde{V}_1}=\frac{V_2}{V_1}\frac{K_a'K_b'+K_b'H_i}{K_b'H_i+H_i^2},\\
&\widetilde{V}_2\left(\frac{\widetilde{K}_1^*}{\widetilde{V}_1}+\frac{\widetilde{K}_2^*}{\widetilde{V}_2}\right)=V_2\left\{\frac{K_1^*}{V_1}\frac{K_aK_b'+K_b'H_i}{K_b'H_i+H_i^2}+\frac{K_2^*}{V_2}\frac{K_bK_b'+K_b'H_i}{K_bK_b'+K_bH_i}\right\},\\
&\frac{\widetilde{K}_1^*}{\widetilde{V}_1}\widetilde{V}_2=V_2\frac{K_1^*}{V_1}\frac{K_aK_b'+K_b'H_i}{K_b'H_i+H_i^2}.
\end{aligned} \tag{7.89}$$

These expressions can be inserted into eqs. (7.85) and (7.86) and the variation of δ_s^* as a function of H_i, or of pH_i, can be simulated by computer. If the expression for $\widetilde{V}_2/\widetilde{V}_1$ is

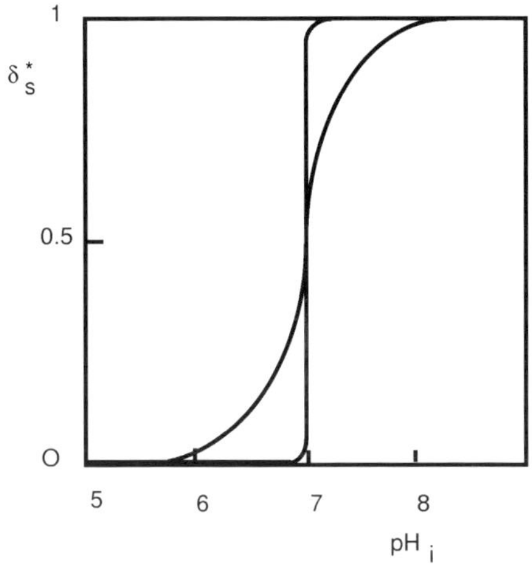

Fig. 7.12. Change of the fixed charge density as a function of the local pH. The steepness of the transition depends on the value of K_1^* when the ratio K_2^*/K_1^* is held constant. The transition becomes extremely steep when the value of K_1^* becomes very small. Adapted from ref. [10].

inserted into eq. (7.85), the variation of δ_s^* *versus* pH_i may display extremely steep co-operativity for certain pH values. The existence of this abrupt transition of charge density is the consequence of the opposite pH-dependences of pectin methyl esterase and of wall loosening enzymes (endotransglycosylase). The situation is thus similar to that encountered for monocyclic cascades, and the abrupt transition is a property of the system, not a property of an individual enzyme. For constant values of K_2^*/K_1^* the steepness of the transition depends on the value of K_1^*. Changing this value does not alter the symmetry of the curve about its midpoint and does not bring about any pH-shift (Fig. 7.12). However, changing the value of the ratio V_2/V_1 generates a shift of the transition and a moderate decrease of its steepness. If $\widetilde{V}$ and $\widetilde{K}^*$ both are pH-dependent, as assumed in eq. (7.86), the system may still display the steep transition shown in Fig. 7.12, and the conclusions derived for this figure still hold.

The interest of this simplified, perhaps oversimplified, model is that it allows one to make a number of predictions that can be tested experimentally. These predictions are the following:

- pectin methyl esterase should be controlled by cations and protons;
- pectin methyl esterase should be responsible for the building up, *in situ*, of the $\Delta\Psi$ between the inside and outside of the wall;
- there should exist an ionic control of wall loosening enzymes;
- the pectin methyl esterase – wall loosening enzyme system should display strong co-operativity of its response to slight changes of pH, thus controlling the local ion concentration and the activity of the enzymes involved in cell wall extension and building up.

7.3.3.2. Primary cell wall as a complex auto-regulated polyelectrolyte system

The aim of the present section is to offer experimental evidence for both the validity (as an approximation) and the usefulness of the simplified model presented above.

7.3.3.2.1. Ionic control of pectin methyl esterase. The above model relies in part on the existence of a strict ionic control of pectin methyl esterase. Pectin methyl esterase from soybean cell walls is a monomeric protein of 33 kDa molecular mass [24–27]. The protein can be purified to homogeneity from soybean cell wall fragments, obtained from isolated cells in sterile culture. Nearly complete solubilization of the enzyme can be obtained by simply raising the salt concentration up to 1 M. There is little doubt that this enzyme requires cations for full activity. In the absence of cations, pectin methyl esterase is nearly totally inactive. However, at high cation concentration, the enzyme becomes inhibited. Moreover pectin methyl esterase is maximally active at a pH value close to neutrality. Above and below this value, the activity falls off. The optimum cation concentration depends on the pH of the reaction medium. At the optimum pH, the optimum cation concentration is low, but this optimum increases as the pH is decreased.

The reaction mechanism of the enzyme process involves a nucleophilic attack of the enzyme on the carboxyl of the substrate, the subsequent release of methanol, followed by an attack of water and the formation of an acid [26]. A number of experiments suggest that the metal ion does not interact with the enzyme, but with the negatively charged groups of pectin. The activation and inhibition of the enzyme reaction by cations, such as sodium or calcium, can be mimicked by methylene blue which carries a positive charge and can be stacked on the polyanion [26]. Numerous studies have shown that, in natural pectins, there exists "blocks" of carboxylate groups that tend to trap cations, but also trap enzyme molecules that are positively charged. Thus the binding of cations to the "blocks" tends to release the enzyme molecules initially bound to these regions of the polyanion. This release indeed results in an enhancement of the reaction rate, for more enzyme molecules become available for the reaction. Kinetic results, which will be not discussed here, suggest that metal binding to the "blocks" is a non co-operative process. Moreover, a methylated unit can undergo demethylation only if the neighbouring units are negatively charged. This is understandable, for pectin methyl esterase should bind to these negative charges. In the presence of an excess of metal ions, these negative charges will be neutralized by cations, thus preventing the binding of enzyme molecules in the vicinity of the methyl groups of pectin. The consequence of this situation is indeed a decrease of the enzyme reaction rate as the metal concentration is increased. Kinetic studies suggest this second type of metal binding to pectin to be positively co-operative [26].

7.3.3.2.2. The building up and control of plant cell wall electrostatic potential by pectin methyl esterases. As previously mentioned, it is possible to estimate the $\Delta\Psi$ value of cell wall fragments by monitoring proton efflux when increasing the ionic strength in the bulk phase. As we know the volume of the bulk phase and that of the cell wall fragments, one can estimate, from the measurement of the proton efflux, the local proton concentration and the electrostatic potential difference between the inside and the outside of the wall. Since solubilized pectin methyl esterase is apparently more active under neutral pH conditions and relatively "high" cation concentrations, one would expect that, after incubation of the walls under the conditions where the enzyme is maximally active, the $\Delta\Psi$ of this organelle would increase. This is exactly what is found. Cell wall fragments are prepared at pH 5 from clumps of sycamore or soybean cells grown *in vitro*. A series of samples of cell wall fragments are pre-incubated for 15 minutes at a pH higher, or lower, than 5, then

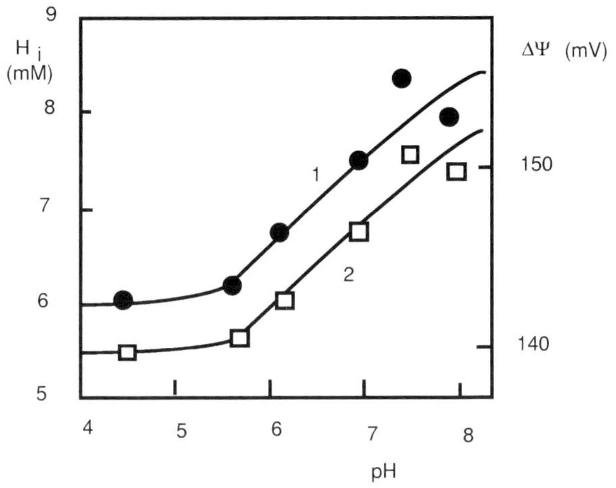

Fig. 7.13. Variation of the local proton concentration and of the ΔΨ of the cell wall as a function of the pH of the incubation mixture. 1 – Proton concentration. 2 – ΔΨ values. See text. Adapted from ref. [7].

washed and brought back to pH 5. Another sample is subjected to the same treatment, but with a pre-incubation at pH 5. It plays the part of a control. Whatever the pre-incubation, the proton efflux is always measured under the same experimental conditions. From this proton efflux, the local proton concentration in the wall and the ΔΨ values are plotted as a function of the pH of the pre-incubation mixture (Fig. 7.13).

From these results it is obvious that stimulating the activity of pectin methyl esterase brings about an enhancement of the ΔΨ values of the wall. This implies that pectin methyl esterase is involved in the build-up of the electrostatic potential of the wall. This conclusion can be confirmed by measuring the variations of the rate (at constant pH) and of ΔΨ as a function of the sodium concentration. Under identical experimental conditions, the concentration of NaCl that brings about the maximum reaction rate also results in the maximum of ΔΨ values. This result can hardly be a coincidence and one must conclude that pectin methyl esterase is responsible for the build up of the ΔΨ value of the wall.

Although the local pH within cell wall fragments suspended in water may be of the order of 2, it is unlikely that these values ever occur *in vivo* during cell wall expansion and build up, for counter ions other than protons are present in the extending wall. Therefore the local pH cannot be as low as two. These results leave no doubt as to the participation of pectin methyl esterase to the ΔΨ value of the wall. The changes of ΔΨ values observed in the above experiments are not very large. This is quite understandable since most pectins in the cell wall fragments have already been demethylated by pectin methyl esterase during the time required to break the cells and prepare cell wall fragments. The changes of ΔΨ are expected to be much larger *in vivo.*

These results thus offer a coherent physico-chemical basis for the mechanism of plant cell wall extension. When the ΔΨ value of the wall is large, the local proton concentration is large as well. Pectin methyl esterase is thus relatively inactive under these conditions and the cell wall extends under the influence of wall loosening enzymes (see below) and turgor pressure. But this extension, in turn, results in the decrease of the electrostatic potential of the wall since the pectins incorporated in the extending wall are methylated. As the wall extends, the density of fixed charges falls off, thus giving rise to an increase of the local pH and activation of pectin methyl esterase. This scheme, however, embodies an apparent difficulty, for proton and cation concentrations affect pectin methyl esterase in two opposing ways. Namely, "high" proton concentrations bring about an inhibition of pectin methyl esterase whereas "high" metal ion concentrations result in an enhanced enzyme activity. Moreover, as a change of fixed charge density results in a simultaneous increase, or decrease, of proton and metal ion concentrations, the effects of protons and cations would be expected to be mutually antagonistic. In fact, this difficulty is only apparent, for it will be noted that the metal ion concentration, which gives a maximum of the rate, is much lower for neutral than for acidic pH conditions. After the extension of the cell wall has taken place, $\Delta\Psi$ is small and this results in low values of both proton and cation concentrations, which represent favourable conditions for pectin methyl esterase activity. When the fixed charge density increases, both proton and cation concentrations increase, but this still represents favourable conditions for pectin methyl esterase activity, and may thus be viewed as an amplification of its action.

7.3.3.2.3. Ionic control of cell wall autolysis. The theoretical model developed above postulates that $\Delta\Psi$ is the trigger for growth. Large values of $\Delta\Psi$ result in large values of the proton concentration that stimulate the wall loosening enzymes required for local cell wall extension. It is therefore mandatory to obtain direct evidence that "low" pH values are required for maximum activity of wall loosening enzymes. There is in fact little doubt that, during its extension, the cell wall is submitted to a loosening process. Cell wall fragments spontaneously release reducing sugars. This process is enzymatic and can be suppressed if the cell wall fragments are suspended in boiling water. Under the conditions of "high" (0.125) ionic strength, the optimum pH is about 5.5 and is shifted, by about one pH unit, towards lower values at "low" (0.025) ionic strength. Another striking result, is the sharp pH dependence of wall loosening enzymes. Since activating, or inhibiting, pectin methyl esterase by bulk hydroxyl ions, protons or cations results in an increase, or a decrease, of $\Delta\Psi$ and of the local proton concentration, it is then obvious that $\Delta\Psi$ serves as a trigger for the activity of wall loosening enzymes [25]. Moreover, wall loosening enzymes are activated by "high" ionic strength. The attraction of cations in the wall by the fixed negative charges, generated by pectin methyl esterase, stimulate these wall loosening enzymes. Stimulation of pectin methyl esterase activity thus has a dual effect: it increases the local proton concentration and stimulates wall loosening enzymes; it increases the local cation concentration, and this again stimulates these enzymes.

7.3.3.2.4. Co-operativity of the response of the pectin methyl esterase system to changes in local pH during cell growth. Thus, during cell extension, two opposing processes take place. Pectin methyl esterase, which is maximally active under neutral pH conditions, generates the fixed negative charges of the wall and thereby results in a decrease of the local pH. This, in turn, stimulates the wall loosening enzymes associated with the growth process. The extension of the wall generates a decrease of the fixed charge density and

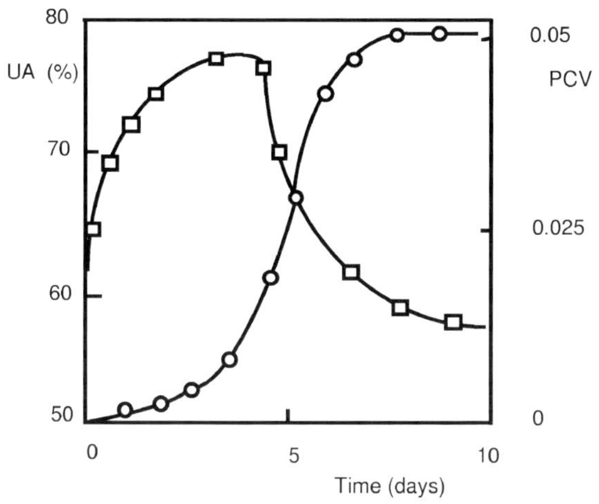

Fig. 7.14. Co-operativity of the response of pectin methyl esterase to changes in local pH. 1 – Variation of unmethylaed uronic acids (UA) during growth of a cell culture. 2 – Variation of the packed cell volume (PCV) of the culture. The first curve expresses the variation of the fixed charge density and the second curve expresses cell wall growth. There is thus a period (up to 5 days) where the fixed charge density and the cell wall volume increase. Adapted from ref. [1].

of the local proton concentration, thus leading to a stimulation of the pectin methyl esterase. The enzyme thus responds to changes of local proton concentration. The results of Fig. 7.14 suggest that this response can be highly co-operative, i.e. minute changes of pH bring about a much larger change in pectin methyl esterase activity, and therefore in the fixed charge density. In fact, if the restoration of the fixed charge density were not co-operative, one would have expected the fixed charge density to remain constant during cell expansion. If, alternatively, pectin methyl esterase did not respond, or responded poorly, to small local pH changes, the fixed charge density should decline continuously during cell growth. Last, if the response of pectin methyl esterase to local changes in proton concentration were co-operative, the charge density should increase together with the cell wall volume. It is possible to check the validity of these predictions experimentally.

As already outlined, the isolated clumps of plant cells in sterile culture represent a good biological system for studying plant growth. When sycamore or soybean cells are transferred to fresh culture medium, they first extend, then divide. One can estimate whether the fixed charge density is increasing, or decreasing, by titrating the methylated and unmethylated uronic acids of the cell wall. If the percent of unmethylated uronic acids, together with the cell wall volume, are plotted as a function of time, one observes that both unmethylated uronic acids and the cell wall volume increase for several days (Fig. 7.14). During this time period the response of pectin methyl esterase to changes of local pH must be co-operative. After this period, the fixed charge density decreases as a function of time.

The ideas that have been discussed above are summarized in Fig. 7.15. The trigger of growth is the $\Delta\Psi$ value. When the electrostatic potential of the cell wall is low, the local

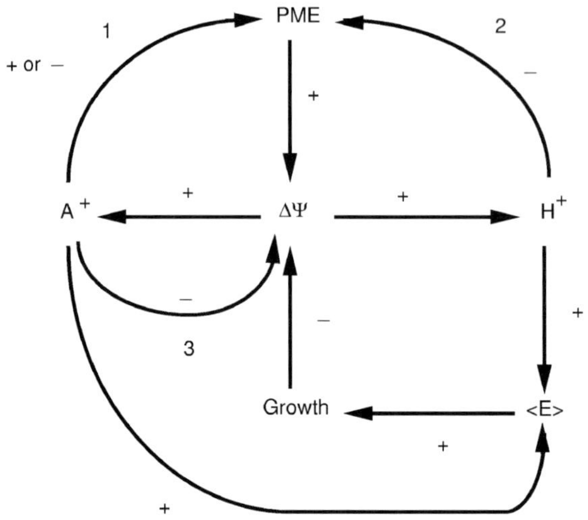

Fig. 7.15. The control of plant cell wall extension and pectin methyl esterase activity by $\Delta\Psi$. The '+' and '−' signs stand for activation and inhibition. See text. From ref. [7].

proton and cation concentration is low as well, and this results in the activation of pectin methyl esterase (loops 1 and 2) which tends to bring about higher $\Delta\Psi$ values. This, in turn, stimulates the attraction of protons and cations in the matrix thus leading to amplification of the electrostatic potential. This amplification is the consequence of two types of events: the shift of the reaction profile towards "high" metal concentrations when the pH drops, and the increase of availability of pectin methyl esterase, which dissociates from the pectin "blocks", as these "blocks" become neutralized by cations. This probably results in a concentration of protons and cations that is optimal for the activity of glucanases involved in cell wall loosening and extension. If the metal ion (and the proton) concentration in the wall were too high, the control of the process would be achieved through the apparent inhibition of pectin methyl esterase by an excess of the metal. As this inhibition is exerted through the binding of the metal to the polyanion (loop 3), it results in a decrease of $\Delta\Psi$ [7].

7.4. Sensing, memorizing and conducting signals by polyelectrolyte-bound enzymes

It was implicitly assumed, in the previous sections, that enzyme reactions were taking place in a polyelectrolyte matrix, under quasi-equilibrium conditions. This is certainly not always the case, and one may therefore wonder which types of events can be expected to take place under conditions where "slow" diffusion of charged substrate and product occurs, together with their electric repulsion by the fixed negative charges of the matrix.

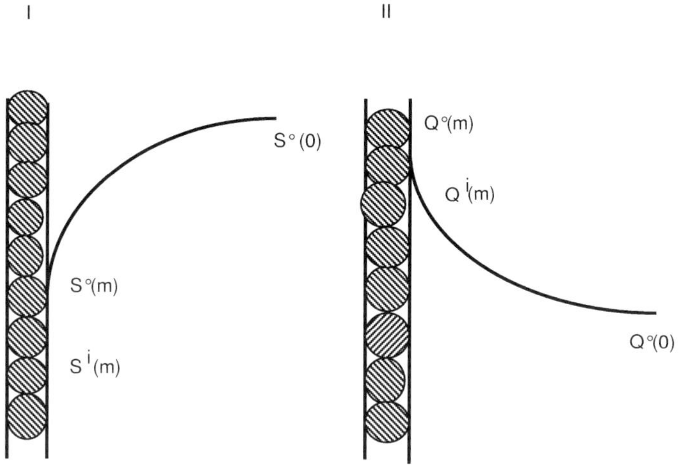

Fig. 7.16. Gradients of substrate and product of an enzyme reaction taking place on a negatively charged surface. I – Gradient of substrate. II – Gradient of product. $S^\circ(0)$ and $Q^\circ(0)$ are the concentrations of substrate and product in the reservoir, $S^i(m)$ and $Q^i(m)$ their concentrations in the charged membrane, and $S^\circ(m)$ and $Q^\circ(m)$ the corresponding concentrations in the immediate vicinity of the membrane. See text.

7.4.1. Diffusion of charged substrate and charged product of an enzyme reaction

Let us consider the simple enzyme reaction

$$E \xleftarrow{S} (ES, EPQ) \longrightarrow EQ \underset{Q}{\xleftrightarrow{}} E$$
$$\searrow P$$

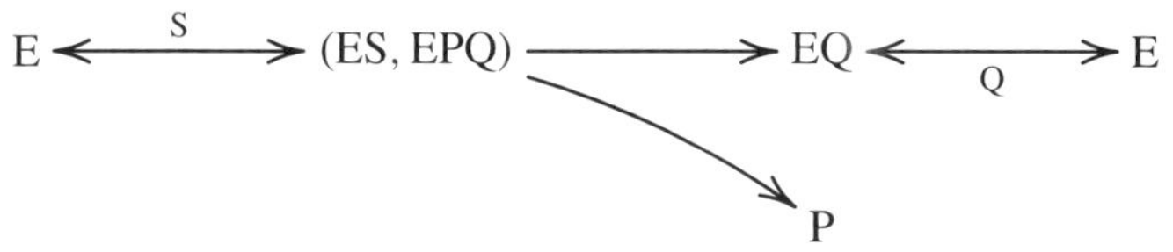

taking place on a negatively charged surface. Moreover, it is assumed that both the substrate S and the product Q are negatively charged with valences $z' = -z$ and $\lambda z' = -\lambda z$ (λ is a positive number such that the product λz is an integer), respectively. Moreover, the diffusion of S and Q is assumed to display some resistance in such a way that a coupling between diffusion and enzyme reaction takes place (Fig. 7.16). The situation is somewhat similar, although more complex than the one we have already considered in chapter 3.

The diffusion of the substrate is indeed expressed by the Nernst–Planck equation (see chapter 5), namely

$$J_s = -D\frac{\partial S^\circ(x)}{\partial x} - DS^\circ(x)\frac{zF}{RT}\frac{\partial \Psi^\circ(x)}{\partial x}, \tag{7.90}$$

where J_s is the diffusion flow of the substrate, $S^\circ(x)$ the substrate concentration at a distance x, $\Psi^\circ(x)$ the electric potential at the same distance, D the diffusion constant, F, R and T the Faraday constant, the gas constant and the absolute temperature, respectively.

If the origin of the diffusion process is located at a distance m from the membrane, then $x \in [0, m]$. Moreover we assume, as previously, that the potential varies linearly as a function of the distance. Thus one has

$$-D\frac{\partial \Psi^{\circ}(x)}{\partial x} = -\frac{D}{m}\{\Psi^{\circ}(0) - \Psi^{\circ}(m)\} = -\frac{D}{m}\Delta\Psi^{\circ}. \tag{7.91}$$

In eqs. (7.90) and (7.91) the superscript $^{\circ}$ refers to the outside of the charged membrane. Setting, as previously,

$$\alpha = \frac{D}{m}\frac{zF}{RT}\Delta\Psi^{\circ}, \tag{7.92}$$

the Nernst–Planck equation can be rewritten as

$$J_{\mathrm{s}}\,\mathrm{d}x = -D\,\mathrm{d}S^{\circ}(x) - \alpha S^{\circ}(x)\,\mathrm{d}x \tag{7.93}$$

and integrated. One finds

$$x = -\frac{D}{\alpha}\ln\frac{J_{\mathrm{s}} + \alpha S^{\circ}(x)}{J_{\mathrm{s}} + \alpha S^{\circ}(0)} \tag{7.94}$$

or

$$\frac{J_{\mathrm{s}} + \alpha S^{\circ}(x)}{J_{\mathrm{s}} + \alpha S^{\circ}(0)} = \exp\left(-\frac{\alpha x}{D}\right). \tag{7.95}$$

Since

$$\frac{\alpha x}{D} = \frac{x}{m}\frac{zF\Delta\Psi^{\circ}}{RT}, \tag{7.96}$$

the solution of the Nernst–Planck equation is

$$J_{\mathrm{s}} = \tilde{h}_{\mathrm{d}}\{\widetilde{S}^{\circ}(0) - S^{\circ}(x)\} \tag{7.97}$$

with

$$\tilde{h}_{\mathrm{d}} = \frac{\dfrac{D}{m}\dfrac{zF}{RT}\Delta\Psi^{\circ}}{1 - \exp\left(-\dfrac{zF\Delta\Psi^{\circ}}{RT}\dfrac{x}{m}\right)} \tag{7.98}$$

and

$$\widetilde{S}^{\circ}(0) = S^{\circ}(0)\exp\left(-\frac{zF\Delta\Psi^{\circ}}{RT}\frac{x}{m}\right). \tag{7.99}$$

The diffusion equation (7.97) is written in the same form as the diffusion equation of a neutral molecule under steady state conditions, but these equations are basically different. In the case of the diffusion of a neutral molecule under steady state, the concentration gradient is linear. In the case of the diffusion of an ion, the gradient cannot be linear for when $S^{\circ}(x)$ varies, both $\widetilde{S}^{\circ}(0)$ and $\tilde{h}_{\mathrm{d}}$ vary as well. However, as we shall see later, it is advantageous to write the diffusion equation of an ion in the form of expression (7.97).

7.4.2. Electric partition of ions and Donnan potential under gobal nonequilibrium conditions

We have already seen that, under equilibrium conditions, the partition coefficient for both the substrate S and the charged product Q is given by

$$\Pi_{\mathrm{e}} = \left(\frac{\overline{S}^{\mathrm{o}}}{\overline{S}^{\mathrm{i}}}\right)^{1/z'} = \left(\frac{\overline{Q}^{\mathrm{o}}}{\overline{Q}^{\mathrm{i}}}\right)^{1/\lambda z'} = \exp\left(\frac{F\Delta\overline{\Psi}_{\mathrm{D}}}{RT}\right) \tag{7.100}$$

where, as usually, $z' = -z$ and $\Delta\overline{\Psi}_{\mathrm{D}}$ is the electrostaic potential difference between the outside and the inside of the charged matrix. The present symbolism has been slightly modified relative to that used so far. The superscripts 'o' and 'i' refer to the outside and the inside of the matrix and the bar above the symbols specifies that the system is in equilibrium. In principle both the partition coefficient and the electrostatic potential difference are meaningful only for equilibrium conditions. But one may wonder whether these concepts of partition coefficient and Donnan potential can be extrapolated to global nonequilibrium conditions.

Let us consider for instance the flow of the substrate. Its concentration in the reservoir is $S^{\mathrm{o}}(0)$. In the immediate vicinity of the charged membrane, its concentration is $S^{\mathrm{o}}(m)$, and inside the membrane it is now $S^{\mathrm{i}}(m)$. The substrate flow can be pictured as

$$\longrightarrow S^{\mathrm{o}}(0) \underset{h_{\mathrm{d}}}{\overset{h_{\mathrm{d}}}{\longleftrightarrow}} S^{\mathrm{o}}(m) \underset{k'h_{\mathrm{d}}}{\overset{kh_{\mathrm{d}}}{\longleftrightarrow}} S^{\mathrm{i}}(m) \longrightarrow$$

The coefficients k and k' express the extent of the repulsion of the charged substrate by the fixed charges of the matrix [28,29], namely

$$k = \exp\left(-\frac{F\Delta\Psi}{2RT}\right), \qquad k' = \exp\left(\frac{F\Delta\Psi}{2RT}\right), \tag{7.101}$$

where $\Delta\Psi = \Psi^{\mathrm{o}} - \Psi^{\mathrm{i}} > 0$. Therefore, one must have $0 < k < 1$ and $k' > 1$. In all generality, $\Delta\Psi$ is not a Donnan potential for the system is not in equilibrium but it is of interest to determine the conditions where it can be approximated to a Donnan potential. One may thus determine the expression for the flow of substrate, namely

$$J_{\mathrm{s}} = \frac{h_{\mathrm{d}}k}{1 + k + k'} S^{\mathrm{o}}(0). \tag{7.102}$$

If $\Delta\Psi$ is such that there is a strong repulsion of the substrate, one must have

$$\exp\left(\frac{F\Delta\Psi}{2RT}\right) \gg 1 + \exp\left(-\frac{F\Delta\Psi}{2RT}\right). \tag{7.103}$$

This expression implies that the rate constant k' is much larger than the rate constant k. More specifically, $k' \gg 1 + k$.

Under these conditions eq. (7.102) reduces to

$$J_s = h_d \frac{k}{k'} S^o(0). \tag{7.104}$$

Comparison of expressions (7.102) and (7.104) implies that the step of electric repulsion of the substrate is in fast equilibrium relative to the other steps of the flow. This means that the substrate transport scheme can be reexpressed as

$$\longrightarrow S^o(0) \underset{h_d f'}{\overset{h_d}{\rightleftarrows}} X \xrightarrow{\;h_d f\;}$$

with

$$X = \overline{S}^o(m) + \overline{S}^i(m) \tag{7.105}$$

and

$$\begin{aligned} f &= \frac{\overline{S}^i(m)}{\overline{S}^o(m) + \overline{S}^i(m)} = \frac{1}{1 + \Pi_e^{z'}}, \\ f' &= \frac{\overline{S}^o(m)}{\overline{S}^o(m) + \overline{S}^i(m)} = \frac{\Pi_e^{z'}}{1 + \Pi_e^{z'}}. \end{aligned} \tag{7.106}$$

In these expressions, Π_e is the partition coefficient because the concentrations $\overline{S}^o(m)$ and $\overline{S}^i(m)$ are in fact the equilibrium concentrations. Therefore, the flow of substrate corresponding to the last model above is now

$$J_s = \frac{h_d}{1 + \Pi_e^{z'}} S^o(0). \tag{7.107}$$

Since it has been assumed that the electrostatic repulsion of the substrate is very strong, $\Pi_e^{z'} \gg 1$, and eq. (7.107) reduces to

$$J_s = \frac{h_d}{\Pi_e^{z'}} S^o(0). \tag{7.108}$$

The two kinetic models presented above are thus equivalent when the electrostatic repulsion of the charged substrate by the surface of the membrane is large. Therefore, eqs. (7.102), (7.104), (7.107) and (7.108) are equivalent. This implies that

$$\frac{k'}{k} = \Pi_{\mathrm{e}}^{z'} = \exp\left(\frac{F\Delta\Psi}{RT}\right) \tag{7.109}$$

and thus proves that $\Delta\Psi$ in this equation is nearly identical to the Donnan potential $\Delta\overline{\Psi}_{\mathrm{D}}$.

7.4.3. Coupling between diffusion, reaction and electric partition of the substrate and the product

The net flow of the substrate which reaches the enzyme molecules, within the matrix, can then be expressed as

$$J_{\mathrm{s}} = \tilde{h}_{\mathrm{d}}\left\{\widetilde{S}^{\mathrm{o}}(0) - \Pi^{z'} S^{\mathrm{i}}(m)\right\}, \tag{7.110}$$

and the net flow of the product Q which leaves the enzyme molecules is

$$J_{\mathrm{Q}} = \tilde{h}_{\mathrm{d}}\left\{\Pi_{\mathrm{e}}^{\lambda z'} Q^{\mathrm{i}}(m) - \widetilde{Q}^{\mathrm{o}}(0)\right\}, \tag{7.111}$$

where $Q^{\mathrm{i}}(m)$ is the product concentration within the matrix and $\widetilde{Q}^{\mathrm{o}}(0)$ the apparent concentration of product Q in the reservoir. The expression of the enzyme reaction rate is now

$$v_{\mathrm{e}} = \frac{V S^{\mathrm{i}}(m)/K_{\mathrm{S}}}{1 + S^{\mathrm{i}}(m)/K_{\mathrm{S}} + Q^{\mathrm{i}}(m)/K_{\mathrm{Q}}}, \tag{7.112}$$

where K_{S} and K_{Q} are the Michaelis constants of the substrate and of the product, and V the maximum rate. It is convenient to rewrite these expressions in dimensionless form. Setting

$$\begin{aligned} s_{\mathrm{o}} &= \frac{\widetilde{S}^{\mathrm{o}}(0)}{K_{\mathrm{S}}}, & s_{\sigma} &= \frac{S^{\mathrm{i}}(m)}{K_{\mathrm{S}}}, \\ q_{\mathrm{o}} &= \frac{\widetilde{Q}^{\mathrm{o}}(0)}{K_{\mathrm{Q}}}, & q_{\sigma} &= \frac{Q^{\mathrm{i}}(m)}{K_{\mathrm{Q}}}, \\ h_{\mathrm{d}}^{*} &= \frac{\tilde{h}_{\mathrm{d}} K_{\mathrm{S}}}{V}, & h_{\mathrm{d}}^{*\prime} &= \frac{\tilde{h}_{\mathrm{d}} K_{\mathrm{Q}}}{V}, \end{aligned} \tag{7.113}$$

one may write a first equation that couples substrate diffusion, electric partition coefficient and enzyme reaction. This equation is

$$h_{\mathrm{d}}^{*}\left(s_{\mathrm{o}} - \Pi_{\mathrm{e}}^{z'} s_{\sigma}\right) - \frac{s_{\sigma}}{1 + s_{\sigma} + q_{\sigma}} = 0, \tag{7.114}$$

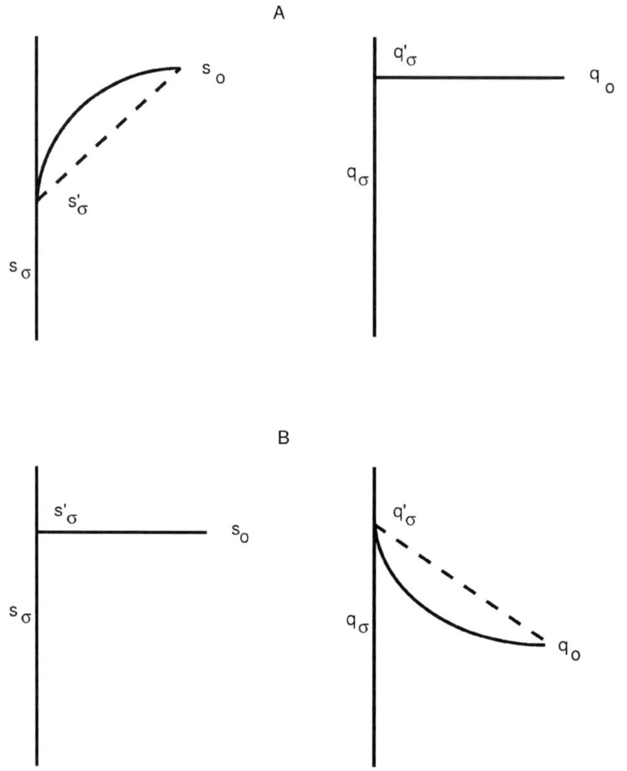

Fig. 7.17. Gradients of substrate and product in the vicinity of the surface. A – The substrate diffuses slowly and the product diffuses rapidly. B – The substrate diffuses rapidly and the product diffuses slowly. The dotted lines represent a linear approximation of the nonlinear gradients. See text.

and it has two implications: first, there exists a steady state for s_σ and a quasi-equilibrium for q_σ, which is nearly identical to q_o; second, as $\widetilde{S}^o(0)$ and $\tilde{h}_d$ are defined for the distance $x = 0$, they are equal to $S^o(0)$ and h_d, respectively. This means that S diffuses slowly whereas Q diffuses rapidly, and that the nonlinear gradient of S is approximated by a linear one. This situation is illustrated in Fig. 7.17.

If, alternatively, the product diffuses slowly whereas the substrate diffuses rapidly, one can write the following equation

$$\frac{s_\sigma}{1+s_\sigma+q_\sigma} - h_d^{*\prime}\left(\Pi_e^{\lambda z'} q_\sigma - q_o\right) = 0, \tag{7.115}$$

which expresses the coupling between product diffusion, product repulsion and enzyme reaction rate. In this case it is assumed that s_σ is in quasi-equilibrium and therefore that $s_\sigma = s_o$. Moreover, the nonlinear gradient of product in the vicinity of the surface is approximated to a linear gradient. This is illustrated in Fig. 7.17.

Equations (7.114) and (7.115), which are mutually incompatible, are both expressed in terms of the internal substrate and product concentrations, s_σ and q_σ. They can also be expressed in terms of the corresponding external concentrations, s'_σ and q'_σ, in the vicinity of the membrane. One has

$$\Pi_{\mathrm{e}}^{z'} = \frac{S^{\mathrm{o}}(m)/K_{\mathrm{S}}}{S^{\mathrm{i}}(m)/K_{\mathrm{S}}} = \frac{s'_\sigma}{s_\sigma}, \qquad \Pi_{\mathrm{e}}^{\lambda z'} = \frac{Q^{\mathrm{o}}(m)/K_{\mathrm{Q}}}{Q^{\mathrm{i}}(m)/K_{\mathrm{Q}}} = \frac{q'_\sigma}{q_\sigma}, \tag{7.116}$$

and, therefore, the two coupling equations assume the form

$$h_{\mathrm{d}}^{*}(s_{\mathrm{o}} - s'_\sigma) - \frac{s'_\sigma/\Pi_{\mathrm{e}}^{z'}}{1 + s'_\sigma/\Pi_{\mathrm{e}}^{z'} + q'_\sigma/\Pi_{\mathrm{e}}^{\lambda z'}} = 0 \tag{7.117}$$

and

$$\frac{s'_\sigma/\Pi_{\mathrm{e}}^{z'}}{1 + s'_\sigma/\Pi_{\mathrm{e}}^{z'} + q'_\sigma/\Pi_{\mathrm{e}}^{\lambda z'}} - h_{\mathrm{d}}^{*\prime}(q'_\sigma - q_{\mathrm{o}}) = 0. \tag{7.118}$$

If, for simplicity, one assumes that $z' = \lambda = 1$, one has

$$\Pi_{\mathrm{e}} = \frac{\Delta^-}{S^{\mathrm{o}}(m) + Q^{\mathrm{o}}(m)}, \tag{7.119}$$

where Δ^- is still the fixed charge density. Moreover, if one aims at studying, under nonequilibrium conditions, the effect of substrate or product on the partition coefficient Π_{e}, one has to hold on the advancement of the reaction at a fixed value, i.e. to maintain the ratio

$$\rho = \frac{Q^{\mathrm{o}}(m)}{S^{\mathrm{o}}(m)} \tag{7.120}$$

constant. The expression of Π_{e} can be rewritten as

$$\Pi_{\mathrm{e}} = \frac{\Delta^-}{K_{\mathrm{S}}} \frac{1}{1+\rho} \frac{1}{s'_\sigma}, \tag{7.121}$$

or as

$$\Pi_{\mathrm{e}} = \frac{\Delta^-}{K_{\mathrm{Q}}} \frac{\rho}{1+\rho} \frac{1}{q'_\sigma}. \tag{7.122}$$

Setting

$$\delta_{\mathrm{s}} = \frac{\Delta^-}{K_{\mathrm{S}}} \frac{1}{1+\rho}, \qquad \delta_{\mathrm{q}} = \frac{\Delta^-}{K_{\mathrm{Q}}} \frac{\rho}{1+\rho}, \tag{7.123}$$

one finds

$$\Pi_{\mathrm{e}} = \frac{\delta_{\mathrm{s}}}{s'_{\sigma}} = \frac{\delta_{\mathrm{q}}}{q'_{\sigma}}. \tag{7.124}$$

δ_s and δ_q are the fixed charge densities scaled with respect to the Michaelis constants of the substrate and of the product. One can then replace s'_σ and q'_σ in eqs. (7.117) and (7.118) by their expressions derived from (7.124). One finds

$$h_{\mathrm{d}}^{*} s_{\mathrm{o}} \Pi_{\mathrm{e}}^{3} - h_{\mathrm{d}}^{*} \delta_{\mathrm{s}} \Pi_{\mathrm{e}}^{2} + \left\{ h_{\mathrm{d}}^{*} s_{\mathrm{o}} (\delta_{\mathrm{s}} + \delta_{\mathrm{q}}) - \delta_{\mathrm{s}} \right\} \Pi_{\mathrm{e}} - h_{\mathrm{d}}^{*} \delta_{\mathrm{s}} (\delta_{\mathrm{s}} + \delta_{\mathrm{q}}) = 0 \tag{7.125}$$

and

$$h_{\mathrm{d}}^{*\prime} q_{\mathrm{o}} \Pi_{\mathrm{e}}^{3} - h_{\mathrm{d}}^{*\prime} \delta_{\mathrm{q}} \Pi_{\mathrm{e}}^{2} + \left\{ h_{\mathrm{d}}^{*\prime} q_{\mathrm{o}} (\delta_{\mathrm{s}} + \delta_{\mathrm{q}}) + \delta_{\mathrm{s}} \right\} \Pi_{\mathrm{e}} - h_{\mathrm{d}}^{*\prime} \delta_{\mathrm{q}} (\delta_{\mathrm{s}} + \delta_{\mathrm{q}}) = 0. \tag{7.126}$$

These two equations are indeed incompatible. The first expresses the effect of changing the scaled substrate concentration in the reservoir (at constant advancement of the reaction) on the partition coefficient, in a coupled system involving the "slow" diffusion of the substrate, its repulsion and its consumption during an enzyme reaction. The second equation expresses the effect of the scaled product concentration in the reservoir (again at constant advancement of the reaction) on a partition coefficient, in a coupled system involving the "slow" diffusion of the product, its repulsion and its inhibitory effect on the enzyme reaction. These equations are, at first sight, very similar, but in fact they display a major difference that distinguishes them. The first equation (7.125) can be shown to have one real positive root only, whereas the second (7.126) can have three [31–33]. This conclusion has some important implications. It means that, for the same value of the product concentration in the reservoir, the partition coefficient can display three different values. Two of them correspond to stable steady states of the system, whereas the last one pertains to an unstable state. This means that the partition coefficient will assume different values depending on whether the present product concentration in the reservoir has been reached via an increase or a decrease of the concentration. In other words, the electrostatic repulsion effects exerted by a negatively charged membrane can generate sensing properties of this membrane. The present situation is thus somewhat similar to that already discussed in chapter 3.

7.4.4. Conduction of ionic signals by membrane-bound enzymes

It has been assumed thus far that the bound enzyme system was partly coupled, i.e. it was postulated that diffusion of the substrate, or of the product, was taking place, but not the diffusion of both substrate and product. If it is now considered that both ligands diffuse, one has, under steady state conditions

$$\begin{aligned} f(s_\sigma, q_\sigma) &= 0 = J_{\mathrm{si}}^{*} - \Pi_{\mathrm{e}}^{z'} J_{\mathrm{so}}^{*} - v_{\mathrm{e}}, \\ g(s_\sigma, q_\sigma) &= 0 = v_{\mathrm{e}} - \Pi_{\mathrm{e}}^{\lambda z'} J_{\mathrm{qo}}^{*} + J_{\mathrm{qi}}^{*}, \end{aligned} \tag{7.127}$$

where

$$J_{so}^{*} = h_{d}^{*} s_{\sigma}, \qquad J_{si}^{*} = h_{d}^{*} s_{o},$$
$$J_{qo}^{*} = h_{d}^{*\prime} q_{\sigma}, \qquad J_{qi}^{*} = h_{d}^{*\prime} q_{o}. \tag{7.128}$$

The dynamic behaviour of the system in the immediate vicinity of the steady state can be studied by the so-called phase plane technique [34]. The principles of this technique will be briefly described below. If s and q are the scaled substrate and product concentrations, one has

$$\frac{ds}{dt} = \frac{ds_{\sigma}}{dt} + \frac{dx_1}{dt} = \frac{dx_1}{dt},$$
$$\frac{dq}{dt} = \frac{dq_{\sigma}}{dt} + \frac{dx_2}{dt} = \frac{dx_2}{dt}. \tag{7.129}$$

If the deviations relative to the steady state are small enough they can be expanded in Taylor series and the nonlinear terms can be neglected. Thus one has

$$\frac{d}{dt}\begin{bmatrix} x_1 \\ x_2 \end{bmatrix} = \begin{bmatrix} \partial f/\partial s & \partial f/\partial q \\ \partial g/\partial s & \partial g/\partial q \end{bmatrix}\begin{bmatrix} x_1 \\ x_2 \end{bmatrix}. \tag{7.130}$$

The characteristic equation of this dynamic system is

$$D^2 - T_j D + \Delta_j = 0, \tag{7.131}$$

where D is the differential operator d/dt, T_j and Δ_j the trace and the determinant of the Jacobian matrix of the system. Its general solution is

$$x_1 = c_{11}\, e^{\lambda_1 t} + c_{12}\, e^{\lambda_2 t},$$
$$x_2 = c_{21}\, e^{\lambda_1 t} + c_{22}\, e^{\lambda_2 t}, \tag{7.132}$$

where c_{11}, c_{12}, c_{21} and c_{22} are the integration constants, λ_1 and λ_2 the time constants of the system, i.e. the roots of the characteristic equation. The dynamic behaviour of the system and its temporal organization rely on the respective values of T_j, Δ_j and $T_j^2 - 4\Delta_j$. The parabola

$$T_j^2 - 4\Delta_j = 0, \tag{7.133}$$

and the corresponding axes define six regions in the (T_j, Δ_j) plane (Fig. 7.18). Moreover, one has

$$T_j = \lambda_1 + \lambda_2, \qquad \Delta_j = \lambda_1 \lambda_2. \tag{7.134}$$

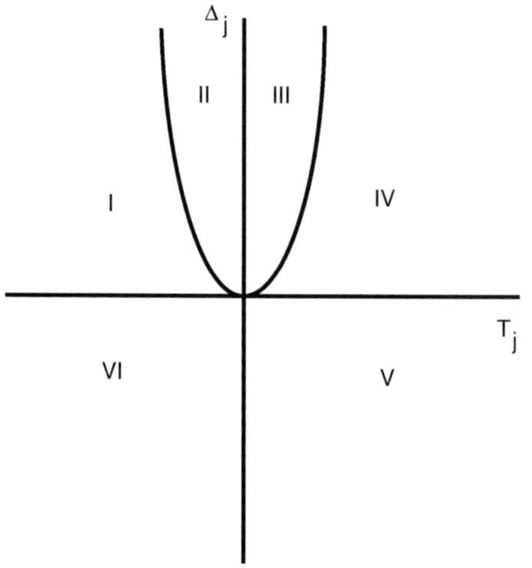

Fig. 7.18. The (Δ_j, T_j) plane. See text.

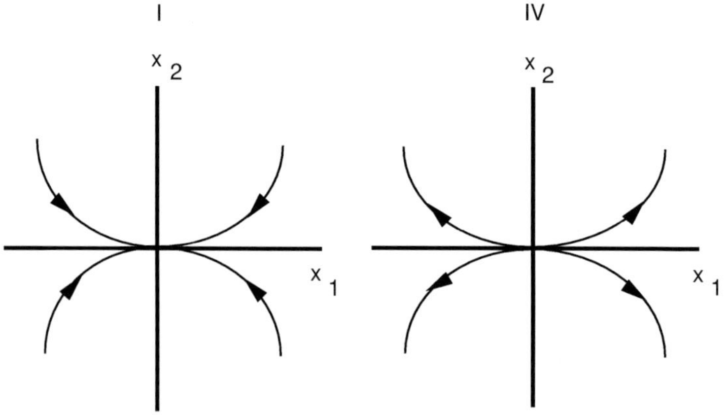

Fig. 7.19. Stable and unstable node of a metabolic cycle. I – Stable node. IV – Unstable node.

In region I

$$T_j < 0, \qquad \Delta_j > 0, \qquad T_j^2 - 4\Delta_j > 0, \tag{7.135}$$

hence the two roots of the characteristic equation are real and negative. The system displays a stable node. It returns back to its initial steady state when perturbed form this steady state. Moreover, this return is monotonic (Fig. 7.19). In region IV the situation is symmetrical, namely,

$$T_j > 0, \qquad \Delta_j > 0, \qquad T_j^2 - 4\Delta_j > 0. \tag{7.136}$$

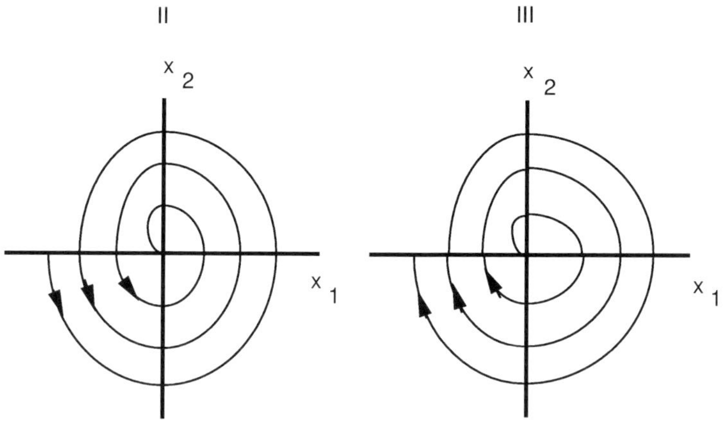

Fig. 7.20. Stable and unstable focus of a metabolic cycle. II – Stable focus. III – Unstable focus.

The two roots are real and positive. The system has an unstable node. It tends to diverge monotonically from its steady state (Fig. 7.19).

In region II

$$T_j < 0, \qquad \Delta_j > 0, \qquad T_j^2 - 4\Delta_j < 0, \tag{7.137}$$

and the two roots are complex with a negative real part, namely,

$$\lambda_{1,2} = \frac{T_j}{2} \pm i\omega \tag{7.138}$$

with

$$\omega = \frac{1}{2}\sqrt{4\Delta_j - T_j^2}. \tag{7.139}$$

The general solution of the dynamic system is now

$$\begin{aligned} x_1 &= e^{(T_j/2)t}\left(c_{11}\,e^{i\omega t} + c_{12}\,e^{-i\omega t}\right), \\ x_2 &= e^{(T_j/2)t}\left(c_{21}\,e^{i\omega t} + c_{22}\,e^{-i\omega t}\right). \end{aligned} \tag{7.140}$$

Hence the system displays damped oscillations because $T_j < 0$. When the system is perturbed from its initial steady state, it returns back to the same steady state in a damped periodic manner. It is said to display a stable focus (Fig. 7.20). In region III, one has

$$T_j > 0, \qquad \Delta_j > 0, \qquad T_j^2 - 4\Delta_j < 0, \tag{7.141}$$

and the roots are complex with a positive real part. The system has an unstable focus. It spontaneously drifts away from its initial steady state, describing amplified oscillations (Fig. 7.20).

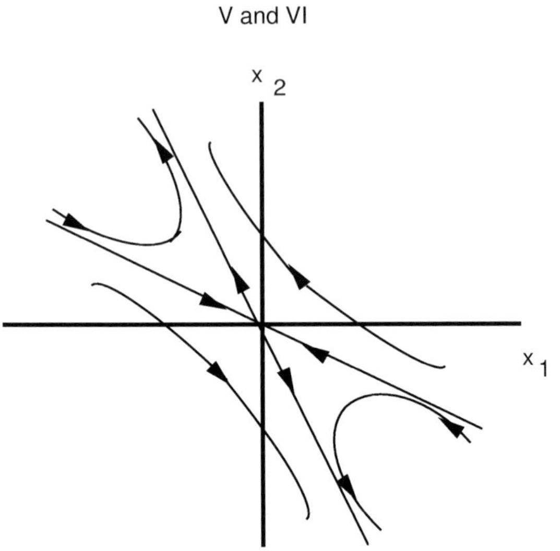

Fig. 7.21. Saddle point of a metabolic cycle.

In regions VI and V

$$\Delta_\mathrm{j} < 0, \qquad T_\mathrm{j}^2 - 4\Delta_\mathrm{j} > 0,$$
$$T_\mathrm{j} > 0 \text{ (region V)}, \qquad T_j < 0 \text{ (region VI)}, \tag{7.142}$$

and the two roots are real and opposite in sign. The system is unstable and exhibit a saddle point (Fig. 7.21).

A particularly important case occurs when

$$T_\mathrm{j} = 0, \qquad \Delta_\mathrm{j} > 0, \qquad T_\mathrm{j}^2 - 4\Delta_\mathrm{j} < 0, \tag{7.143}$$

and the system displays sustained oscillations. Its trajectory in the phase plane is an ellipse (Fig. 7.22). A random perturbation generates self-organization of the system, which spontaneously oscillates. Such system has been referred to as a "dissipative structure" [35].

It is of interest to apply these general considerations to the polyanion-bound enzyme system, in order to know what kind of behaviour this system can display. The elements of the Jacobian matrix of the dynamic system can be shown to take the form

$$\begin{aligned}
\frac{\partial f}{\partial s} &= -z'\Pi_\mathrm{e}^{z'-1}\frac{\partial \Pi_\mathrm{e}}{\partial s}J_\mathrm{so}^* - \Pi_\mathrm{e}^{z'}\frac{\partial J_\mathrm{so}^*}{\partial s} - \frac{\partial v_\mathrm{e}}{\partial s},\\
\frac{\partial g}{\partial s} &= \frac{\partial v_\mathrm{e}}{\partial s} - J_\mathrm{qo}^*\lambda z'\Pi_\mathrm{e}^{\lambda z'-1}\frac{\partial \Pi_\mathrm{e}}{\partial s},\\
\frac{\partial f}{\partial q} &= -z'\Pi_\mathrm{e}^{z'-1}\frac{\partial \Pi_\mathrm{e}}{\partial q}J_\mathrm{so}^* - \frac{\partial v_\mathrm{e}}{\partial q},\\
\frac{\partial g}{\partial q} &= \frac{\partial v_\mathrm{e}}{\partial q} - \lambda z'\Pi_\mathrm{e}^{\lambda z'-1}\frac{\partial \Pi_\mathrm{e}}{\partial q}J_\mathrm{qo}^* - \Pi_\mathrm{e}^{\lambda z'}\frac{\partial J_\mathrm{qo}^*}{\partial q}.
\end{aligned} \tag{7.144}$$

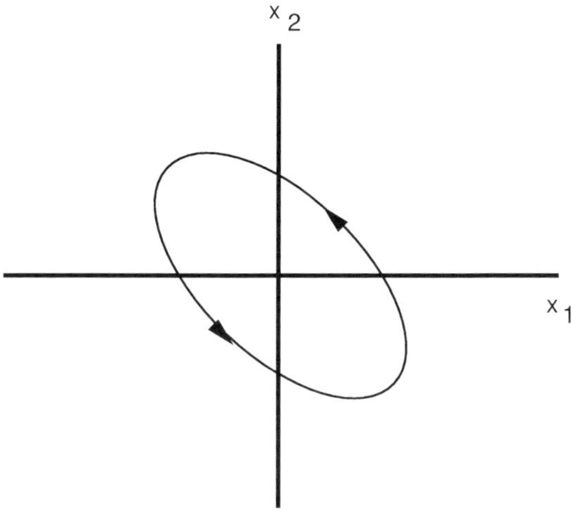

Fig. 7.22. Limit cycle of a metabolic cycle.

The expressions of the trace and of the Jacobian matrix are rather complicated and will not be given here. One can show that, independently of the partition coefficient value, $\Delta_j > 0$ and $T_j^2 - 4\Delta_j > 0$. However, the sign of the trace T_j can change depending on the value of the electric partition coefficient. To simplify the calculations, let us assume that $z' = \lambda = 1$. Then one has

$$T_j = -\frac{\partial \Pi_e}{\partial s}(J^*_{so} + J^*_{qo}) - \Pi_e\left(\frac{\partial J^*_{so}}{\partial s} + \frac{\partial J^*_{qo}}{\partial q}\right) - \frac{\partial v_e}{\partial s} + \frac{\partial v_e}{\partial q}. \tag{7.145}$$

It is obvious from this expression that, if the electrostatic repulsion effect is weak,

$$\Pi_e \approx 1, \qquad \frac{\partial \Pi_e}{\partial s} \approx 0, \qquad \frac{\partial v_e}{\partial s} > 0, \qquad \frac{\partial v_e}{\partial q} < 0, \tag{7.146}$$

then $T_j < 0$ and the system can have only a stable node. However, if there is significant electrostatic repulsion

$$\Pi_e > 1, \qquad \frac{\partial \Pi_e}{\partial s} < 0, \qquad \frac{\partial v_e}{\partial s} > 0, \qquad \frac{\partial v_e}{\partial q} < 0, \tag{7.147}$$

and the trace may become positive. The system then has an unstable node. It spontaneously tends to diverge from its initial steady state.

If this mathematical condition is applied to the plant cell surface, it means that, in a region of the primary cell wall of high fixed charge and enzyme densities (a "cluster"), the local substrate and product concentrations may increase, in this limited region of space, at the expense of the substrate and product present in the reservoir. The immediate consequence of this local increase in concentration, is the propagation, by diffusion, of this perturbation to other regions of the cell surface. This situation may be viewed as a simple device that allows the conduction of a chemical signal. It is evident that this local amplification of a chemical signal cannot go on indefinitely. As the local concentration of substrate and product, in the cluster of fixed charges, increases, the local value of the partition coefficient decreases. This, in turn, results in a decrease of the trace T_j of the Jacobian matrix, which may turn negative. The system becomes stabilized and the propagation of the chemical perturbation stops. Moreover the local amplification and the propagation of a chemical signal is accompanied by a local depolarization of the wall, which propagates.

7.5. Complexity of biological polyelectrolytes and the emergence of novel functions

When an enzyme is randomly distributed in a polyelectrolyte that does not display any spatial organization of its fixed charges, the intrinsic properties of the bound enzyme are apparently altered relative to its behaviour in free solution. These alterations include a shift of the pH-profile of the bound enzyme and, if the substrate is an ion, an apparent co-operativity of the reaction. This co-operativity is positive if the charges of the matrix and of the substrate have the same sign. Alternatively, this co-operativity will be negative if they are opposite in sign. Last but not least, all these effects are modulated or suppressed if the ionic strength is increased.

It is very likely, however, that most biological polyelectrolytes do not display this lack of organization, i.e. a random distribution of the fixed charges. Thus, plant cell walls display some sort of fuzzy organization that can be estimated roughly on the basis of statistical criteria. This complex organization introduces considerable flexibility and novelty in the response of cell wall-bound enzymes. Thus, for instance, an acid phosphatase is bound to primary plant cell walls. One would have expected this monomeric enzyme to display apparent positive co-operativity, owing to the fact that the cell wall is a polyanion and the substrate of the enzyme (an anion) is repelled by the fixed charges of the matrix. Precisely the opposite is observed, and this is due to the fact that the fixed negative charges and the phosphatase molecules are more or less clustered in the cell wall. Hence the complexity and the fuzzy organization generate extremely rich kinetic behaviour that cannot be predicted from a simple study of the phosphatase in free solution, or even from the study of the phosphatase randomly bound to a polyelectrolyte devoid of fuzzy organization, i.e. a polyelectrolyte displaying randomly distributed charges.

Another important functional aspect, brought about by the complexity of the organization of the polyelectrolyte–enzyme system, is an enhancement of the enzyme reaction rate. This enhancement is directly associated with the variance of the spatial distribution of the fixed charge density within the matrix. The larger this statistical parameter and the larger the reaction rate is expected to be. As the variance expresses the lack of uniformity of the fixed charge density, it expresses, in a way, the complexity of the spatial distribution of fixed charges.

Another interesting consequence of this complex organization is the possible existence of a conduction of chemical signals within biological polyelectrolytes. If the fixed charges are non-uniformly distributed in space, as occurs with plant cell walls, the local concentration of charged substrate and product of an enzyme reaction can become amplified, thus leading to the propagation, by diffusion, of these charged ligands within the cell wall. Since the electrostatic partition coefficient depends on the concentration of these ligands, the whole process will be equivalent to the propagation of a depolarization. Again, these events can be expected to take place only if the fixed charges are non-uniformly distributed within the polyelectrolyte matrix. They are therefore a consequence of the complexity of this polyelectrolyte.

References

[1] Ricard, J. (1989) Modulation of enzyme catalysis in organized biological systems. A physico-chemical approach. Catalysis Today 5, 275–384.

[2] Maurel, P. and Douzou, P. (1976) Catalytic implications of electrostatic potentials: the lytic activity of lysozyme as a model. J. Mol. Biol. 102, 253–264.

[3] Goldstein, L., Levin, Y. and Katchalsky, E. (1964) A water-insoluble polyanionic derivative of trypsin. II. Effect of the polyelectrolyte carrier on the kinetic behaviour of the bound trypsin. Biochemistry 3, 1913–1919.

[4] Ricard, J., Noat, G., Crasnier, M. and Job, D. (1981) Ionic control of immobilized enzymes. Kinetics of acid phosphatase bound to plant cell walls. Biochem. J. 56, 477–487.

[5] Noat, G., Crasnier, M. and Ricard, J. (1980) Ionic control of acid phosphatase activity in plant cell walls. Plant Cell Environ. 3, 225–229.

[6] Engasser, J.M. and Horvath, C. (1975) Electrostatic effects on the kinetics of bound enzymes. Biochem. J. 145, 431–435.

[7] Ricard, J., Mulliert, G., Kellershohn, N. and Giudici-Orticoni, M.T. (1994) Dynamics of enzyme reactions and metabolic networks in living cells. A physico-chemical approach. Progress Mol. Subcell. Biol. 13, 1–80.

[8] Kendall, M.G. (1948) The Advanced Theory of Statistics. Charles Griffin, London.

[9] Ricard, J., Kellershohn, N. and Mulliert, G. (1989) Spatial order as a source of kinetic cooperativity in organized bound enzyme systems. Biophys. J. 56, 477–487.

[10] Ricard, J. and Noat, G. (1986) Electrostatic effects and the dynamics of the enzyme reactions at the surface of plant cells. I. A theory of the ionic control of a complex multi-enzyme system. Eur. J. Biochem. 155, 183–190.

[11] Goldberg, R. (1984) Changes in the properties of cell wall pectin methyl esterase along the *Vigna radiata* hypocotyl. Physiol. Plant. 61, 58–63.

[12] Crasnier, M., Moustacas, A.M. and Ricard, J. (1985) Electrostatic effects and calcium ion concentration as modulators of acid phosphatase bound to plant cell walls. Eur. J. Biochem. 151, 187–190.

[13] Fry, S.C. (1993) Loosening the ties. Current Biol. 3, 355–357.

[14] Fry, S.C., Smith, R.C., Renwick, K.F., Martin, D.J., Hodge, S.K. and Matthews, K.J. (1992) Xyloglucan endotransglycosylase, a new wall loosening activity from plants. Biochem. J. 282, 821–828.

[15] Smith, R.C. and Fry, S.C. (1991) Endotransglycosylation of xyloglucans in plant cell suspension cultures. Biochem. J. 279, 529–535.

[16] McDougall, G.J. and Fry, S.C. (1980) Xyloglucan oligosaccharides promote growth and activate cellulase: evidence for a role of cellulase in cell suspension. Plant Physiol. 93, 1042–1048.

[17] Nishitani, K. and Tominaga, R. (1992) Endo-xyloglucan transferase, a novel class of glycosyl transferase that catalyzes transfer of a segment of xyloglucan molecule to another xyloglucan molecule. J. Biol. Chem. 267, 21058–21064.

[18] Farkas, V., Sulova, Z., Stratilova, E., Hanna, R. and McLachlan, G. (1992) Cleavage of xyloglucan by nasturtium seed xyloglucanase and transglycosylation to xyloglucan subunit oligosaccarides. Arch. Biochem. Biophys. 298, 365–370.

[19] McQueen-Mason, S. and Cossgrove, D.J. (1994) Disruption of hydrogen bonding between plant cell wall polymers by proteins that induce wall extension. Proc. Natl. Acad. Sci. USA 91, 6574–6578.

[20] Crasnier, M., Noat, G. and Ricard, J. (1980) Purification and molecular properties of acid phosphatase from sycamore cell walls. Plant Cell Environ. 3, 217–224.

[21] Dussert, C., Rasigni, M., Palmari, J., Rasigni, A. and Llebaria, A. (1987) Minimal spanning tree analysis of biological structures. J. Theor. Biol. 125, 317–323.

[22] Dussert, C., Mulliert, G., Kellershohn N., Ricard J., Giordani, R., Noat, G., Palmari, J., Rasigni, M., Llebaria, A. and Rasigni, G. (1989) Molecular organization and clustering of cell wall-bound enzymes as a source of kinetic apparent co-operativity. Eur. J. Biochem. 185, 281–290.

[23] Ricard, J., Kellershohn, N. and Mulliert, G. (1992) Dynamic aspects of long distance functional interactions between membrane bound enzymes. J. Theor. Biol. 156, 1–40.

[24] Moustacas, A.M., Nari, J., Noat, G., Crasnier, M., Borel, M. and Ricard, J. (1986) Electrostatic effects and the dynamics of enzyme reactions at the surface of plant cells. 2. The role of pectin methyl esterase in the modulation of electrostatic effects in soybean cell walls. Eur. J. Biochem. 155, 191–197.

[25] Nari, J., Noat, G., Diamantidis, G., Woudstra, M. and Ricard, J. (1986) Electrostatic effects and the dynamics of enzyme reactions at the surface of plant cells. 3. Interplay between limited cell wall autolysis, pectin methyl esterase activity and electrostatic effects in soybean cell walls. Eur. J. Biochem. 155, 199–202.

[26] Nari, J., Noat, G. and Ricard, J. (1991) Pectin methyl esterase, metal ions and plant cell wall extension. Hydrolysis of pectin by plant cell wall pectin methyl esterase. Biochem. J. 279, 342–350.

[27] Moustacas, A.M., Nari, J., Borel, M., Noat, G. and Ricard, J. (1991) Pectin methyl esterase, metal ions and plant cell wall extension. The role of metal ion in plant cell wall extension. Biochem. J. 279, 351–354.

[28] Hill, T.L. (1977) Free Energy Transduction in Biology. Academic Press, New York.

[29] Hill, T.L. and Chen, Y. (1970) Cooperative effects in models of steady state transport across membranes. III. Simulation of potassium ion transport in nerve. Proc. Natl. Acad. Sci. USA 66, 607–611.

[30] Cha, S. (1968) A simple method for the derivation of rate equations for enzyme-catalyzed reactions under the rapid equilibrium assumption or combined assumptions of equilibrium and steady state. J. Biol. Chem. 243, 820–825.

[31] Ricard, J. and Noat, G. (1984) Enzyme reactions at the surface of living cells. I. Electric repulsion effects of charged ligands and recognition of signals from the external milieu. J. Theor. Biol. 109, 555–568.

[32] Ricard, J. and Noat, G. (1984) Enzyme reactions at the surface of living cells. II. Destabilization in the membrane and conduction of signals. J. Theor. Biol. 109, 571–580.

[33] Ricard, J. (1987) Dynamics of multienzyme reactions, cell growth and perception of ionic signals from the external milieu. J. Theor. Biol. 128, 253–278.

[34] Pavlidis, T. (1973) Biological Oscillators: Their Mathematical analysis. Academic Press, New York.

[35] Nicolis, G. and Prigogine, I. (1997) Self-Organization in Nonequilibrium Systems. From Dissipative Structures to Order through Fluctuations. John Wiley and Sons, New York.

CHAPTER 8

Dynamics and motility of supramolecular edifices in the living cell

The supramolecular edifices that are present in a living cell are not always permanent entities. During cell life, they may appear, disappear, grow, shrink, bend, slide and may thus display many different types of distorsion and motion. The appearance and disappearance of the mitotic spindle, the motion of motor proteins along a microtubule, the contraction of a myofibril, are some examples, among many others, of these dynamic events. A common feature displayed by these dynamic processes is the need for energy consumption. The aim of the present chapter is to describe very briefly some of the biological events related to tubulin and actin polymerization–depolymerization as well as to muscle contraction, and to discuss the kinetics and thermodynamics of these events in the frame of a model of conversion of chemical energy.

8.1. Tubulin, actin and their supramolecular edifices

This discussion on the dynamics of multimolecular edifices will be restricted to microtubules and actin filments which are the major components of the cytoskeleton of eukaryotic cells.

8.1.1. Tubulin and microtubules

Tubulin is a heterodimer consisting of two tightly linked polypeptides called α- and β-tubulin. This protein dimer is present in virtually all eukaryotic cells. Tubulin can undergo polymerization and then forms microtubules. A microtubule is a hollow tube consisting of 13 protofilaments each composed of alternating α- and β-tubulin. Moreover, a microtubule is not a permanent entity of the cell. It is continuously growing and shrinking through the polymerization of tubulin and depolymerization of the microtubule. The two ends of a microtubule are not functionally equivalent. One end, called “plus”, grows thanks to a polymerization process, whereas the other end, called “minus”, shrinks owing to the depolymerization of this supramolecular structure (Fig. 8.1).

As we shall see in another section, the polymerization–depolymerization process is not a simple event and is under the control of guanosine triphosphate (GTP) hydrolysis and exchange. As a matter of fact, both α- and β-tubulin bear a site that can bind GTP. Whereas the GTP bound to α-tubulin cannot be exchanged or hydrolyzed, the GTP bound to β-tubulin can be hydrolyzed to GDP (guanosine diphosphate) and this GDP can, in turn, be exchanged for another molecule of GTP present in the reaction mixture. *In vitro*, in the presence of a metal ion and GTP, tubulin can undergo polymerization, but the rate of this polymerization process is not constant.

Before polymerization starts in earnest there is a nucleation phase which corresponds to a lag of the polymerization process. The nucleation is followed by an elongation phase and

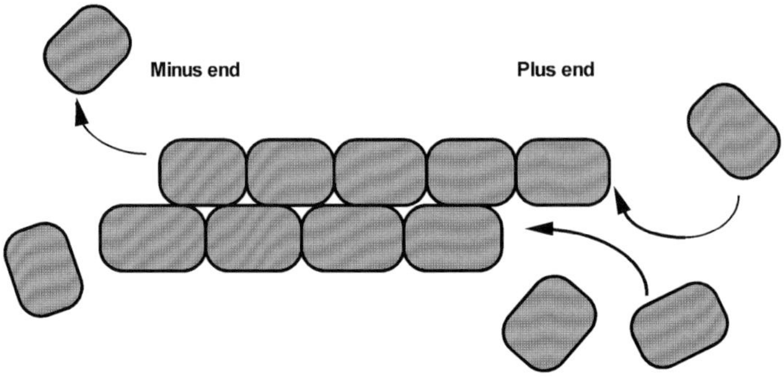

Fig. 8.1. Schematic picture of polymerization–depolymerization of tubulin and actin filaments. See text.

$$\ldots \text{TTT - GTP} \rightleftarrows \ldots \text{TTT - GDP - P}_i \rightleftarrows \ldots \text{TTT - GDP} + \text{P}_i$$

$$\updownarrow \qquad\qquad \updownarrow \qquad\qquad \updownarrow$$

$$\text{T - GTP} \qquad\qquad \text{T - GDP - P}_i \qquad\qquad \text{T - GDP}$$

$$\text{GDP} \qquad\qquad \text{GTP}$$

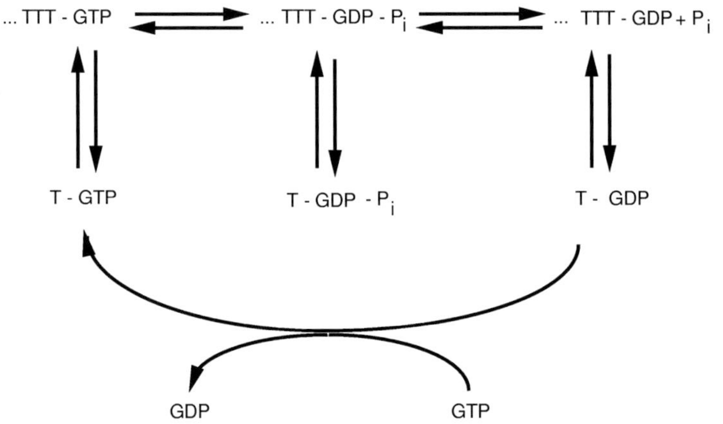

Fig. 8.2. GTP hydrolysis and exchange on a microtubule. M stands for monomer, P_i for inorganic phosphate.

finally by a steady state in which the length of each microtubule does not vary significantly. This type of situation shows that the binding of a tubulin molecule to another one is a less probable event than the binding of the tubulin dimer to the plus end of a microtubule. In fact, the polymerization of tubulin dimers takes place with polymers that have bound a GTP on the β-site. The microtubule grows at the plus end as long as GTP is bound to the β-sites. If GTP is hydrolyzed to GDP, the microtubule tends to disassemble until the GDP of the β-site has not been exchanged for another GTP molecule present in the reaction mixture (Fig. 8.2).

In vivo, microtubules originate from a cell structure, referred to as the centrosome, sitting next to the nucleus. The centrosome is thus the site of nucleation of microtubules and the polymer elongates in the cytoplasm by polymerization of tubulin at the plus end of the microtubule. The mitotic spindle of a dividing cell consists of microtubules. Different drugs interfere with the stability of this spindle. Thus colchicine and colcemid cause the rapid disappearance of the mitotic spindle, thereby abolishing cell division. As a matter of fact,

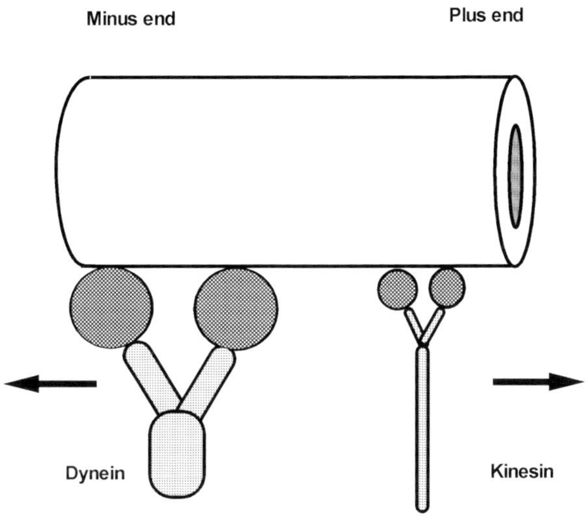

Fig. 8.3. Motor proteins slide along a microtubule. See text.

these drugs bind to tubulin, thus preventing the polymerization process. The microtubule then disassembles without incorporating new tubulin molecules. Another drug, taxol, has a different effect. It binds to a microtubule and stabilizes the overall edifice. Cell division is then blocked. All these chemicals act as anticancer drugs.

Various proteins are associated with microtubules. They are called MAPs (microtubule associated proteins). Owing to the diverse functions of microtubules, there are many different kinds of MAPs and their molecular mass ranges from 50 kDa to 300 kDa. These MAPs serve to speed up the nucleation process and to stabilize microtubules against disassembly. Other proteins called kinesins and dyneins slide along the microtubules (Fig. 8.3). Dyneins slide towards the minus end and kinesins towards the plus end of the microtubule. These proteins are thus defined as motor proteins. They play an important role in the transport of various organelles in the cytoplasm. The transport of synaptic vesicles along axons is brought about by kinesins. These motor proteins are made up of a globular head which binds to the microtubule and a tail that can associate with the organelle to be transported (Fig. 8.3). The globular domain, which slides along the mcrotubule, binds ATP. Thus the energy required for the motion of these motor proteins originates from ATP hydrolysis [1–12].

8.1.2. Actin, actin filaments and myofibrils

Globular actin (G-actin) is present in all eukaryotic cells. The protein consists of a single polypeptide chain of 375 aminoacids. Its three-dimensional structure has been determined [13]. G-actin polymerizes as a left-handed helix of 5.9 nm pitch [14] called an actin filament (F-actin). Like microtubules, actin filaments are polarized, i.e. the two ends are not functionally equivalent. The plus end is a fast growing structure and the minus end a slowly

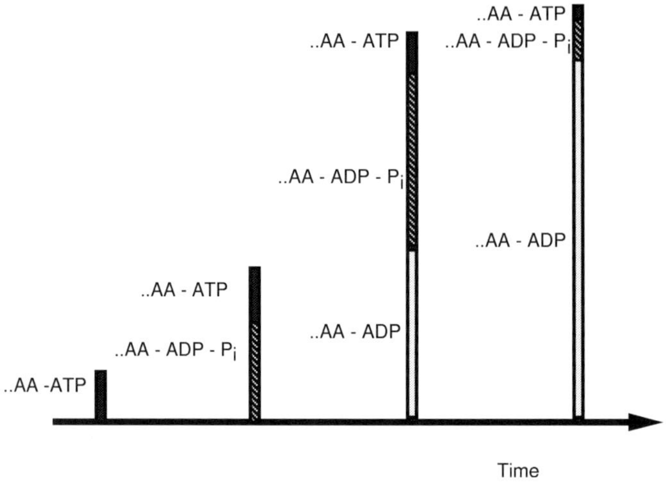

Fig. 8.4. The three states of an actin filament. See text.

disassembling edifice [15,16]. As actin filaments are often associated with myosin heads (see below), the difference between the plus and the minus ends is not only functional but also morphological. The plus end has an arrowhead and the minus end a barbed appearance. The two ends are thus called "pointed" and "barbed", respectively. Actin filaments are usually shorter, thinner and more flexible than microtubules. In the animal cell, they are usually located at the periphery, near the plasma membrane. Actin filaments rarely occur in isolation but rather in bundles that contribute to give animal cells their shape. Roughly speaking, actin polymerizes by the same mechanism as that described for tubulin except that GTP has been replaced by ATP. ATP bound to actin filaments can undergo hydrolysis and when ADP is bound to the polymer, the polymer tends to disassemble. An actin filament may exist in a F-ATP state, a F-ADP-P_i state and a F-ADP state (Fig. 8.4).

Bound ADP can be exchanged for ATP, but the exchange process, unlike what happens with tubulin, is extremely slow. This property helps maintain monomeric actin at a high level in the cytoplasm. As with tubulin, the polymerization of actin requires the presence of mono- or divalent cations. Various drugs interfere with the polymerization–depolymerization process. This is the case for cytochalasins and phalloidins, extracted from fungi. Cytochalasins prevent G-actin from polymerizing because they bind to the plus end of the filament whereas phalloidins bind to F-actin, thus preventing its disassembly. *In vivo* actin polymerization is tightly regulated by proteins, called thymosin and profilin, that can bind to actin and inhibit the polymerization process. This helps understanding why, in living cells, the concentration of G-actin is so high [17,18].

Muscle contains large amounts of actin. Fibers of skeletal muscles are very large cells formed by the fusion, during development, of many separate cells. The nuclei of these fibers sit next to the plasma membrane and most of the cytoplasm is made up of myofobrils which are the contractile elements of the fibers. Each myofibril consists of a sequence of tiny units called sarcomeres (Fig. 8.5). These sarcomeres represent the supramolecular

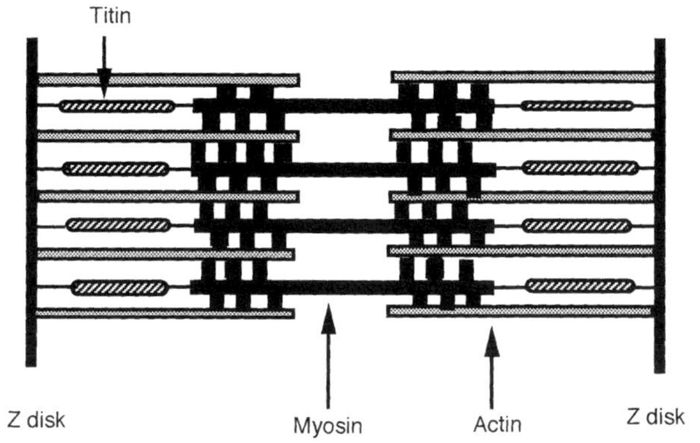

Fig. 8.5. Schematic picture of a skeletal muscle sarcomere. See text.

Fig. 8.6. Schematic picture of a myosin molecule. See text.

machinery reponsible for muscle contraction. The myofibrils display a chain of dark and light bands. Sitting in the middle of each light band is a dense line, called the Z line, or Z disk. The region delimited by two Z disks is a sarcomere. Each sarcomere is a highly organized assembly of filaments that are partly overlapping (Fig. 8.5). Thin actin filaments occupy the light, and part of the dark bands. The actin filaments are stuck to the Z disks by their plus ends. The dark bands are occupied by thick filaments of myosin.

Myosins are motor proteins of different types. Myosin II of skeletal muscles is made up of two heavy chains (each comprising about 2000 aminoacid residues) and four light chains. The N-termini of the two heavy chains constitute the globular heads that bear actin binding sites. The rest of the dimeric molecule constitutes a rodlike tail (Fig. 8.6). The tail is thus made up of two coiled-coil α-helices driven by the association of regularly spaced hydrophobic aminoacids. As we shall see later, myosin heads can bind to actin filaments. Next to this region is an ATP-binding site. Once ATP is bound to this site, it is hydrolyzed and this brings about the spontaneous self-assembly of several myosin molecules that form a thick bipolar filament of about 15 to 20 molecules. The thick filaments in the dark band of a sarcomere consist of these bundles of myosin molecules. Myosin heads at the surface of the bundle are oriented in all directions. In the centre of this bundle of filaments there is a bare zone displaying no myosin head. Myosins I are probably more primitive forms of myosins and possess only one polypeptide chain. The conserved region of the myosin molecule is the globular head that interact with actin, binds and hydrolyzes ATP. Other accessory proteins are present in the myofibril. One such is titin. This protein is composed of about 27 000 aminoacids and spans half the sarcomere, from the Z-disk to the centre of the sarcomere. It probably plays a role in maintaining the elasticity of the sarcomere and displays protein kinase activity [22].

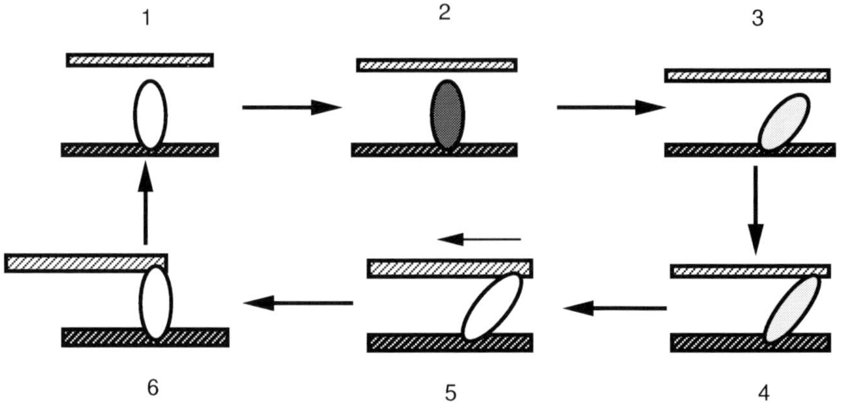

Fig. 8.7. The myosin walk along an actin filament. Myosin head binds ATP (2). ATP hydrolysis generates a hinge motion of the head (3) that binds to actin (4). The release of ADP generates a force exerted on actin (5) and the myosin head takes its initial position (6 and 1).

In muscle, myosin II molecules walk towards the plus end of actin. This walk, as for any motor protein, is unidirectional and requires ATP consumption. The mechanism of the walk along actin fibers is schematically pictured in Fig. 8.7. At the beginning of the process, a myosin head is tightly bound to an actin filament (6). During the first step, an ATP molecule binds to the myosin head and this reduces the affinity of myosin for actin (2). ATP is then being hydrolyzed, but ADP and phosphate remain bound to the globular head of myosin. The hydrolysis of ATP produces a large conformational rearrangement of the myosin head that slides 5 nm away from its initial position on actin (3). In the next step, the myosin head binds loosely to a new region of actin, thus producing the release of phosphate (4). During the last step, which is the force-generating step, the release of ADP results in a return of the myosin head to its initial conformation. But, as the head is bound to actin, this produces a sliding relative to the thick bundle of myosin molecules (5, 6). All these events produce the contraction of the sarcomere [19–22].

8.2. Dynamics and thermodynamics of tubulin and actin polymerization

"Linear" aggregation of tubulin and actin most probably requires conformation changes of the elementary "bricks" of these edifices (i.e., actin and tubulin). These conformation changes have been studied in the case of actin [23–27]. They take part, in particular, in the process of nucleation (Fig. 8.8). The first step of this process is probably an "activation" of the monomer through the binding of ATP and a metal ion which leads to a conformation change of G-actin [25,26]. A dimer is formed, then a trimer appears that undergoes an isomerization process leading to a helical structure. The polymerization process requires the existence of the trimer in its helical structure [23,24]. In the course of the elongation process, activated monomers are polymerized on this helical structure.

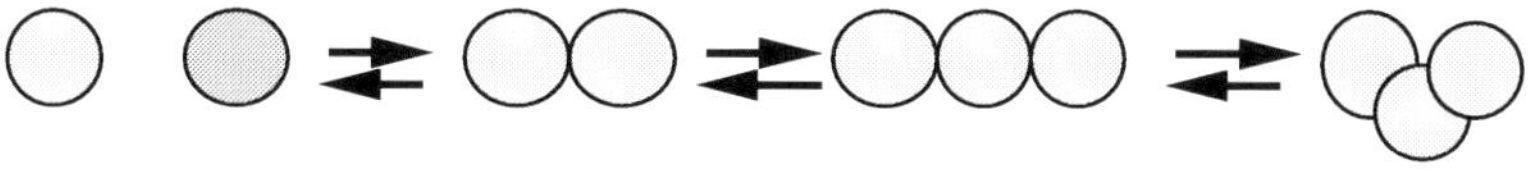

Fig. 8.8. Actin polymerization requires the formation of a trimer in helical conformation.

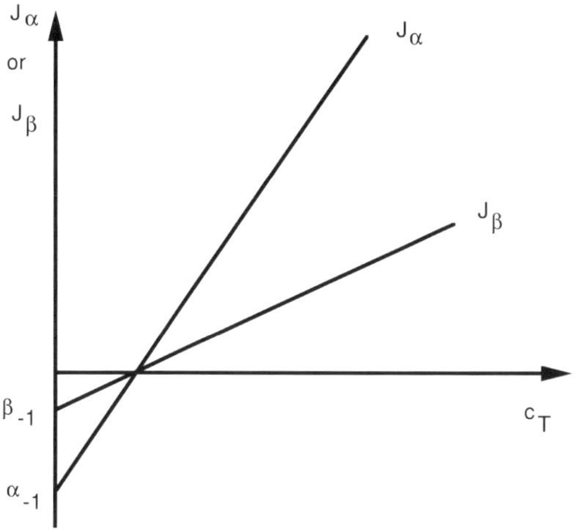

Fig. 8.9. Polymerization flow of an equlibrium polymer as a function of the total concentration of monomers. See text.

As outlined previously, fast polymerization at the plus end of the actin filament requires hydrolysis of ATP. A monomer which has bound ATP can be polymerized at the plus end of the actin filament and the ATP is then hydrolyzed to ADP and phosphate. The phosphate is then released from the actin unit, whereas the ADP remains bound to the protein. Depending on the relative rates of polymerization, ATP hydrolysis and phosphate release, one may expect actin filaments to be capped with actin units bearing ADP and phosphate. But the actin filaments may also carry no cap. If the rate of polymerization is very fast, the actin filaments should bear an actin–ATP cap. If the polymerization rate is slower, the actin filaments should bear an actin–ADP–phosphate cap and if the polymerization is even slower, then all the actin units should bear bound ADP. This is precisely what has been found experimentally (see Fig. 8.4).

The aim of the present section is to discuss simple theoretical models of polymerization of actin and tubulin.

8.2.1. Equilibrium polymers

Let us first consider the simplest situation, namely, that of a "linear" polymer $A_DA_D \ldots A_D$ A_D, for instance, an actin filament with an ADP boud to every actin unit. Let us assume that the polymer is in thermodynamic equilibrium with monomers A_D that may be bound to (and released from) the $\alpha(+)$ and $\beta(-)$ ends of this polymer (Fig. 8.9). The flow, J_α, of actin molecules that bind to the plus (α) end is

$$J_\alpha = \alpha_1 c_T - \alpha_1, \tag{8.1}$$

and similarly

$$J_\beta = \beta_1 c_T - \beta_{-1}. \tag{8.2}$$

In these expressions c_T is the total concentration of monomers in solution (for simplicity an actin molecule is considered a monomer). As the two ends are different, $\alpha_1 \neq \beta_1$ and $\alpha_{-1} \neq \beta_{-1}$. Moreover, the plus end is the site of polymerization, so one must expect that $\alpha_1 > \beta_1$ and $\alpha_{-1} < \beta_{-1}$. Thus, plotting J_α and J_β as a function of c_T should yield linear plots, as shown in Fig. 8.9 [28,29].

These expressions show that J_α and J_β may be positive or negative depending on the value of c_T. There exists a critical concentration, called $\bar{c}_T$, where the two flows are separately null (Fig. 8.9). For this concentration, referred to as the critical concentration, the polymer is in equilibrium with its medium. Below and above this concentration, the polymer is shrinking or growing.

8.2.2. *Drug effects on equilibrium polymers*

Actin filaments and microtubules are extremely sensitive to various drugs. This is the case, for instance, for microtubules which spontaneously tend to disassemble in the presence of colchicine. This drug binds to the isolated monomer and to the α- and to the β-ends of the polymer. The situation is schematized in the model of Fig. 8.10. Colchicine is assumed to bind to the free monomer (affinity constant K) and to the last subunit of the polymer at the α- and β-ends (affinity constants K'_α and K'_β).

In fact, there is no experimental proof that colchicine is bound only to the "last" subunit. This is a simplification made in the model to make it tractable [29]. The processes of ligand (colchicine) binding are assumed to be fast relative to the other steps. The rate constants a_1 and b_1 are the rate constants for the binding of a liganded monomer to the α- and β-ends of the polymer. These two rate constants are extremely small and, in most cases, negligible.

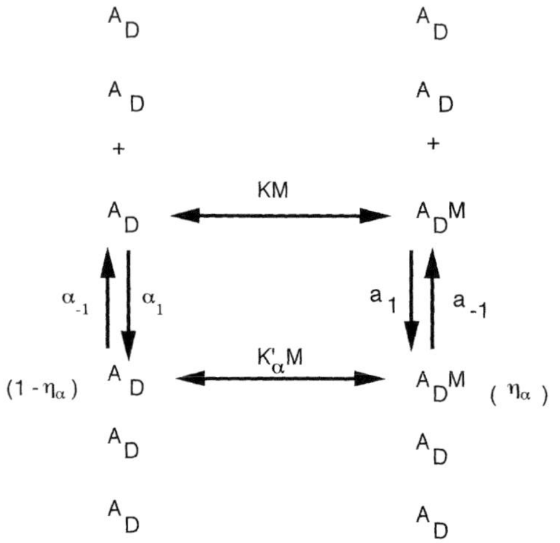

Fig. 8.10. Drug binding on an equilibrium polymer. See text.

The reverse rate constants, however, are not small. Thus the ligand (colchicine) prevents, to a large extent, the binding of colchicine-liganded monomers to the α- and β-ends of the polymer, but does not impede the dissociation of this liganded subunit from the α- or β-end of the polymer. If c_{M} is the concentration of M (colchicine), c_{A} is the concentration of free, unliganded, monomer, c_{AM} is the concentration of the liganded monomer, c_{T} is the total concentration of monomer, then one has

$$c_{\mathrm{A}} = \frac{c_{\mathrm{T}}}{1 + K c_{\mathrm{M}}}, \qquad c_{\mathrm{AM}} = \frac{c_{\mathrm{T}} K c_{\mathrm{M}}}{1 + K c_{\mathrm{M}}}. \tag{8.3}$$

The fraction of the liganded monomers at the α-end of the polymer is

$$\eta_\alpha = \frac{K'_\alpha c_{\mathrm{M}}}{1 + K'_\alpha c_{\mathrm{M}}}, \tag{8.4}$$

and, similarly, the fraction of liganded monomers at the β-end of the polymer is

$$\eta_\beta = \frac{K'_\beta c_{\mathrm{M}}}{1 + K'_\beta c_{\mathrm{M}}}. \tag{8.5}$$

The fractions of α- and β-ends that have not bound M are thus $1 - \eta_\alpha$ and $1 - \eta_\beta$, respectively. The net flow of aggregation (or of decay) of the α-end is thus

$$J_\alpha = \alpha_1 c_{\mathrm{A}}(1 - \eta_\alpha) + a_1 c_{\mathrm{AM}} \eta_\alpha - \alpha_{-1}(1 - \eta_\alpha) - a_{-1}\eta_\alpha. \tag{8.6}$$

Under equilibrium conditions one has

$$\begin{aligned} &\alpha_1 \bar{c}_{\mathrm{A}}(1 - \eta_\alpha) - \alpha_{-1}(1 - \eta_\alpha) = 0, \\ &a_1 \bar{c}_{\mathrm{AM}} \eta_\alpha - a_{-1}\eta_\alpha = 0, \end{aligned} \tag{8.7}$$

where $\bar{c}_{\mathrm{A}}$ and $\bar{c}_{\mathrm{AM}}$ are the equilibrium concentration of free and liganded monomers, respectively. From these equations, it follows that

$$\bar{c}_{\mathrm{A}} = \frac{\alpha_{-1}}{\alpha_1}, \qquad \bar{c}_{\mathrm{AM}} = \frac{a_{-1}}{a_1}. \tag{8.8}$$

These ratios represent the dissociation constants of the unliganded and liganded subunits from the α-end of the polymer. As the rate constant a_1 is extremely small, the second term in eq. (8.6) can usually be neglected. One must stress, however, that the first term in the second equation of (8.7), which contains a_1 as well, cannot be dropped, for then $\bar{c}_{\mathrm{AM}}$ would have to be extremely large. This means that the equilibrium of the polymer that has bound ligands at both ends is shifted towards free subunits, and this polymer spontaneously tends to disassemble. The affinity constants α_1/α_{-1} and a_1/a_{-1} are in fact related. Thermodynamics demands that

$$K\left(\frac{\alpha_{-1}}{\alpha_1}\right) = K'_\alpha\left(\frac{a_{-1}}{a_1}\right), \tag{8.9}$$

or

$$\frac{\alpha_1}{\alpha_{-1}} = \frac{K}{K'_\alpha} \frac{a_1}{a_{-1}}. \tag{8.10}$$

Experimental studies show that

$$\frac{\alpha_1}{\alpha_{-1}} > \frac{a_1}{a_{-1}}, \tag{8.11}$$

and this implies that

$$K > K'_\alpha. \tag{8.12}$$

The affinity of M is much larger for the free subunit than for the α-end of the polymer. Similar reasoning applies to the β-end of this polymer.

Coming back to eq. (8.6) and taking advantage of expressions (8.3) and (8.4), the flow J_α can be rewritten in explicit form as

$$J_\alpha = \frac{c_T(\alpha_1 + a_1 K K'_\alpha c_M^2)}{(1 + K c_M)(1 + K'_\alpha c_M)} - \frac{\alpha_{-1} + a_{-1} K'_\alpha c_M}{1 + K'_\alpha c_M}. \tag{8.13}$$

If there is no ligand in the medium, $c_M = 0$ and this expression reduces to eq. (8.1). If, alternatively, c_M is very large, eq. (8.13) reduces to

$$J_\alpha = \frac{c_T(\alpha_1 + a_1 K K'_\alpha c_M^2)}{K K'_\alpha c_M^2} - \alpha_{-1}. \tag{8.14}$$

Since a_1 is very small, α_1 will never be negligible relative to $a_1 K'_\alpha c_M$. Thus, by measuring the J_α flow at very low and very high concentrations of M, the intercepts of the $J_\alpha = f(c_T)$

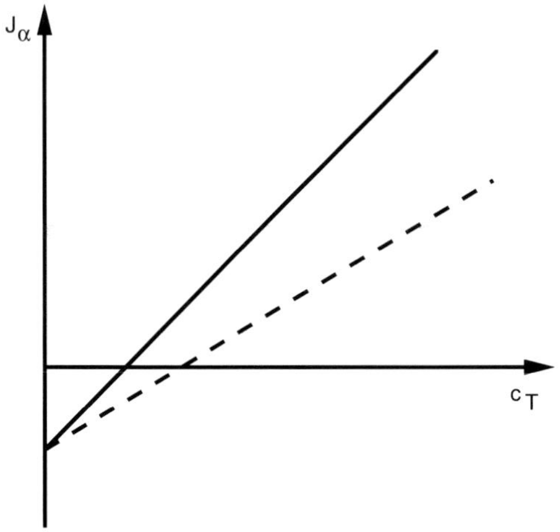

Fig. 8.11. Inhibition of the polymerization flow of an equilibrium polymer. The dotted line pertains to the inhibited flow.

plots yield the first order rate constant of polymer dismantling. At low c_M values, one obtains the constant α_{-1}, and at high c_M values one has the constant a_{-1}. Similar reasoning would hold for the β-end of the polymer. The $J_\alpha = f(c_T)$ plots are shown in Fig. 8.11. As previously, the plots should intersect the x axis. When the concentration of the inhibitor M is increased, the slope of the plots decreases but the intercept will either become more and more negative if $\alpha_{-1} > a_{-1}$, or decline in absolute value if $a_{-1} > \alpha_{-1}$.

The model that has been presented above, and which is mostly derived from the work of Hill [29], is interesting but has no general value for different drugs may have different mode of action. The model has thus to be modified in order to accomodate different experimental results.

8.2.3. Treadmilling and steady state polymers

The growth process of a polymer, such as F-actin, usually does not take place under equilibrium conditions but under steady state. This means that monomers that have bound ATP (for F-actin) or GTP (for tubulin) bind to the α- and β-ends of the polymer. These monomers are designated below A_T. The units of the bulk polymer still have one molecule of ADP (or GDP) bound and are termed, as previously, A_D. When an A_T monomer binds to the α- or to the β-end, it hydrolyzes its ATP. The system is thus not under quasi-equilibrium conditions but, possibly, in steady state. As actin filaments are polarized, the properties of the plus and of the minus ends are different. This implies that the G-actin units will move along the polymer and this phenomenon is called treadmilling (Fig. 8.12).

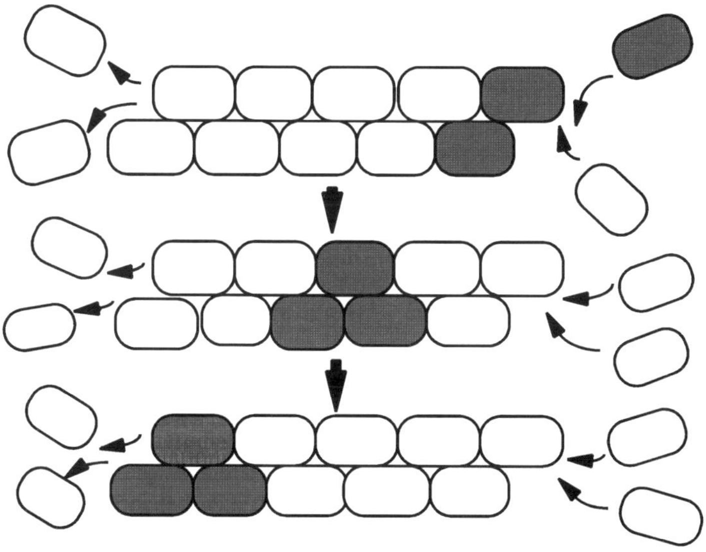

Fig. 8.12. Treadmilling of a steady state polymer. Labelled monomers (hatched in the figure) move along the steady state polymer.

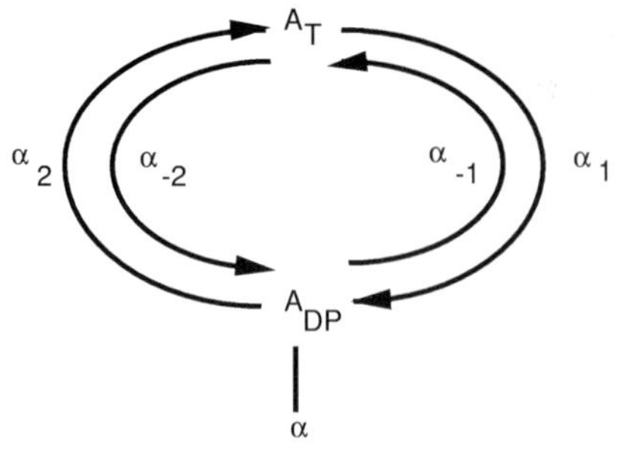

Fig. 8.13. Kinetic scheme of the polymerization–depolymerization process at the α-end of a steady state polymer. See text.

We have already outlined that growing actin filaments are capped by actin units bearing ADP plus phosphate [16]. We shall assume now, for simplicity, that only one actin unit bearing these ligands is sitting at the plus and minus ends of the polymer. These actin units are termed below A_{DP}. The kinetic scheme of the dynamic polymerization–depolymerization process is shown in Fig. 8.13.

At the α- and β-ends of the actin filaments, two types of events are taking place. The first event is the hydrolysis of ATP bound to actin monomers and the binding of this monomer to the α- or β-end of the polymer. The corresponding rate constants are α_1 and α_{-1} for the first end and β_1 and β_{-1} for the second one. Thus one has

$$\alpha + A_T \underset{\alpha_{-1}}{\overset{\alpha_1}{\longleftrightarrow}} \alpha - A_{DP}$$

and

$$\beta + A_T \underset{\beta_{-1}}{\overset{\beta_1}{\longleftrightarrow}} \beta - A_{DP}$$

In these equilibria, α and β still represent the α-end and the β-end of the polymer. The two constants of each couple, namely, α_1 and β_1 as well as α_{-1} and β_{-1}, must be different since the two ends of the polymer are different. The equilibrium constants, however, must be the same. This means that

$$\frac{\alpha_1}{\alpha_{-1}} = \frac{\beta_1}{\beta_{-1}}. \tag{8.15}$$

Moreover, other processes are taking place at the two ends of the polymer, namely the release of an actin unit from the ends of the polymer together with the release of ADP (D) and phosphate (P) and their replacement on the actin monomer by an ATP (T) molecule. These events can be pictured as

$$\alpha - A_{DP} \longleftrightarrow A + D + P + \alpha$$

$$A + T \longleftrightarrow A_T$$

at the α-end and as

$$\beta - \mathrm{A_{DP}} \longleftrightarrow \mathrm{A} + \mathrm{D} + \mathrm{P} + \beta$$

$$\mathrm{A} + \mathrm{T} \longleftrightarrow \mathrm{A_T}$$

at the β-end. Therefore, the overall process at the two ends of the polymer is

$$\mathrm{T} + \alpha - \mathrm{A_{DP}} \underset{\alpha_{-2}}{\overset{\alpha_2}{\longleftrightarrow}} \mathrm{A_T} + \mathrm{D} + \mathrm{P} + \alpha$$

and

$$\mathrm{T} + \beta - \mathrm{A_{DP}} \underset{\beta_{-2}}{\overset{\beta_2}{\longleftrightarrow}} \mathrm{A_T} + \mathrm{D} + \mathrm{P} + \beta$$

Again, one must have $\alpha_2 \neq \beta_2$ and $\alpha_{-2} \neq \beta_{-2}$, but the two equilibrium constants must be the same, i.e.

$$\frac{\alpha_2}{\alpha_{-2}} = \frac{\beta_2}{\beta_{-2}}. \tag{8.16}$$

If c_{AT} is the concentration of the A_{T} monomer, the rate of addition of monomers at the ends of the same polymer is

$$\begin{aligned} \frac{\mathrm{d}n_\alpha}{\mathrm{d}t} &= (\alpha_1 + \alpha_{-2})c_{\mathrm{AT}} - (\alpha_{-1} + \alpha_2), \\ \frac{\mathrm{d}n_\beta}{\mathrm{d}t} &= (\beta_1 + \beta_{-2})c_{\mathrm{AT}} - (\beta_{-1} + \beta_2), \end{aligned} \tag{8.17}$$

where n_α and n_β represent the number of monomers added to the α-end and to the β-end of the polymer. The steady state flow of monomers within the polymer, or the rate of displacement of a monomer within this polymer, is

$$\widetilde{J} = \frac{\mathrm{d}n_\alpha}{\mathrm{d}t} = -\frac{\mathrm{d}n_\beta}{\mathrm{d}t}. \tag{8.18}$$

$\widetilde{J}$ is thus the treadmill flow. If $\tilde{c}_{\mathrm{AT}}$ is the steady state concentration of the monomer, then one must have

$$(\alpha_1 + \alpha_{-2})\tilde{c}_{\mathrm{AT}} - (\alpha_{-1} + \alpha_2) = -(\beta_1 + \beta_{-2})\tilde{c}_{\mathrm{AT}} + (\beta_{-1} + \beta_2) \tag{8.19}$$

or

$$\tilde{c}_{\mathrm{AT}} = \frac{\alpha_2 + \beta_2 + \alpha_{-1} + \beta_{-1}}{\alpha_1 + \beta_1 + \alpha_{-2} + \beta_{-2}}. \tag{8.20}$$

The expression of the treadmill flow is thus

$$\widetilde{J}=\frac{(\alpha_1+\alpha_{-2})(\alpha_2+\beta_2+\alpha_{-1}+\beta_{-1})-(\alpha_{-1}+\alpha_2)(\alpha_1+\beta_1+\alpha_{-2}+\beta_{-2})}{\alpha_1+\beta_1+\alpha_{-2}+\beta_{-2}}, \tag{8.21}$$

which can be rewritten as

$$\widetilde{J}=\frac{\alpha_1\beta_{-1}-\alpha_{-1}\beta_1+\alpha_{-2}\beta_2-\alpha_2\beta_{-2}+\alpha_1\beta_2-\alpha_2\beta_1+\alpha_{-2}\beta_{-1}-\alpha_{-1}\beta_{-2}}{\alpha_1+\beta_1+\alpha_{-2}+\beta_{-2}}. \tag{8.22}$$

From eqs. (8.15) and (8.16) one has

$$\alpha_1\beta_{-1}=\alpha_{-1}\beta_1, \qquad \alpha_2\beta_{-2}=\alpha_{-2}\beta_2, \qquad \frac{\alpha_1\alpha_2}{\alpha_{-1}\alpha_{-2}}=\frac{\beta_1\beta_2}{\beta_{-1}\beta_{-2}}, \tag{8.23}$$

and eq. (8.22) reduces to

$$\widetilde{J}=\frac{\alpha_1\beta_2-\alpha_2\beta_1+\alpha_{-2}\beta_{-1}-\alpha_{-1}\beta_{-2}}{\alpha_1+\beta_1+\alpha_{-2}+\beta_{-2}}, \tag{8.24}$$

which can be rearranged to

$$\widetilde{J}=\frac{(\alpha_1\beta_2-\alpha_2\beta_1)\left(1-\dfrac{\alpha_{-1}\alpha_{-2}}{\alpha_1\alpha_2}\right)}{\alpha_1+\beta_1+\alpha_{-2}+\beta_{-2}}. \tag{8.25}$$

Moreover, a comparison of the above equilibria leads to the new equilibrium

$$\mathrm{T} \underset{\alpha_{-1}\alpha_{-2}}{\overset{\alpha_1\alpha_2}{\rightleftarrows}} \mathrm{D}+\mathrm{P}$$

and the same equilibrium would be obtained from the ratio $\beta_1\beta_2/\beta_{-1}\beta_{-2}$. The thermodynamic significance of these ratios is thus

$$\frac{\alpha_1\alpha_2}{\alpha_{-1}\alpha_{-2}}=\frac{\beta_1\beta_2}{\beta_{-1}\beta_{-2}}=\exp\frac{\mu_\mathrm{T}-\mu_\mathrm{D}-\mu_\mathrm{P}}{RT}. \tag{8.26}$$

It is important to stress that

$$X=\mu_\mathrm{T}-\mu_\mathrm{D}-\mu_\mathrm{P} \tag{8.27}$$

is not the standard free energy of ATP hydrolysis, but the affinity, or force, that drives the treadmill flow. Equation (8.25) can be rearranged to

$$\widetilde{J} = \frac{(\alpha_1\beta_2 - \alpha_2\beta_1)(1 - \mathrm{e}^{-X/RT})}{\alpha_1 + \beta_1 + \alpha_{-2} + \beta_{-2}}. \tag{8.28}$$

This equation, derived by Hill [31], has two important implications. First, it shows that if the force X is nil, no treadmilling can take place. Second, if $\alpha_1 = \beta_1$ and $\alpha_2 = \beta_2$ the polymer is not polarized and again no treadmilling is expected to occur. Thus nonequilibrium thermodynamics offers the proof that treadmilling relies on the polarity of actin filaments and on energy consumption.

As shown in Fig. 8.13, the conversion of $\mathrm{A_T}$ to α-$\mathrm{A_{DP}}$ and back constitutes a cycle. It is therefore worth considering how much free energy is required to keep this cycle running. The reasoning is basically the same for the α-end and the β-end of the polymer. For the first half of this cycle, one has

$$\mu_{\mathrm{AT}} = \mu^{\circ}_{\mathrm{AT}} + RT \ln \bar{c}_{\mathrm{AT}} = \mu_{\mathrm{ADP}}, \tag{8.29}$$

where $\bar{c}_{\mathrm{AT}}$ (the equilibrium concentration of A_{T}) is equal to α_{-1}/α_1. Therefore,

$$\mu^{\circ}_{\mathrm{AT}} + RT \ln \frac{\alpha_{-1}}{\alpha_1} = \mu_{\mathrm{ADP}} \tag{8.30}$$

and

$$RT \ln \frac{\alpha_1}{\alpha_{-1}} = \mu^{\circ}_{\mathrm{AT}} - \mu_{\mathrm{ADP}}. \tag{8.31}$$

For the second half of the cycle, one has

$$\mu_{\mathrm{ADP}} + \mu_{\mathrm{T}} = \mu_{\mathrm{AT}} + \mu_{\mathrm{D}} + \mu_{\mathrm{P}}. \tag{8.32}$$

Moreover, for global equilibrium conditions

$$\mu_{\mathrm{AT}} = \mu^{\circ}_{\mathrm{AT}} + RT \ln \bar{c}_{\mathrm{T}} = \mu^{\circ}_{\mathrm{AT}} + RT \ln \frac{\alpha_2}{\alpha_{-2}}, \tag{8.33}$$

and eq. (8.32) above becomes

$$\mu_{\mathrm{ADP}} + \mu_{\mathrm{T}} = \mu^{\circ}_{\mathrm{AT}} + RT \ln \frac{\alpha_2}{\alpha_{-2}} + \mu_{\mathrm{D}} + \mu_{\mathrm{P}}, \tag{8.34}$$

and can be rearranged to

$$RT \ln \frac{\alpha_2}{\alpha_{-2}} = \mu_{\mathrm{ADP}} - \left(\mu^{\circ}_{\mathrm{AT}} - X\right). \tag{8.35}$$

The significance of the energy costs defined by eqs. (8.31) and (8.35) is shown in Fig. 8.14.

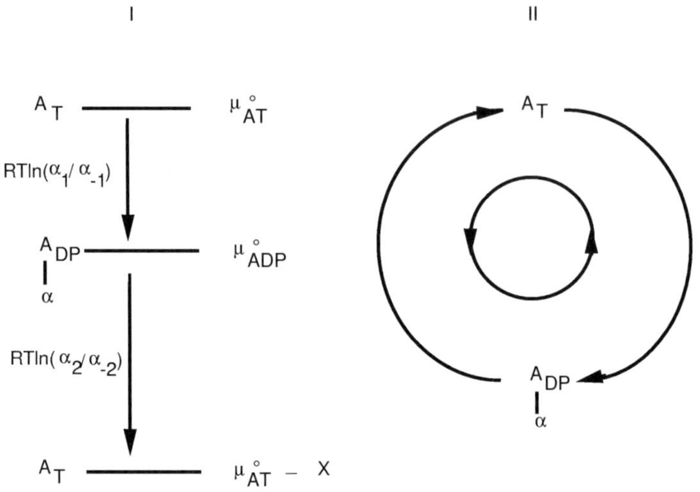

Fig. 8.14. Energetics of the A_T, α-A_{DP} cycle. I – Energetics of the treadmill flow. II – The A_T, α-A_{DP} cycle. See text.

The hydrolysis of ATP on the monomer and the subsequent binding of this monomer to the α-end of the actin filament is associated with an energy cost of μ_{ADP} (Fig. 8.14). Alternatively, the release of the actin unit of the α-end that bears ADP+phosphate and the replacement of these ligands by ATP has an energy cost of $\mu^{\circ}_{AT} - X$ (Fig. 8.14). The energy required to keep the cycle running is thus the sum of expressions (8.31) and (8.35), i.e.

$$RT \ln \frac{\alpha_1 \alpha_2}{\alpha_{-1}\alpha_{-2}} = X. \tag{8.36}$$

8.2.4. Drug action on steady state polymers

In this section, we shall assume that the polymer is in a steady state and that a ligand, which plays the role of an inhibitor, can be bound to the free monomer and to the two ends of the polymer. For simplicity, we assume that the last unit only, at the α-end and the β-end of the polymer, binds M and that all the units of the bulk polymer bind an ADP molecule [29,30]. The relevant model is shown in Fig. 8.15.

In this model which, in a different form, has been worked out by Hill [29], the rate constants a_1 and a_{-2} are extremely small. At the α-end of the polymer two "cycles" are running and pertain to the reversible conversion of α-A_D to A_T and α-A_DM to A_TM. A similar situation, indeed, takes place at the β-end of the polymer. This means that forces are keeping the cycles running. At the α-end the first cycle relies upon the two equilibria

$$\alpha + A_T \underset{\alpha_{-1}}{\overset{\alpha_1}{\rightleftarrows}} \alpha - A_D + P$$

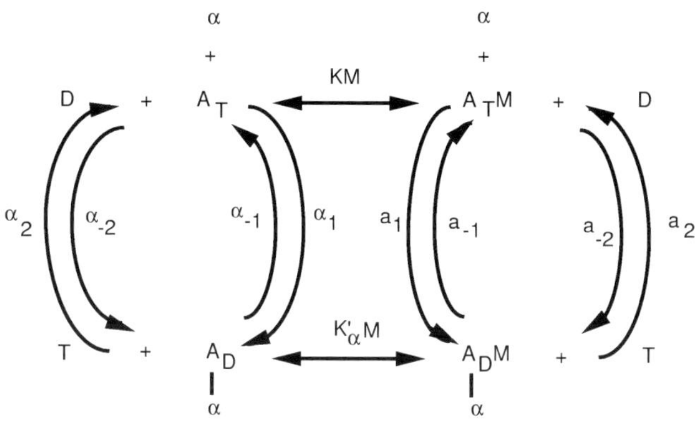

Fig. 8.15. Drug action on a steady state polymer. See text.

$$\mathrm{T} + \alpha - \mathrm{A_D} \underset{\alpha_{-2}}{\overset{\alpha_2}{\longleftrightarrow}} \mathrm{A_T} + \mathrm{D} + \alpha$$

and the second cycle upon

$$\alpha + \mathrm{A_T M} \underset{a_{-1}}{\overset{a_1}{\longleftrightarrow}} \alpha - \mathrm{A_D M} + \mathrm{P}$$

$$\alpha - \mathrm{A_D M} + \mathrm{T} \underset{a_{-2}}{\overset{a_2}{\longleftrightarrow}} \mathrm{A_T M} + \mathrm{D} + \alpha$$

Moreover, these two cycles are not independent, for thermodynamics demands that

$$\frac{a_1}{a_{-1}} = \frac{K'_\alpha}{K}\frac{\alpha_1}{\alpha_{-1}}, \qquad \frac{a_2}{a_{-2}} = \frac{K}{K'_\alpha}\frac{\alpha_2}{\alpha_{-2}}. \tag{8.37}$$

From the equilibria discussed in the previous section, one has

$$\begin{aligned} RT\ln\frac{\alpha_1}{\alpha_{-1}} &= \mu^\circ_{\mathrm{AT}} - (\mu_{\mathrm{AD}} + \mu_{\mathrm{P}}), \\ RT\ln\frac{\alpha_2}{\alpha_{-2}} &= \mu_{\mathrm{AD}} + \mu_{\mathrm{T}} - (\mu^\circ_{\mathrm{AT}} + \mu_{\mathrm{D}}). \end{aligned} \tag{8.38}$$

Thus one can write, from eq. (8.37)

$$\begin{aligned} RT\ln\frac{a_1}{a_{-1}} &= \mu^\circ_{\mathrm{AT}} - (\mu_{\mathrm{AD}} + \mu_{\mathrm{P}}) + RT\ln\frac{K'_\alpha}{K}, \\ RT\ln\frac{a_2}{a_{-2}} &= \mu_{\mathrm{AD}} + \mu_{\mathrm{T}} - (\mu^\circ_{\mathrm{AT}} + \mu_{\mathrm{D}}) - RT\ln\frac{K'_\alpha}{K}. \end{aligned} \tag{8.39}$$

Then the force driving the first cycle is

$$RT \ln \frac{\alpha_1 \alpha_2}{\alpha_{-1} \alpha_{-2}} = \mu_T - \mu_D - \mu_P = X, \tag{8.40}$$

and, similarly, the force driving the second is

$$RT \ln \frac{a_1 a_2}{a_{-1} a_{-2}} = \mu_T - \mu_D - \mu_P = X. \tag{8.41}$$

Thus, as expected, it is the same force, namely, the hydrolysis of ATP, that drives the two cycles. Relationships (8.40) and (8.41) have an interesting implication. Because the rate constants a_1 and a_{-2} are very small and the two cycles are driven by the same force, the two other rate constants must be large. In other words, thermodynamics shows that the ligand M should induce the dismantling of the polymer.

The flow of addition of monomers to the α-end of the polymer is

$$\begin{aligned} J_\alpha &= (\alpha_1 + \alpha_{-2}) c_{AT} (1 - \eta_\alpha) - (\alpha_{-1} + \alpha_2)(1 - \eta_\alpha) \\ &\quad + (a_1 + a_{-2}) c_{ATM} \eta_\alpha - (a_2 + a_{-1}) \eta_\alpha, \end{aligned} \tag{8.42}$$

where c_{AT} is the concentration of monomers bearing ATP, c_{ATM} is the concentration of monomers that bear both ATP and the drug M. η_α has the same significance as it had in section 8.2.2. We have already mentioned that the rate constants a_1 and a_{-2} are very small but, for most drugs, the rate constants α_{-2}, α_{-1} and a_{-1} are small as well. Therefore, a good approximation of the flow is

$$J_\alpha = \alpha_1 c_{AT} (1 - \eta_\alpha) - \alpha_2 (1 - \eta_\alpha) - a_2 \eta_\alpha. \tag{8.43}$$

Taking advantage of eqs. (8.3) and (8.4), one obtains an explicit expression of J_α, namely,

$$J_\alpha = \frac{\alpha_1 c_T}{(1 + K c_M)(1 + K'_\alpha c_M)} - \frac{\alpha_2 + a_2 K'_\alpha c_M}{1 + K'_\alpha c_M}. \tag{8.44}$$

As expected, the ligand M will decrease the slope of the $J_\alpha = f(c_T)$ plot. It can also alter the value of its intercept. Last but not least, it may be of interest to know what is the critical concentration of monomers, c_T, that results in a flow of addition of monomers to the α-ends of the polymers equal to zero. This critical concentration, $\bar{c}_T$, is readily derived from eq. (8.44) above by setting $J_\alpha = 0$. One thus finds

$$\bar{c}_T = \frac{(1 + K c_M)(\alpha_2 + a_2 K'_\alpha c_M)}{\alpha_1}, \tag{8.45}$$

and the critical concentration increases as the drug concentration increases.

8.3. *Molecular motors and the statistical physics of muscle contraction*

It has been possible in recent years, through the use of sophisticated physical techniques, to study individual motor molecules, to measure their motion and the force they exert on subcellular structures [28]. Moreover, physical models of muscular contraction have been proposed that allow one to grasp, in terms of statistical physics, the logic of this complex process. The physical model that will be presented below is largely derived from that developed some years ago by Hill [32,33]. As shown in section 8.1.2 of this chapter, the actin–myosin complex, which will be termed "cross-bridge", can exist in, at least, six states. These states are schematically pictured in Fig. 8.16.

In state 1 the myosin head is unliganded and not attached to actin. In state 2 it binds ATP (T) but is still unattached to actin. In state 3, ATP is hydrolyzed but ADP and phosphate (D) remain bound to myosin which is still unattached to actin. ATP hydrolysis induces a hinge motion of the myosin head. During the next step of the process (state 4) myosin binds to actin (A). The binding occurs at the nearest actin site. As each actin unit bears an actin site, each site is thus separated from its neighbour by a fixed distance d. The binding of a myosin head to its nearest actin site may involve a slight bending of the myosin head and thus a storage of energy in the cross-bridge, for the myosin head can exert a force on the actin filament (Fig. 8.16). This important question will be addressed later. The release of ADP and phosphate (state 5) from the myosin head produces a second hinge motion of the myosin head that brings it back to its initial orientation relative to actin. However, as myosin is still attached to actin, the actin filaments move relative to the myosin thick filaments and this results in the shortening of the sarcomere. It is thus in state 5 that the greatest force is exerted on actin filaments. The reorientation of the myosin head relative to the actin filament leaves the cross-bridge in state 6. This state can also store some energy for, owing to the fact that many myosin heads are attached to the same actin filament,

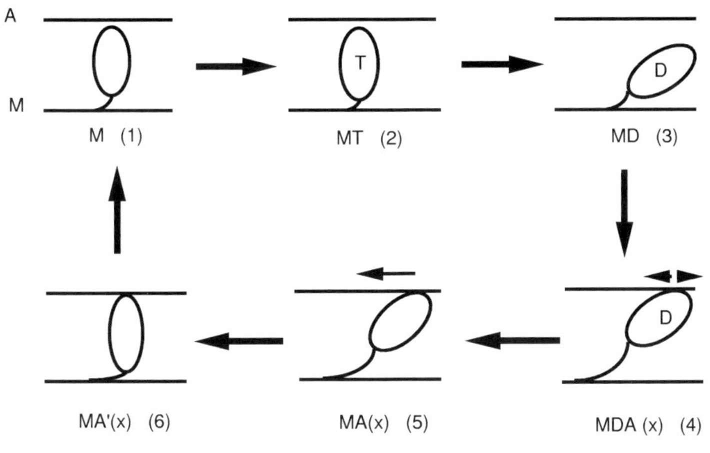

Fig. 8.16. The six states of a cross-bridge. See also Fig. 8.7. Hydrolysis of ATP (T) induces an hinge motion of the myosin head that binds to actin and the release of ADP (D) produces the force-generating step.

the final orientation of each head may not be the one associated with the lowest energy level. The six states considered above are designated M, MT, MD, MDA(x), MA(x) and MA$'(x)$. Three states in six pertain to attached cross-bridges. Moreover most of the free energy that will be converted into work is stored in state MA(x).

Let us come now to the thermodynamic significance of the results considered above. A myosin molecule in state MDA(x) can bind to an actin site without any bending of its head and this will be called the "equilibrium position" of the head. This situation is certainly the most frequent one. But the binding process may also require the bending of the head, and this implies that some free energy be stored in the corresponding cross-bridge. Moreover, each myosin head which is not in equilibrium position will exert a force on the actin filament but the resultant of these forces will be nil. The two extreme actin sites that may accomodate the same myosin head are separated by the distance d and the quilibrium position is at the distance $x = 0$ (Fig. 8.17).

The Helmholtz free energy stored in a cross-bridge thus depends on the distance x. Moreover, one must have $-d/2 \leqslant x \leqslant d/2$. The simplest expression of the Helmholtz free energy, A_i, of cross-bridge i is thus

$$A_i = A_i^\circ + \frac{1}{2}\alpha x^2. \tag{8.46}$$

A_i° is the Helmoltz free energy of the equilibrium position for cross-bridge i, α is a constant. The partition function, Q, for a population of cross-bridges is thus

$$Q = \sum_{i=1}^{n} \exp\left(-\frac{A_i}{k_\mathrm{B}T}\right). \tag{8.47}$$

In the case of $x \neq 0$ (eq. (8.46)), the term $Q(A_i)$ of this partition function that pertains to cross-bridge i is

$$Q(A_i) = \exp\left\{-\frac{A_i^\circ + \alpha x^2/2}{k_\mathrm{B}T}\right\}. \tag{8.48}$$

This expression represents the probability, for cross-bridge i, to bind to an actin site at a distance x from the corresponding equilibrium position. The maximum probability, $Q(A_i)$, is indeed obtained for $x = 0$, and decreases for $x \to \pm d/2$ (Fig. 8.17). Moreover, eq. (8.46) shows that the Helmoltz free energy increases as $x \to \pm d/2$ (Fig. 8.17).

In classical fluid thermodynamics, the Gibbs–Duhem relationship is written as

$$\mathrm{d}A = -S\,\mathrm{d}T - p\,\mathrm{d}V, \tag{8.49}$$

where S is the entropy, T is the absolute temperature, p is the pressure and V is the volume. In the present case, the term $-p\,\mathrm{d}V$ should be replaced by $F\,\mathrm{d}x$, where F is the force. Therefore, one must have, for cross-bridge i

$$\mathrm{d}A_i = -S_i\,\mathrm{d}T + F_i\,\mathrm{d}x \tag{8.50}$$

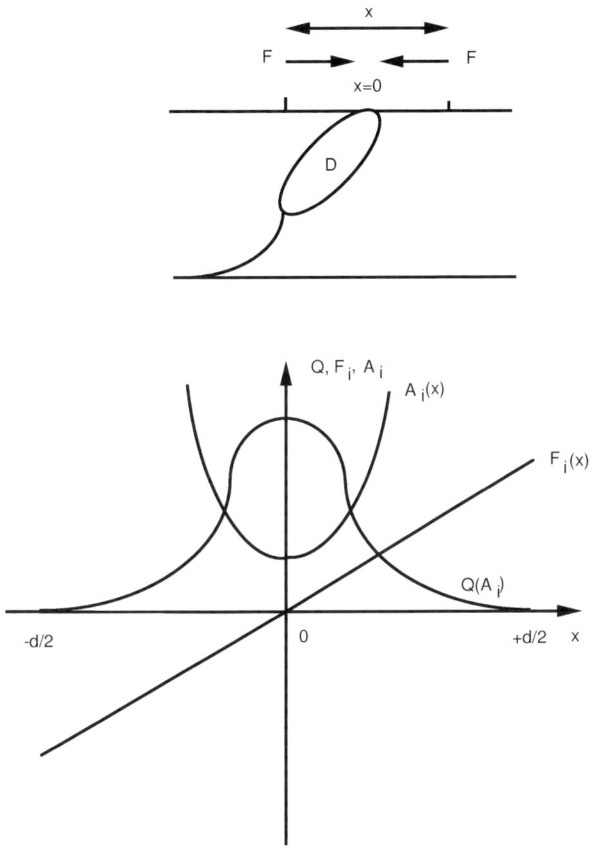

Fig. 8.17. Statistical mechanics of the binding of myosin head to actin. Top – The myosin head bearing an ADP bound may bind actin at the equilibrium position ($x = 0$) or at a position different from equilibrium. Bottom – If the binding takes place at the equilibrium position the partition function, Q, has its maximum value, the Helmoltz free energy, A_i, is minimum and the force exerted on actin filament, F_i, is nil. The Helmoltz free energy increases and the force becomes different from zero if binding takes place above or below the equilibrium position.

and

$$F_i = \left(\frac{\partial A_i}{\partial x}\right)_T. \tag{8.51}$$

If this relationship is applied to eq. (8.46), one can derive the expression of the force as a function of the distance x

$$F_i = \alpha x. \tag{8.52}$$

The magnitude of the force varies linearly as a function of the distance x. Equation (8.48) shows that there is an equal probability, for different cross-bridges, to bind to actin sites

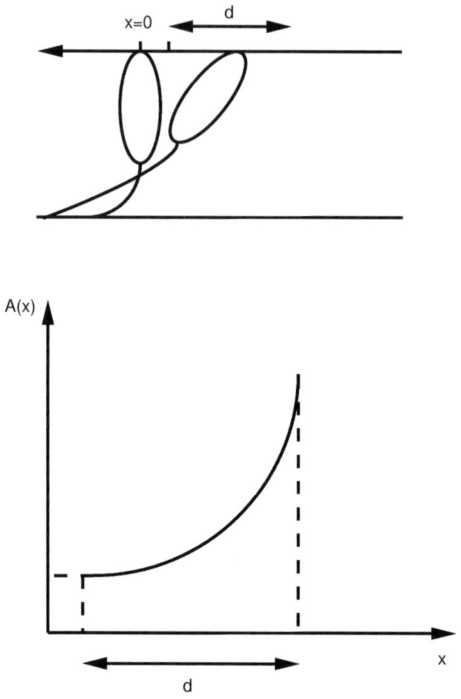

Fig. 8.18. In state MA(x) the mean position of the myosin head has a positive value. Top – The mean position of the myosin head is positive. This implies that the Helmoltz free energy of a population of cross-bridges is large and that the mean force exerted on the actin filaments is positive. Bottom – The Helmoltz free energy of a population of cross-bridges is of necessity above its minimum value. This state generates a force.

located at equally distant positive and negative x values. This means that the resultant force (Fig. 8.17) is equal to zero

$$\sum_i p_i Q_i = 0, \tag{8.53}$$

where Q_i is the term of the partition function pertaining to cross-bridge i and p_i is the corresponding frequency of cross-bridges in state i. In other words, this is the probability for a cross-bridge to be in state i. Therefore, as previously alluded to, a population of cross-bridges in state MDA(x), bound to the same actin filament, will exert a resultant null-force on this filament. The same conclusion will also apply to state MA$'$(x).

If the cross-bridge is in state MA(x), the myosin head is not in its equilibrium position. The corresponding value of x is positive (Fig. 8.18). The cross-bridge has thus stored free energy and exerts a force on the actin filament, for the myosin head tends to reach its equilibrium position, i.e. precisely the position which is reached in state MA$'$(x). Thus if we consider a population of attached cross-bridges, only one state (state MA(x)) in three is able to generate a force and a work, because the site of attachment of myosin to actin is at a distance x different from that of the equilibrium position ($x = 0$). Conversely, the other attached states, namely MDA(x) and MA$'$(x), cannot generate a force because the average position of their myosin head is nearly identical to the corresponding equilibrium position.

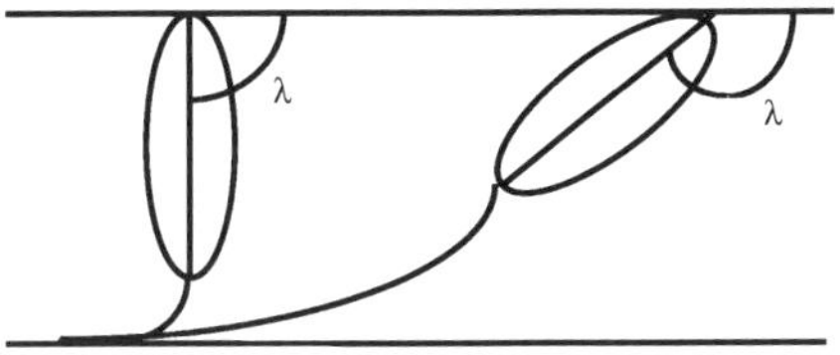

Fig. 8.19. The Helmoltz free energy stored in a cross-bridge depends on the value of the λ angle. See text.

For operational reasons it is advantageous to consider, as we have done so far, that the force F_i and the corresponding Helmholtz free energy, A_i, to depend on the distance x. However, x will not, in general, be the "natural" independent variable of which F_i and A_i depend. It is more sensible to consider that this "natural" variable is the angle λ, as shown in Fig. 8.19. It is, however, possible to show that expressing A_i and F_i as a function of this angle λ is equivalent to expressing them as a function of the distance x. The Helmoltz free energy A_i can also be rewritten as

$$A_i = A_i^\circ + \frac{1}{2}\alpha_\lambda(\lambda - \lambda_0)^2. \tag{8.54}$$

λ_0 is then the angle formed by the actin filament and the attached myosin head in the equilibrium position (Fig. 8.19). The distance x is indeed a function of the angle λ

$$x = f(\lambda), \tag{8.55}$$

which can be expanded in Taylor series

$$x = \left(\frac{\partial x}{\partial \lambda}\right)_{\lambda_0}(\lambda - \lambda_0) + \cdots. \tag{8.56}$$

If the difference between λ and λ_0 is small enough, one can derive the expression of $\lambda - \lambda_0$ from this equation. Inserting this expression into eq. (8.54) leads to

$$A_i = A_i^\circ + \frac{1}{2}\alpha_\lambda\left(\frac{\partial x}{\partial \lambda}\right)_{\lambda_0}^{-2} x^2, \tag{8.57}$$

which is equivalent to eq. (8.46). There is, therefore, no harm in expressing the thermodynamic functions F_i and A_i as a function of the distance x.

The conversion of these states can be pictured by the kinetic scheme of Fig. 8.20. In this diagram the transitions from one state to another are governed by transition probabilities. Some of these transition probabilities depend on the distance x already referred to, but others do not. We have already seen that the mean Helmholtz free energy for states 4 and 6 is independent of the distance x. Therefore, one has

$$\langle A_4(x)\rangle = \langle A_4\rangle, \qquad \langle A_6(x)\rangle = \langle A_6\rangle. \tag{8.58}$$

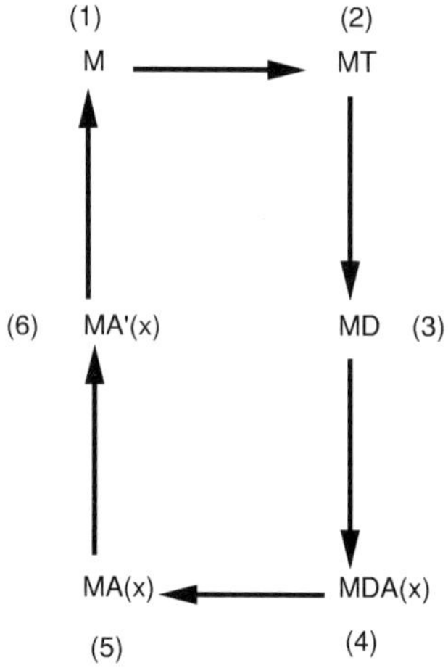

Fig. 8.20. Formal scheme of the six states involved in muscle contraction. See text.

If we consider the transitions of a population of cross-bridges, the ratios of forward and backward transition probabilities are formally equivalent to equilibrium constants in classical enzyme kinetics. Thus, for the model of Fig. 8.20, one has

$$
\begin{aligned}
\frac{\alpha_{12}}{\alpha_{21}} &= \exp\frac{A_1 + \mu_T - A_2}{k_B T} = K_{12}, \\
\frac{\alpha_{23}}{\alpha_{32}} &= \exp\frac{A_2 - A_3}{k_B T} = K_{23}, \\
\frac{\alpha_{34}}{\alpha_{43}} &= \exp\frac{A_3 - \langle A_4\rangle}{k_B T} = K_{34}, \\
\frac{\alpha_{45}(x)}{\alpha_{54}(x)} &= \exp\frac{\langle A_4\rangle - \langle A_5(x)\rangle - \mu_D}{k_B T} = K_{45}(x), \\
\frac{\alpha_{56}(x)}{\alpha_{65}(x)} &= \exp\frac{\langle A_5(x)\rangle - \langle A_6\rangle}{k_B T} = K_{56}(x), \\
\frac{\alpha_{61}}{\alpha_{16}} &= \exp\frac{\langle A_6\rangle - A_1}{k_B T} = K_{61}.
\end{aligned}
\tag{8.59}
$$

Note that, in these expressions, the "ligand" actin does not appear explicitly because it is part of the cross-bridge structure. As previously, μ_T and μ_D are the chemical potentials of ATP and ADP+phosphate. These expressions apply to an average cross-bridge. The cost of free energy required to move this average cross-bridge along the actin filament is thus obtained from the product

$$
K_{12}K_{23}K_{34}K_{45}(x)K_{56}(x)K_{61} = \exp\frac{\mu_T - \mu_D}{k_B T}, \tag{8.60}
$$

and is, indeed, equal to $\mu_T - \mu_D$ [34].

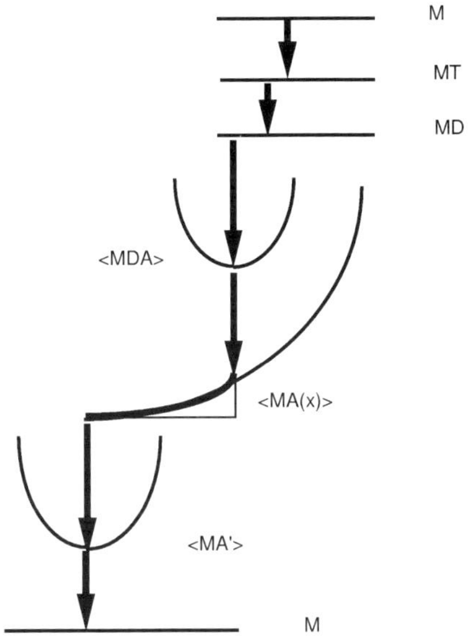

Fig. 8.21. Random walk of cross-bridges. This scheme pictures the decrease of energy during the cycle of the the preceding figure. Only the ideal mean state $\langle \mathrm{MA}(x) \rangle$ can generate a work and a force for reasons expressed in Figs. 8.17 and 8.18.

If one considers now the sequence of Fig. 8.20, the cross-bridges follow a biased random walk from state M back to the same state M. The general trend of the walk is downhill and the corresponding transition probabilities are the α's of eq. (8.59). In Fig. 8.21 hypothetical random walks of cross-bridges are shown. As mentioned previously, the states $\mathrm{MDA}(x)$ and $\mathrm{MA}'(x)$ of an individual cross-bridge can generate a force but the resultant of these forces is nil. If one considers an ideal mean cross-bridge, only the ideal state $\langle \mathrm{MA}(x) \rangle$ can generate a force and do work. The efficiency of the contraction will then be equal to

$$\rho = \frac{\langle A_5(x) \rangle}{\mu_\mathrm{T} - \mu_\mathrm{D}}. \tag{8.61}$$

It is thus a state, not a transition between states, that generate a force.

Contraction does not necessarily generate work. If the force exerted on actin by the cross-bridge is just balanced by an external force, F^{ext}, then no work will be performed and the contraction will be isometric. It may also happen that the contraction is isotonic, i.e. the rate of contraction is constant. We are now in a position to describe the walk of the myosin heads along the actin filaments, towards their plus ends. As this process is not taking place in solution, the usual rules of chemical kinetics do not apply. It is thus important to define a number of simplifying assumptions that will allow one to express in quantitative terms the dynamics of the walk of the motor protein. Let us consider an ensemble of attached cross-bridges in the overlap zone that associates myosin thick filaments. We have already outlined that the probability of the myosin head to bind to an actin site is distributed, as a function of the distance (eq. (8.48)), about a maximum value obtained for $x = 0$. As $x \to \pm d/2$ this probability approaches zero. Thus, in this model, the density of the actin sites is assumed to vary periodically along an actin filament and this reasoning can be extended to all the actin filaments of the overlap zone. If we consider, in this overlap zone,

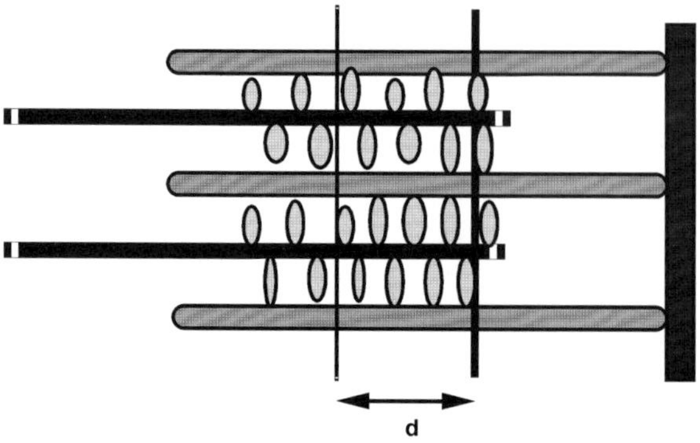

Fig. 8.22. The "flow" of cross-bridges that enter and leave a definite region of the overlap zone. The thickness of the overlap zone is d. See text.

all the cross-bridges located at a given x value, the subensemble of cross-bridges located within a region defined by a short displacement, $x + \mathrm{d}x$, will appear to gain and lose cross-bridges continuously as the actin filaments move relative to the thick myosin filaments (Fig. 8.22). There is apparently a flow of cross-bridges that enter and leave this region of the overlap zone.

Let p_i be the probability of finding, in the subensemble, cross-bridges in state i. Depending on the state considered, this probability may depend only on time, $p_i(t)$, or it may depend on both the time and the distance, $p_i(t, x)$. States M, MT, MD will depend on time only, and also states $\langle \mathrm{MDA}(x)\rangle$ and $\langle \mathrm{MA}'(x)\rangle$ for one has

$$\langle \mathrm{MDA}(x)\rangle = \langle \mathrm{MDA}\rangle, \qquad \langle \mathrm{MA}'(x)\rangle = \langle \mathrm{MA}'\rangle, \tag{8.62}$$

as we have already seen. $\langle \mathrm{MA}(x)\rangle$ in contrast will depend on the distance x, so three equations will depend on both the time and the distance. For any of these equations one will have [34]

$$\frac{\mathrm{d}p}{\mathrm{d}t} = \left(\frac{\partial p}{\partial t}\right) + \left(\frac{\partial p}{\partial x}\right)\left(\frac{\partial x}{\partial t}\right). \tag{8.63}$$

These equations are:

$$\begin{aligned}
\left(\frac{\partial p_4}{\partial t}\right) + \left(\frac{\partial p_4}{\partial x}\right)\left(\frac{\partial x}{\partial t}\right) &= \alpha_{34} p_3 + \alpha_{54}(x) p_5 - \left\{\alpha_{43} + \alpha_{45}(x)\right\} p_4, \\
\left(\frac{\partial p_6}{\partial t}\right) + \left(\frac{\partial p_6}{\partial x}\right)\left(\frac{\partial x}{\partial t}\right) &= \alpha_{16} p_1 + \alpha_{56}(x) p_5 - \left\{\alpha_{61} + \alpha_{65}(x)\right\} p_6, \\
\left(\frac{\partial p_5}{\partial t}\right) + \left(\frac{\partial p_5}{\partial x}\right)\left(\frac{\partial x}{\partial t}\right) &= \alpha_{45}(x) p_4 + \alpha_{65}(x) p_6 - \left\{\alpha_{54}(x) + \alpha_{56}(x)\right\} p_5,
\end{aligned} \tag{8.64}$$

and depend on both t and x. The others, for instance,

$$\frac{\mathrm{d}p_2}{\mathrm{d}t} = \alpha_{12}p_1 + \alpha_{32}p_3 - (\alpha_{21} + \alpha_{23})p_2 \tag{8.65}$$

depend on t only. In principle at least, these equations can be solved to obtain $p_i(t,x)$ for i and $t > 0$. These differential equations, however, become much simpler if the length of the muscle fiber is held constant ($\partial x/\partial t = 0$) and if the p_i do not vary in time ($\partial p_i/\partial t = 0$). The contraction is then steady and isometric. The mucle fiber will exert a force without producing any work. The set of differential equations degenerates into algebraic equations and one has

$$\begin{aligned} 0 &= \alpha_{61}p_6 + \alpha_{21}p_2 - (\alpha_{12} + \alpha_{16})p_1 \\ &\vdots \\ 0 &= \alpha_{34}p_3 + \alpha_{54}(x)p_5 - \left\{\alpha_{43} + \alpha_{45}(x)\right\}p_4 \\ &\vdots \end{aligned} \tag{8.66}$$

This system, together with

$$\sum_i p_i = 1 \tag{8.67}$$

can be solved for the p_i.

Another situation where the differential equations (8.64) become simpler is that obtained for a steady isotonic contraction. Then, $\partial x/\partial t = -v =$ constant and the three equations (8.64) become

$$\begin{aligned} -v\frac{\mathrm{d}p_4}{\mathrm{d}x} &= \alpha_{34}p_3 + \alpha_{54}(x)p_5 - \left\{\alpha_{43} + \alpha_{45}(x)\right\}p_4, \\ -v\frac{\mathrm{d}p_6}{\mathrm{d}x} &= \alpha_{16}p_1 + \alpha_{56}(x)p_5 - \left\{\alpha_{61} + \alpha_{65}(x)\right\}p_6, \\ -v\frac{\mathrm{d}p_5}{\mathrm{d}x} &= \alpha_{45}(x)p_4 + \alpha_{65}(x)p_6 - \left\{\alpha_{54}(x) + \alpha_{56}(x)\right\}p_5, \end{aligned} \tag{8.68}$$

thus allowing the possibility of solving the system.

If, as already outlined, the force $F_5(x)$ exerted by cross-bridges in state 5 can be measured experimentally, the mean force (per cross-bridge) exerted on the actin filament at time t by the cross-bridges in the subensemble, at the distance x, is then

$$F(t,x) = p_5(t,x)F_5(x). \tag{8.69}$$

$p_5(t,x)$ is obtained by solving the differential equations (see above). The mean force (per cross-bridge) exerted by all the cross-bridges in the interval d is thus

$$F(t) = \frac{1}{d}\int_{-d/2}^{+d/2} F(t,x)\,\mathrm{d}x. \tag{8.70}$$

8.4. Dynamic state of supramolecular edifices in the living cell

As shown in the above sections of this chapter, several supramolecular edifices of the eukaryotic cell are continuously being built up and dismantled. Probably the most conspicuous example of this situation is offered by the mitotic spindle, which appears at certain stages of the cell division cycle and disappears afterwards. A similar situation occurs for actin filaments that grow and shrink depending on the experimental conditions. From a physical viewpoint, there are two basic reasons for this situation. On the one hand, these supramolecular edifices are polarized, i.e. the two ends of the same filament have different structures. There is a plus and a minus end. Moreover, these edifices are not in equilibrium. Their very existence requires that some free energy be continuously spent in order to avoid their shrinking. This energy stems from GTP hydrolysis for microtubules, and from ATP hydrolysis in the case of actin filaments. The dynamic state of these edifices has two important implications. First, it allows one to understand why they could dismantle in the presence of certain drugs such as colchicine. Second, it offers a possible explanation to the existence of periodic oscillations of microtubule polymerization [15]. As microtubules are growing at the plus end, and are shrinking at the minus end, one can intuitively understand that the polymerization process be periodic under certain conditions.

Another feature of the dynamic state of cell substructures is the existence and activity of motor proteins, such as myosin and kinesin. These motors convert the energy of ATP hydrolysis into mechanical work. Thus kinesin can walk along microtubules carrying membrane vesicles, for example. As already mentioned, there now exists some devices that allow one to study individual motor proteins at work [28,35–37]. In so doing, one is able to measure forces and distances travelled by these individual molecular motors. An interesting feature of, at least, some of these motors is their behaviour as processive enzymes. Processive enzymes, such as polymerases, and ATP-dependent mechanoenzymes, may complete several iterations before releasing the substrate or the product. Therefore the number of iterations, and the time required for these iterations, may vary greatly from enzyme molecule to enzyme molecule, within a homogeneous population of the same molecular species. In the case of mechanoenzymes, such as kinesin, the distance travelled may also vary from enzyme molecule to enzyme molecule [28,36]. One may therefore wonder about the biological significance, if any, of the statistical distribution of the distance travelled by individual motor proteins belonging to the same molecular species, and how some local experimental conditions, for instance ATP concentration, may affect this distribution. These questions are far from being solved at the moment.

Moreover, single myofibrils may display oscillations that appear for well defined conditions of ADP and phosphate concentrations [38]. Physical models, which will not be considered here, have been proposed that explain this periodic behaviour [39–41].

There is thus little doubt that the cell substructures display highly dynamic events that have often been described in the current scientific literature but are not always understood in terms of physical chemistry.

References

[1] Dustin, P. (1984) Microtubules, 2nd edition. Springer-Verlag, New York.
[2] Amos, L.A. and Baker, T.S. (1979) The three-dimensional structure of tubulin protofilaments. Nature 279, 607–612.
[3] Sullivan, K.F. (1988) Structure and utilization of tubulin isotypes. Annu. Rev. Cell Biol. 4, 687–716.
[4] de Brabander, M. (1986) Microtubule dynamics during the cell cycle: the effect of taxol and nocodazole on the microtubular system of PtK_2 cells at different stages of the mitotic cycle. Int. Rev. Cytol. 101, 215–274.
[5] Mandelkow, E.M. and Mandelkow, E. (1992) Microtubule oscillations. Cell Motil. Cytoskeleton 22, 235–244.
[6] Bergen, L.G. and Borisy, G.G. (1980) Head-to-tail polymerization of microtubules *in vitro*. Electron microscope analysis of seeded assembly. J. Cell Biol. 84, 141–150.
[7] McIntosh, J.R. and Euteneur, U. (1984) Tubulin hooks as probes for microtubule polarity: an analysis of the method and evaluation of the data on microtubule polarity in the mitotic spindle. J. Cell Biol. 88, 525–533.
[8] Mazia, D. (1984) Centrosomes and mitotic poles. Exp. Cell Res. 88, 525–533.
[9] Erickson, H.P. and O'Brien, E.T. (1992) Microtubule dynamic instability and GTP hydrolysis. Annu. Rev. Biophys. Biomol. Struct. 21, 145–166.
[10] Mitchison, T.J. and Kirschner, M.W. (1984) Dynamic instability of microtubule growth. Nature 312, 237–242.
[11] Olmsted, J.B. (1986) Microtubule-associated proteins. Annu. Rev. Cell Biol. 2, 421–457.
[12] Vale, R.D., Reese, T. and Sheetz, M.P. (1985) Identification of a novel force-generating protein, kinesin, involved in microtubule-based motility. Cell 42, 39–50.
[13] Kabsch, W., Mannherz, H.G., Suck, D., Pai, E.F. and Holmes, K.C. (1990) Atomic structure of actin: DNase I complex. Nature 347, 37–44.
[14] Oosawa, F. and Asakura, S. (1975) Thermodynamics of the Polymerization of Proteins. Academic Press, New York.
[15] Carlier, M.F. (1989) Role of nucleotide hydrolysis in the dynamics of actin filaments and microtubules. Int. Rev. Cytol. 115, 139–170.
[16] Carlier, M.F. (1991) Actin: protein structure and filament dynamics. J. Biol. Chem. 266, 1–4.
[17] Cooper, J.A. (1997) Effects of cytochalasin and phalloidin on actin. J. Cell Biol. 105, 1473–1478.
[18] Fecheimer, M. and Zigmond, S.H. (1993) Focusing on unpolymerized actin. J. Cell Biol. 123, 1–5.
[19] Holmes, K.C., Popp, D., Gebhard, W. and Kabsch, W. (1990) Atomic model of the actin filament. Nature 347, 44–49.
[20] Pollard, T.D. (1997) The myosin-crossbridge problem. Cell 48, 909–910.
[21] Rayment, I., Rypniewski, W.R., Smidt-Base, K., et al. (1993) Three-dimensional structure of myosin subfragment I: a molecular motor. Science 261, 50–58.
[22] Means, A.R. (1998) The clash in titin. Nature 385, 846–847.
[23] Frieden, C. (1985) Actin and tubulin polymerization: The use of kinetic methods to determine mechanism. Annu. Rev. Biophys. Biophys. Chem. 14, 189–210.
[24] Pollard, T.D. and Cooper, J.A. (1986) Actin and actin-binding proteins. A critical evaluation of mechanisms and functions. Annu. Rev. Biochem. 55, 987–1035.
[25] Carlier, M.F., Pantaloni, D. and Korn, E.D. (1986) Fluorescence measurements of the binding of cations to high-affinity and low-affinity sites on ATP-G-actin. J. Biol. Chem. 261, 10778–10784.
[26] Estes, J.E., Selden, L.A. and Gershman, L.C. (1987) Tight binding of divalent cations to monomeric actin. Binding kinetics supports a simplified model. J. Biol. Chem. 262, 4952–4957.
[27] Carlier, M.F., Pantaloni, D. and Korn, E.D. (1986) The effects of Mg at the high-affinity and low-affinity sites on the polymerization of actin and associated ATP hydrolysis. J. Biol. Chem. 261, 10785–10792.

[28] Schnitzer, M.J. and Block, S.M. (1995) Statistical kinetics of processive enzymes. Cold Spring Harbour Quant. Biol. 60, 793–802.

[29] Hill, T.L. and Kirschner, M. (1983) Regulation of microtubule and actin filament assembly–disassembly by associated small and large molecules. Int. Rev. Cytol. 84, 185–234.

[30] Hill, T.L. (1981) Steady state head-to-tail polymerization of actin or microtubules. II. Two-state and three-state kinetic cycles. Biophys. J. 33, 353–371.

[31] Hill, T.L. (1980) Bioenergetic aspects and polymer length distribution in steady state head-to-tail polymerization of actin or microtubules. Proc. Natl. Acad. Sci. USA 77, 4803–4807.

[32] Hill, T.L. (1974) Theoretical formalism for the sliding filament model of contraction of striated muscle. Part I. Prog. Biophys. Molec. Biol. 28, 267–340.

[33] Hill, T.L. (1975) Theoretical formalism for the sliding filament model of contraction of striated muscle. Part II. Prog. Biophys. Molec. Biol. 29, 105–159.

[34] Hill, T.L. (1977) Free Energy Transduction in Biology. Academic Press, New York.

[35] Block, S.M. (1995) Nanometers and piconewtons: The macromolecular mechanics of kinesin. Trends Cell Biol. 5, 169–171.

[36] Block, S.M., Goldstein, L.S.B. and Schnapp, B.J. (1990) Bead movement by single kinesin molecules studied with optical tweezers. Nature 348, 348–350.

[37] Svoboda, K., Mitra, P.P. and Block, S.M. (1994) Fluctuation analysis of motor protein movement and single enzyme kinetics. Proc. Natl. Acad. Sci. USA 91, 11782–11796.

[38] Yasuda, K., Shindo, Y. and Ishiwata, S. (1996) Spontaneous oscillations of sarcomeres in skeletal myofibrils under isotonic conditions. Biophys J. 70, 1823–1829.

[39] Chauwin, J.F., Adjari, A. and Prost, J. (1994) Force-free motion in asymmetric structures: a mechanism without diffusive steps. Europhys. Lett. 27, 421–426.

[40] Jülicher, F. and Prost, J. (1995) Cooperative molecular motors. Phys. Rev. Lett. 75, 2618–2621.

[41] Jülicher, F. and Prost, J. (1997) Spontaneous oscillations of collective molecular motors. Phys. Rev. Lett. 78, 4510–4513.

CHAPTER 9

Temporal organization of metabolic cycles and structural complexity: oscillations and chaos

In chapter 4 we have assumed that enzyme reactions take place in the cell under steady state conditions. There are, however, many experimental results that show beyond doubt this is far from always being the case. Many metabolic networks and cycles in fact display a temporal organization. In other words, they display periodic, or aperiodic, oscillations. Numerous review articles and even books [1–5] have been published on this topic, and the aim of the present chapter is certainly not to offer a new comprehensive review of the abundant literature on the subject. It is rather to discuss whether the compartmentalization and the supramolecular complexity of the living cell can generate temporal organization of certain important metabolic processes. Besides the experimental results that have shown that many metabolic processes display oscillations, theoretical models have also been developed to explain these results. To the best of our knowledge, these models explain the features of these oscillations through the individual properties of the enzyme reactions and through the topology of the network of biochemical processes. At the moment, no enzyme reaction can be considered *per se* an oscillatory process. It is *a system* of several reactions that may possibly display oscillations. The peroxidase reaction is sometimes considered oscillatory. It is in fact not the peroxidase reaction alone that is being studied under the conditions where oscillations take place, but a system comprising the enzyme reaction coupled to a complex chain of free radicals that propagate. In biology, periodic or aperiodic oscillations are thus emergent processes that can only be detected at the supramolecular level.

After a brief overview of two biological processes that have left a mark in the field, namely glycolytic and calcium oscillations, the question will be addressed for a simple open metabolic cycle, to define the minimal condition that generates oscillations. We shall then consider how compartmentalization and electrostatic repulsion of the reagents of a chemical reaction may generate sustained oscillations. Last, we shall study how the complexity of the organization of supramolecular edifices in the cell may lead to periodic and chaotic oscillations of some important biological processes.

9.1. *Brief overview of the temporal organization of some metabolic processes*

In this section, we shall very briefly review some salient features of two types of biological phenomena, namely glycolytic oscillations and calcium spiking.

9.1.1. *Glycolytic oscillations*

The existence of glycolytic oscillations has been observed in cell free extracts, in intact yeast cells as well as in intact myocytes (see, for instance, ref. [5]). Thus, yeast cell free extracts display sustained oscillations of NADH absorbance, or fluorescence, after periodic additions of glucose. Phosphofructokinase is considered as a source of oscillations [1,3]. Theoretical studies have shown that more complex oscillatory behaviour, such as bi-periodicity and chaos, may be expected to occur if the enzymes are coupled in series and activated by their respective products [6–8]. Depending on the periodicity of glucose injection in the reactor containing cell free extract, NADH fluorescence can display periodic, quasi-periodic or chaotic patterns.

In a single intact yeast cell, the microfluorimetric detection of NADH fluorescence has been observed to be periodic with a fairly regular period. Moreover, a population of cells displays a remarkable synchronization of their dynamic properies [5]. Acetaldehyde is probably the coupling agent in a population of yeast cells. Oscillations of NADH fluorescence in yeast cells appear to be coupled to the plasma membrane potential. In fact, ATP which is generated by the glycolytic process drives the proton translocating ATPase, thus resulting in the build-up of a membrane potential that can be monitored through the fluorescence of rhodamine [5].

9.1.2. Calcium spiking

It was observed several years ago, that the addition of physiological concentrations of a hormone to an intact animal cell resulted in a sequence of calcium spikes [9]. Calcium spiking occurs in many different animal cells such as hepatocytes [9,10], endothelial cells [11], fibroblasts [12], etc. This process displays a pattern that has been encountered in nearly all animal cells investigated so far. The main features of spiking can be summarized as follows [13]. Spiking takes place above a sharp threshold of hormone concentration. Below that threshold, no change in the level of calcium can be observed. The frequency of spiking is more or less proportional to the hormone concentration. Above the hormone threshold, the rise time, the amplitude and the duration of spikes are nearly independent of the hormone concentration. This suggests that calcium spiking is due to the excitation of a bistable system whose nature is still hypothetical. Many cells undergo spiking in the absence of external calcium. This shows that the source of calcium is intracellular.

Calcium spikes are reminiscent of the action potential of neurons. The sharpness of the threshold and the all-or-none response of the system suggest that the molecular phenomena underlying calcium spiking are a positive feedback and co-operativity. Co-operativity generates a threshold and positive feedback amplifies a stimulus above the threshold. Two models have been proposed to explain the kind of results already mentioned. The first model postulates that inositol 1,4,5-triphosphate (IP_3), triggered by a receptor, leads to calcium release from the endoplasmic reticulum. The increase of calcium in turn stimulates the synthesis of IP_3 thus generating a positive feedback [14–16]. This model is called the ICC model (IP_3 calcium cross-coupling). The second model postulates that the increase of the calcium level induced by IP_3 opens calcium channels. The positive feedback originates from calcium itself which enhances its own release [17]. This model is called the CICR model (calcium-induced calcium release).

9.2. Minimum conditions required for the emergence of oscillations in a model metabolic cycle

9.2.1. The model

Let us consider the simplest open metabolic cycle. This cycle consists of two antagonistic enzyme reactions, two reagents, S_1 and S_2, an input and an output of matter. If v_1 and v_2 are the rates of the two enzyme reactions, and v_i and v_o the input and output of matter, the dynamic system that expresses the time evolution of the concentration of the reagents S_1 and S_2 can be written in matrix form as

$$\frac{\mathrm{d}}{\mathrm{d}t}\begin{bmatrix} S_1 \\ S_2 \end{bmatrix} = \begin{bmatrix} v_i - v_1 + v_2 \\ v_1 - v_2 - v_o \end{bmatrix}. \tag{9.1}$$

If an explicit form of enzyme rates v_1 and v_2 is to be taken into account, for instance, if both enzymes follow the reaction rates

$$v_1 = \frac{V_1 S_1}{K_1 + S_1}, \qquad v_2 = \frac{V_2 S_2}{K_2 + S_2}, \tag{9.2}$$

there is an obvious advantage in expressing these rates in dimensionless form so as to decrease the number of parameters. If the model metabolic cycle were in steady state, the two enzymes would follow Michaelis–Menten kinetics. Setting

$$\begin{aligned} &\alpha = \frac{S_1}{K_1}, \qquad \beta = \frac{S_2}{K_2}, \\ &\nu_1 = \frac{v_1}{V_1}, \qquad \nu_2 = \frac{v_2}{V_1}, \qquad \nu_i = \frac{v_i}{V_1}, \qquad \nu_o = \frac{v_o}{V_1}, \\ &\varepsilon = \frac{K_1}{K_2}, \qquad \theta = \frac{V_1}{K_1}t, \qquad \lambda = \frac{V_2}{V_1}, \end{aligned} \tag{9.3}$$

the dynamic system then assumes the form

$$\frac{\mathrm{d}}{\mathrm{d}\theta}\begin{bmatrix} \alpha \\ \beta \end{bmatrix} = \begin{bmatrix} 1 & 0 \\ 0 & \varepsilon \end{bmatrix}\begin{bmatrix} \nu_i - \nu_1 + \nu_2 \\ \nu_1 - \nu_2 - \nu_o \end{bmatrix}. \tag{9.4}$$

The elements of the transformation matrix that appear in this expression are thus defined by the normalization procedure. If the rates v_1 and v_2 are defined as expressed in eq. (9.2), the dynamic system (9.1) contains nine parameters whereas the corresponding normalized system (9.4) has only seven independent parameters.

9.2.2. Steady states of a model metabolic cycle

Let us now consider the two functions

$$\begin{aligned} u_1(\alpha, \beta) &= \nu_i - \nu_1(\alpha, \beta) + \nu_2(\alpha, \beta), \\ u_2(\alpha, \beta) &= \nu_1(\alpha, \beta) - \nu_2(\alpha, \beta) - \nu_o. \end{aligned} \tag{9.5}$$

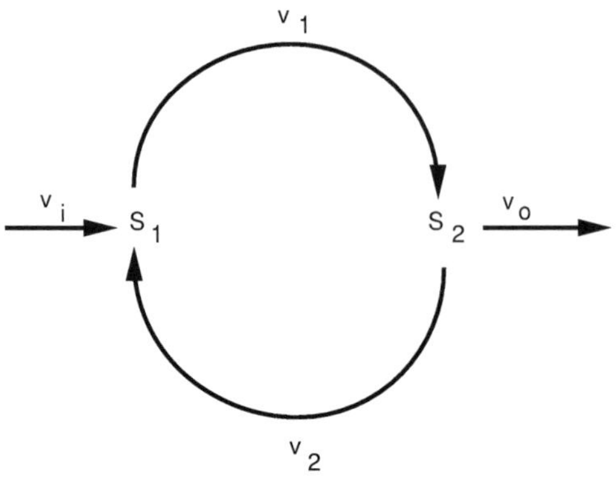

Fig. 9.1. A simple open metabolic cycle. Two enzymes catalyse antagonistic reactions.

If, in the scheme of Fig. 9.1, the output v_o is first order in reagent S_2, then

$$\nu_0 = \frac{kK_2}{V_1}\beta = \mu\beta, \tag{9.6}$$

where k is the first order rate constant for the output of S_2 and μ is the corresponding dimensionless constant. If the system is in steady state, then

$$\begin{aligned} &\nu_i - \nu_1(\alpha,\beta) + \nu_2(\alpha,\beta) = 0, \\ &\nu_1(\alpha,\beta) - \nu_2(\alpha,\beta) - \mu\beta = 0. \end{aligned} \tag{9.7}$$

From these equations, it necessarily follows that the steady state concentration of β, β^*, is

$$\beta^* = \nu_i/\mu. \tag{9.8}$$

The steady state concentration, α^*, is thus a solution of the equation

$$\nu_1(\alpha,\beta^*) - \nu_2(\alpha,\beta^*) - \mu\beta^* = 0. \tag{9.9}$$

As ν_i is assumed to be constant, eq. (9.8) shows there is one steady state value of β. But eq. (9.9) leaves open the possibility of several steady states for α. Multiple steady states, if they exist, are defined by paired values

$$(\alpha_1^*,\beta_1^*),\ (\alpha_2^*,\beta_2^*),\ \ldots. \tag{9.10}$$

In fact, the existence, or the impossibility, of multiple steady states depends on the expression of the individual reaction rates. Thus, if the two enzymes under steady state follow Michaelis–Menten kinetics, one has, in dimensionless form

$$\nu_1 = \frac{\alpha}{1+\alpha}, \qquad \nu_2 = \frac{\lambda\beta}{1+\beta}. \tag{9.11}$$

The parameter λ has already been defined (eq. (9.3)). Therefore, eq. (9.9) assumes the form

$$\frac{\alpha}{1+\alpha} - \frac{\lambda\beta^*}{1+\beta^*} - \mu\beta^* = 0. \tag{9.12}$$

The steady state expression of α, α^*, can thus be derived from this equation and one finds

$$\alpha^* = \frac{\mu\beta^* + \lambda\beta^*/(1+\beta^*)}{1 - \mu\beta^* - \lambda\beta^*/(1+\beta^*)}. \tag{9.13}$$

This expression shows that a steady state for α exists only if

$$1 > \mu\beta^* + \frac{\lambda\beta^*}{1+\beta^*}. \tag{9.14}$$

Let us now consider the two functions

$$\begin{aligned} Y_1(\alpha) &= v_1(\alpha) = \frac{\alpha}{1+\alpha}, \\ Y_2(\beta) &= \mu\beta^* + \frac{\lambda\beta^*}{1+\beta^*}. \end{aligned} \tag{9.15}$$

Comparison of eqs. (9.8) and (9.9) shows that, under steady state

$$\mu\beta^* = v_{\mathrm{i}}. \tag{9.16}$$

The expression of Y_1 above is thus the rate of consumption of α (or the rate of production of β) and the expression of Y_2 is the rate of production of α (or the rate of consumption of β). Therefore, a steady state for α will be obtained if

$$Y_1(\alpha) = Y_2(\beta), \tag{9.17}$$

which is indeed equivalent to eq. (9.12). If Y_1 is plotted as a function of α, one obtains a rectangular hyperbola. Plotting Y_2 *versus* α on the contrary yields a straight line parallel to the α axis (Fig. 9.2).

This figure makes it obvious that there may exist one steady state only, or no steady state if expression (9.14) is not fulfilled. Let us now assume that the system is perturbed in such a way that

$$Y_2(\beta) > Y_1(\alpha). \tag{9.18}$$

This means that the rate of production of α becomes larger than its rate of consumption. Therefore, α accumulates and $Y_1(\alpha)$ increases until the expression (9.17) is fulfilled. Alternatively, if

$$Y_1(\alpha) > Y_2(\beta), \tag{9.19}$$

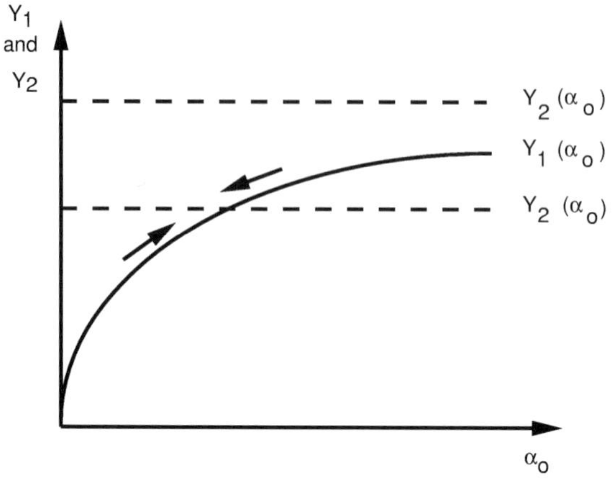

Fig. 9.2. Stable steady state of the open metabolic cycle. If the two enzymes follow Michaelis–Menten kinetics with respect to their own substrate and if they are unaffected by the other substrate, only one stable steady state is possible.

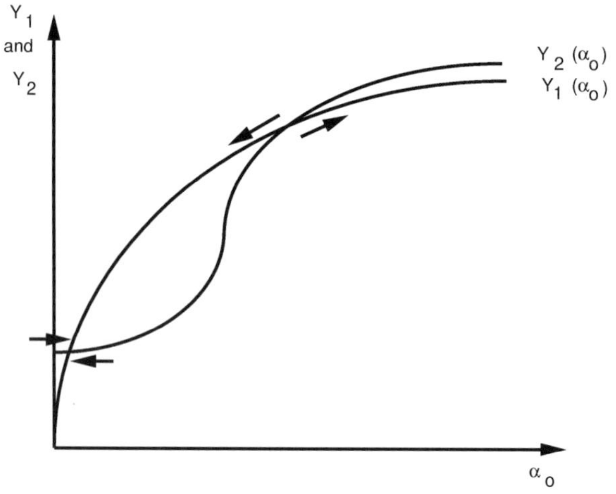

Fig. 9.3. Multiple steady states for the open metabolic cycle. If one reaction rate (v_2) depends on both substrates and if the corresponding kinetics is complex, the system may have several steady states.

the rate of consumption of α becomes larger than its rate of production, thus leading to a decrease of α. As α falls off the rate $Y_1(\alpha)$ decreases and the steady state, expressed by eq. (9.15), spontaneously readjusts. Thus the steady state, if it exists, should be stable. If slightly perturbed, the system should drift back to its initial steady state.

The above reasoning leaves open the possibility that multiple steady states may exist for the simple metabolic cycle of Fig. 9.1, but this should imply that one of the Y functions, for instance Y_2, be a function of both α and β. If the rate process $v_1(\alpha)$ is hyperbolic, the second one, $v_2(\alpha, \beta)$, should display, relative to α (but at constant β) the complex behaviour shown in Fig. 9.3. This situation would generate two steady states, the first stable and the second unstable. Owing to its complexity, this situation, although possible, is not to be often encountered in nature. It would imply that the enzyme E_2 is "allosterically" activated by, at least, two molecules of its reaction products. This would require an extremely complex enzyme, bearing, at least, two regulatory sites distinct from the catalytic site.

The conclusion of this analysis is that multiple steady states of a simple metabolic cycle, taking place in a homogeneous phase, can occur. But this would require at least one enzyme of the cycle to have a complex structure and to follow a rather unusual kinetics. It will be shown later that these multiple steady states are easily obtained if the cell milieu is not homogeneous and if there is an electric repulsion of the reagents by the fixed negative charges of a membrane.

9.2.3. Stability analysis of the model metabolic cycle

Stability analysis of the cycle can be performed with the phase-plane technique, as already discussed (chapter 7). The time evolution of normalized concentrations α and β can be expressed as

$$\frac{\mathrm{d}\alpha}{\mathrm{d}t} = \frac{\mathrm{d}\alpha^*}{\mathrm{d}t} + \frac{\mathrm{d}x_\alpha}{\mathrm{d}t}, \qquad \frac{\mathrm{d}\beta}{\mathrm{d}t} = \frac{\mathrm{d}\beta^*}{\mathrm{d}t} + \frac{\mathrm{d}x_\beta}{\mathrm{d}t}, \tag{9.20}$$

where x_α and x_β are deviations relative to the steady state concentrations α^* and β^*. Thus one has

$$\frac{\mathrm{d}\alpha}{\mathrm{d}t} = \frac{\mathrm{d}x_\alpha}{\mathrm{d}t}, \qquad \frac{\mathrm{d}\beta}{\mathrm{d}t} = \frac{\mathrm{d}x_\beta}{\mathrm{d}t}. \tag{9.21}$$

Close to a steady state, when x_α and x_β are small, the time evolution of these deviations may be studied through the linear variational system

$$\frac{\mathrm{d}}{\mathrm{d}t}\begin{bmatrix} x_\alpha \\ x_\beta \end{bmatrix} = \begin{bmatrix} \partial u_1^*/\partial\alpha & \partial u_1^*/\partial\beta \\ \partial u_2^*/\partial\alpha & \partial u_2^*/\partial\beta \end{bmatrix}\begin{bmatrix} x_\alpha \\ x_\beta \end{bmatrix}. \tag{9.22}$$

As already outlined (see chapter 7), the dynamic behaviour of the system will depend on the values of the trace T_j and of the determinant of the Jacobian matrix

$$J = \begin{bmatrix} \partial u_1^*/\partial\alpha & \partial u_1^*/\partial\beta \\ \partial u_2^*/\partial\alpha & \partial u_2^*/\partial\beta \end{bmatrix}. \tag{9.23}$$

In these equations, u_1^* and u_2^* are defined, as previously, by expressions (9.5). The trace and the determinant of this Jacobian matrix are thus

$$\begin{aligned} T_\mathrm{j} &= \frac{\partial u_1^*}{\partial\alpha} + \frac{\partial u_2^*}{\partial\beta}, \\ \Delta_\mathrm{j} &= \frac{\partial u_1^*}{\partial\alpha}\frac{\partial u_2^*}{\partial\beta} - \frac{\partial u_2^*}{\partial\alpha}\frac{\partial u_1^*}{\partial\beta}. \end{aligned} \tag{9.24}$$

If Y_1 is a function of α only, and Y_2 a function of β only, then

$$T_{\rm j} = -\left(\frac{\partial \nu_1}{\partial \alpha} + \frac{\partial \nu_2}{\partial \beta} + \mu\right), \qquad \Delta_{\rm j} = \mu \frac{\partial \nu_1}{\partial \alpha}. \tag{9.25}$$

Let us first consider the case of two enzymes that follow Michaelis–Menten kinetics, then

$$\nu_1 = \frac{\alpha}{1+\alpha}, \qquad \nu_2 = \frac{\lambda \beta}{1+\beta}, \tag{9.26}$$

and the corresponding derivatives are

$$\frac{\partial \nu_1}{\partial \alpha} = \frac{1}{(1+\alpha)^2}, \qquad \frac{\partial \nu_2}{\partial \beta} = \frac{\lambda}{(1+\beta)^2}. \tag{9.27}$$

Going back to eq. (9.25), these derivatives make it obvious that $T_{\rm j} < 0$ and $\Delta_{\rm j} > 0$. Moreover, $T_{\rm j}^2 > 4\Delta_{\rm j}$, so the model cycle displays a stable node. If perturbed from its steady state, it drifts monotonically back to the same steady state.

If now a nonlinear term in substrate concentration is introduced into the reaction rate equation ν_2 so that

$$\nu_2 = \frac{\lambda \beta}{1+\beta+\varsigma\beta^2}, \tag{9.28}$$

the enzyme reaction process is then said to be inhibited by excess substrate. The relevant derivative, $\partial \nu_2/\partial \beta$, assumes the form

$$\frac{\partial \nu_2}{\partial \beta} = \frac{\lambda(1-\varsigma\beta^2)}{(1+\beta+\varsigma\beta^2)^2}, \tag{9.29}$$

and may adopt positive or negative values. If, for instance

$$\frac{\lambda(1-\varsigma\beta^2)}{(1+\beta+\varsigma\beta^2)^2} = -\left\{\frac{1}{(1+\alpha)^2} + \mu\right\}, \tag{9.30}$$

then $T_{\rm j} = 0$ and $\Delta_{\rm j} > 0$. The system can thus display sustained oscillations and is thus a temporal "dissipative structure" [18].

The general conclusions that can be drawn from this analysis are straightforward. The first conclusion is that these effects are expected to occur if the system is under nonequilibrium conditions. Spontaneous temporal organization implies a dissipation of matter and energy. The second conclusion is that oscillatory dynamics requires, at least, the existence of a nonlinear term in the equations. In the case of a homogeneous reaction phase, which was precisely the case considered above, the nonlinear term is introduced into the system by one of the enzymes. These oscillations require that one of the enzymes display rather complex dynamics (inhibition by excess substrate for instance). This complex dynamics reflects a complex structure of the enzyme, probably the existence of, at least, two sites in interaction. "Simple" enzymes that follow Michaelis–Menten kinetics cannot be expected to generate this type of behaviour. The last conclusion is that the properties of the overall system are qualitatively novel with respect to those of any of the individual enzymes. Thus, although the overall system may display oscillatory dynamics, none of the enzymes displays this type of behaviour and none can be considered as oscillatory.

9.3. *Emergence of a temporal organization generated by compartmentalization and electric repulsion effects*

In the previous section, we have treated dynamic processes occurring in a homogeneous bulk phase. We shall now study the dynamics of a similar model metabolic cycle taking place in a heterogeneous phase, at the interface between a bulk and a polyanionic phase, a membrane or a cell wall for instance.

9.3.1. The model

The model we are dealing with is shown in Fig. 9.4. It corresponds to the simplest open metabolic cycle in heterogeneous phase. Two antagonistic enzyme reactions convert S_1 into S_2 and back. The first reaction takes place in the bulk phase and the second in the polyanionic phase. For simplicity, the two enzymes are assumed to follow Michaelis–Menten kinetics. Moreover, one assumes either that both S_1 and S_2 are monoanions, or

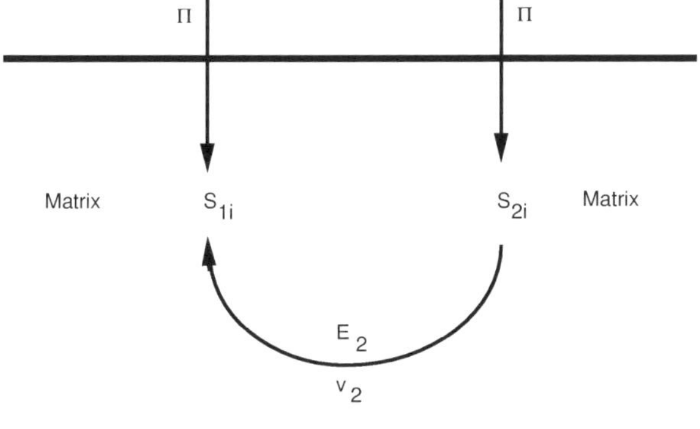

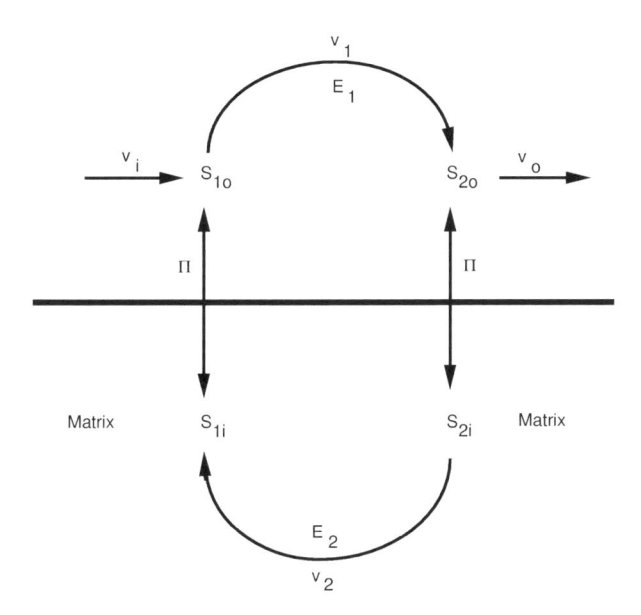

Fig. 9.4. Open metabolic cycle taking place at the surface of a negatively charged matrix. See text. Adapted from ref. [22].

that one of these reagents is a monoanion and the other a neutral molecule. The anions are subject to electric partitioning between the bulk and the polyanionic phases. The local density of fixed negative charges in the matrix is Δ. As previously, the enzyme reaction rates are termed v_1 and v_2. The input and output of matter are called v_i and v_o. Both are assumed to take place in the bulk phase. The process of electrostatic partitioning between the inside and outside of the matrix is assumed to be fast relative to all the other processes. Electrostatic partitioning may therefore be considered as taking place under quasi-equilibrium conditions whereas the whole system is not. For simplicity, the two enzyme reactions are assumed to be nearly irreversible. This will always be the case if the two antagonistic enzyme reactions require the participation of a reagent X_1, for the first reaction $S_1 + X_1 \rightarrow S_2 + X'_1$, and of a second reagent X_2, for the second reaction $S_2 + X_2 \rightarrow S_1 + X'_2$. If X_1 and X_2 are saturating, the two enzyme reactions will be irreversible and the reagents X_1 and X_2 will cancel out from the rate equations. The main assumption of the model is that of a fast equilibrium between the bulk and the local concentrations of the reagents. This allows one to express a problem of heterogeneous enzyme dynamics in terms of equations that are formally similar (but not physically equivalent) to equations that describe a dynamic process occurring in a bulk phase. Last but not least, the enzyme molecules and the fixed negative charges are assumed to be randomly distributed in the matrix.

For a dilute bulk phase, one can define a partition coefficient, Π, from the ratios of the concentrations of anions, or cations, in the bulk and matrix phases. One thus has

$$\Pi = \frac{S_{1o}}{S_{1i}} = \frac{S_{2o}}{S_{2i}} = \exp\frac{F\Delta\Psi}{RT},$$
$$\Pi = \frac{H_i}{H_o} = \frac{A_i}{A_o} = \exp\frac{F\Delta\Psi}{RT}, \tag{9.31}$$

where the subscipts 'i' and 'o' refer to the inside and the outside of the matrix. A is the concentration of a cation, $\Delta\Psi$ is the electrostatic potential difference, F, R and T have their usual significance (see chapter 7). From expressions (9.31) and the electroneutrality conditions, one can rewrite this electrostatic partition coefficient as

$$\Pi = \frac{\sqrt{\Delta^2 + 4(S_{1o} + S_{2o})^2} + \Delta}{2(S_{1o} + S_{2o})} = \frac{2(S_{1o} + S_{2o})}{\sqrt{\Delta^2 + 4(S_{1o} + S_{2o})^2} - \Delta}. \tag{9.32}$$

Combining eqs. (9.31) and (9.32) allows one to express the local (in the polyanion) concentrations S_1, S_2 and H as a function of Δ. One has

$$S_{1i} = \frac{S_{1o}}{2(S_{1o} + S_{2o})}\left\{\sqrt{\Delta^2 + 4(S_{1o} + S_{2o})^2} - \Delta\right\},$$
$$S_{2i} = \frac{S_{2o}}{2(S_{1o} + S_{2o})}\left\{\sqrt{\Delta^2 + 4(S_{1o} + S_{2o})^2} - \Delta\right\},$$
$$H_i = \frac{H_o}{2(S_{1o} + S_{2o})}\left\{\sqrt{\Delta^2 + 4(S_{1o} + S_{2o})^2} + \Delta\right\}. \tag{9.33}$$

9.3.2. The dynamic equations of the system and the sensitivity coefficients

If J_{io} and J'_{io} are the flows of S_1 and S_2 from the inside to the outside of the matrix, J_{oi} and J'_{oi} the flows from the outside to the inside of the same matrix, one has

$$\frac{dS_{1o}}{dt} = v_i - v_1 + J_{io} - J_{oi},$$
$$\frac{dS_{1i}}{dt} = v_2 + J_{oi} - J_{io}. \tag{9.34}$$

Similarly, one should also have

$$\frac{dS_{2o}}{dt} = v_1 + J'_{io} - J'_{oi} - v_o,$$
$$\frac{dS_{2i}}{dt} = J'_{oi} - J'_{io} - v_2. \tag{9.35}$$

Therefore one has

$$\frac{dS_{1o}}{dt} + \frac{dS_{1i}}{dt} = v_i - v_1 + v_2,$$
$$\frac{dS_{2o}}{dt} + \frac{dS_{2i}}{dt} = v_1 - v_2 - v_o. \tag{9.36}$$

Moreover, one can write

$$\frac{dS_{1i}}{dt} = \left(\frac{\partial S_{1i}}{\partial S_{1o}}\right)\frac{dS_{1o}}{dt} + \left(\frac{\partial S_{1i}}{\partial S_{2o}}\right)\frac{dS_{2o}}{dt},$$
$$\frac{dS_{2i}}{dt} = \left(\frac{\partial S_{2i}}{\partial S_{1o}}\right)\frac{dS_{1o}}{dt} + \left(\frac{\partial S_{2i}}{\partial S_{2o}}\right)\frac{dS_{2o}}{dt}, \tag{9.37}$$

and inserting these expressions into eq. (9.36) yields

$$\left\{1 + \left(\frac{\partial S_{1i}}{\partial S_{1o}}\right)\right\}\frac{dS_{1o}}{dt} + \left(\frac{\partial S_{1i}}{\partial S_{2o}}\right)\frac{dS_{2o}}{dt} = v_i - v_1 + v_2,$$
$$\left(\frac{\partial S_{2i}}{\partial S_{1o}}\right)\frac{dS_{1o}}{dt} + \left\{1 + \left(\frac{\partial S_{2i}}{\partial S_{2o}}\right)\right\}\frac{dS_{2o}}{dt} = v_1 - v_2 - v_o. \tag{9.38}$$

These equations have to be compatible with the mass balance equation

$$\frac{dn_{1o}}{dt} + \frac{dn_{2o}}{dt} = 0, \tag{9.39}$$

where n_{1o} and n_{2o} are the mole numbers of reagents S_1 and S_2 outside the matrix. Since v_i and v_o are constant and must, of necesity, be equal, one obtains summing up the two equations of (9.38)

$$\left\{1+\left(\frac{\partial S_{1i}}{\partial S_{1o}}\right)+\left(\frac{\partial S_{2i}}{\partial S_{1o}}\right)\right\}\frac{dS_{1o}}{dt} + \left\{1+\left(\frac{\partial S_{1i}}{\partial S_{2o}}\right)+\left(\frac{\partial S_{2i}}{\partial S_{2o}}\right)\right\}\frac{dS_{2o}}{dt} = v_i - v_o = 0, \tag{9.40}$$

and this expression should be equivalent to the mass balance equation (9.39). This condition requires that

$$\left(\frac{\partial S_{1i}}{\partial S_{1o}}\right)+\left(\frac{\partial S_{2i}}{\partial S_{1o}}\right)=\left(\frac{\partial S_{1i}}{\partial S_{2o}}\right)+\left(\frac{\partial S_{2i}}{\partial S_{2o}}\right). \tag{9.41}$$

It is therefore essential to demonstrate the validity of this relationship. It is worth noting that the partial derivatives that appear in these equations are equivalent to sensitivity coefficients (chapter 3) for they express how the local concentrations of the reagents, S_1 and S_2, vary on changing the bulk concentrations. One can thus derive the expressions of these four sensitivity coefficients yielding

$$\begin{aligned}
\sigma_{11} = \frac{\partial S_{1i}}{\partial S_{1o}} &= \frac{\sqrt{\Delta^2+4(S_{1o}+S_{2o})^2}-\Delta}{2(S_{1o}+S_{2o})} \times \left\{1+S_{1o}\frac{\Delta}{S_{1o}+S_{2o}}\frac{1}{\sqrt{\Delta^2+4(S_{1o}+S_{2o})^2}}\right\},\\
\sigma_{12} = \frac{\partial S_{1i}}{\partial S_{2o}} &= \frac{\sqrt{\Delta^2+4(S_{1o}+S_{2o})^2}-\Delta}{2(S_{1o}+S_{2o})} \times S_{1o}\frac{\Delta}{S_{1o}+S_{2o}}\frac{1}{\sqrt{\Delta^2+4(S_{1o}+S_{2o})^2}},\\
\sigma_{21} = \frac{\partial S_{2i}}{\partial S_{1o}} &= \frac{\sqrt{\Delta^2+4(S_{1o}+S_{2o})^2}-\Delta}{2(S_{1o}+S_{2o})} \times S_{2o}\frac{\Delta}{S_{1o}+S_{2o}}\frac{1}{\sqrt{\Delta^2+4(S_{1o}+S_{2o})^2}},\\
\sigma_{22} = \frac{\partial S_{2i}}{\partial S_{2o}} &= \frac{\sqrt{\Delta^2+4(S_{1o}+S_{2o})^2}-\Delta}{2(S_{1o}+S_{2o})} \times \left\{1+S_{2o}\frac{\Delta}{S_{1o}+S_{2o}}\frac{1}{\sqrt{\Delta^2+4(S_{1o}+S_{2o})^2}}\right\}.
\end{aligned} \tag{9.42}$$

Simple inspection of these equations shows that the relationship (9.41) is of necessity valid. Equation (9.40) thus reduces to

$$\frac{dS_{1o}}{dt}+\frac{dS_{2o}}{dt}=0, \tag{9.43}$$

which is equivalent to the mass balance equation (9.39). It should be noted that the sensitivity coefficients are expressed relative to bulk concentrations of the reagents and to the fixed charge density. This is also true for the partition coefficient. One can therefore wonder whether there exists some relationships between the sensitivity and partition coefficients. Algebraic manipulation of eq. (9.32) yields

$$\Pi = \frac{\sqrt{\Delta^2 + 4(S_{1o} + S_{2o})^2} - \Delta}{2(S_{1o} + S_{2o})} + \frac{\Delta}{S_{1o} + S_{2o}} = \Pi^{-1} + \frac{\Delta}{S_{1o} + S_{2o}}, \tag{9.44}$$

which may be reexpressed as

$$\frac{\Delta}{S_{1o} + S_{2o}} = \frac{\Pi^2 - 1}{\Pi}. \tag{9.45}$$

Inserting this expression into eq. (9.32) leads to

$$\frac{\sqrt{\Delta^2 + 4(S_{1o} + S_{2o})^2}}{2(S_{1o} + S_{2o})} = \frac{\Pi^2 + 1}{2\Pi}, \tag{9.46}$$

and one thus finds

$$\frac{1}{\sqrt{\Delta^2 + 4(S_{1o} + S_{2o})^2}} = \frac{1}{S_{1o} + S_{2o}} \frac{\Pi}{\Pi^2 + 1}. \tag{9.47}$$

Inserting these expressions into eq. (9.42) allows one to reexpress them in a more compact and elegant form, namely

$$\begin{aligned} \sigma_{11} &= \Pi^{-1}\left(1 + \frac{S_{1o}}{S_{1o} + S_{2o}} \frac{\Pi^2 - 1}{\Pi^2 + 1}\right), & \sigma_{22} &= \Pi^{-1}\left(1 + \frac{S_{2o}}{S_{1o} + S_{2o}} \frac{\Pi^2 - 1}{\Pi^2 + 1}\right), \\ \sigma_{12} &= \Pi^{-1} \frac{S_{1o}}{S_{1o} + S_{2o}} \frac{\Pi^2 - 1}{\Pi^2 + 1}, & \sigma_{21} &= \Pi^{-1} \frac{S_{2o}}{S_{1o} + S_{2o}} \frac{\Pi^2 - 1}{\Pi^2 + 1}. \end{aligned} \tag{9.48}$$

As mentioned above, these sensitivity coefficients are not independent and one has the following summation properties

$$\begin{aligned} &\sigma_{11} + \sigma_{21} = \sigma_{22} + \sigma_{12} = \frac{2\Pi}{\Pi^2 + 1}, \\ &\sigma_{11} + \sigma_{22} - (\sigma_{12} + \sigma_{21}) = 2\Pi^{-1}. \end{aligned} \tag{9.49}$$

As before, there is an advantage in expressing concentrations and rates in dimensionless form. One may thus set

$$\begin{aligned} &\alpha_o = \frac{S_{1o}}{K_1}, \quad \beta_o = \frac{S_{2o}}{K_2}, \quad \delta = \frac{\Delta}{K_2}, \quad \varepsilon = \frac{K_1}{K_2}, \quad \lambda = \frac{V_2}{V_1}, \\ &\nu_i = \frac{v_i}{V_1}, \quad \nu_1 = \frac{v_1}{V_1}, \quad \nu_2 = \frac{v_2}{V_1}, \quad \nu_o = \frac{v_o}{V_1}, \quad \theta = \frac{V_1}{K_1} t, \end{aligned} \tag{9.50}$$

where K_1 and K_2 are the K_m of the two enzymes and V_1 and V_2 are the two maximum rates. The partition coefficient then assumes the form

$$\Pi = \frac{\sqrt{\delta^2 + 4(\varepsilon\alpha_o + \beta_o)^2} + \delta}{2(\varepsilon\alpha_o + \beta_o)} = \frac{2(\varepsilon\alpha_o + \beta_o)}{\sqrt{\delta^2 + 4(\varepsilon\alpha_o + \beta_o)^2} - \delta}, \tag{9.51}$$

and the sensitivity coefficients can be rewritten as

$$\sigma_{11} = \Pi^{-1}\left(1 + \frac{\varepsilon\alpha_o}{\varepsilon\alpha_o + \beta_o}\frac{\Pi^2 - 1}{\Pi^2 + 1}\right), \qquad \sigma_{22} = \Pi^{-1}\left(1 + \frac{\beta_o}{\varepsilon\alpha_o + \beta_o}\frac{\Pi^2 - 1}{\Pi^2 + 1}\right),$$

$$\sigma_{12} = \Pi^{-1}\frac{\varepsilon\alpha_o}{\varepsilon\alpha_o + \beta_o}\frac{\Pi^2 - 1}{\Pi^2 + 1}, \qquad \sigma_{21} = \Pi^{-1}\frac{\beta_o}{\varepsilon\alpha_o + \beta_o}\frac{\Pi^2 - 1}{\Pi^2 + 1}. \tag{9.52}$$

The dynamic system (9.38) can thus be expressed in matrix form as

$$\frac{d}{d\theta}\begin{bmatrix}\alpha_o \\ \beta_o\end{bmatrix} = \Omega^{-1}\begin{bmatrix}1 + \sigma_{22} & -\sigma_{12} \\ -\varepsilon\sigma_{21} & (1 + \sigma_{11})\varepsilon\end{bmatrix}\begin{bmatrix}v_i - v_1 + v_2 \\ v_1 - v_2 - v_o\end{bmatrix}, \tag{9.53}$$

where Ω is equal to

$$\Omega = (1 + \sigma_{11})(1 + \sigma_{22}) - \sigma_{12}\sigma_{21} = \frac{(\Pi + 1)^3}{\Pi(\Pi^2 + 1)}. \tag{9.54}$$

It is now of interest to look at the behaviour of the system when the normalized charge density δ is zero. Then $\Pi = 1$, $\sigma_{11} = \sigma_{22} = 1$, $\sigma_{12} = \sigma_{21} = 0$ and the differential system (9.53) becomes

$$\frac{d}{d\theta}\begin{bmatrix}\alpha_o \\ \beta_o\end{bmatrix} = \begin{bmatrix}1/2 & 0 \\ 0 & \varepsilon(1/2)\end{bmatrix}\begin{bmatrix}v_i - v_1 + v_2 \\ v_1 - v_2 - v_o\end{bmatrix}. \tag{9.55}$$

This system is different from expressions (9.4) because the reagents, in the present case, are equally distributed in two phases. This is expressed in eq. (9.55) by the coefficient 1/2 that appears in the transformation matrix. Since the medium was assumed to be homogeneous in the case of the system (9.4), this coefficient was lacking.

9.3.3. Local stability of the system

Under steady state conditions one has

$$u_1^* = v_i - v_1^* + v_2^* = 0, \qquad u_2^* = v_1^* - v_2^* - v_o = 0, \tag{9.56}$$

with

$$v_o = \mu\beta_o^*, \tag{9.57}$$

which implies that

$$\beta_o^* = \frac{\nu_i}{\mu}. \tag{9.58}$$

Thus, there can only be one steady state for β_o. μ has the meaning it had in section 9.2.2. Local stability analysis is carried out as already discussed in section 9.2.2. One can define functions $F_1(\alpha_o, \beta_o)$ and $F_2(\alpha_o, \beta_o)$ as

$$\begin{bmatrix} F_1(\alpha_o, \beta_o) \\ F_2(\alpha_o, \beta_o) \end{bmatrix} = \frac{d}{d\theta} \begin{bmatrix} \alpha_o \\ \beta_o \end{bmatrix}. \tag{9.59}$$

This is equivalent to

$$\begin{aligned} F_1 &= \Omega^{-1}(1+\sigma_{22})u_1 - \Omega^{-1}\sigma_{12}u_2, \\ F_2 &= -\varepsilon\Omega^{-1}\sigma_{21}u_1 + \varepsilon\Omega^{-1}(1+\sigma_{11})u_2. \end{aligned} \tag{9.60}$$

The local stability of the differential system (9.53) depends on the trace and the determinant of the matrix

$$J = \begin{bmatrix} \partial F_1^*/\partial\alpha_o & \partial F_1^*/\partial\beta_o \\ \partial F_2^*/\partial\alpha_o & \partial F_2^*/\partial\beta_o \end{bmatrix}. \tag{9.61}$$

The elements of this matrix can be obtained by differentiating the steady state values of F_1 and F_2 with respect to α_o and β_o. Thus, for instance,

$$\begin{aligned} \frac{\partial F_1^*}{\partial\alpha_o} &= \frac{\partial\Omega^{*-1}}{\partial\alpha_o}(1+\sigma_{22}^*)u_1^* + \Omega^{*-1}\frac{\partial\sigma_{22}^*}{\partial\alpha_o}u_1^* + \Omega^{*-1}(1+\sigma_{22}^*)\frac{\partial u_1^*}{\partial\alpha_o} \\ &\quad - \frac{\partial\Omega^{*-1}}{\partial\alpha_o}\sigma_{12}^*u_2^* - \Omega^{*-1}\frac{\partial\sigma_{12}^*}{\partial\alpha_o}u_2^* - \Omega^{*-1}\sigma_{12}^*\frac{\partial u_2^*}{\partial\alpha_o}. \end{aligned} \tag{9.62}$$

Under steady state conditions, $u_1^* = u_2^* = 0$, so this expresson reduces to

$$\frac{\partial F_1^*}{\partial\alpha_o} = \Omega^{*-1}(1+\sigma_{22}^*)\frac{\partial u_1^*}{\partial\alpha_o} - \Omega^{*-1}\sigma_{12}^*\frac{\partial u_2^*}{\partial\alpha_o}. \tag{9.63}$$

In these expressions and in the following, the starred symbols mean that the system is in steady state. The same reasoning as above could be applied to the other elements of the Jacobian matrix (9.61). This means that this matrix assumes the form

$$J = \Omega^{*-1}\begin{bmatrix} 1+\sigma_{22}^* & -\sigma_{12}^* \\ -\varepsilon\sigma_{21}^* & \varepsilon(1+\sigma_{11}^*) \end{bmatrix}\begin{bmatrix} \partial u_1^*/\partial\alpha_o & \partial u_1^*/\partial\beta_o \\ \partial u_2^*/\partial\alpha_o & \partial u_2^*/\partial\beta_o \end{bmatrix}. \tag{9.64}$$

The elements of the J matrix are obtained from this relationship. One finds

$$\frac{\partial F_1^*}{\partial\alpha_o} = \Omega^{*-1}\left\{(1+\sigma_{22}^*)\frac{\partial u_1^*}{\partial\alpha_o} - \sigma_{12}^*\frac{\partial u_2^*}{\partial\alpha_o}\right\},$$

$$\frac{\partial F_1^*}{\partial \beta_o} = \Omega^{*-1}\left\{(1+\sigma_{22}^*)\frac{\partial u_1^*}{\partial \beta_o} - \sigma_{12}^*\frac{\partial u_2^*}{\partial \beta_o}\right\},$$

$$\frac{\partial F_2^*}{\partial \alpha_o} = -\varepsilon\Omega^{*-1}\left\{\sigma_{21}^*\frac{\partial u_1^*}{\partial \alpha_o} - (1+\sigma_{11}^*)\frac{\partial u_2^*}{\partial \alpha_o}\right\},$$

$$\frac{\partial F_2^*}{\partial \beta_o} = -\varepsilon\Omega^{*-1}\left\{\sigma_{21}^*\frac{\partial u_1^*}{\partial \beta_o} - (1+\sigma_{11}^*)\frac{\partial u_2^*}{\partial \beta_o}\right\}. \tag{9.65}$$

As the partial derivatives of the functions u_1^* and u_2^* are obtained by differentiation of eq. (9.56) with respect to α_o or to β_o, one has

$$\frac{\partial u_1^*}{\partial \alpha_o} = \frac{\partial v_2^*}{\partial \alpha_o} - \frac{\partial v_1^*}{\partial \alpha_o}, \qquad \frac{\partial u_1^*}{\partial \beta_o} = \frac{\partial v_2^*}{\partial \beta_o} - \frac{\partial v_1^*}{\partial \beta_o},$$

$$\frac{\partial u_2^*}{\partial \alpha_o} = \frac{\partial v_1^*}{\partial \alpha_o} - \frac{\partial v_2^*}{\partial \alpha_o}, \qquad \frac{\partial u_2^*}{\partial \beta_o} = \frac{\partial v_1^*}{\partial \beta_o} - \frac{\partial v_2^*}{\partial \beta_o} - \mu. \tag{9.66}$$

It can be seen that

$$\frac{\partial u_2^*}{\partial \alpha_o} = -\frac{\partial u_1^*}{\partial \alpha_o}, \qquad \frac{\partial u_2^*}{\partial \beta_o} = -\frac{\partial u_1^*}{\partial \beta_o} - \mu. \tag{9.67}$$

Therefore, a single function u^*, for instance u_1^*, is sufficient to describe the dynamics of the system in the vicinity of a steady state. These relationships allow one to reexpress eq. (9.65) in a more compact form, namely,

$$\frac{\partial F_1^*}{\partial \alpha_o} = \Omega^{*-1}\{1+\sigma_{12}^*+\sigma_{22}^*\}\frac{\partial u_1^*}{\partial \alpha_o},$$

$$\frac{\partial F_2^*}{\partial \alpha_o} = -\varepsilon\Omega^{*-1}\{1+\sigma_{11}^*+\sigma_{21}^*\}\frac{\partial u_1^*}{\partial \alpha_o},$$

$$\frac{\partial F_1^*}{\partial \beta_o} = \Omega^{*-1}\left\{(1+\sigma_{22}^*+\sigma_{12}^*)\frac{\partial u_1^*}{\partial \beta_o} + \mu\sigma_{12}^*\right\},$$

$$\frac{\partial F_2^*}{\partial \beta_o} = -\varepsilon\Omega^{*-1}\left\{(1+\sigma_{11}^*+\sigma_{21}^*)\frac{\partial u_1^*}{\partial \beta_o} + (1+\sigma_{11}^*)\mu\right\}. \tag{9.68}$$

Writing for simplicity

$$\frac{\varepsilon\alpha_o^*}{\varepsilon\alpha_o^*+\beta_o^*} = \rho^*, \tag{9.69}$$

one can express Ω^{*-1} and the sensitivity coefficients as a function of the steady state partition coefficient Π^* and of the parameter ρ^*. One thus finds

$$\Omega^{*-1}(1+\sigma_{11}^*+\sigma_{21}^*) = \Omega^{*-1}(1+\sigma_{22}^*+\sigma_{12}^*) = \frac{\Pi^*}{1+\Pi^*},$$

$$\mu\Omega^{*-1}\sigma_{12}^{*}=\rho^{*}\frac{\Pi^{*}-1}{(\Pi^{*}+1)^{2}},$$
$$\mu\Omega^{*-1}(1+\sigma_{11}^{*})=\frac{\Pi^{*2}+1}{(\Pi^{*}+1)^{2}}\mu+\rho^{*}\frac{\Pi^{*}-1}{(\Pi^{*}+1)^{2}}\mu, \tag{9.70}$$

and expression (9.68) can be written as

$$\frac{\partial F_1^*}{\partial\alpha_o}=\frac{\Pi^*}{\Pi^*+1}\frac{\partial u_1^*}{\partial\alpha_o},$$
$$\frac{\partial F_2^*}{\partial\alpha_o}=-\varepsilon\frac{\Pi^*}{\Pi^*+1}\frac{\partial u_1^*}{\partial\alpha_o},$$
$$\frac{\partial F_1^*}{\partial\beta_o}=\frac{\Pi^*}{\Pi^*+1}\frac{\partial u_1^*}{\partial\beta_o}+\rho^*\frac{\Pi^*-1}{(\Pi^*+1)^2}\mu,$$
$$\frac{\partial F_2^*}{\partial\beta_o}=-\varepsilon\frac{\Pi^*}{\Pi^*+1}\frac{\partial u_1^*}{\partial\beta_o}-\varepsilon\frac{\Pi^{*2}+1}{(\Pi^*+1)^2}\mu-\varepsilon\rho^*\frac{\Pi^*-1}{(\Pi^*+1)^2}\mu. \tag{9.71}$$

There is an advantage in expressing the trace and the determinant of the Jacobian matrix in terms of the derivatives of ν_1^* and ν_2^* with respect to α_o and β_o. One thus has

$$T_j=\frac{\Pi^*}{\Pi^*+1}\left\{\left(\frac{\partial\nu_2^*}{\partial\alpha_o}-\frac{\partial\nu_1^*}{\partial\alpha_o}\right)+\varepsilon\left(\frac{\partial\nu_1^*}{\partial\beta_o}-\frac{\partial\nu_2^*}{\partial\beta_o}\right)\right\}$$
$$-\varepsilon\frac{\Pi^{*2}+1}{(\Pi^*+1)^2}\mu-\varepsilon\rho^*\frac{\Pi^*-1}{(\Pi^*+1)^2}\mu \tag{9.72}$$

and

$$\Delta_j=\varepsilon\frac{\Pi^*(\Pi^{*2}+1)}{(\Pi^*+1)^3}\left(\frac{\partial\nu_1^*}{\partial\alpha_o}-\frac{\partial\nu_2^*}{\partial\alpha_o}\right)\mu. \tag{9.73}$$

These expressions make it clear what conditions can qualitatively alter the dynamic behaviour of the system in the vicinity of a steady state. If there is no repulsive effect exerted on the reagents, $\Pi^*=1$, and if ν_1 is a function of α_o only, ν_2 a function of β_o only, and if the two enzymes follow Michaelis–Menten kinetics, then $\partial\nu_2^*/\partial\alpha_o=\partial\nu_1^*/\partial\beta_o=0$. Under these conditions

$$T_j=-\frac{1}{2}\left(\frac{\partial\nu_1^*}{\partial\alpha_o}+\varepsilon\frac{\partial\nu_2^*}{\partial\beta_o}\right)-\frac{1}{2}\varepsilon\mu,$$
$$\Delta_j=\frac{1}{4}\varepsilon\mu\frac{\partial\nu_1^*}{\partial\alpha_o}. \tag{9.74}$$

As the two enzymes are assumed to follow Michaelis–Menten kinetics, $\partial\nu_1^*/\partial\alpha_o>0$ and $\partial\nu_2^*/\partial\beta_o>0$. Under these conditions, $T_j<0$, $\Delta_j>0$ and $T_j^2-4\Delta_j>0$. The system can only display a stable node. A condition required to generate dynamic behaviour qualitatively different from the present one, is offered by the existence of the electrostatic repulsion exerted on substrate β, in such a way that the rate ν_2 is now a function of both α_o and β_o, namely $\nu_2(\alpha_o, \beta_o)$. Then the signs of T_j and Δ_j can be changed, thus leading to qualitatvely different dynamics in the vicinity of a steady state. To illustrate this point, let us consider the two reaction rates

$$\nu_1 = \frac{\alpha_o}{1+\alpha_o}, \qquad \nu_2 = \frac{\lambda\beta_o}{\Pi+\beta_o}. \tag{9.75}$$

Differentiating the first function with respect to α_o and the second with respect to α_o and β_o leads to

$$\begin{aligned}
\frac{\partial \nu_1^*}{\partial \alpha_o} &= \frac{1}{(1+\alpha_o)^2},\\
\frac{\partial \nu_2^*}{\partial \alpha_o} &= -\lambda\frac{\beta_o(\partial\Pi^*/\partial\alpha_o)}{(\Pi^*+\beta_o)^2},\\
\frac{\partial \nu_2^*}{\partial \beta_o} &= \lambda\frac{\Pi^*-\beta_o(\partial\Pi^*/\partial\beta_o)}{(\Pi^*+\beta_o)^2}.
\end{aligned} \tag{9.76}$$

Moreover, since $\partial\nu_1^*/\partial\beta_o = 0$, eqs. (9.72) and (9.73) take the form

$$\begin{aligned}
T_j &= -\frac{\Pi^*}{\Pi^*+1}\left\{\frac{1}{(1+\alpha_o)^2} + \varepsilon\frac{\lambda\Pi^*}{(\Pi^*+\beta_o)^2}\right\} - \varepsilon\frac{\Pi^{*2}+1}{(\Pi^*+1)^2}\mu - \varepsilon\rho^*\frac{\Pi^*-1}{(\Pi^*+1)^2}\mu,\\
\Delta_j &= \varepsilon\frac{\Pi^*(\Pi^{*2}+1)}{(\Pi^*+1)^3}\left\{\frac{1}{(1+\alpha_o)^2} + \frac{\lambda\beta_o}{(\Pi^*+\beta_o)^2}\frac{\partial\Pi^*}{\partial\alpha_o}\right\}.
\end{aligned} \tag{9.77}$$

In the expression for T_j the terms in $\partial\Pi^*/\partial\alpha_o$ and $\partial\Pi^*/\partial\beta_o$ cancel out because

$$\frac{\partial\Pi^*}{\partial\alpha_o} = \varepsilon\frac{\partial\Pi^*}{\partial\beta_o}. \tag{9.78}$$

Moreover, the expression for $\partial\Pi^*/\partial\alpha_o$ is always negative because

$$\frac{\partial\Pi^*}{\partial\alpha_o} = -\frac{\varepsilon\delta\left\{\sqrt{\delta^2+4(\varepsilon\alpha_o+\beta_o)^2}+\delta\right\}}{2(\varepsilon\alpha_o+\beta_o)^2\sqrt{\delta^2+4(\varepsilon\alpha_o+\beta_o)^2}}. \tag{9.79}$$

Therefore, the value of Δ_j can become negative, thus resulting in a complete change of the dynamic properties in the vicinity of a steady state. The stable node is replaced by a saddle point. This change in dynamics is not an intrinsic property of the enzymes, which are still Michaelian, but a consequence of the electrostatic repulsion effects of the reagents by the fixed charges of the membrane.

9.3.4. *Electrostatic repulsion effects and multiple steady states*

It was shown in the previous sections that, even with Michaelian enzymes, electric repulsion effects of charged reagents can dramatically alter the dynamic behaviour of a simple metablic cycle. We shall now see whether these electric repulsion effects may not generate multistability for enzyme reactions that intrinsically follow Michaelis–Menten kinetics. If, as assumed previously, the equation of the first enzyme process taking place in the bulk phase is

$$\nu_1 = \frac{\alpha_o}{1+\alpha_o}, \tag{9.80}$$

and if that of the second enzyme buried in the polyanionic matrix is

$$\nu_2 = \frac{\lambda\beta_o}{\Pi+\beta_o}, \tag{9.81}$$

then, the two functions Y_1 and Y_2 (see section 9.2.2) will be functions of α_o (see eq. (9.51)). These two functions assume the form

$$\begin{aligned} Y_1(\alpha_o) &= \nu_1(\alpha_o) = \frac{\alpha_o}{1+\alpha_o}, \\ Y_2(\alpha_o) &= \mu\beta_o^* + \frac{\lambda\beta_o^*}{\Pi+\beta_o^*}. \end{aligned} \tag{9.82}$$

$Y_2(\alpha_o)$ is an increasing function of α_o for one has

$$\frac{\partial Y_2(\alpha_o)}{\partial\alpha_o} = -\frac{\lambda\beta_o^*}{(\Pi+\beta_o^*)^2}\frac{\partial\Pi}{\partial\alpha_o}, \tag{9.83}$$

and $\partial\Pi/\partial\alpha_o$ is of necessity negative (eq. (9.79)). Moreover, the curve for the variation of Y_2 as a function of α_o may possess an inflection point and therefore may display a sigmoidal shape. Thus, two cases are of potential interest. If

$$\lim Y_2(\alpha_o) > \lim Y_1(\alpha_o) \quad (\text{for } \alpha_o \to \infty), \tag{9.84}$$

then the system can have either no steady state or two steady states one of which is stable and the other unstable (Fig. 9.5). If, alternatively,

$$\lim Y_2(\alpha_o) < \lim Y_1(\alpha_o) \quad (\text{for } \alpha_o \to \infty), \tag{9.85}$$

then the system can display three steady states if $Y_2(\alpha_o)$ is sigmoidal. One steady state is unstable, the other two aré stable (Fig. 9.6).

If the steady state values of α_o, α_o^*, are plotted as a function of β_o^* one obtains a curve that displays hysteresis. The same type of situation is observed if the electric partition coefficient is plotted as a function of β_o^* (Fig. 9.6). This multistability is generated by electric

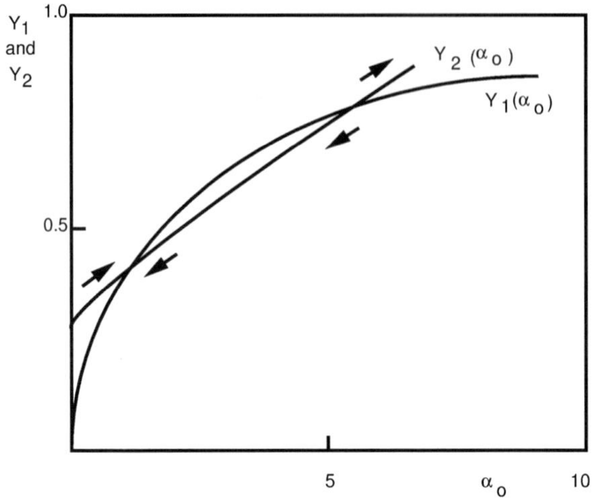

Fig. 9.5. Multiple steady states for the metabolic cycle of Fig. 9.4. The two enzymes are assumed to follow simple Michaelis–Menten kinetics but, owing to the electrostatic repulsion in the matrix, the two enzyme reactions become dependent on both substrates and two steady states (one stable, one unstable) may exist. Adapted from ref. [22].

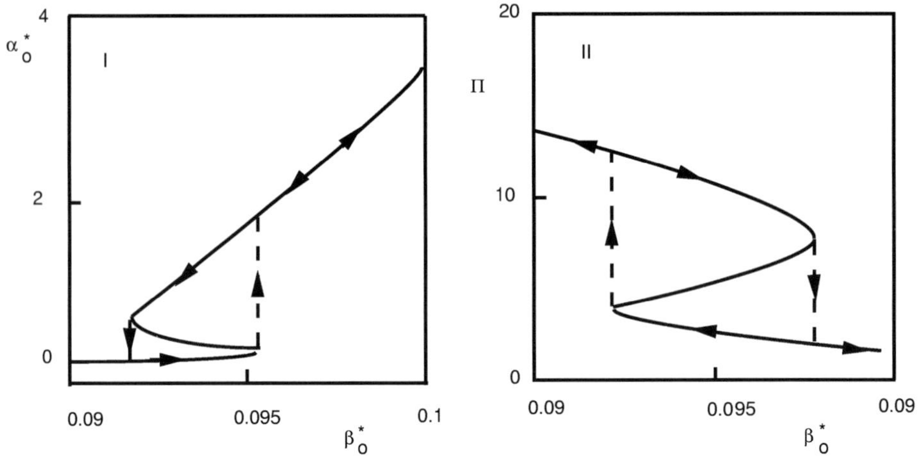

Fig. 9.6. Hysteresis of a charged reaction intermediate resulting in hysteresis of the electric partition coefficient. I – When the concentration of the intermediate S_2 is varied, the steady state concentration of S_1 displays hysteresis. II – Under these conditions the electric partition coefficient, Π, varies as well. Adapted from ref. [22].

repulsion effects and is not an intrinsic property of the enzymes. Since β_o^* is equivalent to the balance between the input and the output of matter in the system, changing this balance may generate hysteresis of the steady state concentration of S_1, or hysteresis of the electric partition coefficient, Π. The electric repulsion of mobile charges by the membrane is thus sensitive, not only to the value of a signal (the balance between the input and the output of matter), but also to the direction of change of the signal. In other words the system is now able to store short-term memory [19]. These multiple steady states have already been found in current biochemical literature [20,21]. But such properties are usually considered to be the consequence of nonlinear terms in the expression of the rate law of one of the enzyme processes. In the present case, as already outlined, multiple steady states appear as a consequence of electric partitioning effects exerted by the membrane [22–24].

9.3.5. pH-effects and the oscillatory dynamics of bound enzyme systems

When the bulk concentration of a charged substrate is varied, the electrostatic coefficient varies as well. This means that the local proton concentration varies, and the enzyme embedded in the charged matrix may be sensitive to this local pH change. More specifically, the local pH rises as the reagent concentration is increased. The pH effects on enzymes may be rather complex but, for reasons that will appear later, we shall assume that the rate falls off under alkaline pH conditions. Moreover, we shall postulate, for simplicity, that the reagent S_1 (or α) is neutral, whereas the other one, S_2 (or β), is a monoanion. In addition to the dimensionless variables and parameters, which we have already defined, one has to express the normalized bulk proton concentration, γ_b, as

$$\gamma_b = \frac{H_o}{K_b}, \tag{9.86}$$

where K_b is the base ionization constant of the enzyme–substrate complex. γ_b is roughly constant if the volume of the bulk phase is much larger than that of the matrix. If not, γ_b can be held constant by maintaining H_o constant. If it is the V_{max} which is sensitive to "high" pH values, then

$$\widetilde{V}_2 = V_2 \frac{H_i}{K_b + H_i} = V_2 \frac{H_o \Pi}{K_b + H_o \Pi} = V_2 \frac{\gamma_b \Pi}{1 + \gamma_b \Pi}, \tag{9.87}$$

where $\widetilde{V}_2$ is the apparent V_{max} of the second enzyme reaction. Using these definitions, one has

$$\begin{aligned} \nu_1(\alpha_o) &= \frac{\alpha_o}{1+\alpha_o}, \\ \nu_2(\beta_o) &= \lambda \frac{\gamma_b \Pi}{1 + \gamma_b \Pi} \frac{\beta_o}{\Pi + \beta_o} = \lambda \zeta(\beta_o) \xi(\beta_o). \end{aligned} \tag{9.88}$$

The two functions $\zeta(\beta_o)$ and $\xi(\beta_o)$ are thus

$$\zeta(\beta_o) = \frac{\gamma_b \Pi}{1 + \gamma_b \Pi}, \qquad \xi(\beta_o) = \frac{\beta_o}{\Pi + \beta_o}. \tag{9.89}$$

Since the reagent S_2 is charged whereas S_1 is not, the expression of Π is now

$$\Pi = \frac{\sqrt{\delta^2 + 4\beta_o^2} + \delta}{2\beta_o} = \frac{2\beta_o}{\sqrt{\delta^2 + 4\beta_o^2} - \delta}. \tag{9.90}$$

The first derivative of the function $\zeta(\beta_o)$ is

$$\frac{\partial \zeta(\beta_o)}{\partial \beta_o} = \frac{\gamma_b}{(1 + \gamma_b \Pi)^2} \frac{\partial \Pi}{\partial \beta_o}, \tag{9.91}$$

and $\partial \Pi / \partial \beta_o$ is equal to

$$\frac{\partial \Pi}{\partial \beta_o} = -\frac{\delta\left(\sqrt{\delta^2 + 4\beta_o^2} + \delta\right)}{2\beta_o^2 \sqrt{\delta^2 + 4\beta_o^2}}, \tag{9.92}$$

and can have only negative values. Therefore the function $\zeta(\beta_o)$ is monotonically decreasing. Its two limits are

$$\begin{aligned} &\lim \zeta(\beta_o) = 1 \quad (\text{for } \beta_o \to 0), \\ &\lim \zeta(\beta_o) = \frac{\gamma_b}{1 + \gamma_b} \quad (\text{for } \beta_o \to \infty). \end{aligned} \tag{9.93}$$

Similarly, the first derivative of the function $\xi(\beta_o)$ is

$$\frac{\partial \xi(\beta_o)}{\partial \beta_o} = \frac{\Pi - \beta_o(\partial \Pi / \partial \beta_o)}{(\Pi + \beta_o)^2}. \tag{9.94}$$

Since $\partial \Pi / \partial \beta_o$ is negative, the function $\xi(\beta_o)$ is of necessity increasing. Moreover, it displays an inflection point. It follows from these results that the function

$$v_2(\beta_o) = \lambda \zeta(\beta_o) \xi(\beta_o) \tag{9.95}$$

may increase first, may reach a maximum, then falls off towards an asymptotic value equal to $\lambda \gamma_b / (1 + \gamma_b)$. This type of behaviour is illustrated in Fig. 9.7. Thus, although the enzyme E_2 follows Michaelis–Menten kinetics in free solution, it may appear inhibited by excess substrate when bound to a polyanionic matrix.

In order to derive the expression of the dynamic system that governs the behaviour of the metabolic cycle, one has to derive first the expression of the sensitivity coefficients for the reagent S_1. One has thus

$$\begin{aligned} \sigma_{11} &= 1, \qquad \sigma_{12} = 0, \qquad \sigma_{21} = 0, \\ \sigma_{22} &= \Pi^{-1}\left(1 + \frac{\Pi^2 - 1}{\Pi^2 + 1}\right) = \frac{2\Pi}{\Pi^2 + 1}, \end{aligned} \tag{9.96}$$

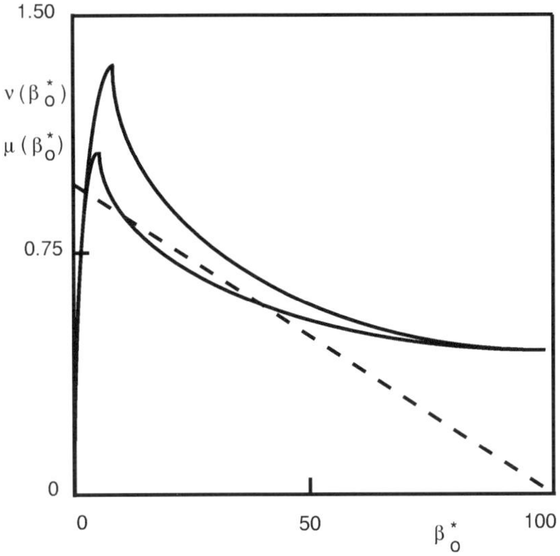

Fig. 9.7. Existence of three steady states generated by the sensitivity of the enzyme to changes of pH values. When the concentration of intermediate S_2 is varied, the local proton concentration changes and the enzyme burried in the matrix may become inhibited, thus leading to three steady states. Adapted from ref. [22].

and the expression of the dynamic system is then

$$\frac{d}{d\theta}\begin{bmatrix}\alpha_o\\ \beta_o\end{bmatrix} = \Omega^{-1}\begin{bmatrix}1+\sigma_{22} & 0\\ 0 & 2\varepsilon\end{bmatrix}\begin{bmatrix}u_1\\ u_2\end{bmatrix} \tag{9.97}$$

with

$$\Omega = 2(1+\sigma_{22}) = 2\frac{(\Pi+1)^2}{\Pi^2+1} \tag{9.98}$$

and

$$\begin{aligned} u_1 &= \nu_i - \nu_1(\alpha_o) + \nu_2(\beta_o),\\ u_2 &= \nu_1(\alpha_o) - \nu_2(\beta_o) - \mu\beta_o. \end{aligned} \tag{9.99}$$

Here, ν_2 is not a function of α_o, for α is uncharged and Π does not depend on this variable. The differential system above can be rewritten as

$$\begin{aligned} \frac{d\alpha_o}{d\theta} &= F_1(\alpha_o, \beta_o) = \frac{1}{2}\{\nu_i - \nu_1(\alpha_o) + \nu_2(\beta_o)\},\\ \frac{d\beta_o}{d\theta} &= F_2(\alpha_o, \beta_o) = \varepsilon\frac{\Pi^2+1}{(\Pi+1)^2}\{\nu_1(\alpha_o) - \nu_2(\beta_o) - \mu\beta_o\}. \end{aligned} \tag{9.100}$$

Steady state values of the variables α and β are solutions of the system

$$\begin{aligned}
&\nu_{\mathrm{i}} - \nu_1(\alpha_{\mathrm{o}}) + \nu_2(\beta_{\mathrm{o}}^*) = 0,\\
&\nu_1(\alpha_{\mathrm{o}}) - \nu_2(\beta_{\mathrm{o}}^*) - \mu\beta_{\mathrm{o}}^* = 0,
\end{aligned} \tag{9.101}$$

which implies that

$$\beta_{\mathrm{o}}^* = \frac{\nu_{\mathrm{i}}}{\mu}. \tag{9.102}$$

As expected, there exists only one steady state value of β_{o}. The steady state value (or values) of α_{o}, α_{o}^*, is (or are) a solution (or solutions) of the equation

$$\frac{\alpha_{\mathrm{o}}}{1+\alpha_{\mathrm{o}}} = \lambda \frac{\gamma_{\mathrm{b}}\Pi^*}{1+\gamma_{\mathrm{b}}\Pi^*} \frac{\beta_{\mathrm{o}}^*}{\Pi^* + \beta_{\mathrm{o}}^*} + \mu\beta_{\mathrm{b}}^*, \tag{9.103}$$

where, as previously, Π^* is the value of the partition coefficient when $\beta_{\mathrm{o}} = \beta_{\mathrm{o}}^*$. The solution of this equation is

$$\alpha_{\mathrm{o}}^* = \frac{\mu\beta_{\mathrm{o}}^* + \lambda \dfrac{\gamma_{\mathrm{b}}\Pi^*}{1+\gamma_{\mathrm{b}}\Pi^*} \dfrac{\beta_{\mathrm{o}}^*}{\Pi^* + \beta_{\mathrm{o}}^*}}{1 - \mu\beta_{\mathrm{o}}^* - \lambda \dfrac{\gamma_{\mathrm{b}}\Pi^*}{1+\gamma_{\mathrm{b}}\Pi^*} \dfrac{\beta_{\mathrm{o}}^*}{\Pi^* + \beta_{\mathrm{o}}^*}}. \tag{9.104}$$

Equations (9.103) and (9.104) have two implications. Because the left-hand side of eq. (9.103) is hyperbolic with respect to α_{o} and the right-hand side of the same equation is independent of α_{o}, there can exist only one intersection of ν_1 and ν_2 when they are plotted as a function of α_{o}. Therefore, the system can have only one stable steady state. The second implication is that, in order to have a steady state for α_{o}, one must have

$$1 - \mu\beta_{\mathrm{o}}^* > \lambda \frac{\gamma_{\mathrm{b}}\Pi^*}{1+\gamma_{\mathrm{b}}\Pi^*} \frac{\beta_{\mathrm{o}}^*}{\Pi^* + \beta_{\mathrm{o}}^*}. \tag{9.105}$$

Defining the conditions for a steady state more precisely requires prior knowledge of the roots of the equation

$$\mu(\beta_{\mathrm{o}}^*) - \nu(\beta_{\mathrm{o}}^*) = 0, \tag{9.106}$$

where

$$\begin{aligned}
&\mu(\beta_{\mathrm{o}}^*) = 1 - \mu\beta_{\mathrm{o}}^*,\\
&\nu(\beta_{\mathrm{o}}^*) = \lambda \frac{\gamma_{\mathrm{b}}\Pi^*}{1+\gamma_{\mathrm{b}}\Pi^*} \frac{\beta_{\mathrm{o}}^*}{\Pi^* + \beta_{\mathrm{o}}^*}.
\end{aligned} \tag{9.107}$$

The roots of eq. (9.106) correspond to the intersection of these two curves. Depending on the value of the normalized fixed charge density one may have one or three roots (Fig. 9.7).

The local stability of the system in the vicinity of a steady state requires, as previously, prior knowledge of the elements of the Jacobian matrix J (eq. (9.61)). One has, from eq. (9.97)

$$\begin{aligned} J &= \begin{bmatrix} \partial F_1^*/\partial\alpha_o & \partial F_1^*/\partial\beta_o \\ \partial F_2^*/\partial\alpha_o & \partial F_2^*/\partial\beta_o \end{bmatrix} \\ &= \Omega^{*-1}\begin{bmatrix} 1+\sigma_{22}^* & 0 \\ 0 & 2\varepsilon \end{bmatrix}\begin{bmatrix} \partial u_1^*/\partial\alpha_o & \partial u_1^*/\partial\beta_o \\ \partial u_2^*/\partial\alpha_o & \partial u_2^*/\partial\beta_o \end{bmatrix} \end{aligned} \qquad (9.108)$$

where u_1^* and u_2^* are defined as expressed in eq. (9.99) and Ω^* is the value of Ω when $\beta_o = \beta_o^*$. The elements of the J matrix may be determined from eq. (9.108)

$$\begin{aligned} &\frac{\partial F_1^*}{\partial\alpha_o} = -\frac{1}{2}\frac{\partial v_1^*}{\partial\alpha_o}, && \frac{\partial F_1^*}{\partial\beta_o} = \frac{1}{2}\frac{\partial v_1^*}{\partial\beta_o}, \\ &\frac{\partial F_2^*}{\partial\alpha_o} = \varepsilon\frac{\Pi^{*2}+1}{(\Pi^*+1)^2}\frac{\partial v_1^*}{\partial\alpha_o}, && \frac{\partial F_2^*}{\partial\beta_o} = -\varepsilon\frac{\Pi^{*2}+1}{(\Pi^*+1)^2}\left(\frac{\partial v_2^*}{\partial\beta_o}+\mu\right). \end{aligned} \qquad (9.109)$$

The local stability of the enzyme system depends on the trace and the determinant of the Jacobian matrix as well as on the discriminant $T_j^2 - 4\Delta_j$. One thus has

$$\begin{aligned} T_j &= -\left\{\frac{1}{2}\frac{\partial v_1^*}{\partial\alpha_o} + \varepsilon\frac{\Pi^{*2}+1}{(\Pi^*+1)^2}\left(\frac{\partial v_2^*}{\partial\beta_o}+\mu\right)\right\}, \\ \Delta_j &= \frac{\varepsilon}{2}\frac{\Pi^{*2}+1}{(\Pi^*+1)^2}\mu\frac{\partial v_1^*}{\partial\alpha_o}, \\ T_j^2 - 4\Delta_j &= \left\{\frac{1}{2}\frac{\partial v_1^*}{\partial\alpha_o} + \varepsilon\frac{\Pi^{*2}+1}{(\Pi^*+1)^2}\left(\frac{\partial v_2^*}{\partial\beta_o}+\mu\right)\right\}^2 - 2\varepsilon\frac{\Pi^{*2}+1}{(\Pi^*+1)^2}\mu\frac{\partial v_1^*}{\partial\alpha_o}. \end{aligned} \qquad (9.110)$$

These expressions show that a necessary, although not sufficient, condition for obtaining sustained oscillations of α_o and β_o is $\partial v_2^*/\partial\beta_o < 0$. Then the trace T_j can be negative or nil and the discriminant can be negative. Sustained oscillations, which may occur, are thus the consequence of an inhibition of the enzyme reaction rate v_2 by the "alkaline" local pH conditions generated by an increase of the values of the normalized concentrations β_o. As usual, these temporal oscillations can be described as a time-evolution of normalized concentrations α_o and β_o or as trajectories in the phase-plane (α_o, β_o). These oscillations are shown in Fig. 9.8.

Moreover, as the concentrations of S_2 oscillate, the expression of the electric partition coefficient Π should oscillate as well, for its value depends on β_o. This means that the electric repulsion effects exerted by the membrane on the mobile anions is not constant but varies periodically with time. Similarly, since the sensitivity coefficient σ_{22} can be expressed in terms of the electrostatic partition coefficient Π, one may expect the sensitivity coefficient to vary periodically. These periodic variations are shown in Fig. 9.9. Moreover, close to a critical value of the normalized fixed charge density δ where the system has no steady state, a slight variation of this δ value results in a dramatic change in the period of oscillations, or even to suppression of the oscillations.

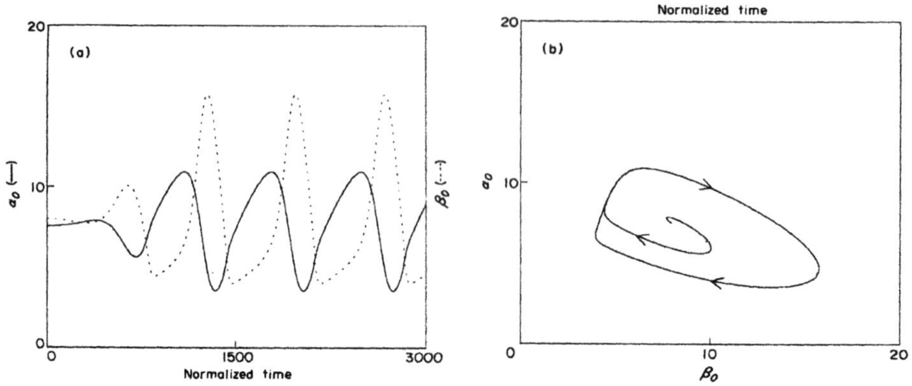

Fig. 9.8. Sustained oscillations of the two substrates of the metabolic cycle of Fig. 9.4. (a) – The oscillations display a phase-shift. (b) – Limit cycle in the phase plane. From ref. [22].

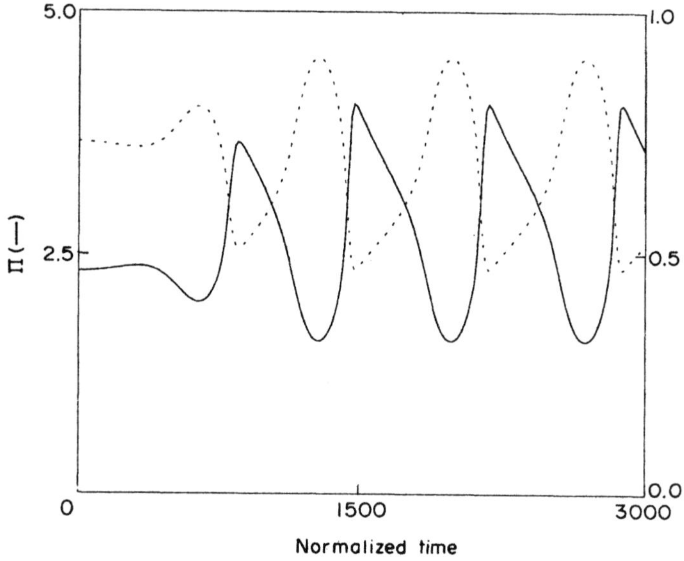

Fig. 9.9. Periodic oscillations of the electric partition and sensitivity coefficient. Full line: partition coefficient. Dotted line: sensitivity coefficient σ_{22}. From ref. [22].

Thus, both the existence of multiple steady states and the oscillatory dynamics do not appear, in the present case, to be the consequence of the properties of an individual enzyme, but rather of the spatial organization and of the complexity of the living cell. Compartmentalization of the cell and the occurrence of metabolic processes at the surface of membranes may generate instabilities and oscillations that rely solely on the existence of this compartmentalization.

9.4. *Periodic and aperiodic oscillations generated by the complexity of the supramolecular edifices of the cell*

In chapter 7 we described a model, the two-state model, which attempts at understanding the plant cell wall extension. We shall now analyse a new version of this model considered in a dynamic, and thermodynamically open, perspective. Let us first recall, very briefly, some biological results that serve as a basis for the model. In primary plant cell walls, cellulose microfibrils are interconnected by xyloglucan chains. Cell extension requires both the breaking of hydrogen bonds that associate cellulose microfibrils and xyloglucans, and the splitting and making of $\beta(1 \rightarrow 4)$ bonds of xyloglucans [22]. This process is enzymatic and requires the participation of a specific transglycosylase [26]. Unzipping xyloglucans, as well as breaking the $\beta(1 \rightarrow 4)$ bonds, results in a loosening of the wall. This loosening is localized and takes place "in mosaic". It allows the limited and localized extension of the wall under the influence of the turgor pressure. This extension process requires a "low" local pH, which is generated by a high local density of fixed negative charges of demethylated pectins. A pectin methyl esterase demethylates methylated pectins and generates the fixed negative charges. When the cell wall extends, neutral methylated pectins are incorporated into the wall. Thus, in the course of the wall extension, two antagonistic processes take place: the creep of cellulose microfibrils associated with the incorporation of methylated pectins in the wall; the demethylation of methylated pectins by pectin methyl esterase. The first process results in a decrease of the fixed charge density whereas the second leads to an increase of this charge. The first process is activated by acid pH conditions whereas the second requires neutral or slightly alkaline local pHs. As these two processes take place under open thermodynamic conditions and are associated with cell growth, one may wonder whether they can not lead to periodic, or aperiodic (chaotic) oscillations of the growth rate.

9.4.1. The model

The model is based on four basic ideas that are presented below. First, a limited region of the wall that undergoes an extension process (growth "in mosaic") is in a destabilized state, owing to unzipping of xyloglucans from cellulose microfibrils. This region can be considered a collection of micro-domains. Second, each micro-domain can be viewed as a complex supramolecular edifice of a constant volume, comprising either highly methylated, or poorly methylated pectins. These micro-domains exist in only two states, called X and Y. Each state, however, usually comprises many different sub-states with different degrees of methylation (or of different charge density). X represents a micro-domain of high charge density and Y a micro-domain of low charge density. The view that a clear distinction can be made between state X and state Y is only an ideal approximation that allows to make the model tractable. Third, the extension of the wall is accompanied by the incorporation of methylated pectins in some micro-domains. As the region of the wall is extending and the density of the glucidic material in the domains is roughly constant, this means that acidic pectins "leave" these micro-domains. This process is thus equivalent to the conversion of X into Y. Fourth, pectin methyl esterase tends to demethylate pectins in the extending region of the wall and this leads to the conversion of Y into X.

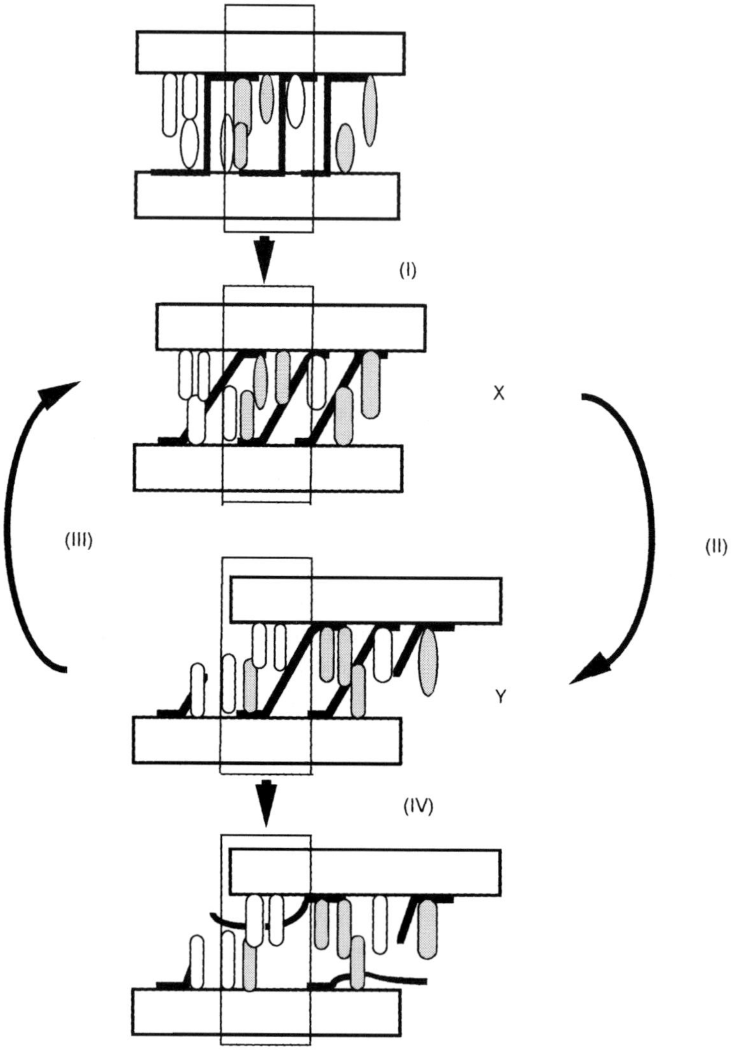

Fig. 9.10. Simplified representation of the events that take place during plant cell wall extension. This scheme is a dynamic version of the two-state model of chapter 7. The black lines represent xyloglucan molecules that associate two cellulose microfibrils. The grey and the white "balloons" are the acidic and methylated pectins, respectively. See text. Adapted from ref. [25].

The model based on these ideas is shown in Fig. 9.10 [25]. It is an extension to open conditions of the two-state model discussed in chapter 7. During the first step (I) of this complex process a small region of the wall is destabilized by the unzipping of xyloglucans from the cellulose microfibrils. This process is driven by a high turgor pressure and the corresponding limited region is assumed to be extremely rich in negatively charged pectins.

A micro-domain of a fixed volume, is defined in this region and is referred to as X. The rate of the process leading to the formation of this destabilized micro-domain is v_i. In this X micro-domain, methylated pectins (as well as other glucidic material) are incorporated and endotransglycosylases catalyse the breaking of $\beta(1 \rightarrow 4)$ bonds, thus allowing the cellulose microfibrils to slide and the corresponding region to expand. After the sliding has occurred, new $\beta(1 \rightarrow 4)$ bonds are formed thanks to the same enzyme. As the small region of the wall extends, methylated molecules are incorporated in the micro-domain X, which is thus transformed into a micro-domain Y. The rate of conversion of X into Y is v_1 (step II). Demethylation of methylated pectins in this micro-domain converts Y back to X. The rate of the corresponding process is v_2 (step III). Step III is thus an enzymic process controlled by pectin methyl esterase. But micro-domain Y may also undergo further unzipping of xyloglucans and destabilization (step IV). The corresponding destabilization rate is v_o. The overall process is thus that already described in chapter 7 but with two additional processes of destabilization (steps I and IV) that make this process thermodynamically open (Figs. 9.10 and 9.12), and with the assumption that states X and Y usually encompass sub-states of different degrees of methylation.

The same process should also take place at different rates in different regions of the wall, in such a way that the rate, which is measured experimentally, is the mean of individual rates of many different regions of the wall. Metal ions and protons are indeed present in the wall, and their concentration varies form micro-domain to micro-domain, in relation to the local variation of fixed charge density of pectins. Both pectin methyl esterases and endotransglycosylases seem to be ionically bound to the wall. This suggests that these enzymes may circulate within the cell wall if the corresponding local charge density is small, but they may become temporarily stuck in certain regions of the wall if the local fixed charge density is large.

9.4.2. The basic enzyme equations

It is sensible, at least as a first approximation, to assume that the rate of extension, v_1, of a region encompassing many micro-domains of constant volume X, follows a hyperbolic kinetics relative to the density of negatively charged pectins (or to the density of negatively charged micro-domains in this micro-region). If X represents this density one has

$$v_1 = \frac{V_1 X/\overline{K}_1}{1 + (X/\overline{K}_1)}, \tag{9.111}$$

where the maximum rate, V_1, is proportional to the local concentration of endotransglycosylase in the micro-domains X and $\overline{K}_1$ is the apparent half-saturation constant. It is implicitly assumed, in this scheme, that the local density of acidic pectins is roughly proportional to that of xyloglucans. The symbols X and X thus represent a negatively charged micro-domain and the density of these micro-domains (or the density of negatively charged pectins) in the corresponding region of the wall. The rate v_1 can also be viewed as the rate of conversion of micro-domains X into micro-domains Y. Moreover, it is implicitly assumed that metal binding to acidic pectins does not significantly affect the enzyme reaction rate.

Other regions of the wall contain partly methylated pectins (or Y micro-domains). The total density of these Y micro-domains is referred to as Y_{T}. In the cell wall methylated pectins, there exist "blocks" of carboxyl groups, termed Y^-, and carboxyl groups adjacent to the methyl groups to be cleaved, called Y'^-. The total local densities of the "blocks" and of the charged regions adjacent to the methyl groups are designated by Y_{T}^- and by $Y_{\mathrm{T}}'^-$, respectively. Moreover, it was shown previously [27] that metal ions bind non co-operatively to the Y^- "blocks", but co-operatively to the Y'^- regions. The local densities that have not bound cations are thus

$$
\begin{aligned}
Y^- &= \frac{Y_{\mathrm{T}}^-}{(1+K_{\mathrm{a}}A)^m}, \\
Y'^- &= \frac{Y_{\mathrm{T}}'^-}{1+K_1K_2\ldots K_nA^n}.
\end{aligned}
\tag{9.112}
$$

In these expressions, A is the local concentration of a metal ion, K_{a} is the intrinsic binding constant of the metal ion to a negatively charged group of a "block" and $K_1, K_2, \ldots, K_n$ the affinity constants of the same cation with respect to the charged regions adjacent to the methyl groups. There is an advantage in expressing Y'^- in terms of the geometric mean of the various affinity constants $K_1, K_2, \ldots, K_n$. This geometric mean is

$$
K_{\mathrm{a}}' = (K_1K_2\ldots K_n)^{1/n},
\tag{9.113}
$$

and therefore

$$
Y'^- = \frac{Y_{\mathrm{T}}'^-}{1+(K_{\mathrm{a}}'A)^n}.
\tag{9.114}
$$

As will be seen later the use of this geometric mean allows one to express the dynamic equations in dimensionless form. The binding isotherms of cation A to the two regions Y^- and Y'^- are

$$
\bar{\nu}_m = \frac{mK_{\mathrm{a}}AY_{\mathrm{T}}^-}{1+K_{\mathrm{a}}A}, \qquad \bar{\nu}_n = \frac{n(K_{\mathrm{a}}'A)^nY_{\mathrm{T}}'^-}{1+(K_{\mathrm{a}}'A)^n}.
\tag{9.115}
$$

Moreover, one has

$$
Y_{\mathrm{T}} = Y_{\mathrm{T}}^- + Y_{\mathrm{T}}'^-.
\tag{9.116}
$$

Thus, if the pectin concentration varies, the densities of $\mathrm{Y}_{\mathrm{T}}^-$ and $\mathrm{Y}_{\mathrm{T}}'^-$ should vary in a constant ratio $\alpha = Y_{\mathrm{T}}^-/Y_{\mathrm{T}}'^-$. It therefore follows that

$$
\begin{aligned}
Y^- &= \frac{\alpha}{1+\alpha}\frac{Y_{\mathrm{T}}}{(1+K_{\mathrm{a}}A)^m}, & Y'^- &= \frac{1}{1+\alpha}\frac{Y_{\mathrm{T}}}{1+(K_{\mathrm{a}}'A)^n}, \\
\bar{\nu}_m &= \frac{mK_{\mathrm{a}}A}{1+K_{\mathrm{a}}A}\frac{\alpha}{1+\alpha}Y_{\mathrm{T}}, & \bar{\nu}_n &= \frac{n(K_{\mathrm{a}}'A)^n}{1+(K_{\mathrm{a}}'A)^n}\frac{1}{1+\alpha}Y_{\mathrm{T}}.
\end{aligned}
\tag{9.117}
$$

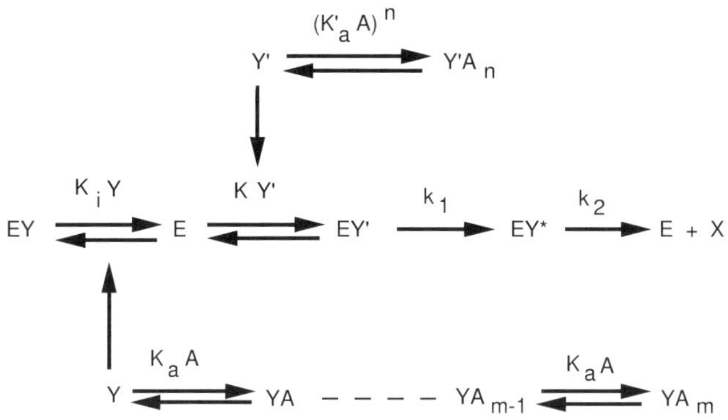

Fig. 9.11. Simplified kinetic scheme of pectin methyl esterase. The enzyme may be either trapped in the "blocks" Y^-, and is then unavailable to the catalytic process, or may be bound to the charged regions Y'^- adjacent to the methyl groups, and then takes part to the chemical process of demethylation. Cation A is bound non-co-operatively to the "blocks" and co-peratively to the regions Y'^-. From ref. [25].

As X micro-domains contain many more fixed negative charges than the Y micro-domains, metal ions should be bound preferentially to X. Let us call X_T the total density of X micro-domains. The binding isotherm of the metal ion to X is then

$$\bar{\nu}_w = \frac{wK''_a A}{1 + K''_a A} X_T, \tag{9.118}$$

where w is the number of binding sites on X and K''_a the affinity constant of the metal for these sites, which are all considered equivalent and independent.

A simplified model for the action of pectin methyl esterase is shown in Fig. 9.11. In this model it is postulated that the enzyme molecules are trapped by the "blocks", for the enzyme is positively charged. But the enzyme can also bind to the fixed negative charges, in the vicinity of the methyl groups to be cleaved. The first mode of binding is abortive for the enzyme reaction whereas the second is productive.

This model allows one to understand the mode of action of metal ions, already described in chapter 7. There is no evidence that metal ions interact with the enzyme. In fact they interact only with the charged regions of pectins [27]. At low concentrations, metal ions neutralize the fixed negative charges of the "blocks". They therefore release the enzyme molecules, which were initially trapped by the "blocks" and which thereby become available for the enzyme reaction. In this concentration range, metal ions thus behave as activators of the reaction. But at higher concentrations, cations also bind to the charged groups located in the vicinity of the methyl groups and prevent enzyme binding to the charged regions. Therefore, at higher concentrations, metal ions behave as inhibitors [27]. The corresponding reaction rate that accomodates these experimental findings is

$$v_2 = \frac{k_1 E_T K Y'^-}{1 + \dfrac{k_1 + k_2}{k_2} K Y'^- + K_i Y^-}, \tag{9.119}$$

where E_{T} is the total concentration of pectin methyl esterase, K is the affinity constant of the enzyme for the Y'^{-} region and K_{i} is the affinity constant of the enzyme for the "blocks" Y^{-}. Setting for simplicity

$$\widetilde{K} = K\frac{1}{1+\alpha}, \qquad \widetilde{K}_{\mathrm{i}} = K_{\mathrm{i}}\frac{\alpha}{1+\alpha}, \tag{9.120}$$

the enzyme reaction rate assumes the form

$$v_2 = \frac{k_1 E_{\mathrm{T}} \dfrac{\widetilde{K} Y_{\mathrm{T}}}{1+(K_{\mathrm{a}}' A)^n}}{1+\dfrac{k_1+k_2}{k_2}\dfrac{\widetilde{K} Y_{\mathrm{T}}}{1+(K_{\mathrm{a}}' A)^n}+\dfrac{\widetilde{K}_{\mathrm{i}} Y_{\mathrm{T}}}{(1+K_{\mathrm{a}} A)^m}}. \tag{9.121}$$

The concentration of free metal ion which appears in this expression is thus equal to

$$A = A_{\mathrm{T}} - \bar{\nu}_w - (\bar{\nu}_m + \bar{\nu}_n), \tag{9.122}$$

where A_{T} is the total cation concentration. This expression assumes a simple form if $\alpha \approx 1$, namely,

$$A = A_{\mathrm{T}} - \frac{w K_{\mathrm{a}}'' A}{1+K_{\mathrm{a}}'' A} X_{\mathrm{T}} - \left\{ \frac{m K_{\mathrm{a}} A}{1+K_{\mathrm{a}} A} + \frac{n (K_{\mathrm{a}}' A)^n}{1+(K_{\mathrm{a}}' A)^n} \right\} \frac{Y}{2}. \tag{9.123}$$

As the number of fixed negatively charged groups of X is much larger than that of Y, one must have

$$w > m + n, \tag{9.124}$$

and the difference

$$r = w - (m+n), \tag{9.125}$$

represents the number of methyl groups cleaved by pectin methyl esterase.

During the extension of a limited region of the wall, methylated pectins "enter" the micro-domains arbitrarily defined in this extending region, and acidic pectins "leave" these micro-domains. Thus X micro-domains appear to be formed. Conversely, when pectin methyl esterase is maximally active, methylated pectins become negatively charged and Y micro-domains are generated from X. The differential equations that govern the dynamics of these events can be written as

$$\frac{\mathrm{d}X}{\mathrm{d}t} = v_{\mathrm{i}} - v_1 + v_2, \qquad \frac{\mathrm{d}Y_{\mathrm{T}}}{\mathrm{d}t} = v_1 - v_2 - k Y_{\mathrm{T}}, \tag{9.126}$$

where v_{i} is the rate of "production" of X and k is the output rate of Y. The overall process can be represented by the open cycle of Fig. 9.12.

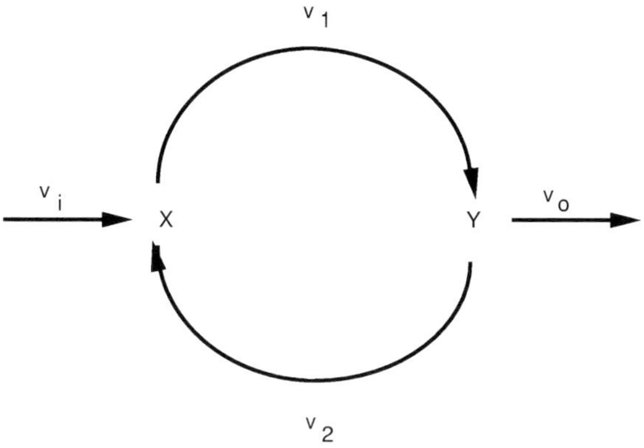

Fig. 9.12. Formal dynamic scheme of the events of Fig. 9.10. In this scheme X and Y represent the second and third states shown in Fig. 9.10. They are associated with micro-regions rich in acidic and methylated pectins, respectively.

As always, there is an advantage in expressing these equations in dimensionless form. Setting

$$
\begin{aligned}
& x = \frac{X}{\overline{K}_1}, \qquad y = \widetilde{K}\frac{k_1+k_2}{2k_2}Y_T, \qquad z = K_a A, \\
& \tau = V_1\overline{K}_1 t, \qquad \mu = \frac{K_a}{\widetilde{K}}\frac{k_2}{k_1+k_2}, \qquad \eta = \frac{\widetilde{K}_i}{\widetilde{K}}\frac{k_2}{k_1+k_2}, \\
& \lambda = \frac{k_1 E_T}{V_1}\frac{k_2}{k_1+k_2}, \qquad \sigma = \frac{k}{\widetilde{K}V_1}\frac{k_2}{k_1+k_2}, \qquad \rho = \frac{\widetilde{K}}{\overline{K}_1}\frac{k_1+k_2}{k_2}, \\
& z_o = K_a A_T, \qquad \varepsilon_1 = \frac{K_a'}{K_a}, \qquad \varepsilon_2 = \frac{K_a''}{K_a}, \qquad \nu_i = \frac{v_i}{V_1},
\end{aligned}
\tag{9.127}
$$

allows to reduce the number of variables and parameters from 19 to 15. One can rewrite the differential system (9.126) as

$$
\begin{aligned}
\frac{dx}{d\tau} &= \nu_i - \nu_1(x) + \nu_2(x,y), \\
\frac{dy}{d\tau} &= \rho\{\nu_1(x) - \nu_2(x,y) - \sigma y\},
\end{aligned}
\tag{9.128}
$$

with

$$
\begin{aligned}
\nu_1(x) &= \frac{x}{1+x}, \\
\nu_2(x,y) &= \frac{\lambda y}{1+(\varepsilon_1 z)^n + \left\{1+\eta\frac{1+(\varepsilon_1 z)^n}{(1+z)^m}\right\}y}.
\end{aligned}
\tag{9.129}
$$

Moreover, the conservation equation (9.123) can be rewritten in dimensionless form as

$$z = z_o - \mu\left\{\rho\frac{w\varepsilon_2 z}{1+\varepsilon_2 z}x + \left[\frac{mz}{1+z} + \frac{n(\varepsilon_1 z)^n}{1+(\varepsilon_1 z)^n}\right]y\right\}. \tag{9.130}$$

Therefore, at a fixed value of z_o the value of z depends on both x and y which gives eq. (9.129) a markedly nonlinear character. Equations (9.129) and (9.130) express in mathematical terms the view that, as y (the methylated pectin) is increased, the corresponding rate of pectin methyl esterase activity increases at first and then falls off for high y values. In fact, as y increases, z decreases (eq. (9.130)). This means that the free metal concentration has become very small. The enzyme pectin methyl estease is then trapped by the fixed negative charges of acidic pectins, thus explaining that the enzyme reaction levels off. All these results have been obtained experimentally [27]. A steady state of the system is defined by the values, x^* and y^*, of the variables such that

$$\begin{aligned} F_1^* &= \nu_i - \nu_1(x) + \nu_2(x, y^*) = 0, \\ F_2^* &= \rho\{\nu_1(x) - \nu_2(x, y^*) - \sigma y^*\} = 0, \end{aligned} \tag{9.131}$$

which leads in turn to

$$x^* = \frac{\nu_i + \nu_2(x^*, y^*)}{1 - \nu_i - \nu_2(x^*, y^*)}, \qquad y^* = \frac{\nu_i}{\sigma}. \tag{9.132}$$

9.4.3. Homogeneous population of elementary oscillators

Within the frame of the present model, the primary cell wall can be viewed as a collection of micro-domains that differ in their local charge densities. It is now of interest to know whether any of these micro-domains may behave as an elementary oscillator. This can be achieved, as previously, by stability analysis of the present model. One can show that the trace T_j and the determinant Δ_j of the relevant Jacobian matrix of the system have the form

$$\begin{aligned} T_j &= -\left\{\frac{\partial v_1^*}{\partial x} + \rho\sigma\right\} + \left\{\frac{\partial v_2^*}{\partial x} - \rho\frac{\partial v_2^*}{\partial y}\right\}, \\ \Delta_j &= \rho\sigma\left\{\frac{\partial v_1^*}{\partial x} - \frac{\partial v_2^*}{\partial x}\right\}. \end{aligned} \tag{9.133}$$

A necessary, although not sufficient, condition for the system to display sustained oscillations is

$$\Delta_j > 0, \qquad T_j > 0, \qquad T_j^2 - 4\Delta_j < 0. \tag{9.134}$$

$\partial v_1^*/\partial x$ is of necessity positive, but $\partial v_2^*/\partial x$ and $\partial v_2^*/\partial y$ may be either positive or negative. In fact, a necessary condition for the appearance of oscillations is $\partial v_2^*/\partial y < 0$. The

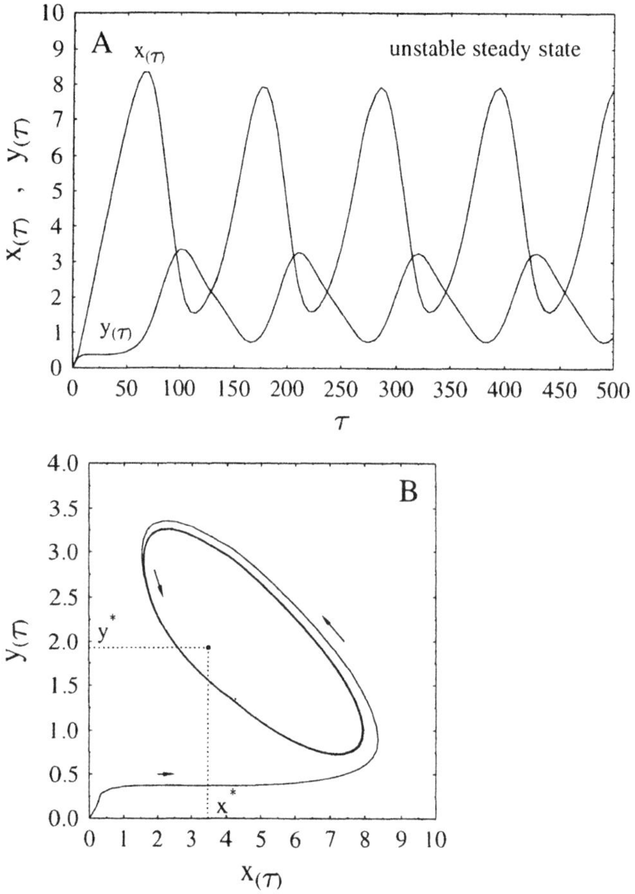

Fig. 9.13. Sustained oscillatory behaviour of acidic and methylated pectins during extension of cell wall micro-region. A – Temporal variations of the two variables. B – Phase-plane representation. From ref. [25].

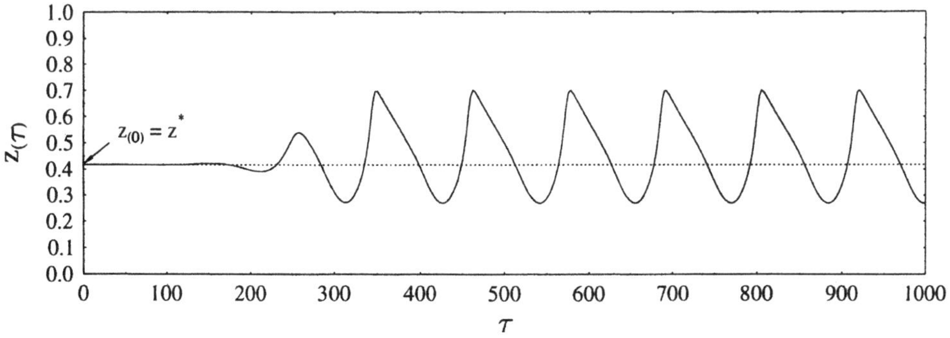

Fig. 9.14. Sustained variations of cation concentration in a cell wall micro-region. See text. From ref. [25].

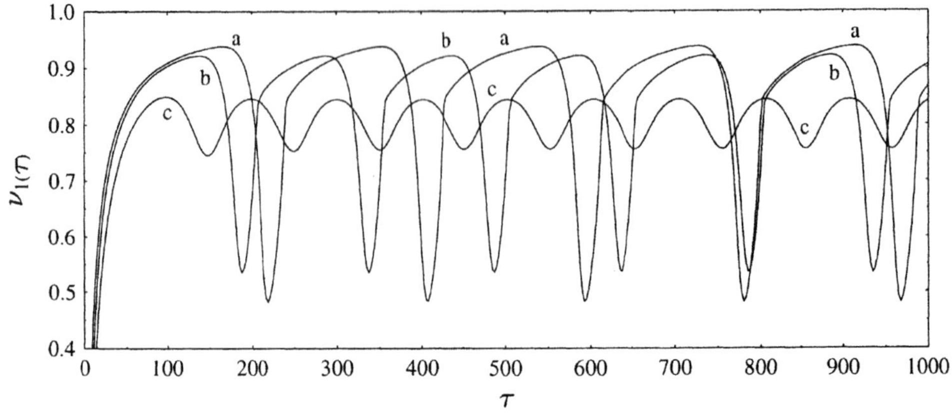

Fig. 9.15. Influence of slight differences of charge density in the "blocks" on the periodic growth rate of a cell wall micro-region. The period and the amplitude of the oscillation dramatically depends on this charge density. From ref. [25].

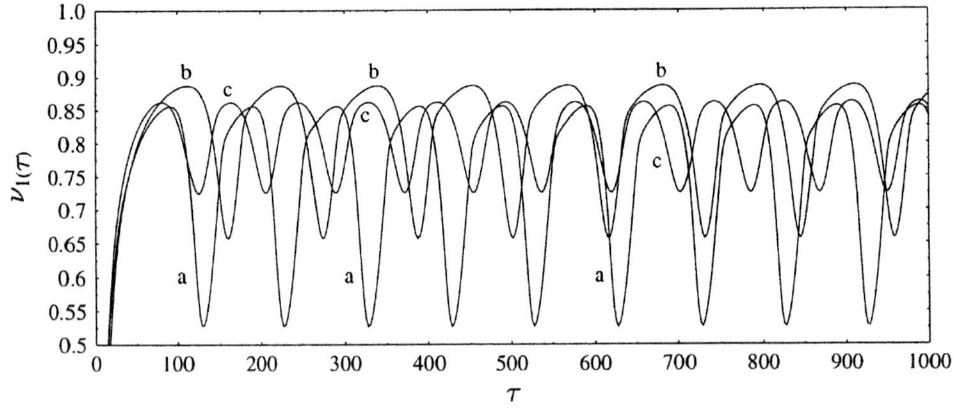

Fig. 9.16. Influence of the negatively charged groups of methylated pectins on the periodic growth of a cell wall micro-domain. The period and the amplitude of this oscillation depends also on this charge density. From ref. [25].

decrease of the pectin methyl esterase reaction rate for high pectin concentrations, or densities, results from a decrease of free metal concentration. In Fig. 9.13 is shown an example of oscillatory dynamics of the density of unmethylated and methylated pectins in a homogeneous population of micro-domains.

ν_1 may be viewed as the local growth rate associated with a homogeneous population of micro-domains and indeed displays oscillations (see Figs. 9.15 and 9.16). In Fig. 9.13(B) a phase plane representation of the oscillatory pattern of Fig. 9.13(A) is shown. The free metal ion concentration may also display sustained oscillations in these micro-domains. These oscillations are shown in Fig. 9.14.

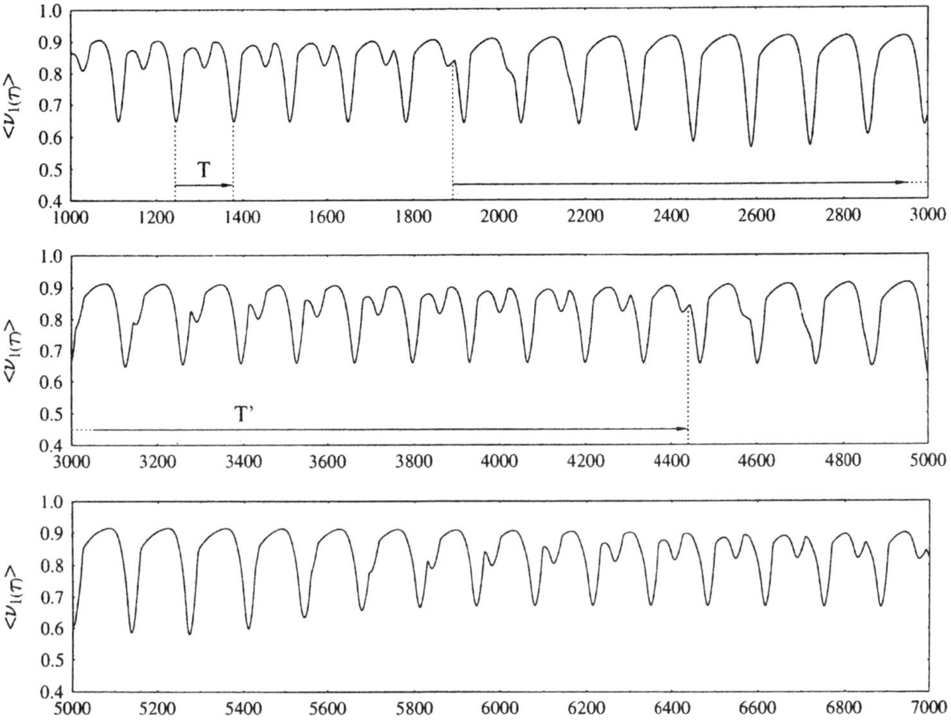

Fig. 9.17. Complex periodic behaviour of a population of micro-regions. The amplitude of the oscillations varies, biperiodicities appear and disappear and the process displays at least two different periodicities. From ref. [25].

9.4.4. Periodic and "chaotic" behaviour of the overall growth rate

The oscillations of the growth rate of the region of the wall associated with micro-domains possess an interesting property: they are sensitive to differences in local charge density of the "blocks" and of the regions located in the vicinity of the methyl groups to be cleaved. This means that the population of micro-domains is probably not homogeneous. Changes in the charge numbers m and n alter both the amplitude and the period of the oscillations. This is illustrated in Figs. 9.15 and 9.16.

From an experimental viewpoint, it is only possible to measure the extension of a wall bearing the micro-domains that behave as oscillators. But as the fixed charge density is likely to vary from place to place, one may expect the resulting oscillations of the growth rate to display a rather complex dynamic regime. Such a regime is simulated in Fig. 9.17. Thus, in the example of this figure, one can observe mono- and bi-periodicities as well as two markedly different periods. Moreover, since growth is taking place "in mosaic" many regions of the wall are not extending whereas others are. This is understandable on the basis of the above model. In order to extend, a limited region of the wall should possess both the strategic transglycosylase and pectin methyl esterase. If either, or both, of these enzymes is, or are, lacking in the corresponding micro-domains, the growth process cannot

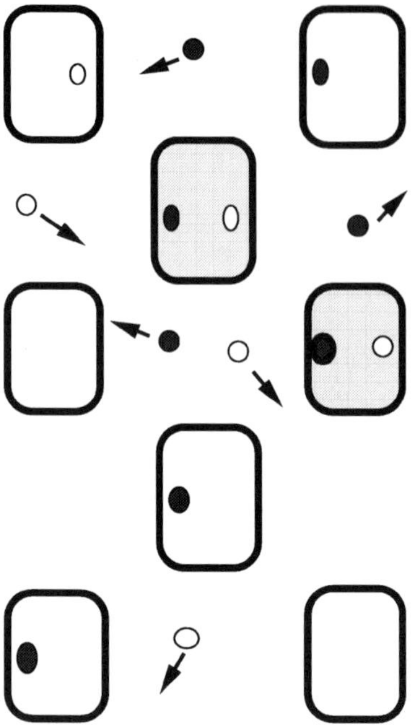

Fig. 9.18. Population of "active" and "inactive" micro-domains. The two types of enzymes involved in the growth process are depicted by black and void circles. These enzymes circulate in the wall. Extension of a micro-region of the wall occurs when these two types of enzymes are simultaneously present in a corresponding micro-domain. In this figure only two micro-domains are assumed to be "active".

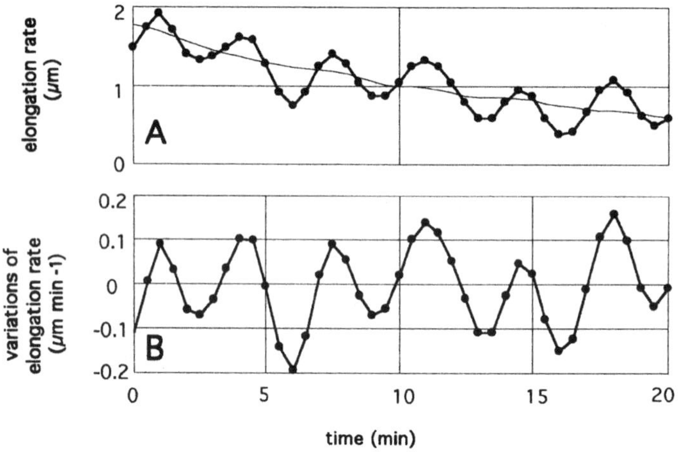

Fig. 9.19. Periodic oscillations of the growth rate of segments of *Vigna radiata*. A – Oscillations of the growth rate about the trend of the time series. The decline of the trend indicates that the growth rate decreases. B – Stationary oscillations about the trend. From ref. [31].

take place. Moreover, these enzymes are apparently ionically bound to the wall. Therefore, depending on the fixed local negative charge density, the enzymes may be bound in the micro-domains of the extending region, or they may circulate from micro-domain to micro-domain. This is understandable in view of the periodic variation of cation concentration in the micro-domains of the wall. This represents a rationale for the view that extension may start in a micro-domain, stops after a while, and then starts extending again. Moreover, if extension starts in several micro-domains that have identical, or very similar, charge densities, then the observed oscillations of the growth rate will be roughly periodic. But if, after a while, extension stops in these micro-domains because one (or two) of the enzymes responsible for the growth process has (or have) moved elsewhere, and if extension starts in different micro-domains with different charge densities, then the overall dynamics will appear aperiodic or "chaotic". The model thus predicts that the dynamics of growth may be periodic or aperiodic and may turn from periodic to aperiodic, or *vice versa*. An example of such aperiodic, or "chaotic", behaviour is shown in the next section. The model of the growth process, viewed as a mosaic of elementary oscillators that are functioning in a random manner because the enzymes involved in these oscillations are randomly circulating in the wall, is depicted in Fig. 9.18. These theoretical considerations can indeed be applied to real experimental data, and this precisely the aim of the next section.

9.4.5. Periodic and aperiodic oscillations of the elongation rate of plant cells

It is now possible, thanks to specific experimental devices called auxanometers, to monitor the continuous extension of segments of roots, stems, or coleoptiles [28–31]. If these segments are appropriately chosen, their elongation is strictly proportional to the mean extension rate of the cells. If the elongation of segments of Mung bean (*Vigna radiata*) is continuously monitored, one may observe oscillations of the growth rate about the trend of the time-series. In some cases, the oscillations are periodic (Fig. 9.19) and one can express the differences between the actual rate and the trend as a function of time (Fig. 9.19).

When the oscillations are periodic, their period is comprised between 4 and 8 minutes. This conclusion can be reached by averaging the time-series with itself, shifted by a given time-length. If the time-length is equal to half-a-period, the oscillations tend to disappear. They reappear again, if the time-length coincides with the period. This is precisely what is shown in Fig. 9.20. Thus, for these data, if the curve-shift is 3.5 minutes the oscillations are dramatically attenuated. If the cirve-shift is 7 minutes the oscillations reappear (Fig. 9.20). This means that the mean period of the oscillations is 7 minutes.

For most cases the oscillations appear aperiodic or "chaotic". This is illustrated in Fig. 9.21. It may even occur that the oscillation regime of the same hypocotyl segment changes during the growth process (Fig. 9.21).

Some general ideas can be derived from the theoretical and experimental results presented. Biological oscillations may indeed originate from sophisticated properties of some specific enzymes. But we believe that, in most cases, these oscillations originate rather from the complexity of cell organization. Thus, compartmentalization of the cell milieu by a polyelectrolyte membrane may generate biological oscillations, even if the enzymes that take part in a metabolic cycle at the interface between a membrane and the cell milieu follow simple Michaelis–Menten kinetics. In the case of oscillations of the growh rate of plant cells, these oscillations are not due to a specific property of pectin methyl esterase for this enzyme follows classical Michaelis–Menten kinetics relative to artificial substrates [27]. The oscillations of growth rate are due to the complex structure of the cell

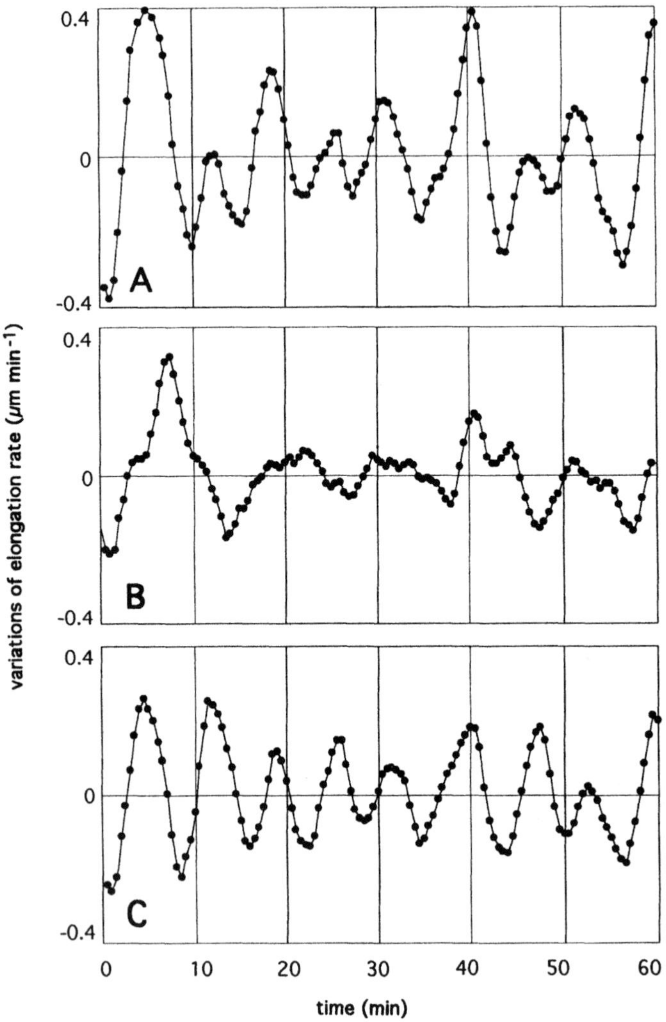

Fig. 9.20. Estimation of the mean period of an oscillation. The initial curve is averaged with itself after a shift of 3.5 minutes (B) and 7 minutes (C). In the present case the mean period is thus 7 minutes. From ref. [31].

wall and of the pectins. Thus, the apparent activation and inhibition of pectin methyl esterase by cations is understandable because the fixed charges of pectins are clustered in "blocks", or are located in the vicinity of the methyl groups. Surprisingly, the emergence of a "chaotic" regime of oscillations stems from a non-random distribution of fixed charges in the cell wall. These charges are unevenly distributed within the micro-domains of the cell wall and it is this complex, fuzzy organization that generates the chaotic regime of oscillations.

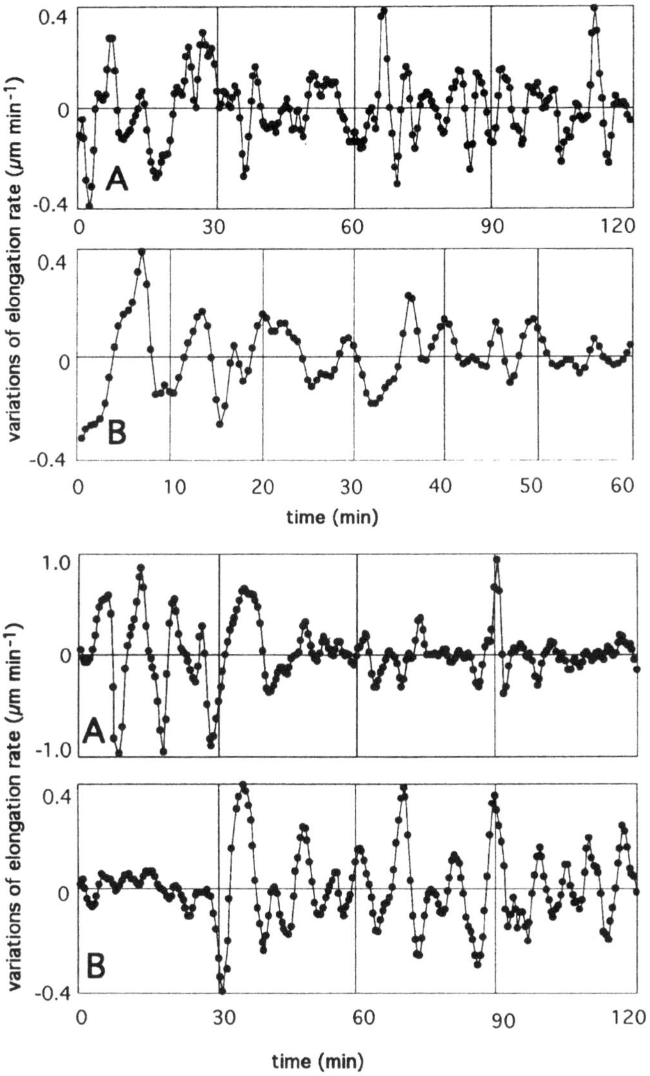

Fig. 9.21. Chaotic oscillations of the growth rate and change of periodicity. A and B (top) – Chaotic oscillations. A and B (bottom) – Change of periodicity. From ref. [31].

9.5. ATP synthesis and active transport induced by periodic electric fields

About two decades ago, Witt et al. [32] reported that chloroplasts subjected to an electric field can synthesize ATP in the dark. Since then, some interesting studies have been devoted to both ATP synthesis and transport of ions and molecules induced by periodic electric fields [33–40]. Thus, Na, K-ATPase of human erythrocytes can respond to periodic electric fields, and ATP-dependent sodium, potassium and rubidium pumping activities have been detected. The optimal field strength for activating both sodium and potassium pumps is 20 $\mathrm{V\,cm^{-1}}$, which corresponds to a $\Delta\Psi$ of 12 mV, or to a transmembrane electric field of 24 $\mathrm{kV\,cm^{-1}}$. The stimulated cation pumping activity does not require ATP hydrolysis. Moreover, cation pumping can be inhibited by the usual inhibitors ouabain, oligomycin and vanadate [41]. Similarly ATP-synthase anchored in a membrane and submitted to periodic electric fields can synthesize ATP [42,43]. The usual inhibitors of this enzyme arrest ATP synthesis obtained in these conditions. In fact, periodic electric fields result in an increase of the equilibrium constant of ADP–ATP phosphorylation–dephosphorylation, which is multiplied by a factor of 10^5. The yield of ATP synthesis, however, is small, less than one molecule synthesized per enzyme molecule and per electric pulse. This yield increases in the presence of dithiothreitol. Indeed, ATP synthesis takes place in the total absence of electron transfer process.

The mechanism of ATP synthesis by periodic fields has been studied in detail. In fact, membrane proteins may interact with electric fields for different, but interrelated, reasons. At neutral pH, Lys, Arg and His are positively charged, whereas Asp and Glu are negatively charged. These positive and negative charges often constitute dipoles. The peptide unit of an α-helix constitutes a dipole of 3.5 Debyes. An α-helix made up of n peptide units thus constitutes a dipole of $3.5n$ Debyes. β-sheets, however, have little dipole moments. Modified (for instance phosphorylated) aminoacids carry charges (negative charges if phosphorylated). Water molecules are often aligned within the protein structure and form dipoles as well. All these features give membrane proteins properties permitting them interact with electric fields.

When submitted to periodic fields, a membrane protein may undergo electroconformational changes that can alter its properties. If K is the equilibrium constant between two of these conformational states, one has

$$\left(\frac{\partial \ln K}{\partial E_{\mathrm{e}}}\right) = \frac{\Delta M}{RT}, \tag{9.135}$$

where E_{e} is the effective transmembrane electric field equal to $E_{\mathrm{e}} = -\partial\Psi/\partial x$ and ΔM is the difference between the final and initial states. Thus, in the case of ATP-synthase, the successive binding steps of ADP and phosphate, followed by the phosphorylation of ADP, are all exergonic. It is the release of ATP which is endergonic (see chapter 5). The electroconformational change of the enzyme precisely allows the release of ATP. This requires in turn that the ΔM value for the final state (the state of the free enzyme) be positive and larger than that of the initial state (that of the E-ATP complex), in such a way that the final state be stabilized relative to the initial state. Then the step of ATP desorption will become exergonic (Fig. 9.22).

Tsong [33] has studied a simple theoretical model that fulfils the requirements of a pump enforced by a periodic electric field. This model is formally identical to those already discussed in chapter 5. But here L is a neutral molecule, not an ion. In this model, the ligand binding site of P_1 for L faces the cytoplasm, and that of P_2 faces the bulk medium. Moreover, one has to assume that the molar electric moment of P_2 is larger than that of P_1.

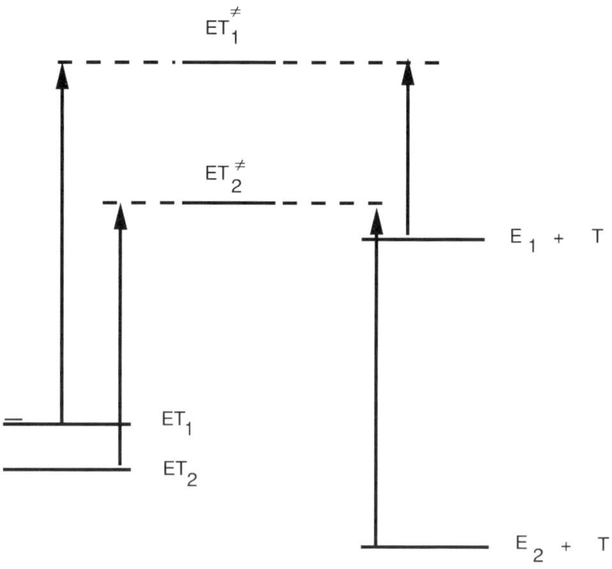

Fig. 9.22. Free energy diagram showing how an electric field may change a chemical process. In the example chosen, ATP (T) does not dissociate from ATP-synthase because the final level ($E_1 + T$) is higher than the initial one (ET_1). A transmembrane electric field may stabilize the final state ($E_2 + T$) more effectively than the initial one (ET_2).

If the system is submitted to a sine wave of an electric field, then the concentrations of P_1, P_2, P_1L and P_2L may start oscillating and this allows the transport of ligand L against a concentration gradient. This model has been worked out in detail by Westerhoff [44] and Chen [45] and will not be discussed further here.

Although these model studies are no doubt interesting from a physico-chemical viewpoint, one may wonder about their biological relevance. One may, in particular, raise the point of the existence of periodic electric fields in living cells. We believe there exists in the cell the physical conditions required for the existence of such periodic fields. We have seen, in the present chapter, that even a simple open metabolic cycle taking place at the interface between a charged membrane and a bulk medium can generate sustained oscillations of the electrostatic partition coefficient Π if, at least, one of the intermediates of the cycle is charged. The periodic variation of Π implies the existence of a periodic variation of $\Delta\Psi$, i.e. of the electric field. It then becomes sensible to think that ATP may be synthesized owing to the periodic electric fields generated in the membranes, thanks to periodic biochemical systems. If this were the case, ATP would be synthesized in the living cell by three different types of mechanisms: the coupling between endergonic ATP synthesis and a scalar exergonic biochemical reaction (this is what takes place in glycolysis); the coupling between ATP synthesis and the vectorial transport of ions across a membrane (this is the case for oxidative and photosynthetic phoshorylation); the coupling between ATP synthesis and a periodic electric field generated by a periodic biochemical process, taking place at the interface between a membrane and the bulk medium.

9.6. Some functional advantages of complexity

We have seen, in the previous section, that the periodic behaviour of metabolic processes, taking place at the interface between a membrane and a bulk phase, can generate a periodic electric field that could drive the active pumping of ions or molecules, or could allow the synthesis of ATP in the absence of any electron transfer process. Since these effects would not be expected to occur if the metabolic process were in steady state, there is an obvious functional advantage, for the living cell, to display these periodic metabolic processes. It has also been outlined by Ross and his associates [46] that periodic biochemical events display reduced energy dissipation and therefore an improved yield relative to the same process taking place under steady state conditions.

Ross and Schell [46] have compared the efficiency of a proton pump, working in a periodic and in a steady state mode. As already outlined (chapter 5) a proton pump transfers protons across a membrane, against an electrochemical potential gradient and at the expense of ATP consumption. This proton transfer is accompanied by a temporal variation of the membrane potential according to the equation

$$C\frac{\mathrm{d}\Delta\Psi}{\mathrm{d}t} = I_{\mathrm{p}} + I_{\mathrm{i}}, \tag{9.136}$$

where C is the capacitance of the membrane, I_{p} and I_{i} are the contributions of the proton (I_{p}) and the other ions (I_{i}) to the current, respectively. If the proton pump operates in the steady state mode, that is if the ATP concentration is held at a constant value, and if now a brief impulse of excess ATP is given to the overall system, then $\Delta\Psi$ will display a rather moderate oscillatory relaxation to a steady state. The total current ($I_{\mathrm{p}} + I_{\mathrm{i}}$), however, will display damped oscillations of larger amplitudes than those observed for the relaxation of $\Delta\Psi$, and these oscillations last longer.

For any scalar or vectorial process, one can define the energy dissipation, D, as [46]

$$D = T\frac{\mathrm{d}S}{\mathrm{d}t}, \tag{9.137}$$

or as

$$D = A\frac{\mathrm{d}\xi}{\mathrm{d}t}, \tag{9.138}$$

where S is the entropy, ξ is the advancement of the reaction and A is the affinity. Energy dissipation is thus equivalent to the product of the affinity of a process by its flow. The efficiency, η, of the proton pump may thus be defined as

$$\eta = \frac{\langle(\mu_{\mathrm{H_o}} - \mu_{\mathrm{H_i}})J_{\mathrm{H}} + \Delta\Psi I_{\mathrm{p}}\rangle}{\langle A_{\mathrm{a}} J_{\mathrm{a}}\rangle}, \tag{9.139}$$

where $\mu_{\mathrm{H_o}}$ and $\mu_{\mathrm{H_i}}$ are the electrochemical potentials of the proton outside and inside the membrane, J_{H} is the proton flow, I_{p} is the current generated by the proton flow, A_{a} is the affinity of ATP hydrolysis and J_{a} is the flow of ATP hydrolysis. The numerator and the denominator of this expression are averaged over a given period of time. Indeed, the efficiency coefficient can be determined whether the pump is working in the oscillatory or in the steady state mode. If the efficiency coefficient is measured in the oscillatory mode, the average has to be effected over one period. One may then compare the efficiency in the steady state mode (η_{ss}) with the efficiency in the oscillatory mode (η_{os}). If the ratio η_{os}/η_{ss} is plotted as a function of the ratio ω/ω_o, where ω is the frequency of ATP perturbation and ω_o the autonomous relaxation frequency, the ratio η_{os}/η_{ss} may display values higher or lower than unity depending on the value of the ratio ω/ω_o. Thus certain definite experimental conditions result in an efficiency which is better if the pump works in the periodic mode. A similar and perhaps clearer conclusion is obtained with glycolysis. The comparison of the efficiency of glycolysis, working in the oscillatory and in the steady state modes, can be made by computing the ratio $(D_{os} - D_{ss})/|D_{ss}|$, where D_{os} is the dissipation in the oscillatory mode and D_{ss} the dissipation in the steady state mode. When this ratio is plotted as a function of the total adenine concentration ($A_T = \text{ATP} + \text{ADP} + \text{AMP}$), the values obtained are negative and decrease as a function of adenine concentration. Under many different experimental conditions, the value of the ratio $(D_{os} - D_{ss})/|D_{ss}|$ is always negative, indicating that the oscillatory mode is associated with a lower dissipation than the corresponding steady state mode. In fact, reduction of dissipation means that the ATP/ADP ratio should be larger under oscillatory than under steady state conditions. Several simulations performed by Ross and Schell [46] show this is indeed true. The value of this ratio can be doubled in the oscillatory mode. These conclusions have led several biochemists (see, for instance, ref. [5]) to conclude that metabolic oscillations should be the rule rather than the exception *in vivo*.

Some general ideas, which have already been considered in this chapter, will now be discussed again in the perspective of the advantage that a complex non-homogeneous structure offers to the living cell. Even the simple compartmentalization of the intracellular space by a polyanionic membrane can generate effects that are unthinkable in a homogeneous phase with enzymes that follow Michaelis–Menten kinetics. A simple metabolic cycle, taking place at the interface between the membrane and the bulk phase, can display multiple steady states, even if the enzymes obey simple Michaelis–Menten kinetics. When the electric partition coefficient Π is plotted as a function of β_o^* (see section 9.3.4) the resulting curve may display three steady states. As β_o^* represents the material balance between the input and the output of matter, this means that the heterogeneous system is able to sense not only the value of this balance, but also whether the balance is varying in favour of the input or output of matter. Thus, electric repulsion effects coupled to biochemical reactions can play the part of an elementary biosensor. A simple metabolic cycle taking place at the interface between a charged membrane and a bulk phase may experience electric repulsion of the reagents by the fixed charges of the membrane, and this can generate sustained oscillations of the reagent concentrations, of the electric partition coefficient and of the membrane electric field. Thus, in a way, compartmentalization can drive reactions such as active transport and ATP synthesis that would never occur if the system were in steady state. Attempts to monitor the $\Delta\Psi$ values of membranes usually yield values that do not vary as a function of time. Thus the resting value of a neuron is about -70 mV and the $\Delta\Psi$ of a mitochondrial membrane is about -200 mV. Since these values are roughly constant, one may ask whether these results are not in contradiction with the existence of periodic electric fields in biological membranes. It has been outlined, however, that there is no contradiction between the physical prediction of oscillatory electric fields and the apparent constant value of $\Delta\Psi$ measured experimentally [33]. As a matter of fact, the $\Delta\Psi$ values are time-average and space-average values. Electrostatic interactions that are involved in the theoretical developments above are limited to short ranges, for the coulombic interaction has an inverse-square dependence on the distance, and is thus not likely to be detected experimentally.

A much higher degree of complexity is found in the organization of the plant cell wall. This organization has its functional counterpart in the existence of aperiodic, or "chaotic", oscillations of the growth rate and in the possible transitions between a periodic and a "chaotic" regime. If this situation were to be generalized to different biological situations, it would mean that aperiodic oscillations do not originate from lack of organization but rather from an extremely complex and subtle organization. If the Titans, in Greek Mythology, created order out of chaos, physical biochemistry teaches us that chaos can also originate from fuzzy organization and complexity ... It remains to be understood now what could be the functional advantage, if any, of such aperiodic dynamics.

References

[1] Goldebeter, A. and Caplan, S.R. (1976) Oscillatory enzymes. Annu. Rev. Biophys. Bioeng. 5, 469–476.

[2] Goldbeter, A. (1990) Rythmes et Chaos dans les Systèmes Biochimiques et Cellulaires. Masson, Paris.

[3] Goldbeter, A. (1996) Biochemical Oscillations and Biological Rythms. Cambridge University Press, Cambridge.

[4] Hess, B. and Boiteux, A. (1971) Oscillatory phenemena in biochemistry. Annu. Rev. Biochem. 40, 237–258.

[5] Hess, B. (1997) Periodical patterns in biochemical reactions. Quart. Rev. Biophys. 30, 121–176.

[6] Markus, M. and Hess, B. (1984) Transitions between oscillatory modes in a glycolytic model system. Proc. Natl. Acad. Sci. USA 81, 4394–4398.

[7] Markus, M., Kuschmitz, D. and Hess, B. (1984) Chaotic dynamics in yeast glycolysis under periodic substrate input flux. FEBS Lett. 172, 235–238.

[8] Markus, M., Kuschmitz, D. and Hess, B. (1985) Properties of strange attractors in yeast glycolysis. Biophys. Chem. 22, 95–105.

[9] Woods, N.M., Cuthbertson, K.S.R. and Cobbold, P.M. (1986) Repetitive transient rises ion cytoplasmic free calcium in hormone stimulated hepatocytes. Nature 319, 600–602.

[10] Rooney, T.A., Sass, E.J. and Thomas, A.P. (1990) Agonist-induced cytosolic calcium oscillations originate from a specific locus in single hepatocytes. J. Biol. Chem. 265, 10792–10796.

[11] Jacob, R., Meritt, J.E., Hallam, T.J. and Rink, T.J. (1988) Repetitive spikes in cytoplasmic calcium evoked by histamine in human endothelial cells. Nature 335, 40–45.

[12] Harootunian, A.T., Kao, J.P.Y. and Tsien, R.Y. (1988) Agonist-induced calcium oscillations in depolarized fibroblasts and their manipulation by photoreleased IP_3, calcium and calcium buffer. Cold Spring Harbor Quant. Biol. 53, 934–943.

[13] Meyer, T. and Stryer, L. (1991) Calcium spiking. Annu. Rev. Biophys. Biophys. Chem. 20, 153–174.

[14] Maeda, N., Niiobe, M. and Mikoshiba, K. (1990) A cerebellar Purkinje cell marker P400 protein is an inositol 1,4,5-triphosphate ($InsP_3$) receptor protein. Purification and characterization of $InsP_3$ receptor complex. EMBO J. 9, 61–67.

[15] Meyer, T., Holowka, D. and Stryer, L. (1988) Highly cooperative opening of calcium channels by inositol 1,4,5-triphosphate. Science 240, 653–656.

[16] Meyer, T., Wensel, T. and Stryer, L. (1990) Kinetics of calcium channel opening by inositol 1,4,5-triphosphate. Biochemistry 29, 32–37.

[17] Goldbeter, A., Dupont, G. and Berridge, M. (1990) Minimal model for a signal-induced calcium oscillations and for their frequency encoding through protein phosphorylation. Proc. Natl. Acad. Sci. USA 87, 1461–1465.
[18] Nicolis, G. and Progogine, I. (1977) Self-Organization in Nonequilibrium Systems. From Dissipative Structures to Order through Fluctuations. John Wiley and Sons, New York.
[19] Ricard, J., Mulliert, G., Kellershohn, N. and Giudici-Orticoni, M.T. (1994) Dynamics of enzyme reactions and metabolic networks in living cells. A physico-chemical approach. Prog. Mol. Subcell. Biol. 13, 1–80.
[20] Thomas, D., Barbotin, J.N., Hervagault, J.F. and Romette, J.L. (1997) Experimental evidence for a kinetic and electrochemical memory in enzyme membranes. Proc. Natl. Acad. Sci. USA 74, 5313–5317.
[21] Guidi, G.M., Carlier, M.F. and Goldbeter, A. (1998) Bistability in the isocitrate dehydrogenase reaction: an experimentally based theoretical study. Biophys. J. 74, 1229–1240.
[22] Kellershohn, N., Mulliert, G. and Ricard, J. (1990) Dynamics of an open futile cycle at the surface of a charged membrane. I. A simple general model. Physica D 46, 367–374.
[23] Mulliert, G., Kellershohn, N. and Ricard, J. (1990) Dynamics of an open futile cycle at the surface of a charged membrane. II. Multiple steady states and oscillatory behavior generated by electric repulsion effects. Physica D 46, 380–391.
[24] Ricard, J., Kellershohn, N. and Mulliert, G. (1992) Dynamic aspects of long distance functional interactions between membrane-bound enzymes. J. Theor. Biol. 156, 1–40.
[25] Kellershohn, N., Prat, R. and Ricard, J. (1996) Aperiodic ("chaotic") behaviour of plant cell wall extension. I. Nonlinear chemical model of periodic and aperiodic oscillations of cell wall growth. Chaos, Solitons and Fractals 7, 1103–1117.
[26] Fry, S.C. (1993) Loosening the ties. Current Biol. 3, 355–357.
[27] Nari, J., Noat, G. and Ricard, J. (1991) Pectin methyl esterase, metal ions and plant cell wall extension. Hydrolysis of pectin by plant cell wall pectin methyl esterase. Biochem. J. 279 343–360.
[28] Prat, R. (1978) Gradients of growth, spontaneous changes in growth rate and response to auxin of excised segments of *Phaseolus aureus*. Plant Physiol. 62, 75–79.
[29] Prat, R., Geissaz, M.B. and Goldberg, R. (1984) Effects of calcium and magnesium on elongation and proton extrusion of *Vigna radiata* hypocotyl sections. Plant and Cell Physiol. 25, 1459–1467.
[30] Prat, R. and Goldberg, R. (1984) Relationships between auxin- and auxin-induced growth in *Vigna radiata* hypocotyls. Physiol. Plantarum 61, 51–57.
[31] Prat, R., Kellershohn, N. and Ricard, J. (1996) Aperiodic ("chaotic") behaviour of plant cell wall extension. II. Periodic and aperiodic oscillations of the elongation rate of a system of plant cells. Chaos, Solitons and Fractals 7, 1119–1125.
[32] Witt, H.T., Schlodder, E. and Gräber, P. (1976) Membrane-bound ATP synthesis generated by an external electric field. FEBS Lett. 69, 272–276.
[33] Tsong, T.Y. (1990) Electrical modulation of membrane proteins: Enforced conformational oscillations and biological energy and signal transduction. Annu. Rev. Biophys. Biophys. Chem. 19, 83–106.
[34] Astumian, R.D., Chock, P.B., Tsong, T.Y., Chen, Y.D. and Westerhoff, H.V. (1987) Can free energy be transduced from electric noise? Proc. Natl. Acad. Sci. USA 84, 434–438.
[35] Westerhoff, H.V. and Van Dam, K. (1987) Thermodynamics and Control of Biological Free-Energy Transduction. Elsevier, Amsterdam.
[36] Neumann, E. and Katchalsky, E. (1972) Long-lived conformation changes induced by electric impulses in biopolymers. Proc. Natl. Acad. Sci. USA 69, 993–997.
[37] Teissié, J., Knutson, V.P., Tsong, T.Y. and Lane, M.D. (1982) Electric pulse-induced fusion of 3T3 cells in monolayer culture. Science 216, 537–538.
[38] Post, R.L. (1989) Seeds of sodium potassium ATPase. Annu. Rev. Physiol. 51, 1–15.
[39] Tsong, T.Y. and Astumian, R.D. (1987) Electroconformational coupling and membrane protein function. Prog. Biophys. Mol. Biol. 50, 145.
[40] Tsong, T.Y. and Astumian, R.D. (1988) Electroconformational coupling: how membrane-bound ATPase transduces energy from dynamic electric fields. Annu. Rev. Physiol. 50, 273–290.
[41] Serpersu, E.H. and Tsong, T.Y. (1984) Activation of electrogenic Rb transport of (Na,K)-ATPase by an electric field. J. Biol. Chem. 259, 7155–7162.
[42] Teissié, J. (1986) Adenosine 5′-triphosphate synthesis in *Escherichia coli* submitted to a microsecond electric pulse. Bochemistry 25, 368–373.

[43] Teissié, J., Knox, B.E., Tsong, T.Y. and Wehrle, J. (1981) Synthesis of adenosine triphosphate in respiration-inhibited submitochondrial particles induced by microsecond electric pulses. Proc. Natl. Acad. Sci. USA 78, 7473–7477.

[44] Westerhoff, H.V., Tsong, T.Y., Chock, P.B., Chen, Y.D. and Astumian, R.D. (1986) How enzymes can capture and transmit free energy from an oscillating electric field. Proc. Natl. Acad. Sci. USA 83, 4734–4738.

[45] Chen, Y.D. (1987) Asymmetry and external noise-induced free energy transduction. Proc. Natl. Acad. Sci. USA 84, 729–733.

[46] Ross, J. and Schell, M. (1987) Thermodynamic efficiency in nonlinear biochemical systems. Annu. Rev. Biophys. Biophys. Chem. 16, 401–422.

CHAPTER 10

Spatio-temporal organization during the early stages of development

If the events, such as periodic and aperiodic oscillations, that take place in time, were occurring *both* in time and space, then one should observe the emergence of patterns and forms. The aim of the present chapter is to discuss whether the development of a living organism can be understood on the basis of the dynamics of vectorial and scalar molecular processes. The fundamental problem is to offer a physical interpretation of the emergence of patterns and forms in a fertilized egg and in a young embryo. How is it possible to understand that the fertilized egg gives rise to cells that have the same genome, but will display different structures and functions? The structures of genes cannot offer the *direct* explanation of the emergence of patterns and forms out of a homogeneous medium. The only thing a gene does is to direct the synthesis of a protein that may be an enzyme, or a structural protein, or a protein that may act on the expression of another gene. If there is no doubt that genes play an important role in development, they cannot explain the emergence of forms that characterize living organisms.

10.1. Turing patterns

In 1952 [1] Turing was the first to offer a cogent explanation for the emergence of patterns in chemical and biological systems. The basic ideas of Turing were the following:

- There must exist, during early development, gradients of morphogens;
- One of these morphogens enhances its own formation by autocatalysis whereas another one inhibits the action of the first;
- All these events should take place under nonequilibrium conditions and may lead to periodic patterns.

Most biologists did not pay much attention to Turing's seminal paper, probably because his views were purely theoretical. Moreover, no Turing structure had been shown to occur until recently. In 1990, however, experimental evidence for a Turing structure was obtained for the first time [2] and this was confirmed shortly afterwards [3,4]. Several Turing patterns were thus obtained for complex chemical reactions taking place under open thermodynamic conditions in reactors containing polyacrylamide or agarose to decrease the rate of diffusion of the reagents. It thus appears extremely likely that Turing structures also exist in living organisms. The main requirement for the reality of these structures is the existence of gradients of morphogens. It is therefore essential to obtain direct evidence for the existence of these gradients during the early stages of development.

10.2. Positional information and the existence of gradients of morphogens during early development

Small specialized regions of a developing organism play an important role in the developmental process [5–8]. Moreover, it is striking to note that, depending on their spatial localization, cells that originate from the division of the same fertilized egg can have quite different fates. Thus a positional information [6–8] exists, and it is tempting to assume that organizing regions of a multicellular organism are the source of morphogenetic substances, or morphogens, that diffuse within the embryo and are responsible for the emergence of forms and patterns. Today, direct evidence for the existence of these morphogens has been obtained. These morphogenetic waves have been particularly studied in the early development of the *Drosophila* egg [9,10] and in the development of the chick limb [8].

10.2.1. Gradients and the early development of Drosophila egg

A *Drosophila* follicle is surrounded by a layer of follicle cells of somatic origin. The other cells of the follicle are germline in origin. The initial cell has undergone four mitoses, thus leading to sixteen cells. One of these is the oocyte, the others are called the nurse cells (Fig. 10.1). There are thus fifteen nurse cells in the follicle. Cytoplasmic material, namely, proteins and RNA, is transferred from the nurse cells to the oocyte through cytoplasmic connexions that exist between these two types of cells (Fig. 10.1). These cytoplasmic connections exist in a definite region of the oocyte which will become the anterior region of the egg. After fertilization and division of the nucleus, the young embryo is a syncytial blastoderm (Fig.10.2). Then, a membrane appears around each of the nuclei and the embryo becomes a cellular blastoderm (Fig. 10.2).

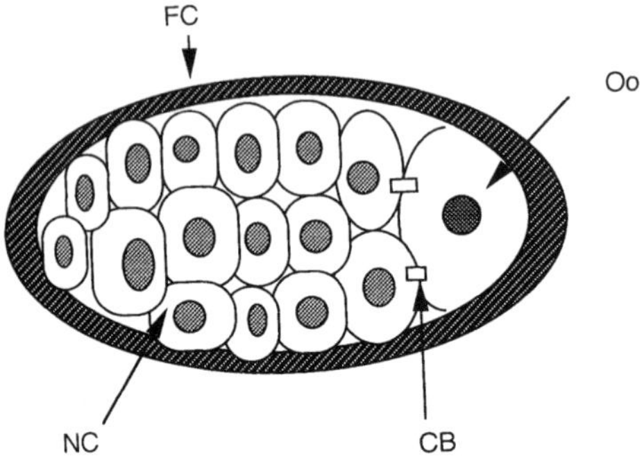

Fig. 10.1. A *Drosophila* follicle. FC: follicle cells; Oo: oocyte; NC: nurse cells; CB: cytoplasmic bridges. See text.

Fig. 10.2. Cellularization of a syncitial blastoderm. Left: syncitial blastoderm. Right: cellular blastoderm.

10.2.1.1. Gradients involve maternal gene products

Early development of the fertilized egg is controlled by maternal genes. A common feature of these maternal genes is that they are expressed prior to fertilization. Gradients are established in the egg and in the young embryo along two axes, namely the anterior–posterior and the dorsal–ventral axes (Fig. 10.3). The gradients consist of proteins and RNA whose concentration varies along the two axes. The anterior–posterior gradient is established soon after fertilization, and the other gradients appear afterwards. There are two kinds of maternal genes: the maternal somatic genes that act on follicle cells, and the maternal germline genes that act on nurse cells and on the oocyte.

Four groups of maternal genes are involved in the development of specific regions of the embryo. The anterior system is involved in the development of the head and thorax. It involves *bicoid* mRNA which is the product of the *bicoid* gene present and transcribed in nurse cells. This *bicoid* mRNA is transported into the oocyte through cytoplasmic connections. Its translation product, the bicoid protein, acts as a transcription regulator and controls the expression of another gene, *hunchback*. The posterior system is involved in the development and segmentation of the abdomen. A gene referred to as *nanos* plays a particularly important role in the process, and the corresponding product acts as a morphogen. The terminal system is responsible for the unsegmented regions of the egg, namely, the acron and the head, as well as the telson and the tail. The initial events take place in the

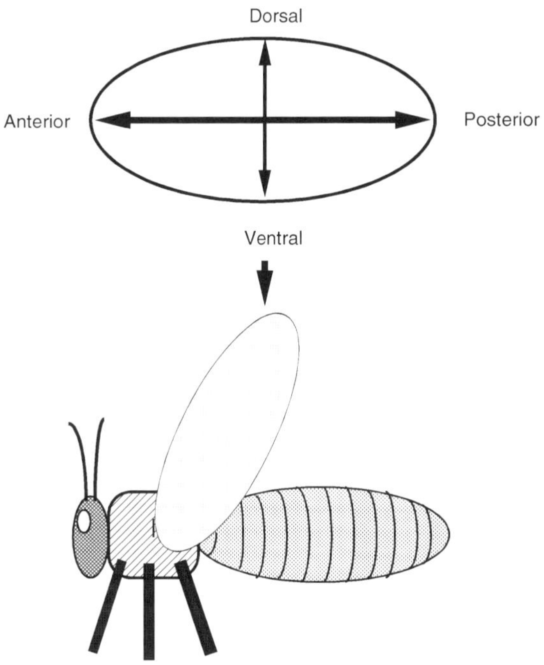

Fig. 10.3. The two axes of the gradients. The gradients are established along two axes: anterior–posterior and dorsal–ventral. These axes correspond to the axes of the adult animal.

follicle cells and result in the activation of a membrane receptor coded for by *torso*. The last system controls the dorsal–ventral development. The process is initiated by a signal originating from a follicle cell in the ventral region of the egg. It is, in turn, recognized by a transmembrane receptor coded for by *Toll.* As a consequence of this process, there appears a gradient of activation of the transcription factor produced by *dorsal.* We shall consider below only the gradients that appear during anterior–posterior and dorsal–ventral development.

10.2.1.2. Gradients of morphogens during anterior–posterior development

Anterior–posterior development has been studied with a technique referred to as the rescue technique. The principle of this technique is the following. Bicoid mutants are devoid of heads. But this can be remedied by injecting the corresponding mutant eggs with cytoplasm originating from the anterior region of wild-type embryos. The same result can be obtained using purified *bicoid* mRNA. Then, the head develops, but usually not in the correct location... These results prove that the components on which *bicoid* acts are ubiquitous. All that is needed is an appropriate concentration of *bicoid* mRNA to trigger the emergence of patterns that characterize anterior regions of the egg. As previously outlined, the *bicoid* gene is transcribed in nurse cells and the corresponding mRNA is transported into the oocyte through cytoplasmic bridges. Translation of this RNA begins after fertilization and a gradient of *bicoid* protein is established between the anterior and posterior regions of the young embryo, whether this embryo is in the syncitial blastoderm, or cellular blastoderm stage (Fig. 10.4).

Above a certain threshold of *bicoid* protein, the protein turns *hunchback* transcription on. This is formally equivalent, or perhaps really equivalent, to a phenomenon of autocatalysis. This allows one to understand that quantitative differences of protein concentration can be converted into different patterns of organization.

The posterior system appears somewhat similar to the anterior one. A maternal mRNA product of the *nanos* gene, is localized at the posterior pole of the young embryo and its corresponding protein tends to diffuse towards the anterior region of the embryo (Fig. 10.5).

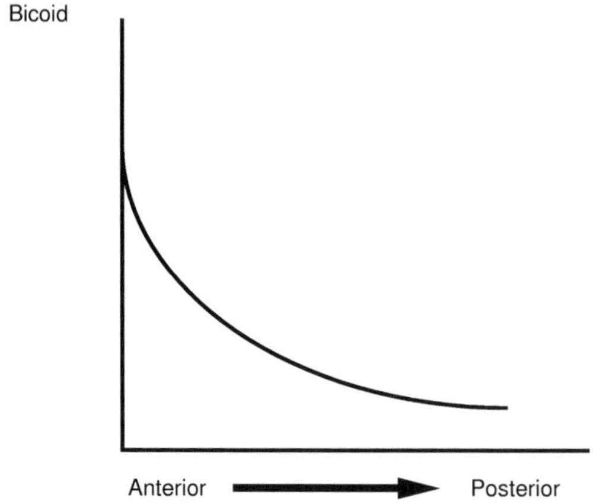

Fig. 10.4. Bicoid gradient in *Drosophila* egg. See text.

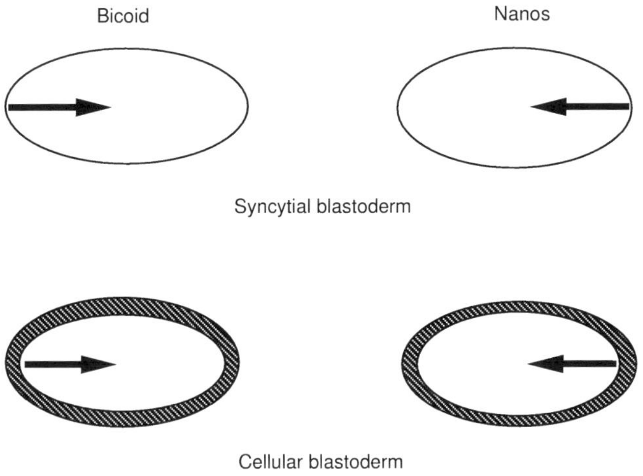

Fig. 10.5. Two gradients along the anterior–posterior axis. See text.

A gradient of *nanos* protein is established when the embryo is in the syncitial and in the cellular blastoderm stages. Both *bicoid* and *nanos* act on the expression of *hunchback* and this gene itself codes for a transcriptional inhibitor. It is required for pattern formation in the anterior region and must be absent for the formation of posterior structure. The interplay between *bicoid* and *nanos* allows the embryo to fulfil these apparently conflicting requirements. The gradient of *bicoid* protein stimulates the synthesis of *hunchback* mRNA in the anterior region, and *nanos* prevents the translation of this mRNA in the posterior region, thus leading to its degradation.

10.2.1.3. Gradients of morphogens during dorsal–ventral development

Dorsal–ventral development begins in the ventral follicle cells. The nature of the initial signal in the follicle is still unclear but, whatever the nature of this signal, it leads, in the outermost layer of the oocyte, to a series of proteolytic cleavages and, in particular, to the cleavage of the *spatzle* product. This cleavage generates a ligand for a membrane receptor coded for by the *Toll* gene. The interaction between the ligand and the *Toll* receptor takes place on the ventral side of the outermost region of the egg. It is probable that the affinity of the ligand for the *Toll* receptor is such that this ligand cannot travel a long way within the egg and remains localized in the ventral region. There exist dominant mutations that confer ventral properties to dorsal regions. The interaction of the ligand with the *Toll* receptor results in activation of the receptor which, after a sequence of events that will not be discussed, leads to a gradient of *dorsal* protein spanning the entire distance from the ventral to the dorsal side of the embryo. As previously, this gradient is established when the embryo is in a syncytial blastoderm stage and is still present in the cellular blastoderm stage (Fig. 10.6). On the ventral side, *dorsal* protein enters the nuclei, but on the dorsal side it remains in the cytoplasm. The gradient of *dorsal* protein becomes steeper as the embryo follows the transition from syncitial blastoderm to cellular blastoderm (Fig. 10.6). *Dorsal*

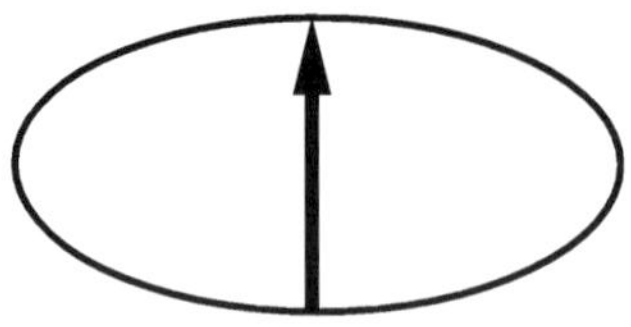

Fig. 10.6. A gradient along the dorsal–ventral axis. See text.

protein activates and represses gene expression. It activates the genes *twist* and *snail* which are required for pattern formation on the ventral side and represses *dpp* and *zen* involved in dorsal pattern formation [11–18].

10.2.2. Gradients and the development of the chick limb

Another quite different example, which is relatively well understood, shows conclusively the existence of morphogenetic waves. It concerns the development of the chick limb (see ref. [8]). In normal development there appear three morphologically distinguishable digits referred to as 2, 3, 4 to stress their homology with the digits of the five-fingered limb. If cells are taken from the posterior region of the limb bud and grafted onto the symmetrical region of a different limb bud, then this bud will develop six digits referred to as 4, 3, 2, 2, 3, 4. The sensible conclusion that can be drawn from these results is that the cells from the posterior region synthesize a substance, which acts as a morphogen and diffuses within the limb bud. At relatively high concentrations, the morphogen induces the formation of digits 2, 3, 4. This morphogen has now been isolated and identified. It appears to be retinoic acid.

Some general conclusions can be derived from this brief overview of some important results:

- gradients of morphogens are unquestionably present in developing organisms;
- at least in some cases (for instance in the case of the *bicoid* gene product), these gradients appear to originate from outside the egg and the embryo, but this does not exclude the possibility that other gradients may be generated by a self-organization process;
- there exist thresholds of morphogens, and this implies that below a threshold the response of the developing organism will be qualitatively different from that observed above the same threshold.

Experimental results such as those briefly described above are indeed essential to grasp the rationale of the physical mechanisms that underlie the dynamics of development. But they cannot give the final answer namely, how the gradients of gene products may generate forms and structures. The aim of the next section is precisely to approach this basic problem.

10.3. The emergence of patterns and forms

On the basis of the Turing equations [1], different models have been proposed to explain the emergence of patterns [19–21]. We shall consider below a model similar to that proposed by Gierer and Meinhardt [22–26] for Meinhardt's model is both sensible and is in direct connection with experimental results. We shall follow, with some modifications, the analytical treatment of this model as proposed by Segel [21] and Granero et al. [27].

10.3.1. The basic model

The model we are considering now is expressed by the two nonlinear partial differential equations

$$\frac{\partial a}{\partial t} = k_{\mathrm{a}}\frac{a^2}{i} - k'_{\mathrm{a}}a + D_{\mathrm{a}}\frac{\partial^2 a}{\partial x^2} + k_{\mathrm{s}},$$
$$\frac{\partial i}{\partial t} = k_{\mathrm{a}}a^2 - k_{\mathrm{i}}i + D_{\mathrm{i}}\frac{\partial^2 i}{\partial x^2}. \tag{10.1}$$

These equations embody a number of experimental results which have already been discussed, together with simple and sensible assumptions. The first equation in eq. (10.1) states that the concentration change of an activating morphogen (or activator) per unit time depends on an autocatalytic term (term in a^2) for the activator production. This can generate thresholds relative to the activator concentration. The autocatalytic term is slowed down by an inhibitory morphogen (or inhibitor) and this corresponds to the expression in $1/i$ (first term of the equation). Moreover, the rate of change of the activator ($\partial a/\partial t$) also depends on the degradation of the activator (term $-k'_{\mathrm{a}}a$). Indeed, a gradient of the activator should exist and the simplest way to explain the existence of this gradient is to assume that the Fick's second law applies ($D_{\mathrm{a}}\partial^2 a/\partial x^2$). But, of course, pure diffusion is by no means the only way to obtain this gradient. Any kind of exchange of the activator with neighbouring cells would fit the model equally well. Last but not least, the rate of change of the activator ($\partial a/\partial t$) is assumed to depend on a small (activator-independent) activator production (term k_{s}). This term is important for it prevents the activator concentration from falling to zero value. Moreover it allows the initiation of the reaction to take place at low activator concentrations. The constant k_{a} is the source density and describes the ability of the cells to perform the autocatalytic reaction. The second equation in eq. (10.1) states that the change of the inhibitor concentration, expressed by $\partial i/\partial t$, is a function of a^2. In other words, the activator stimulates the production of the inhibitor. The change of the inhibitor concentration ($\partial i/\partial t$) also depends on its degradation (term in $-k_{\mathrm{i}}i$) and on the inhibitor gradient (term $D_{\mathrm{i}}\partial^2 i/\partial x^2$). As previously, diffusion is the simplest way to generate a gradient but any other kind of inhibitor exchange with other cells would be acceptable.

These partial nonlinear differential equations can be numerically integrated. This allows one to express the variation of the activator and inhibitor concentrations as a function of both position and time. A condition for the formation of stable patterns is that the inhibitor diffuses more rapidly than the activator and has a shorter life-time ($D_{\mathrm{i}} > D_{\mathrm{a}}$ and $k_{\mathrm{i}} > k_{\mathrm{a}}$). Numerical integrations are indeed useful to illustrate a particular situation, but they cannot replace the analytical study of the equations which is the most rigourous way of studying the problem of the emergence of patterns.

10.3.2. Dimensionless variables

In order to simplify the system of nonlinear differential equations (10.1) one has to define dimensionless variables, namely a dimensionless time, τ, and distance, ξ

$$\tau = tk'_{\mathrm{a}}, \qquad \xi = \frac{x}{\sqrt{D_{\mathrm{a}}/k'_{\mathrm{a}}}}, \tag{10.2}$$

as well as dimensionless activator, α, and inhibitor, β, concentrations

$$\alpha = \frac{a}{a_{\mathrm{o}}}, \qquad \beta = \frac{i}{i_{\mathrm{o}}}. \tag{10.3}$$

a_{o} and i_{o} are concentrations which will be defined more precisely later. Therefore the variables can be expressed as a function of the corresponding normalized variables, giving

$$\begin{aligned} t &= \frac{\tau}{k'_{\mathrm{a}}}, \qquad x = \xi\sqrt{D_{\mathrm{a}}/k'_{\mathrm{a}}}, \\ a &= a_{\mathrm{o}}\alpha, \qquad i = i_{\mathrm{o}}\beta. \end{aligned} \tag{10.4}$$

Moreover, one has

$$\begin{aligned} &\mathrm{d}^2 a = \mathrm{d}^2(a_{\mathrm{o}}\alpha) = a_{\mathrm{o}}\mathrm{d}^2\alpha, \qquad \mathrm{d}^2 i = \mathrm{d}^2(i_{\mathrm{o}}\beta) = i_{\mathrm{o}}\,\mathrm{d}^2\beta, \\ &\mathrm{d}x^2 = d\big(\xi\sqrt{D_{\mathrm{a}}/k'_{\mathrm{a}}}\big)^2 = (D_{\mathrm{a}}/k'_{\mathrm{a}})\,\mathrm{d}\xi^2 \end{aligned} \tag{10.5}$$

Inserting of expressions (10.4) and (10.5) into eq. (10.1) yields

$$\begin{aligned} \frac{\partial\alpha}{\partial\tau} &= \frac{k_{\mathrm{a}}a_{\mathrm{o}}^2}{k'_{\mathrm{a}}i_{\mathrm{o}}}\frac{\alpha^2}{\beta} - \alpha + \frac{\partial^2\alpha}{\partial\xi^2} + \frac{k_{\mathrm{s}}}{k'_{\mathrm{a}}a_{\mathrm{o}}}, \\ \frac{\partial\beta}{\partial\tau} &= \frac{k_{\mathrm{a}}a_{\mathrm{o}}^2}{k'_{\mathrm{a}}i_{\mathrm{o}}}\alpha^2 - \frac{k_{\mathrm{i}}}{k'_{\mathrm{a}}}\beta + \frac{D_{\mathrm{i}}}{D_{\mathrm{a}}}\frac{\partial^2\beta}{\partial\xi^2}. \end{aligned} \tag{10.6}$$

To simplify even more these equations, a_{o} is chosen as to have

$$\frac{k_{\mathrm{s}}}{k'_{\mathrm{a}}a_{\mathrm{o}}} = 1, \tag{10.7}$$

which means that

$$a_{\mathrm{o}} = \frac{k_{\mathrm{s}}}{k'_{\mathrm{a}}}. \tag{10.8}$$

Similarly, i_o is chosen as to have

$$\frac{k_a a_o^2}{k'_a i_o} = \frac{k_i}{k'_a}, \tag{10.9}$$

which implies that

$$i_o = \frac{k_a a_o^2}{k_i}. \tag{10.10}$$

Therefore, equations (10.6) assume the form

$$\begin{aligned}\frac{\partial \alpha}{\partial \tau} &= 1 + \frac{k_a a_o}{k'_a i_o}\frac{\alpha^2}{\beta} - \alpha + \frac{\partial^2 \alpha}{\partial \xi^2}\\ \frac{\partial \beta}{\partial \tau} &= \frac{k_i}{k'_a}\left(\alpha^2 - \beta\right) + \frac{D_i}{D_a}\frac{\partial^2 \beta}{\partial \xi^2}.\end{aligned} \tag{10.11}$$

Setting

$$A = \frac{k_a a_o}{k'_a i_o}, \qquad B = \frac{k_i}{k'_a}, \qquad C = \frac{D_i}{D_a}, \tag{10.12}$$

the fundamental equations (10.6) can be rewritten in compact form as

$$\begin{aligned}\frac{\partial \alpha}{\partial \tau} &= 1 + A\frac{\alpha^2}{\beta} - \alpha + \frac{\partial^2 \alpha}{\partial \xi^2},\\ \frac{\partial \beta}{\partial \tau} &= B\left(\alpha^2 - \beta\right) + C\frac{\partial^2 \beta}{\partial \xi^2}.\end{aligned} \tag{10.13}$$

10.3.3. Stability analysis of temporal organization

Patterns will not be formed if α and β do not vary as a function of the distance. Therefore, in eq. (10.13), let us set

$$\frac{\partial^2 \alpha}{\partial \xi^2} = \frac{\partial^2 \beta}{\partial \xi^2} = 0. \tag{10.14}$$

If the system is in steady state, then

$$\begin{aligned}f(\alpha^*, \beta^*) &= 1 + A\frac{\alpha^{*2}}{\beta^*} - \alpha^* = 0,\\ g(\alpha^*, \beta^*) &= B\left(\alpha^{*2} - \beta^*\right) = 0.\end{aligned} \tag{10.15}$$

In these equations, the starred symbols represent the steady state dimensionless concentrations. These steady state conditons will be fulfilled if

$$\beta^* = \alpha^{*2}, \qquad \alpha^* = 1 + A. \tag{10.16}$$

As usually, stability analysis relies on the Jacobian matrix

$$J = \begin{bmatrix} \partial f/\partial\alpha^* & \partial f/\partial\beta^* \\ \partial g/\partial\alpha^* & \partial g/\partial\beta^* \end{bmatrix} \tag{10.17}$$

its trace, T_j, its determinant, Δ_j, and the discriminant, $T_\mathrm{j}^2 - 4\Delta_\mathrm{j}$, of the corresponding characteristic equation. Differentiating equations (10.15) with respect to α^* and β^* leads to

$$\begin{aligned} &\frac{\partial f}{\partial\alpha^*} = 2A\frac{\alpha^*}{\beta^*} - 1, \qquad \frac{\partial f}{\partial\beta^*} = -A\frac{\alpha^{*2}}{\beta^{*2}}, \\ &\frac{\partial g}{\partial\alpha^*} = 2B\alpha^*, \qquad \frac{\partial g}{\partial\beta^*} = -B. \end{aligned} \tag{10.18}$$

Inserting expressions (10.16) into eq. (10.18) yields

$$\begin{aligned} &\frac{\partial f}{\partial\alpha^*} = \frac{2A}{1+A} - 1 = \frac{A-1}{A+1}, \qquad \frac{\partial f}{\partial\beta^*} = -\frac{A}{(1+A)^2}, \\ &\frac{\partial g}{\partial\alpha^*} = 2B(1+A), \qquad \frac{\partial g}{\partial\beta^*} = -B. \end{aligned} \tag{10.19}$$

The trace of the Jacobian matrix is then

$$T_\mathrm{j} = \frac{A-1}{A+1} - B, \tag{10.20}$$

and its determinant

$$\Delta_\mathrm{j} = -\frac{A-1}{A+1}B + 2B(1+A)\frac{A}{(A+1)^2} = B. \tag{10.21}$$

Δ_j is of necessity positive but T_j can be positive, negative or null, and $T_\mathrm{j}^2 - 4\Delta_\mathrm{j}$ can also be positive or negative. Thus the four following situations can occur:

- $T_\mathrm{j} > 0$, $\Delta_\mathrm{j} > 0$, $T_\mathrm{j}^2 - 4\Delta_\mathrm{j} > 0$, and the system displays an unstable node;
- $T_\mathrm{j} > 0$, $\Delta_\mathrm{j} > 0$, $T_\mathrm{j}^2 - 4\Delta_\mathrm{j} < 0$, and the system has an unstable focus;
- $T_\mathrm{j} < 0$, $\Delta_\mathrm{j} > 0$, $T_\mathrm{j}^2 - 4\Delta_\mathrm{j} > 0$, and the system displays a stable node;
- $T_\mathrm{j} < 0$, $\Delta_\mathrm{j} > 0$, $T_\mathrm{j}^2 - 4\Delta_\mathrm{j} < 0$, and the system has a stable focus.

A particularly simple situation is obtained if A is very large, which means there is strong autocatalysis. Then one has

$$T_\mathrm{j} = 1 - B, \qquad \Delta_\mathrm{j} = B, \qquad T_\mathrm{j}^2 - 4\Delta_\mathrm{j} = B^2 - 6B + 1. \tag{10.22}$$

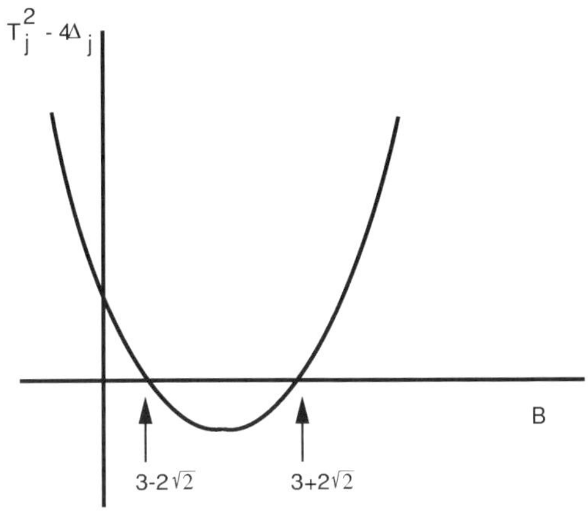

Fig. 10.7. Domain of temporal oscillations in a spatially homogeneous system. See text.

The last equation of (10.22) shows that a plot of $T_{\mathrm{j}}^2 - 4\Delta_{\mathrm{j}}$ as a function of B should yield a parabola (Fig. 10.7). Periodic oscillations of α and β are expected to occur if $T_{\mathrm{j}}^2 - 4\Delta_{\mathrm{j}} < 0$, i.e. if the value of B is located between the two roots of the equation $B^2 - 6B + 1 = 0$, namely, $3 - 2\sqrt{2} < B < 3 + 2\sqrt{2}$.

10.3.4. Stability analysis of spatio-temporal organization

If the dynamic system is expected to generate pattern formation, this means that $\partial^2\alpha/\partial\xi^2$ and $\partial^2\beta/\partial\xi^2$ cannot be neglected in equations (10.13). The dynamic system in the vicinity of a steady state is

$$\begin{aligned}
\frac{\partial x_\alpha}{\partial \tau} &= \frac{\partial f}{\partial \alpha^*} x_\alpha + \frac{\partial f}{\partial \beta^*} x_\beta + \frac{\partial^2 x_\alpha}{\partial \xi^2}, \\
\frac{\partial x_\beta}{\partial \tau} &= \frac{\partial g}{\partial \alpha^*} x_\alpha + \frac{\partial g}{\partial \beta^*} x_\beta + C\frac{\partial^2 x_\beta}{\partial \xi^2},
\end{aligned} \tag{10.23}$$

or, taking advantage of eq. (10.19)

$$\begin{aligned}
\frac{\partial x_\alpha}{\partial \tau} &= \frac{A-1}{A+1} x_\alpha - \frac{A}{(1+A)^2} x_\beta + \frac{\partial^2 x_\alpha}{\partial \xi^2}, \\
\frac{\partial x_\beta}{\partial \tau} &= 2B(1+A)x_\alpha - Bx_\beta + C\frac{\partial^2 x_\beta}{\partial \xi^2}.
\end{aligned} \tag{10.24}$$

In these eqs. (10.23) and (10.24) x_α and x_β represent perturbations relative to the steady state, namely

$$x_\alpha = \alpha - \alpha^*, \qquad x_\beta = \beta - \beta^*. \tag{10.25}$$

The system of partial differential equations (10.24) can be transformed into a system of ordinary differential equations if the perturbation functions x_α and x_β are proportional to their second derivative relative to the dimensionless distance, ξ. This is precisely the case for sine and cosine functions. In the case of the cosine function one has

$$\frac{\partial^2}{\partial \xi^2}\cos(\xi/l) = -\frac{1}{l^2}\cos(\xi/l). \tag{10.26}$$

l is a non-zero constant such that $2\pi l$ is the period of the cosine function. The perturbations x_α and x_β, which are functions of both ξ and τ, are thus defined as

$$\begin{aligned} x_\alpha(\xi,\tau) &= y_\alpha(\tau)\cos(\xi/l),\\ x_\beta(\xi,\tau) &= y_\beta(\tau)\cos(\xi/l). \end{aligned} \tag{10.27}$$

Differentiating x_α and x_β successively with respect to τ and ξ leads to

$$\begin{aligned} \frac{\partial x_\alpha}{\partial \tau} &= \cos(\xi/l)\frac{\partial y_\alpha(\tau)}{\partial \tau}, & \frac{\partial x_\beta}{\partial \tau} &= \cos(\xi/l)\frac{\partial y_\beta(\tau)}{\partial \tau},\\ \frac{\partial x_\alpha}{\partial \xi} &= y_\alpha(\tau)\frac{\partial}{\partial \xi}\cos(\xi/l), & \frac{\partial x_\beta}{\partial \xi} &= y_\beta(\tau)\frac{\partial}{\partial \xi}\cos(\xi/l), \end{aligned} \tag{10.28}$$

and the second derivatives $\partial^2 x_\alpha/\partial\xi^2$ and $\partial^2 x_\beta/\partial\xi^2$ can be obtained from these equations. One has

$$\frac{\partial^2 x_\alpha}{\partial \xi^2} = y_\alpha(\tau)\frac{\partial^2}{\partial \xi^2}\cos(\xi/l), \qquad \frac{\partial^2 x_\beta}{\partial \xi^2} = y_\beta(\tau)\frac{\partial^2}{\partial \xi^2}\cos(\xi/l). \tag{10.29}$$

Inserting expression (10.26) into equation (10.29) leads to

$$\frac{\partial^2 x_\alpha}{\partial \xi^2} = -\frac{y_\alpha}{l^2}\cos(\xi/l), \qquad \frac{\partial^2 x_\beta}{\partial \xi^2} = -\frac{y_\beta}{l^2}\cos(\xi/l). \tag{10.30}$$

As expected, the perturbation functions x_α and x_β are proportional to their second derivatives with respect to ξ. Thus, combining expressions (10.27) and (10.30) yields

$$x_\alpha(\xi,\tau) = -l^2\frac{\partial^2 x_\alpha}{\partial \xi^2}, \qquad x_\beta(\xi,\tau) = -l^2\frac{\partial^2 x_\beta}{\partial \xi^2}. \tag{10.31}$$

Inserting expressions (10.27), (10.28) and (10.29) into eq. (10.24) yields

$$\begin{aligned} \cos(\xi/l)\frac{\partial y_\alpha}{\partial \tau} &= \frac{A-1}{A+1}\cos(\xi/l)y_\alpha - \frac{A}{(1+A)^2}\cos(\xi/l)y_\beta - \cos(\xi/l)\frac{y_\alpha}{l^2},\\ \cos(\xi/l)\frac{\partial y_\beta}{\partial \tau} &= 2B(1+A)\cos(\xi/l)y_\alpha - B\cos(\xi/l)y_\beta - C\cos(\xi/l)\frac{y_\beta}{l^2}. \end{aligned} \tag{10.32}$$

As $\cos(\xi/l)$ can be cancelled from both sides of these equations, the system can be rearranged to

$$\frac{\partial y_\alpha}{\partial \tau} = \left(\frac{A-1}{A+1} - \frac{1}{l^2}\right) y_\alpha - \frac{A}{(1+A)^2} y_\beta,$$
$$\frac{\partial y_\beta}{\partial \tau} = 2B(1+A) y_\alpha - \left(B + \frac{C}{l^2}\right) y_\beta. \tag{10.33}$$

The partial differential equations (10.24) have thus been converted into ordinary linear differential equations. The question which arises now is to determine what value of the wavelength parameter l results in the emergence of patterns. These spatial structures may be periodic, and this can occur even if the dynamic system (10.33) does not display temporal periodicities. For instance, if the system has a saddle point, which is the situation that is going to be studied below, it can exhibit periodic patterns. This is a consequence of equations (10.27), which contain a factor in $\cos(\xi/l)$. Thus, even if $y_\alpha(\tau)$ and $y_\beta(\tau)$ do not vary periodically as a function of time, $x_\alpha(\xi,\tau)$ and $x_\beta(\xi,\tau)$ can vary periodically as a function of the distance. A necessary condition for the emergence of patterns is that the dynamic system (10.33) be unstable. There are in principle two conditions that generate such instability, namely, $T_\mathrm{j} > 0$ or $\Delta_\mathrm{j} < 0$. In the first case the system drifts, either monotonically or periodically, and in the second case it displays a saddle point. We shall now examine either of these two conditions.

The trace and the determinant of the Jacobian matrix associated with the linear system (10.33) are now

$$T_\mathrm{j} = \frac{A-1}{A+1} - B - \frac{C+1}{l^2},$$
$$\Delta_\mathrm{j} = -\left(\frac{A-1}{A+1} - \frac{1}{l^2}\right)\left(B + \frac{C}{l^2}\right) + 2\frac{AB}{1+A}. \tag{10.34}$$

The expression for the determinant can be rearranged to

$$\Delta_\mathrm{j} = B - \left(\frac{A-1}{A+1}C - B\right)\frac{1}{l^2} + \frac{C}{l^4}. \tag{10.35}$$

The first equation of (10.34) shows that if patterns are being formed, they tend to decrease the value of T_j. Thus, for instance, if the system were stable in the absence of pattern formation, it would be even more stable if the patterns were formed. It is, therefore, the second condition above, $\Delta_\mathrm{j} < 0$, that has to be worked out in order to determine which values of l^2 generate pattern formation. Thus, in the following analysis it will be assumed that the spatially homogeneous system is stable ($T_\mathrm{j} < 0$).

One can define a function $F(l^2)$

$$F(l^2) = \frac{l^4 \Delta_\mathrm{j}}{C} = \frac{B}{C} l^4 - \left(\frac{A-1}{A+1}C - B\right)\frac{l^2}{C} + 1, \tag{10.36}$$

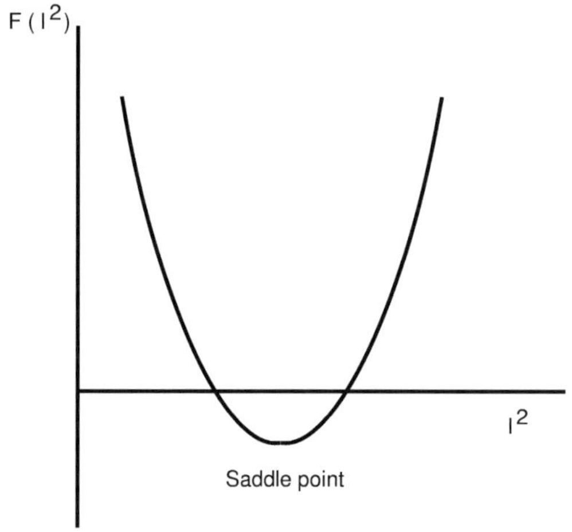

Fig. 10.8. A first condition for the emergence of patterns. Patterns may form if l^2 is comprised between the two roots of the polynomial. See text.

the sign of which is the same as that of Δ_j. Setting

$$U = \frac{B}{C} = \frac{k_\mathrm{i} D_\mathrm{a}}{k'_\mathrm{a} D_\mathrm{i}} \tag{10.37}$$

expression (10.36) becomes

$$F(l^2) = Ul^4 - \left(\frac{A-1}{A+1} - U\right) l^2 + 1. \tag{10.38}$$

The system will be unstable and a pattern will emerge if $\Delta_\mathrm{j} < 0$ or $F(l^2) < 0$. The two roots of eq. (10.38) are

$$l^2 = \frac{1}{2U}\left\{\frac{A-1}{A+1} - U \pm \sqrt{\left(\frac{A-1}{A+1} - U\right)^2 - 4U}\right\}. \tag{10.39}$$

To obtain a saddle point and pattern formation, l^2 must be real and positive and $F(l^2) < 0$. The first condition requires that

$$\left(\frac{A-1}{A+1} - U\right)^2 > 4U, \tag{10.40}$$

or

$$\frac{A-1}{A+1} > U + 2\sqrt{U}. \tag{10.41}$$

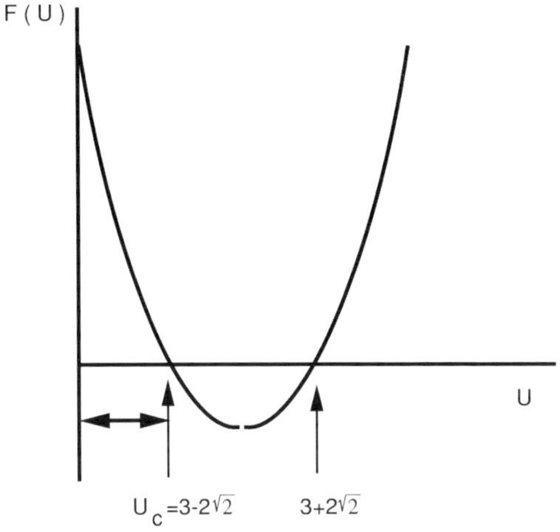

Fig. 10.9. A second condition for the emergence of patterns. Patterns may form if U is smaller than the smaller root of the polynomial. See text.

If the autocatalytic process is very fast (k_a very large relative to k'_a) then $A \gg 1$. Under these conditions, eq. (10.38) reduces to

$$F(l^2) = Ul^4 - (1-U)l^2 + 1. \tag{10.42}$$

As shown in Fig. 10.8, $F(l^2)$ will adopt negative values, and patterns will form if l^2 is comprised between the two roots. It is possible to go a little further in this analysis. As eq. (10.42) must have real roots

$$(1-U)^2 - 4U > 0 \quad \text{or} \quad U^2 - 6U + 1 > 0. \tag{10.43}$$

If one considers the corresponding equation $F(U) = 0$, the two roots are

$$U = \frac{6 \pm \sqrt{36-4}}{2} = 3 \pm 2\sqrt{2}. \tag{10.44}$$

U is indeed positive but must be smaller than unity, otherwise $F(l^2)$ (eq. (10.42)) would have been positive, which is incompatible with the requirement for pattern formation. Thus, the smaller root, $3 - 2\sqrt{2}$, is compatible with the emergence of spatial organization (Fig. 10.9) and one must have

$$0 < U < 3 - 2\sqrt{2}. \tag{10.45}$$

We may now search for the wavelength parameter value that generates the largest instability and therefore clear-cut pattern formation. This value is the value of l^2, l_c^2, that

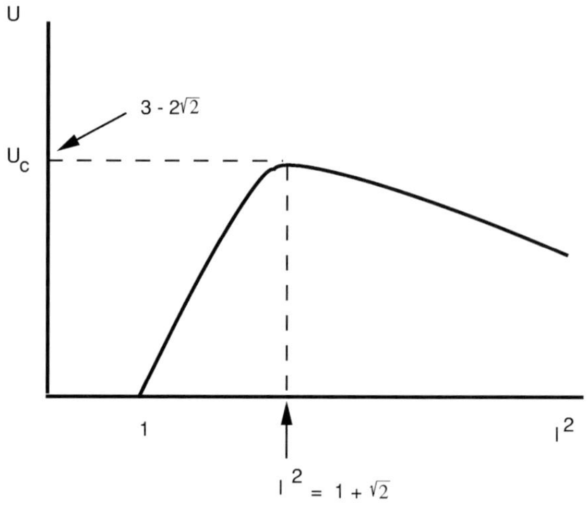

Fig. 10.10. Critical value of l^2 that generates a growing perturbation. See text.

generates the U value, U_c, which is the smaller root of the equation $F(U) = 0$. Thus one has

$$U_c = 3 - 2\sqrt{2}. \tag{10.46}$$

Equation (10.42) can be rearranged to

$$U = \frac{l^2 - 1}{l^4 + l^2}. \tag{10.47}$$

In this form, it becomes obvious that U should increase as a function of l^2, should reach its maximum value $(3 - 2\sqrt{2})$ and decrease afterwards (Fig. 10.10). The maximum of this function can be derived from the expression of $\partial U/\partial(l^2) = 0$. This is indeed equivalent in searching for the roots of the quadratic equation

$$l^4 - 2l^2 - 1 = 0, \tag{10.48}$$

that is

$$l^2 = \frac{2 \pm \sqrt{4+4}}{2} = 1 \pm \sqrt{2}. \tag{10.49}$$

As l^2 is of necessity greater than unity, the critical value, l_c, of the wavelength parameter is obtained from the larger root of eq. (10.48), namely,

$$l_c^2 = 1 + \sqrt{2}. \tag{10.50}$$

Thus, even if the spatially homogeneous system is stable, a periodic perturbation characterized by a wavelength parameter l such that

$$1 < l^2 < 1 + \sqrt{2}. \tag{10.51}$$

will grow exponentially and patterns will emerge from this perturbation.

10.3.5. Emergence of patterns in finite intervals

We have considered so far that the diffusion of the activator and inhibitor is taking place in an unbound medium. However, this is indeed not the case for living systems. One must have

$$x \in [0, L], \tag{10.52}$$

where 0 and L are the boundaries of the system. Moreover, no leak should take place through the boundaries. Therefore, one must have

$$\frac{\partial a}{\partial x} = \frac{\partial i}{\partial x} = 0 \quad \text{at} \quad x = 0 \quad \text{and} \quad x = L. \tag{10.53}$$

Expressed in dimensionless variables, these conditions become

$$\xi \in [0, \lambda] \tag{10.54}$$

and

$$\frac{\partial x_\alpha}{\partial \xi} = \frac{\partial x_\beta}{\partial \xi} = 0 \quad \text{at} \quad \xi = 0 \quad \text{and} \quad \xi = \lambda. \tag{10.55}$$

We know that

$$\frac{\partial x_\alpha}{\partial \xi} = y_\alpha \frac{\partial}{\partial \xi} \cos(\xi/l), \qquad \frac{\partial x_\beta}{\partial \xi} = y_\beta \frac{\partial}{\partial \xi} \cos(\xi/l). \tag{10.56}$$

Moreover, one has

$$\frac{\partial}{\partial \xi} \cos(\xi/l) = -\frac{1}{l} \sin(\xi/l). \tag{10.57}$$

Therefore, conditions (10.55) will be fulfilled if

$$\sin \frac{\xi}{l} = 0 \quad \text{at} \quad \xi = 0 \quad \text{and} \quad \xi = \lambda, \tag{10.58}$$

or

$$\sin \frac{\lambda}{l} = 0, \tag{10.59}$$

and this, in turn, requires that

$$\frac{\lambda}{l} = n\pi \quad (n = 1, 2, \ldots) \tag{10.60}$$

or

$$\lambda = n\pi l \quad (n = 1, 2, \ldots). \tag{10.61}$$

λ has indeed fixed values. Moreover, within the frame of expression (10.61), the value of l that results in a growing perturbation, and therefore in pattern formation, is indeed the largest possible value of l, namely l_c (eq. (10.50)). As λ and π, in eq. (10.61), have fixed values, giving l its maximum value requires that $n = 1$. Therefore, one must have

$$\lambda = \pi l_c. \tag{10.62}$$

Since the period of the cosine function is $2\pi l$, eq. (10.62) means that the growing period will be obtained for a half period. From eq. (10.61) one has

$$\cos\frac{\xi}{l} = \cos\frac{n\pi\xi}{\lambda} \quad (n = 1, 2, \ldots). \tag{10.63}$$

If this expression for $\cos(\xi/l)$ is inserted into eq. (10.33), one has, for each value of n, the system

$$\begin{aligned}
\frac{\partial y_{\alpha,n}}{\partial \tau} &= \left(\frac{A-1}{A+1} - \frac{n^2\pi^2}{\lambda^2}\right) y_{\alpha,n} - \frac{A}{(1+A)^2} y_{\beta,n}, \\
\frac{\partial y_{\beta,n}}{\partial \tau} &= 2B(1+A) y_{\alpha,n} - \left(B + C\frac{n^2\pi^2}{\lambda^2}\right) y_{\beta,n} \quad (n = 1, 2, \ldots).
\end{aligned} \tag{10.64}$$

The trace and determinant of the corresponding Jacobian matrix are

$$\begin{aligned}
T_{\mathrm{j}} &= \frac{A-1}{A+1} - B - \frac{n^2\pi^2}{\lambda^2}(1+C), \\
\Delta_{\mathrm{j}} &= -\left(\frac{A-1}{A+1} - \frac{n^2\pi^2}{\lambda^2}\right)\left(B + C\frac{n^2\pi^2}{\lambda^2}\right) + \frac{2AB}{1+A}.
\end{aligned} \tag{10.65}$$

As π^2 and λ^2 are constants, increasing the value of n results in T_{j} values that decrease and tend to be more and more negative. Conversely, the Δ_{j} values increase and tend to become more and more positive. In other words, higher values of n tend to stabilize the system and leads to the suppression of pattern formation. As shown previously, it is the value $n = 1$ that is most suited for the emergence of spatial structures.

10.4. *Pattern formation and complexity*

Simple biological systems, such as certain viruses, assemble through the direct association of proteins and nucleic acids. In this case, there is no emergence of any novel function because the properties of the virus are, in a way, already present, in a virtual state, in the protein and nucleic acid molecules that constitute this virus.

For most living objects, however, the situation is far more complex, as novel structures, forms and functions emerge that are not already present in the elements that constitute the system. These properties are thus the properties of the system, not of its elements. In order to be explained, and not only described, these emergent properties not only require precise knowledge of the elements of the system, but, above all, are based on their interconnections. Probably the most salient feature of complexity is the fact that each element of the system does not know what is happening in the system as a whole. These elements only respond to the local information they receive. If each element knew what was taking place in the system, it would embody all the complexity of that system. The early development of most living organisms represent an illustration of the complexity of these organisms.

Genes are expressed as products, mRNA and proteins,that may form gradients within the fertilized egg. These gradients may originate from events that take place outside the egg, by transfer of gene products from nurse cells to the egg. But it is also quite possible that, in a number of cases, these gradients emerge by auto-organization through the amplification of a slight deviation from a steady state. These gene products act as morphogens, namely as activators or inhibitors of development.

The central question of the biology of development is to understand how these concentration differences are converted into patterns and forms. Simple models, which are all derived from Turing's seminal proposals, offer sensible answers to this fundamental question. These models are based on the interplay between two morphogens, an activator and an inhibitor, acting under nonequilibrium conditions. The activator displays autocatalysis, which provides the nonlinearity of its response to slight changes in concentration. What is amazing is the simplicity of the ideas that underlie these models and the diversity of the patterns and forms the models can generate [20]. Thus, for instance, the periodical coat patterns of many animals (zebra, tigers, leopards, ...) can be modelled with a surprising precision in this way [20]. A central idea of complexity studies is probably that complexity may emerge from the interconnection of *simple* dynamic events.

References

[1] Turing, A.M. (1952) The chemical basis of morphogenesis. Philos. Trans. Royal Soc. London B 237, 37–72.

[2] Castets, V., Dulos, E., Bouassonade, J. and De Kepper, E. (1990) Experimental existence of a sustained standing Turing-type nonequilibrium chemical pattern. Physical Review Lett. 64, 2953–2956.

[3] De Kepper, P., Pierraud, J.J., Rudovics, B. and Dulos, E. (1995) Experimental study of stationary Turing patterns. Internat. J. Bifurcation and Chaos 4, 1215–1231.

[4] Lee, K.J., McCormick, W.D., Pearson, J.E. and Swinney, H.L. (1994) Experimental observation of self-replicating spots in a reaction–diffusion system. Nature 369, 215–218.

[5] Child, C.M. (1941) Patterns and Problems in Development. University of Chicago Press, Chicago.

[6] Wolpert, L. (1969) Positional information and the spatial pattern of cellular differenciation. J. Theor. Biol. 25, 1–47.

[7] Lewis, J., Slack, J.M.W. and Wolpert, L. (1977) Thresholds in development. J. Theor. Biol. 65, 579–590.

[8] Wolpert, L. (1991) The Triumph of the Embryo. Oxford University Press, Oxford.

[9] Driever, W. and Nusslein-Volhard, C. (1988) A gradient of bicoid protein in *Drosophila* embryos. Cell 54, 83–93.

[10] Driever, W. and Nusslein-Volhard, C. (1988) The bicoid protein determines position in the *Drosophila* embryo in a concentration dependent manner. Cell 54, 95–104.

[11] Lewin, B. (1997) Genes. Oxford University Press, New York.

[12] St. Johnston, D. and Nusslein-Volhard, C. (1992) The origin of pattern and polarity in the *Drosophila* embryo. Cell 68, 201–219.

[13] Anderson, K.V., Bokla, L. and Nusslein-Volhard, C. (1985) Establishment of the dorsal–ventral polarity in the *Drosophila* embryo: the induction of polarity by the *Toll* gene product. Cell 42, 791–798.

[14] Anderson, K.V., Jurgens, G. and Nusslein-Volhard, C. (1985) Establishment of dorsal–ventral polarity in the *Drosophila* embryo: genetic studies on the *Toll* gene product. Cell 42, 779–789.

[15] Ferrandon, D., Elphick, L., Nusslein-Volhard, C. and St. Johnston, D. (1994) Staufen protein associates with the 3′ UTR of bicoid mRNA to form particles that move in a microtubule-dependent manner. Cell 79, 1221–1232.

[16] Roth, S., Stein, D. and Nusslein-Volhard, C. (1989) A gradient of nuclear localization of the dorsal protein determines dorsoventral pattern in the *Drosophila* embryo. Cell 59, 1182–1202.

[17] St. Johnston, D. and Nusslein-Volhard, C. (1992) The origin of pattern and polarity in *Drosophila* embryo. Cell 68, 201–219.

[18] Sprenger, F. and Nusslein-Volhard, C. (1992) Torso receptor activity is regulated by a diffusible ligand produced at the extracellular terminal regions of the *Drosophila* egg. Cell 71, 987–1001.

[19] Harrisson, L.G. (1993) Kinetic Theory of Living Patterns. Cambridge University Press, Cambridge.

[20] Murray, J.D. (1993) Mathematical Biology, 2nd edition. Springer Verlag, Berlin, Heidelberg, New York.

[21] Segel, L.A. (1984) Modeling Dynamic Phenomena in Molecular and Cellular Biology. Cambridge University Press, Cambridge.

[22] Meinhardt, H. (1997) Biological pattern formation as a complex dynamic phenomenon. Internat. J. Bifurcation and Chaos 7, 1–26.

[23] Gierer, A. (1981) Generation of biological patterns and forms: Some physical, mathematical and logical aspects. Prog. Biophys. Molec. Biol. 37, 1–47.

[24] Meinhardt, H. (1983) Cell determination boundaries as organizing regions for secondary embryonic fields. Devel. Biol. 96, 375–385.

[25] Meinhardt, H. (1994) Biological pattern formation – new observations provide support for theoretical predictions. Bioessays 16, 627–632.

[26] Meinhardt, H. and Gierer, A. (1980) Generation and regeneration of sequences of structures during morphogenesis. J. Theor. Biol. 85, 429–450.

[27] Granero, M.I., Porati, A. and Zanacca, D. (1977) A bifurcation analysis of pattern formation in a diffusion governed morphogenetic field. J. Math. Biol. 4, 21–27.

CHAPTER 11

Evolution towards complexity

This chapter is certainly not devoted to the study of prebiotic evolution. Several excellent books [1–4] have recently been published that aptly review what could have been the main steps and transitions in molecular and organismic evolution. This chapter rather attempts to define, in an evolutionary perspective, the various steps that living systems must have passed through in order to reach the present complexity of an eukaryotic cell. As complexity may have many different aspects, the present discussion will be limited to those that have already been considered in the previous chapters.

It is perhaps a peculiarity of biological problems and structures that they can be considered in an evolutionary perspective. Indeed this approach can only be speculative, for it is an account of a series of unique events that occurred a long time ago. But speculative discourse does not necessarily mean unsound disourse... There is little doubt that certain speculations about the origins of life are not sensible because they violate some known physical principle. But others are perfectly reasonable and offer an attractive scenario of what could have happened to lead to more and more complex living organisms. Function is the driving force of evolution. In a neo-Darwinian perspective, mutations affect molecular and supramolecular edifices and selection keeps the structure thas is fittest for a given function. The aim of this chapter is precisely to discuss, from this neo-Darwinian perspective, the different aspects of biological complexity we have previously considered.

11.1. The need for a membrane

It is highly probable that, for quite a long time, molecules able to catalyse chemical reactions and molecules able to replicate their own structure existed on mineral surfaces, thus mimicking a primordial metabolism [5–8]. A first step in complexity may have been reached when these molecules, located in the same area, started acting in a co-ordinated manner. A second, and major, step towards complexity should have been the individuation of the molecules by a membrane that separates them from the external world.

The present relation which we know today between the molecules that store information and replicate (nucleic acids) and those that generate the phenotype (proteins and enzymes) raises a chicken-and-egg paradox, if it is assumed this relation existed even in the prebiotic period. As a matter of fact, enzymes that catalyse chemical reactions require genes in order to be synthesized and genes require enzymes in order to be replicated and expressed. The classical solution of this paradox is to assume that nucleic acids came first and were "charged" with both storage and replication of information and also "endowed" with catalytic activity [9–13]. As could have been expected, it has since been shown that RNAs, called ribozymes, display such catalytic activity [9–13]. It has thus been assumed that an "RNA world" existed in primordial times and that storage of information and expression of a phenotype were carried out by the same molecules, namely RNAs. This hypothesis is unquestionably attractive and cannot probably be avoided, but leads to some serious difficulties. First, contrary to aminoacid synthesis, nucleotide synthesis is extremely difficult to carry out under presumptive prebiotic conditions. Second, isomeric impurities would block polymerization of these nucleotides [14]. To eliminate these potential problems, some authors have assumed that ribose could have been replaced by a simpler molecule, such as glycerol phosphate [15]. Cairns-Smith [16] has suggested that primordial genes were not made up of RNA but of clay. Clay crystals are readily formed in the presence of suitable ions. These crystals grow and then divide, thus mimicking replication. If there are errors in the crystal lattice, and if these errors are replicated, then the process will be equivalent to the storage and replication of information. Clay crystals would have been replaced much later by RNA molecules.

An extremely attractive hypothesis put forward some years ago by Wächterhäuser [7, 8,17] is that of surface metabolism. The basic idea, which constitutes the core of the hypothesis, is the assumption that the primordial chemical reactions were not taking place in a primordial soup [18–20], as thought before, but on mineral surfaces. This view is somewhat reminiscent of Cairns-Smith's proposal. The surface favoured by Wächterhäuser, however, is not clay, but pyrite. As this surface is positively charged, it would allow the binding of anions to the surface, thus decreasing the entropy of the system without impeding migration of the ions on the pyrite surface. This should considerably increase the probability of occurrence of chemical reactions, for which the entropy is known to be a key factor in the control of chemical reactions.

The accuracy of replication raises a problem which, in turn, allows one to predict that replicating molecules should operate in co-operation. If an RNA-template is incubated with a replicase enzyme and activated ribonucleotides, the RNA-primer is replicated many times. A drop of the resulting medium is then transferred to a similar medium devoid of primer. One observes, after a series of transfers, that the length of the replicated RNA has decreased and reached a fixed value [21]. This elegant experiment can be viewed as an example of natural selection exerted at the molecular level. The reason for this result is that short RNA molecules replicate faster than longer ones. If the polynucleotide molecules become too short, however, they are no longer recognized by the replicase. Hence these two antagonistic influences tend to define an optimal length of the replicator. These results show that the only criterion of fitness is the replication rate in a "struggle for life" experiment. Moreover they suggest that primordial replicators could not have stored much information in their molecule. In order to reach a higher order of complexity one has to imagine that different replicators act in a co-ordinated manner. This set of replicators working co-ordinately is referred to as a hypercycle [1,5,6]. A simple hypercycle of four components is shown in Fig. 11.1.

Four replicators $I_1, \ldots, I_4$ are expressed as four primitive replicases, $E_1, \ldots, E_4$, which in the "RNA world" may be identical to the replicators. Each catalyst E_i aids the replication of I_{i+1}. The different replicators serving different functions should not competitively exclude each other. Moreover the integrated system of replicators must compete against less efficient systems [1,5,6]. As outlined by Maynard-Smith and Szathmary [2], any food chain plays the part of a hypercycle. Thus, although the information stored in each replicator is not large, the co-ordination between the replicators within a hypercycle allows one to reach a fairly high degree of sophistication and complexity. Hypercycles, however, cannot be considered primitive individuals in the sense that bacteria are. They just represent populations of molecules acting on mineral surfaces in a co-ordinate manner.

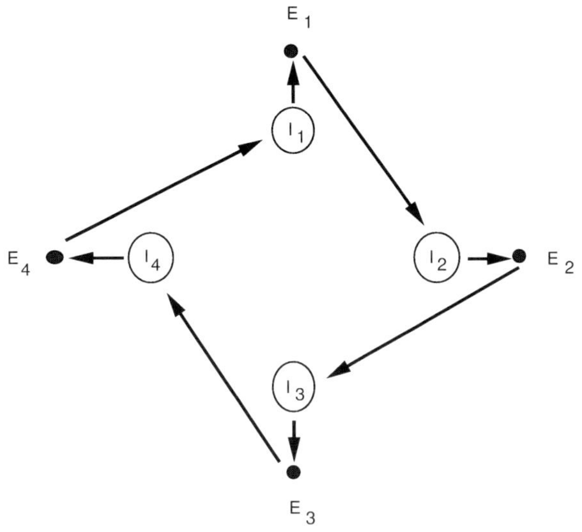

Fig. 11.1. Hypercycle of four components. See text.

It seems that the great leap towards complexity is the isolation of these populations of replicating macromolecules from the external world. It appears logical to consider that there is a selection pressure that favours compartmentalization of a hypercycle, leading to the individuation of a protocell. Let us assume, for instance, that a mutation affects the replicase E_4 which becomes E'_4 (Fig. 11.2). Even if E'_4 is more efficient than E_4 for the replication of I_1, the new hypercycle will be disfavoured relative to the old one because the number of mutated replicase molecules E'_4 is much smaller than in the corresponding wild type. The new hypercycle will then disappear (Fig. 11.2) and the replicase mutation is referred to as altruistic.

Alternatively, if the mutation alters the replicator I_4, which becomes I'_4, the new replicator will be expressed as a new replicase E'_4. If the new replicator I'_4 is a better target for replicase E_3, then the new hypercycle will be favoured relative to the old one and will tend to become dominant (Fig. 11.3). The mutation is defined as selfish.

If a hypercycle subjected to an altruistic mutation is surrounded by a membrane, the new hypercycle will not disappear and may even become dominant. It is therefore logical to conclude that there must exist a selection pressure in favour of the individuation of hypercycles and therefore in favour of the synthesis of a membrane.

The structure of biomembranes is defined by thermodynamic conditions and is associated with a minimum of the energy required for definite organization of lipids. The hydrophobic head of these lipids is in contact with water molecules and the hydrophobic tail is attracted towards the apolar part of other lipid molecules. This means that the most stable structure is a bilayer. Moreover in order to avoid the contact of the edges of the bilayer with water molecules, the supramolecular structure spontaneously adopts a spherical shape and forms vesicles.

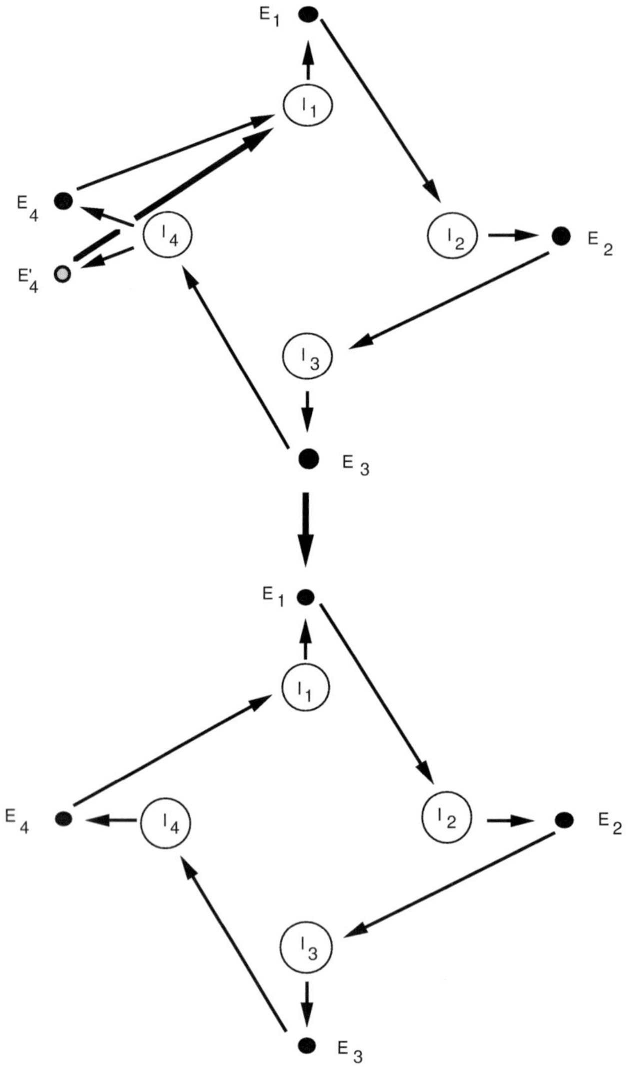

Fig. 11.2. Favourable mutation of a replicase in a hypercycle is not conserved. See text.

Wächterhäuser [7,8,17] has outlined the view that once lipid bilayers are formed on pyrite surfaces, the chemical milieu is changed and acquires a hydrophobic character, thus making polymerization reactions easier. Cellularization may then take place by abstriction form the mineral surface (Fig. 11.4). The incorporation of proteinoids into the membrane would allow the transport of ions and solutes across this membrane. Before the abstriction from the pyrite surface took place, most chemical reactions mimicking proto-metabolism were oxidation–reduction processes involving H_2S, thioesters and more generally iron–

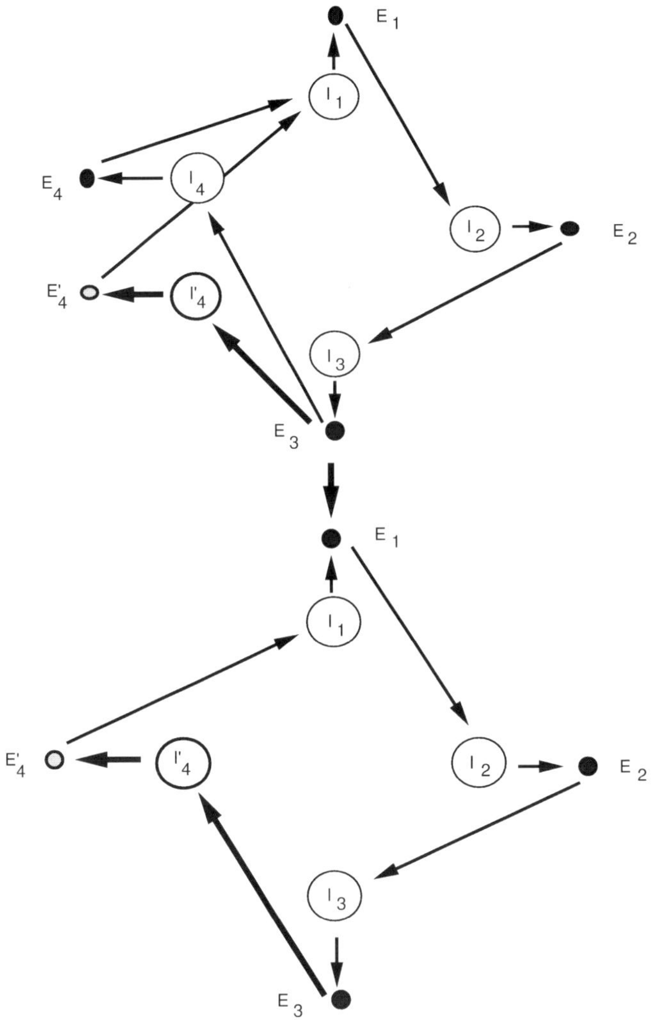

Fig. 11.3. Favourable mutation of a replicator in a hypercycle is conserved. See text.

sulfur compounds. This surface metabolism must have continued within protocells after their individuation [17]. This view of an "iron–sulfur world" is not opposed to the view of an "RNA world". In fact, it has been shown [8] that a RNA-based redox metabolism is likely to have occurred. These protocells probably contained DNA, for a primitive ribonucleotide reductase could convert ribonucleotides to deoxyribonucleotides, thus leading to DNA. The primitive ribo-organisms might have synthesized tetrapyrrols, for the synthesis of the terapyrrol precursor 5-aminolevulinate can be obtained through the reduction of a RNA-bound glutamic acid [8]. These results lead to the view that the primitive ribo-organisms were presumably photosynthetic.

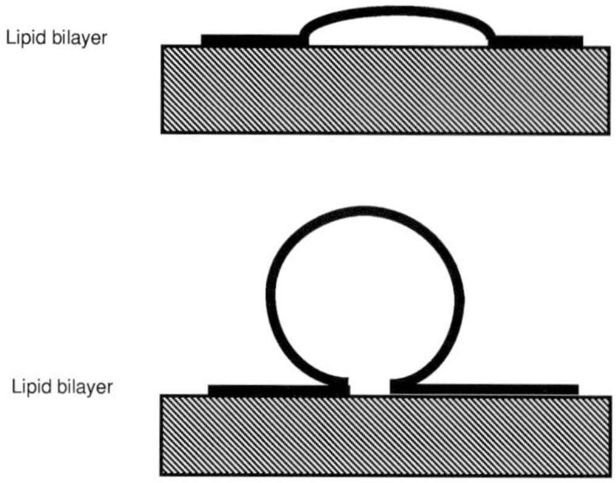

Fig. 11.4. Cellularization by abstriction of a lipid bilayer from a mineral surface. See text.

Protocells may have reacted to changes in temperature by changing their shape. Some shape transformations mimic the process of endocytosis [22]. The reason for these shape changes is that the two lipid layers react differently to changes in temperature of the milieu. These experimental results pave the way to an attractive hypothesis that could allow one to understand how a drift from surface metabolism to real metabolism took place and how the first living systems appeared [7].

A protocell would bind primitive genetic material to its outer surface. The inside of the vesicle would then correspond to the presumptive periplasm of a bacterium. The cell wall is synthesized in this periplasm. Then, probably after a change of external temperature, the vesicle bends in a sort of "endocytosis motion". The two lips fuse, thus leading to the formation of a primordial bacterial cell. The interruptions of the cell wall correspond to Bayer's patches in the bacterium (Fig. 11.5). A variant of this idea is to assume that all these events took place not in an isolated protocell, but on the pyrite surface [23,24].

Once a set of molecules that can replicate and catalyse chemical reactions are individuated as a protocell, the protocells should acquire two essential properties: the ability to propagate favourable mutations; and the property of sensing the external world. The first property is self-evident. A hypercycle bearing a favourable mutation is not necessarily selected whereas a protocell bearing that mutation will be selected. The second property has already been discussed in a different context (chapter 3). This property, however, is far from trivial and is worth being considered again in a slightly different perspective. Diffusion of molecules through channels in the membrane may be a "slow" process that is kinetically coupled to nonlinear chemical reactions within the protocell. As shown in chapter 3, this can lead to two important related effects. A dramatic increase, or decrease, of ligand concentration in the bulk phase can result in a minute change of the concentration of this ligand in the protocell (Fig. 11.6). In a way, the internal concentration is buffered against large changes of concentration in the external milieu.

The second property is the ability for the protocell to sense, not only the present concentration of a chemical in the external milieu, but whether this present concentration has been reached after an increase or a decrease of an earlier concentration. Thus the protocell has stored short-term memory of an event that has taken place in the external world (see chapter 3). As already outlined, it seems that bacterial chemotaxis takes advantage specifically

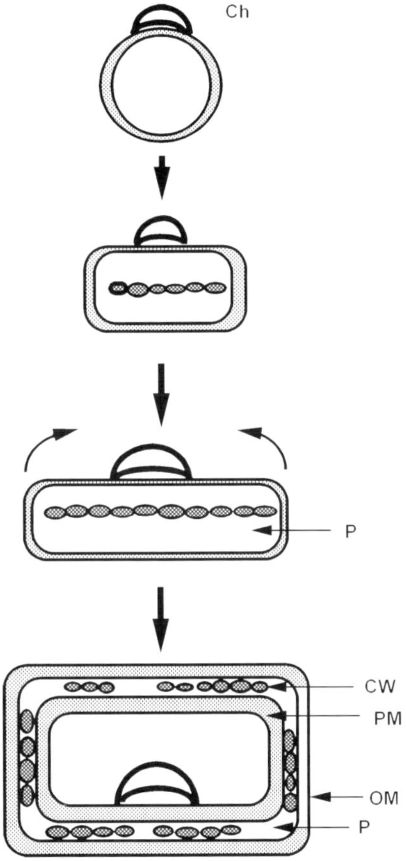

Fig. 11.5. From protocell to primordial bacterial cell. This tentative sequence of events, suggested by Blobel [23], Cavalier-Smith [24], and Maynard-Smith and Szathmary [2] implies that a vesicle bends and the two lips fuse. See text. P: periplasm, CW: cell wall, OM: outer membrane, PM: plasma membrane, Ch: chromosome.

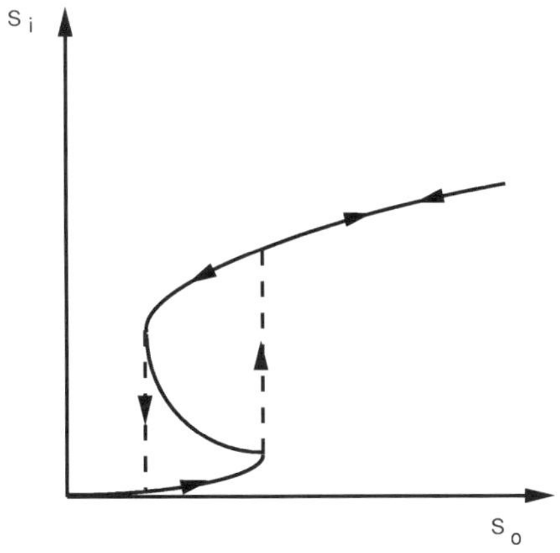

Fig. 11.6. Hysteresis loop of a substrate concentration inside a protocell as a function of the corresponding external concentration. See text.

of a coupling between vectorial and scalar chemical processes. We thus feel that the stage of membrane formation, comprising a set of macromolecules that act in co-ordination, is probably a major transition in evolution, for it gives the resulting protocell the ability to evolve and to sense the external world.

11.2. How to improve the efficiency of metabolic networks in homogeneous phase

Once a protocell evolves, its metabolism may become more and more complex. Even in simple bacterial cells, the metabolic network appears as a system of many connected enzyme-catalysed reactions that take place in a cell milieu which can be considered, as a first approximation, a homogeneous phase. Several important questions can be raised as to the evolution of this complex supramolecular machine. More precisely, these questions are the following. What is the origin of connected metabolic reactions? Are metabolic networks in homogeneous phase efficient supramolecular machines? How can we conceive an increase of their efficiency?

11.2.1. The possible origin of connected metabolic reactions

One may tentatively speculate that the early protocells lived in an environment extremely reach in metabolites [25]. Therefore a connected metabolism was probably not required for the primitive "life" of these "organisms". Metabolism probably emerged as a consequence of a shortage of nutrients in the cell milieu containing the protocells. A sensible idea put forward by Horowitz [25] and Kauffman [3] is that metabolic pathways were built up backwards. Let us assume, for instance, that a primitive catalyst catalyses the conversion of $S_{n-1} \rightarrow S_n$. This means that the catalyst, an enzyme for example, is able to bind S_{n-1} and S_n. The reason for this property is that these two ligands have related structures. But now if a mutation affects the structure of the enzyme which converts S_{n-1} into S_n, the new mutated enzyme is still able to bind S_{n-1} and is also able to bind another substance, S_{n-2}, and to catalyse its conversion into S_{n-1}. Then, through successive mutations, a metabolic network would grow backwards.

11.2.2. The poor efficiency of primitive metabolic networks in homogeneous phase

The dynamics of a complex metabolic network has to be strictly controlled. This control may be effected spontaneously and, in a way, shared by all the enzymes of the pathway. Such a network, however, would be a rather poor energy converter and should suffer from several physical drawbacks. One is associated with the molecular state of the enzymes. These enzymes are different physical entities, and the product of one reaction is the substrate of the reaction that comes next in the reaction sequence. This means that the product of the first reaction has to diffuse to the active site of the enzyme that catalyses the second reaction, and so on. As already outlined, the overall reaction flow is limited by the dilution of the reaction intermediates in the cell compartment and by the diffusion of these intermediates to the active sites of the competent enzymes.

A second physical restriction, probably exerted on the efficiency of primitive metabolic pathways, comes from futile recycling of reaction intermediates. For instance, in present-day metabolism, the two enzymes phosphofructokinase and fructose bisphosphatase catalyse antagonistic enzymatic reactions. The first reaction results in the synthesis of fructose bisphosphate with the consumption of ATP, and in the second reaction fructose bisphosphate is hydrolyzed. This represents a waste of free energy. Last but not least, a linear metabolic sequence of enzymes that follow Michaelis–Menten kinetics tends to adopt a steady state but is unable to display threshold effects. This means that, upon a slight change of concentration of an intermediate, the overall system will show only slight changes in the overall flow. This can be a functional disadvantage, for a metabolic network may have to display a sharp response to a slight change in a metabolic signal.

A potential difficulty raised by the existence of primitive metabolic cycles is that some of the intermediates may be taken away and used in other metabolic processes. Thus, starting the process with n_1 molecules, we will complete the first turn with n_2 molecules of the same intermediate, n_1 being larger than n_2. The cycle will then rapidly run down.

One of the major roles of metabolic networks is to store part of the free energy released by catabolic reactions in energy-rich compounds, such as ATP, and to use these energy-rich metabolites to synthesize complex molecules. If, however, the metabolic network leading to ATP synthesis takes place in a homogeneous phase its yield cannot be large. The reason for the poor yield is that the synthesis of an energy-rich bond has to be coupled with an exergonic reaction that releases more free energy than required for ATP synthesis. Thus the question which is at stake is to understand how bacterial cells may have evolved as to cope with these physical limitations.

11.2.3. How to cope with the physical limitations of a homogeneous phase

Some physical limitations on metabolic efficiency brought about by the existence of a nearly homogeneous phase cannot be suppressed, but others can be circumvented. The dilution of the reaction intermediates before they can reach the active sites of the enzymes may be limited by the small volume of the bacterial cell. This implies that, as cell size increases, as with eukaryotes, simple diffusion of the reaction intermediates towards the active sites of the competent enzymes will no longer provide an adequate physical basis for the efficiency and the control of the eukaryotic metabolic neworks.

To avoid the extinction of simple metabolic cycles [26], one may assume that autocatalysis has been selected to increase the regeneration of certain intermediates. The fine tuning of this process can keep the cycle running.

Futile cycles, which exist in connected metabolism and which represent a waste of energy, can also be used to form cascades that generate strong co-operativity and therefore thresholds (see chapter 3). These thresholds would then be established at the expense of free energy consumption. They would allow the metabolic system to respond in an all-or-none manner to slight changes of the intensity of a stimulus. Another way to cope with futile cycles is precisely to avoid wastage of energy by selection of a temporal organization process. It has been speculated [27], and we find this speculation sensible, that instead of having steady state concentrations of the two intermediates S_1 and S_2 these concentrations oscillate in phase opposition. Then, when the conversion $S_1 \rightarrow S_2$ takes place, the antagonistic reaction $S_2 \rightarrow S_1$ may be virtually inexistant, and *vice versa*. The cell can then avoid energy wastage since the two antagonistic processes will not coexist. This situation may be viewed as a sort of functional compartmentalization [27]. As we have previously shown (chapter 9), sustained oscillations require that at least one enzyme involved in the cycle display remarkable properties. The simplest of these properties is the existence of inhibition by an excess substrate. To observe this phenomenon under realistic physiological substrate concentrations, one has to assume that the enzyme has at least two sites: the standard active site that binds the substrate and is responsible for catalysis; and another site whose occupancy by the substrate blocks the catalytic process. This means that information is transferred from site to site within the enzyme molecule. Therefore this fairly sophisticated dynamic mechanism relies on a complex structure of the enzyme molecule.

Complexification of the structure and function of certain enzymes involved in key metabolic networks can give these networks improved properties and performances. The emergence of enzymes having a multimeric structure and displaying information transfer between identical or different sites can generate complex effects, such as retro-inhibition, that become superimposed on the standard control exerted by the set of enzymes involved in the network.

Two general conclusions can be drawn from this analysis. First, some of the limitations to the efficiency of a metabolic process taking place in a homogeneous phase can be circumvented if selection pressure tends to favour the emergence of complexification of certain enzymes occupying "key" positions in metabolism. Second, the yield of free energy storage cannot be very large in a homogeneous phase.

11.3. The emergence and functional advantages of compartmentalization

The appearance of eukaryotic cells during evolution is associated with the process of compartmentalization. Compartmentalization leads to the formation of specialized organelles such as the nucleus, mitochondria, chloroplasts, ... The conversion of prokaryotes into eukaryotes requires the disappearance of the cell wall that exists in bacteria. The precise reason for the disappearance of the peptidoglycan cell wall is unknown. A possible reason, however, is that an antibiotic was synthesized by a competing prokaryote. The secretion of this antibiotic to the external milieu resulted in the appearance of naked prokaryotic cells. The formation of an endoskeleton then becomes a necessary condition for their viability.

11.3.1. The symbiotic origin of intracellular membranes

As soon as prokaryotic cells lost their cell wall, phagocytosis and endosymbiosis became possible. Mitochondria and chloroplasts of present day cells descend from free-living bacteria [28,29]. There are many arguments in favour of this idea. Mitochondria and chloroplasts contain circular DNA and independent protein-synthesizing machinery, namely polymerases, tRNAs and ribosomes. The machinery is of the bacterial type, not of the eukaryotic type. In the course of evolution, many of the genes of the bacterial symbionts have been transferred to the nucleus. In fact, some nuclear genes of eukaryotic cells bear reminders of their bacterial origin. Their sructure is quite similar to bacterial genes.

Moreover in the organelles there have been changes in the "universal" genetic code. Thus UGA, which is a stop codon in the universal code, specifies tryptophan in animal and yeast mitochondria. The transfer of genes from the organelles to the nucleus resulted in the evolution of a specific transport system between the cytoplasm and the organelles, because proteins synthesized in the cytoplasm have to be transported to the organelles. It is not clear why so many genes have been transferred from mitochondria and chloroplasts to the nucleus. There is perhaps a trend towards simplicity, avoiding the presence of many copies of the same gene in the same cell.

The ancestors of mitochondria and chloroplasts were purple bacteria and cyanobacteria. Purple bacteria cannot perform photosynthesis in the presence of oxygen and cyanobacteria lack the Krebs cycle and are unable to perform normal respiration. It is therefore not a surprise that mitochondria and chloroplasts, which are most probably symbionts, cannot perform photosynthesis and respiration, respectively.

11.3.2. Functional advantages of compartmentalization

In the light of the results of chapter 4, the advantages of compartmentalization are obvious. The existence of mitochondrial and thylakoid membranes with different proton concentrations on the two sides results in ion transport through specific regions of the membrane. The dissipation of the electrochemical gradient is coupled to ATP synthesis. This process is not specific to eukaryotic cells, it already exists in bacteria where ions are transported across the plasma membrane. But this event is more conspicuous in animal and plant cells, for the surface of membrane is much larger in eukaryotic cells.

Another advantage of compartmentalization is that ions and other solutes can be transported against a concentration gradient. This gives the organism a major functional advantage because many cells live in nutrient poor environment. The existence of active transport is thus essential to make these biological systems less dependent on external conditions.

11.4. Evolution of molecular crowding and the different types of information transfer

Bacterial cell has often been viewed as a "bag of enzymes". About 70% of its internal volume is occupied by water. The remaining 30% are filled with protein molecules, tRNAs, mRNAs, ribosomes, etc. Most of the enzyme molecules are physically distinct entities, i.e. they are free in the cytoplasm. Each enzyme is the translation of a message stored in the circular DNA of the bacterial cell, and this message is *in fine* expressed in a definite biological function. This scheme fits perfectly the "central dogma" of molecular biology, which specifies that information is transferred from DNA to protein but never backwards and never from protein to protein. Thus the central dogma paves the way to the popular belief that the functional properties of a living organism are, in a way, defined on the chromosomes by the corresponding structural genes.

However, when the cell becomes compartmentalized some of its compartments are crammed with enzymes, some of which may form multienzyme complexes. A feature that appears common to many of these associated enzymes is that they catalyse consecutive reactions. One of the possible advantages of these multienzyme complexes is the existence of channelling of reaction intermediates from active site to active site within the complex. This allows the cell to cope with the problem of the dilution of the reaction intermediates in a volume which increases as the size of the cell compartment enlarges during evolution. Whatever the size of the cell compartment, the direct transfer of metabolites from site to site optimizes the overall reaction flow.

Many enzymes, however, are associated with membranes, with the cytoskeleton, or with other enzymes that do not catalyse consecutive reactions. Thus channelling cannot be the functional advantage brought about by this type of association. There is little doubt, however, that the tight association of an enzyme with another protein, or with a cell organelle, often results in an alteration of the conformation of the enzyme. In the central dogma, as formulated by Crick [30], the information content of a structural gene and of the corresponding protein is derived from the sequences of base pairs and aminoacids, respectively. But the information may be also be defined in a different manner which is closer to the biological function of the enzyme. As an enzyme in solution exists in different energy states, its information content may be derived from these energy states. If an enzyme is associated with another protein or with a cell organelle, its entropy decreases and its free energy increases. This can give the associated enzyme the ability to catalyse processes that would never take place if the enzyme were free. In other words the functional properties of an enzyme are dramatically altered by its association with another protein, or with a cell organelle. It then becomes impossible to state that the functional properties of an enzyme are fully encoded in the corresponding structural gene, and that information cannot be transferred from protein to protein... But then information has to be defined by the energy levels of the enzyme rather than by its sequence.

It thus appears that, in eukaryotic cells that display overcrowding and protein association, different types of information transfer take place simultaneously. The message of a structural gene, expressed in the structure of an isolated protein, represents the first classical type of information transfer. This process exists in all autonomous living creatures. Beyond this basic process, additional types of information transfer are also present in eukaryotic cells through the interactions that may exist between different proteins. In other words, the supramolecular complexity of the cell modulates and complicates the functional role of a given structural gene. One may expect these effects to become more and more important as cell structure gets more complex.

11.5. Control of phenotypic expression by a negatively charged cell wall

At a certain stage of evolution some cells evolved a rigid cell wall and, after differenciation, led to plants. The cell wall is a rigid envelope that prevents cell motion. But the cell wall is much more than a simple inert skeleton. It can extend, perceive signals and react accordingly. Many different enzymes are buried in the cell wall. They take part in cell wall extension and in the hydrolysis of many organic compounds that then enter the cell. The activity of these enzymes is pH-dependent and the local pH is itself controlled by the fixed charge density of the wall. An enzyme, pectin methyl esterase, demethylates methylated pectins of the wall and generates these fixed negative charges.

The local fixed charge density is important for phenotypic expression because the activity of the enzymes involved in this process are dependent on this charge density. Thus if the substrate of a cell wall enzyme is negatively charged it will tend to be repelled by the fixed charges of the matrix and this electrostatic repulsion effect, superimposed on the intrinsic kinetic properties of the enzyme, mimic positive co-operativity. For enzymes in solution, co-operativity appears as an intrinsic property of an enzyme which is, in a way, a consequence of the multimeric structure of the enzyme and of the information transfer that takes place within the enzyme molecule. In the case of cell wall enzymes, co-operativity is not an intrinsic property of the enzymes, but a property of the enzymes inserted in their complex environment. In other words, the emergence of a complex cell structure such as the cell wall results in complex behaviour of the enzymes sitting in the wall, even if the enzymes themselves are intrinsically simple. It is, we believe, an interesting idea to assume that, in the course of evolution, complexification of the cell milieu can replace the complexification of individual enzyme molecules. These effects can be modulated by the local cation concentration and external ionic strength.

The situation becomes even more complex if the spatial distribution of fixed charges and enzyme molecules is not statistically homogeneous. It has been shown, both theoretically and experimentally, that the same number of enzyme molecules and fixed negative charges in a constant cell wall volume can give rise to different reaction rates depending on the spatial distribution of the charges and of the enzyme molecules. This may appear heterodox but there is little doubt that enzymes can work in quite different ways depending on whether they are in free solution or associated with cell organelles. When the spatial distribution of fixed charges and enzyme molecules is complex, the enzyme reaction rate is complex as well and may display mixed positive and negative co-operativity.

It thus seems that, in the course of evolution, a progressive complexification of the cell architecture progressively replaced the complexification of individual enzyme molecules. As long as the cell milieu was homogeneous (in bacterial cells) the complexification of enzyme molecules was a prerequisite for the complexification of cell metabolism. But as soon as cells became compartmentalized, a dramatic complexification of the chemical machinery of life became possible without having recourse to a complexificaction of enzyme molecules.

11.6. Evolution of the cell structures associated with motion

Many present day cells display spatial supramolecular devices that take part in cell or organelle motion. The aim of this section is to consider briefly the evolution of some of these organelles, namely microtubules, cilia and flagella of eukaryotes. For most bacterial cells, there exists one chromosome only. This chromosome is attached to the cell wall at two points called the origin and the terminus (Fig. 11.7). Replication of the chromosome during cell division starts at the origin and ends at the terminus. As the two replication forks travel along the chromosome, the new origin is carried by the new chromosome. After completion of replication, the new origin becomes attached to the cell wall opposite

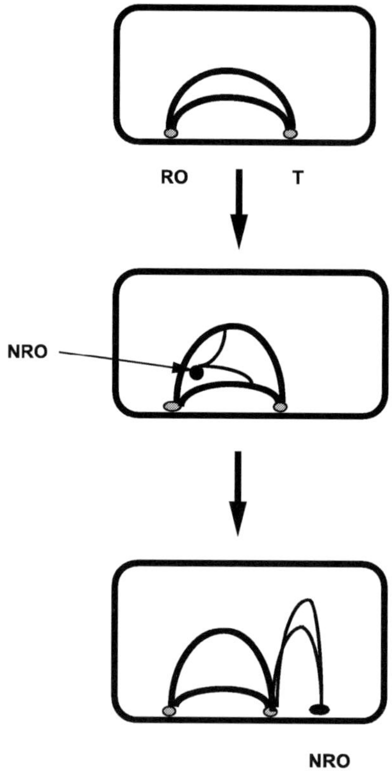

Fig. 11.7. Schematic representation of cell division in bacteria. See text. RO: replication origin, T: terminus, NRO: new replication origin.

the old one. The terminus splits into two parts which correspond to the old and the new terminus (Fig. 11.7). The cell can then divide. This process appears basically different from classical mitosis in which microtubules play a major role in chromosome motion and organization. During mitosis, one observes the formation of a spindle of microtubules from the two centrosomes located at either pole of the cell. Chromosomes are attached to the spindle by their centromeres. After their replication, chromosomes are pulled to the two poles of the cell (Fig. 11.7). Although division of eukaryotic and bacterial cells does not seem to have much in common, one is nevertheless led to think that present day mitosis of eukaryotes originates from a bacterial type of cell division. This is strongly suggested by the existence of pleuromitosis. This mode of cell division is characteristic of some primitive protists. During pleuromitosis two half spindles lie side by side, both attracted by their centromere to the inside of the nuclear membrane. The nuclear membrane is in fact an invagination of the plasma membrane. The two centrosomes are located inside the nucleus, on the nuclear membrane, on the opposite side from the nucleus. The centrosomes have grown microtubules that are attracted to the centromeres (Fig. 11.8). The formation of a bipolar spindle, as it occurs in present day mitosis, require a rotation of the centromeres in such a way they come opposite each other.

It has been suggested that cilia and flagella of eukaryotic cells originate from symbiosis between eukaryotic and prokaryotic cells [28,29]. The validity of this idea is supported by

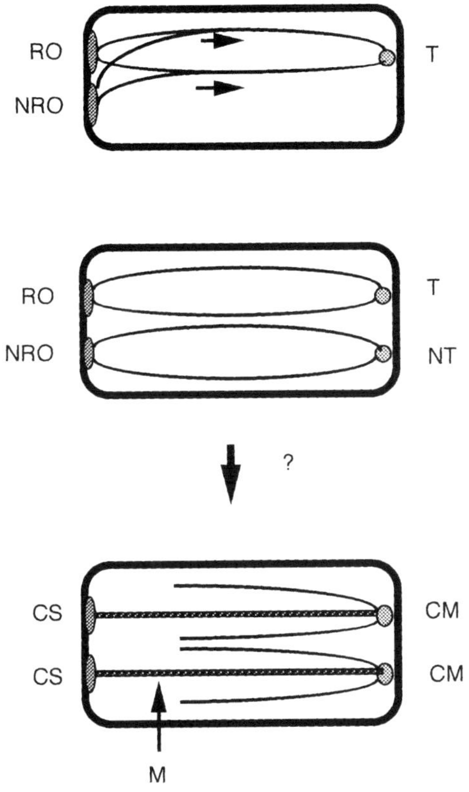

Fig. 11.8. A possible history of mitosis, as suggested by Maynard-Smith and Szathmary [2]. Top and middle – a hypothetical intermediate between bacterial cell division and pleuromitosis. Bottom – pleuromitosis. RO: replication origin, NRO: new replication origin, T: terminus, NT: new terminus, CS: centrosome, CM: centromere, M: microtubule.

a number of observations. Thus spirochaetes (a prokaryote) can bind to the cell membrane of a eukaryotic cell and propel it [31]. This motion requires that the spirochaetes respond in synchrony to an external signal. Moreover, antobodies raised against brain tubulin recognize tubulin-like proteins of spirochaetes. These results as well as others, suggest a symbiotic origin of these organelles involved in cell motion. But this view appears to be in disagreement with other results. For example, the polymerization of the tubulin-like protein of spirochaetes is not inhibited by colchicin [2]. It is therefore not clear whether cilia and flagella of eukaryotic cells stem from symbiosis between eukaryotes and prokaryotes.

11.7. The emergence of temporal organization as a consequence of supramolecular complexity

A host of biological systems display periodic phenomena. This temporal organization is usually explained by complex properties of specific enzymes referred to as "primary oscillators". To the best of our knowledge, this temporal organization of biological events is more frequently encountered with complex than with simple living systems. However the individual enzyme molecules of prokaryotes are no less complex than those of eukaryotes.

Moreover, there is little doubt that, in the course of evolution, a selection pressure has been exerted in favour of the emergence of more and more complex supramolecular structures. These observations lead to the view that the complexity of individual enzyme molecules is probably not the main source of the temporal organization of biological events that have been detected. It appears rather that the supramolecular complexity of eukaryotic cells may represent the main reason for their temporal organization. It may then be the supramolecular complexity of the cell rather than some peculiar property of an enzyme that could be responsible for the oscillations of biochemical processes *in vivo*.

Simple theoretical models which have been considered before confirm this view. A metabolic cycle, involving charged reaction intermediates and taking place at the interface between a charged membrane and a bulk phase, can generate sustained oscillations of the intermediates of the cycle, as well as of the electric potential of the membrane. These oscillations can occur even if all the enzymes of the cycle follow Michaelis–Menten kinetics. The temporal organization of the cycle is therefore not due to some sophisticated property of an enzyme that would play the part of an "elementary oscillator", but rather to the supramolecular organization of the whole system. More specifically, the superposition of the kinetic properties of the enzymes and the electric repulsion of reaction intermediates by the fixed charges of the membrane generates these effects.

It seems that, as certain living systems became more complex, this type of temporal organization is more frequently encountered. One may thus wonder what are the functional avantages of this temporal organization. There seem to be at least three advantages. The first is that the periodic variation of membrane potential drives ATP synthesis [32]. Periodic changes of the membrane potential driven by metabolic oscillations may thus represent a new possible way to synthesize ATP that could have emerged in the course of evolution. The second functional advantage, somewhat related to the first, is that the thermodynamic efficiency of a periodic process is usually larger than that of its steady state counterpart [33]. The third advantage is that a metabolic cycle, taking place at the interface of a membrane, may display multistability and this is an extremely efficient way of controlling the dynamics of the cycle. These functional advantages can help understand why temporal organization of metabolic processes is commonplace in complex living systems. As already outlined [34], periodic behaviour of biochemical processes seems to be the rule rather than the exception. If this idea were valid, metabolic processes under steady state conditions would be more frequent in simple prokaryotic than in eukaryotic cells.

The supramolecular complexity of the cell may also generate aperiodic, or "chaotic" oscillations. This is precisely what occurs during plant cell wall elongation. The growth rate may display periodic oscillations about a trend that may turn aperiodic after a while or *vice versa*. This behaviour is due to the fact that the cell wall is a mosaic of regions which grow at different rates, or even do not grow. The rationale for this process is that any elementary domain of the wall may occur in at least two states: a state in which pectins are demethylated and a state in which they are partially methylated. These two states thus have different charge densities. Conversion between these two states under open thermodynamic conditions can generate sustained oscillations. At least two enzymes (and probably more) are involved in these conversions: a pectin methyl esterase and a trans glucanase. In fact, oscillations of the growth rate of various regions of the wall are not due to the specific properties of either of these enzymes but rather to the complex structure of the cell wall and to the fuzzy organization of the charged and methyl groups. In particular, the frequency and amplitude of the oscillations are extremely sensitive to the local charge densities. Thus, if many different growing regions of the wall have the same charge density the overall growth rate will appear periodic. But if the growing regions have slightly different charge densities, the overall oscillations will be aperiodic, or "chaotic". In the present case, the complexity of oscillations originates from a repeated simplicity, each simple process displaying slight differences with respect to the others. The selection pressure that tends to favour the emergence of more and more complex supramolecular structures also tends to favour the emergence of "chaotic" dynamics. Chaos should not be considered as a strange, exotic situation but rather as the dynamic espression of structural complexity.

11.8. The emergence of multicellular organisms

There are two main types of developments that generate patterns and forms. The first, which has been described for some viruses, consists of the association of protein molecules with a nucleic acid. This association can be viewed as a thermodynamic equilibrium process. Protein molecules spontaneously associate under thermodynamic conditions. A nucleic acid molecule becomes embedded in this supramolecular edifice. The final structure of the virus is thus defined by the structure of the proteins and of the nucleic acid. In a way the shape and the properties of this virus are already present, in a virtual state, in the structure and the properties of these macromolecules. After the molecular association has taken place there is no emergence of any novel property that was not already present in the constituents of the virus. This type of simple morphogenetic process fully justifies a reductionist approach through the concepts and techniques of molecular and structural biology.

It is evident that the situation is completely different with eukaryotic organisms. As soon as the fertilized egg starts dividing and becomes a young embryo, positional information is expressed [35]. This means that different territories of the embryo, which are apparently identical, have quite different fates. This represents a major event in the evolutionary process that allows the appearance on earth of differentiated multicellular organisms. Molecular biologists of development have deciphered, in well chosen biological systems, the sequence of genes that indirectly control the appearance of various patterns and that are characteristic of the developmental process. They have also studied how these genes are turned on and off. Although the knowledge of the genes that become repressed or derepressed during the various stages of development is no doubt essential, it does not provide the final answer to the basic question, namely what are the physical reasons for the emergence of patterns. As a matter of fact, knowledge of the various genes involved and of their mechanism of repression and activation allows one to understand how proteins are formed and how they stop being formed. But this tells us nothing about how the presence of a protein for a limited period of time is associated with the emergence of a definite pattern. One must be aware that true morphogenesis requires that the system be under nonequilibrium conditions, that matter and energy be dissipated and that the whole system be formally equivalent to a dissipative structure [36]. This means that novel properties emerge that cannot be predicted from the sole structure of the elements that constitute the complex system. Hence the reductionist approach does not allow one to tackle the essence of the developmental process.

A major step towards a real understanding of the mechanisms of development is the discovery in young embryos of gradients of gene products (mRNAs and proteins) [37]. This represents the experimenal demonstration of the validity of Turing's ideas put forward as early as 1955 [38] about the intrinsic nature of morphogenetic events. In fact one must realize that the emergence of patterns that define the early steps of development requires the existence of quite stringent conditions. These requirements have probably been met rather recently in the evolutionary process. These conditions have already been mentioned (chapter 10), but will be considered again in a different perspective. Two gradients of morphogens, called activator and inhibitor, must be present in the young embryo. Moreover, they have to be specifically interconnected, the inhibitor has to alter the effect of the activator (see chapter 10). These gradients may be imposed from outside of the fertilized egg, as it is the case for the bicoid mRNA which is synthesized in nurse cells of *Drosophila* and "injected" into the fertilized egg, thus generating a gradient of the morphogen. But this does not exclude that other gradients be the consequence of auto-organization phenomena by amplification of a slight deviation from a steady state. Moreover, the system must respond in a nonlinear fashion to the morphogen concentrations. This is no doubt an important aspect of the model, for the nonlinear response of the system can generate thresholds, i.e. sharp responses to slight changes of morphogen concentrations. In other words nonlinearity can explain that a quantitative difference of morphogen concentration be converted into qualitative differences of structure and form. It is amazing that with diffusion–reaction schemes derived from Turing's basic ideas one can model the periodic and aperiodic patterns found in the living world [39].

Hence the conditions required to generate patterns are extremely restrictive but these conditions are mandatory for explaining the emergence of living creatures that are pluricellular and possess cells that display different functions. It is thus understandable that a selective pressure has been exerted in favour of the emergence of these particular organisms. But it can also be understood that, owing to the specificity of these conditions, pluricellular organisms appeared lately in the history of the living world and many unicellular organisms have never reached the pluricellular state.

11.9. Is natural selection the only driving force of evolution?

Kauffman has presented the view that natural selection is not the only driving force of evolution and does not represent the sole origin of biological order [3,40]. He has put forward the idea that complex systems have the innate property of self-organization and that this tendency is one of the motors of evolution. He has attempted to mimic the evolution of genomes of living organisms by the evolution of random boolean networks in a computer. A random boolean network is an ensemble of connected binary elements that interact randomly. Each element of the network can be turned on or off by one of its neighbours. In general, each element receives K inputs from the elements of the same network. A Boolean network is considered a model of the genome of an organism. Each gene is modelled by an element of a Boolean network and may be turned on or off through an information it receives from the other genes. If each element of a Boolean network receives K inputs, then each element can receive 2^K possible combinations of inputs, since the elements are binary. If $K = N$, the network becomes chaotic. However, Kauffman has observed that when $K = 2$, order spontaneously emerges out of chaos and the network "crystallizes" in a stable state. Kauffman tentatively speculates that the spontaneous evolution of Boolean networks towards ordered systems models the evolution of genomes of living systems. He suggests that each type is the result of the expression of a small number of genes and that these cell types are, in a way, mimicked by the "crystallized states" of the Boolean networks.

These "computer experiments" suggest that the main driving force of evolution is self-organization. In a neo-Darwinian perspective there is little doubt, however, that selection also plays an important role. If, as assumed by Kauffman, "evolution is the marriage of selection and self-organization", it remains to be seen what the respective roles are of these two driving forces of biological evolution.

References

[1] Eigen, M. (1992) Steps Towards Life. Oxford University Press, Oxford.
[2] Maynard-Smith, J. and Szathmary, E. (1995) The Major Transitions in Evolution. Oxford University Press, Oxford.
[3] Kauffman, S.A. (1993) The Origins of Order. Oxford University Press, Oxford.
[4] Maurel, M.C. (1994) Les Origines de la Vie. Syros, Paris.
[5] Eigen, M. (1971) Self-organization of matter and the evolution of biological macromolecules. Naturwiss. 58, 465–523.
[6] Eigen, M. and Schuster, P. (1977) The hypercycle. A principle of natural self-organization. Part A: emergence of the hypercycle. Naturwiss. 64, 541–565.
[7] Wächterhäuser, G. (1988) Before enzymes and templates: theory of surface metabolism. Microbiological Reviews 52, 452–484.
[8] Wächterhäuser, G. (1990) Evolution of the first metabolic cycles. Proc. Natl. Acad. Sci. USA 87, 200–204.
[9] Woese, C.R. (1967) The Genetic Code: the Molecular Basis for Genetic Expression. Harper and Row, New York.
[10] Crick, F.H.C. (1968) The origin of the genetic code. J. Mol. Biol. 38, 367–379.
[11] Orgel, L.E. (1968) Evolution of the genetic apparatus. J. Mol. Biol. 38, 381–393.
[12] Kruger, K., Grabowski, P.J., Zaug, A.J., Sands, J., Gottschling, D.E. and Cech, T.R. (1982) Self-splicing RNA: autoexcision and autocyclisation of the ribosomal intervening sequences of *Tetrahymena.* Cell 31, 147–157.
[13] Zaug, A.J. and Cech, T.R. (1986) The intervening sequence RNA of *Tetrahymena* is an enzyme. Science 231, 470–475.
[14] Joyce, G.F., Schwartz, A.W., Orgel, L.E. and Miller, S.L. (1987) The case of an ancestral genetic system involving simple analogues of the nucleotides. Proc. Natl. Acad. Sci. USA 84, 4398–4402.
[15] Schwartz, A.W. and Orgel, L.E. (1985) Template-directed synthesis of novel, nucleic acid-like structures. Science 228, 585–587.
[16] Cairns-Smith, A.G. (1971) The Life Puzzle. Oliver and Boyd, Edinburgh.
[17] Wächterhäuser, G. (1992) Groundworks for an evolutionary biochemistry: the iron–sulphur world. Prog. Biophys. Mol. Biol. 58, 85–201.
[18] Oparin, A.I. (1965) L'Origine de la Vie sur Terre. Masson, Paris.
[19] Haldane, J.B.S. (1925) The origin of life. Rationalist Annual 1929, 148–169.
[20] Miller, S.L. (1953) A production of aminoacids under possible primitive Earth conditions. Science 117, 528–529.

[21] Spiegelman, S. (1970) Extracellular evolution of replicating molecules. In: F.O. Schmitt (Ed.), The Neuro-Sciences: A Second Study Program. Rockefeller University Press, New York, pp. 927–945.

[22] Lipowski, R. (1991) The conformation of membranes. Nature 349, 475–481.

[23] Blobel, G. (1980) Intracellular membrane topogenesis. Proc. Natl. Acad. Sci. USA 77, 1496–1499.

[24] Cavalier-Smith, T. (1987) The origins of cells: a symbiosis between genes, catalysts and membranes. Cold Spring Harbor Quant. Biol. 52, 805–842.

[25] Horowitz, N.H. (1945) On the evolution of biochemical synthesis. Proc. Natl. Acad. Sci. USA 31, 453–455.

[26] King, G.A.M. (1982) Recycling, reproduction and life's origins. Biosystems 15, 89–97.

[27] Boiteux, A., Hess, B. and Sel'kov, E.E. (1980) Creative functions of instability and oscillations in metabolic systems. Curr. Top. Cell. Regul. 17, 171–203.

[28] Margulis, L. (1970) Origins of Eukaryotic Cells. Yale University Press, New Haven.

[29] Margulis, L. (1981) Symbiosis in Cell Evolution. Freeman, San Francisco.

[30] Crick, F.H.C. (1970) Central dogma of molecular biology. Nature 227, 561–563.

[31] Margulis, L. and Fester, R. (1991) Symbiosis as a Source of Evolutionary Innovation. MIT Press, Cambridge, Mass.

[32] Tsong, T.Y. Electrical modulation of membrane proteins: Enforced conformational oscillations and biological energy, and signal transduction. Annu. Rev. Biophys. Biophys. Chem. 19, 83–106.

[33] Ross, J. and Schell, M. (1987) Thermodynamic efficiency in nonlinear biochemical systems. Annu. Rev. Biophys. Biophys. Chem. 16, 401–422.

[34] Hess, B. (1997) Periodical patterns in biochemical reactions. Quart. Rev. Biophys. 40, 237–258.

[35] Wolpert, L. (1969) Positional information and the spatial pattern of cellular differenciation. J. Theor. Biol. 25, 1–47.

[36] Nicolis, G. and Prigogine, I. (1977) Self-Organization in Nonequilibrium Systems. From Dissipative Structures to Order through Fluctuations. John Wiley and Sons, New York.

[37] Driever, W. and Nusslein-Volhard, C. (1988) A gradient of bicoid protein in *Drosophila* embryos. Cell 54, 83–93.

[38] Turing, A.M. (1952) The chemical basis of morphogenesis. Philos. Trans. Royal Soc. London B 237, 37–72.

[39] Meinhardt, H. (1997) Biological pattern formation as a complex dynamic phenomenon. Internat. J. Bifurcation and Chaos 7, 1–26.

[40] Kauffman, S.A. (1991) Antichaos and adaptation. Scientific American, August, 64–70.

Subject index